Lecture Notes in Computer Science 2068
Edited by G. Goos, J. Hartmanis and J. van Leeuwen

Springer
Berlin
Heidelberg
New York
Barcelona
Hong Kong
London
Milan
Paris
Singapore
Tokyo

Klaus R. Dittrich Andreas Geppert
Moira C. Norrie (Eds.)

Advanced Information Systems Engineering

13th International Conference, CAiSE 2001
Interlaken, Switzerland, June 4-8, 2001
Proceedings

Series Editors

Gerhard Goos, Karlsruhe University, Germany
Juris Hartmanis, Cornell University, NY, USA
Jan van Leeuwen, Utrecht University, The Netherlands

Volume Editors

Klaus R. Dittrich
Andreas Geppert
Universität Zürich
Institut für Informatik
Winterthurerstr. 190, 8057 Zürich, Schweiz
E-mail: {dittrich/geppert}@ifi.unizh.ch

Moira C. Norrie
ETH Zürich
Institute for Information Systems, Department of Computer Science
ETH Zentrum, 8092 Zürich, Switzerland
E-mail: norrie@inf.ethz.ch

Cataloging-in-Publication Data applied for

Die Deutsche Bibliothek - CIP-Einheitsaufnahme

Advanced information systems engineering : 13th international conference ; proceedings / CAiSE 2001, Interlaken, Switzerland, June 4 - 8, 2001. Klaus R. Dittrich ... (ed.). - Berlin ; Heidelberg ; New York ; Barcelona ; Hong Kong ; London ; Milan ; Paris ; Singapore ; Tokyo : Springer, 2001
(Lecture notes in computer science ; Vol. 2068)
ISBN 3-540-42215-3

CR Subject Classification (1998): H.2, H.4-5, H.3, J.1, K.4.3, K.6, D.2, I.2.11

ISSN 0302-9743
ISBN 3-540-42215-3 Springer-Verlag Berlin Heidelberg New York

This work is subject to copyright. All rights are reserved, whether the whole or part of the material is concerned, specifically the rights of translation, reprinting, re-use of illustrations, recitation, broadcasting, reproduction on microfilms or in any other way, and storage in data banks. Duplication of this publication or parts thereof is permitted only under the provisions of the German Copyright Law of September 9, 1965, in its current version, and permission for use must always be obtained from Springer-Verlag. Violations are liable for prosecution under the German Copyright Law.

Springer-Verlag Berlin Heidelberg New York
a member of BertelsmannSpringer Science+Business Media GmbH

http://www.springer.de

© Springer-Verlag Berlin Heidelberg 2001
Printed in Germany

Typesetting: Camera-ready by author, data conversion by Olgun Computergrafik
Printed on acid-free paper SPIN 10781705 06/3142 5 4 3 2 1 0

Preface

Since the late 1980s, the CAiSE conferences have provided a forum for the presentation and exchange of research results and practical experiences within the field of Information Systems Engineering. CAiSE 2001 was the 13th conference in this series and was held from 4th to 8th June 2001 in the resort of Interlaken located near the three famous Swiss mountains – the Eiger, Mönch, and Jungfrau.

The first two days consisted of pre-conference workshops and tutorials. The workshop themes included requirements engineering, evaluation of modeling methods, data integration over the Web, agent-oriented information systems, and the design and management of data warehouses. Continuing the tradition of recent CAiSE conferences, there was also a doctoral consortium. The pre-conference tutorials were on the themes of e-business models and XML application development.

The main conference program included three invited speakers, two tutorials, and a panel discussion in addition to presentations of the papers in these proceedings. We also included a special 'practice and experience' session to give presenters an opportunity to report on and discuss experiences and investigations on the use of methods and technologies in practice.

We extend our thanks to the members of the program committee and all other referees without whom such conferences would not be possible. The program committee, whose members came from 20 different countries, selected 27 high-quality research papers and 3 experience reports from a total of 97 submissions. The topics of these papers span the wide-range of topics relevant to information systems engineering – from requirements and design through to implementation and operation of complex and dynamic systems.

We also take this opportunity to thank all other individuals who helped make this conference possible. These include of course all authors, invited speakers, tutorial presenters, and panel members who found the interest and time in a busy schedule to prepare material and come to Interlaken. We also thank all the organizations who provided financial support. Last but not least, we thank all the individuals involved in the local conference organization whose efforts are mainly behind the scenes, but do not go unnoticed. Finally, we thank the participants and hope that they found the journey to Interlaken worthwhile and that they will be encouraged to return to CAiSE in future years.

March 2001 Moira Norrie

CAiSE 2001 Organization

Advisory Committee

Janis Bubenko Jr.
Kungl. Tekniska Högskolan,
Stockholm,
Sweden

Arne Sölvberg
The Norwegian University of Science
and Technology, Trondheim,
Norway

General Chair

Moira Norrie
ETH Zurich, Switzerland

Program Chair

Klaus R. Dittrich
University of Zurich, Switzerland

Panel and Tutorial Chair

Michel Leonard
University of Geneva,
Switzerland

Workshop and Poster Chair

Stefano Spaccapietra
EPFL Lausanne,
Switzerland

Organizing Committee

Carl A. Zehnder
Andrea Lombardoni
Beat Signer
ETH Zurich,
Switzerland

Program Committee

Witold Abramowicz	University of Poznan, Poland
Hans-Jürgen Appelrath	University of Oldenburg, Germany
Alex Borgida	Rutgers University, USA
Sjaak Brinkkemper	Baan Company R&D, The Netherlands
Janis Bubenko	Royal Institute of Technology and Stockholm University, Sweden
Silvana Castana	University of Milan, Italy
Panos Constantopoulos	University of Crete, Greece
Jan Dietz	Delft University of Technology, The Netherlands
Klaus Dittrich (Chair)	University of Zurich, Switzerland
Mariagrazia Fugini	Politecnico di Milano, Italy
Antonio Furtado	PUC, Rio de Janeiro, Brazil
Andreas Geppert	University of Zurich, Switzerland
Martin Glinz	University of Zurich, Switzerland
Jean-Luc Hainaut	University of Namur, Belgium
Juhani Iivari	University of Oulu, Finland
Stefan Jablonski	University of Erlangen, Germany
Christian S. Jensen	University of Aalborg, Denmark
Manfred Jeusfeld	Tilburg University, The Netherlands
Paul Johannesson	Stockholm University and Royal Institute of Technology, Sweden
Gerti Kappel	Johannes Kepler University of Linz, Austria
Pericles Loucopoulos	UMIST, United Kingdom
Kalle Lyytinen	University of Jyväskylä, Finland
Neil Maiden	City University, England
Michele Missikoff	IASI-CNR, Italy
John Mylopoulos	University of Toronto, Canada
Oscar Nierstrasz	University of Berne, Switzerland
Andreas Oberweis	University of Frankfurt, Germany
Antoni Olivè	Polytechnical University of Catalunya, Spain
Andreas Opdahl	University of Bergen, Norway
Norman Paton	University of Manchester, United Kingdom
Barbara Pernici	Politecnico di Milano, Italy
Alain Pirotte	University of Louvain, Belgium
Klaus Pohl	University of Essen, Germany
Colette Rolland	University of Paris 1, France
Michel Scholl	INRIA, France
Michael Schrefl	Johannes Kepler University of Linz, Austria
Amilcar Sernadas	Technical University of Lisboa, Portugal
Arne Sølvberg	Norwegian Institute of Technology, Norway
Martin Staudt	Swiss Life, Switzerland
Rudi Studer	University of Karlsruhe, Germany
Costantinos Thanos	CNR-IEI, Italy

Aphrodite Tsalgatidou — University of Athens, Greece
Yannis Vassiliou — National Technical University of Athens, Greece
Yair Wand — University of British Columbia, Canada
Benkt Wangler — University of Skoevde, Sweden
Tony Wasserman — Bluestone Software, USA

External Referees

Joseph Barjis
Per Backlund
Stavros J Beis
Roel van den Berg
Terje Brasethvik
Carlos Caleiro
Jaelson Castro
Dariusz Ceglarek
Luiz Marcio Cysneiros
Mohamed Dahchour
Daniela Damm
Ruxandra Domenig
Michael Erdmann
Paula Gouveia
Marco Grawunder
Åsa Grehag
Tony Griffiths
Luuk Groenewegen
Volker Guth
Naji Habra
Arne Harren
Jean Henrard
Olaf Herden
Michael Hess
Patrick Heymans
Holger Hinrichs
Bart-Jan Hommes
Andreas Hotho
Marijke Janssen
Elisabeth Kapsammer
Panos Kardasis
Pawel Jan Klaczyñski
Konstantin Knorr
Manuel Kolp
Minna Koskinen
Gerhard Kramler
Thorsten Lau
Stephan Lechner
Angeliki Lempesi
Kirsten Lenz
Mauri Leppnen
Bin Liu
Jianguo Lu
Paul Mallens
David Massart
Paulo Mateus
Massimo Mecella
Torben B. Pedersen
Nikos Prekas
Guenter Preuner
Jaime Ramos
Johannes Ryser
Motoshi Saeki
Simonas Saltenis
Carina Sandmann
Guido Schimm
Jürgen Schlegelmilch
Kostas Sifakis
Paulo Pinheiro da Silva
Eva Söderström
Anya Sotiropoulou
Marcel Spruit
Steffen Staab
Nenad Stojanovic
Mattias Strand
Gerd Stumme
Xiaomeng Su
Heiko Tapken
Thomas Thalhammer
Dimitris Theotokis
Halvard Traetteberg
Anca Vaduva
Costas Vassilakis
Krzysztof Wêcel
Harald Haibo Xiao
Arian Zwegers

Table of Contents

Invited Talks

Requirements Engineering

Agent-Based Approaches

Workflow Management

Data Models and Design

Reuse and Method Engineering

XML and Information System Integration

Evolution

Conceptual Modelling

Practice and Experience

Flexible Support of Work Processes (Panel)

Project Oxygen: Pervasive, Human-Centric Computing – An Initial Experience

Larry Rudolph

Massachusetts Institute of Technology
Laboratory for Computer Science
200 Technology Square
Cambridge, MA 02139
rudolph@lcs.mit.edu

1 Introduction

For the past six months, I have been integrating several experimental, cutting-edge technologies developed by my colleagues at MIT as part of the MIT LCS/AIL Oxygen project. This paper gives a snapshot of this work-in-progress.

Project Oxygen is a collaborative effort involving many research activities throughout the Laboratory for Computer Science (LCS) and the Artificial Intellegence Laboratory (AIL) at the Massachusetts Institute of Technology (MIT). The Oxygen vision is to bring an abundance of computation and communication within easy reach of humans through natural perceptual interfaces of speech and vision so computation blends into peoples' lives enabling them to easily do tasks they want to do – collaborate, access knowledge, automate routine tasks and their environment. In other words, *pervasive, human-centric computing.*

At first blush, this catch-phrase appears vacuous. Today, computers are certainly pervasive; it is likely, at this moment, you are within 100 meters of a computer. Computers are certainly human-centric; what else can they be? On the other hand, computers are not yet as pervasive as is electricity or water. Although computers perform jobs required by humans, they do not feel human-centric – humans must conform to an unnatural way of communicating and interacting with computers. Finally, the tasks described have little to do with computation; computer-mediated functions is a more accurate term but sounds awkward.

The vision and goals of the Oxygen project are described in detail elsewhere [1,2,3], the purpose here is to show how many maturing technologies can be integrated as a first step towards achieving the Oxygen vision. There are research efforts at other universities and research institutions that roughly share the same vision, however, each institution focuses on integrating their own maturing technologies. Oxygen has a three-pronged approach by dividing the space into three broad categories: the H21, a hand-held device, the N21, an advanced network, and the E21, a sensor-rich environment (see Figure 1).

In what follows, an Oxygen application is described in terms of its human-centric features as well as the required technologies. It is important to keep in

K.R. Dittrich, A. Geppert, M.C. Norrie (Eds.): CAiSE 2001, LNCS 2068, pp. 1–12, 2001.
© Springer-Verlag Berlin Heidelberg 2001

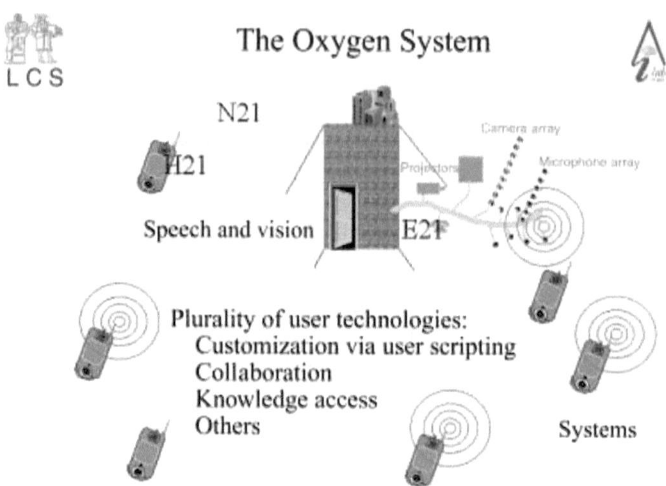

Fig. 1. An overview of the Oxygen Infrastructure, showing the division into three parts: H21, a handheld digital device, N21, the network infrastructure, and E21, the environment infrastructure.

mind that this is just one of many applications and that it is merely a vehicle to explain how many technologies can be integrated and how to create the infrastructure necessary to enable the introduction of context into applications making them more "natural" to use.

The sample application is that of a seminar presentation support system. The next section gives an overview of the application. Section 3, reviews many of the technologies that will go into this application. Section 4, shows how they integrate to form the application. A preliminary programming language and middleware support is described in Section 5.

2 A Computer-Mediated, Seminar Presentation System

This section describes a computer-mediated seminar presentation system. As you read through the description, compare it to how presentations are given today. Although a laptop with programs like Powerpoint or Freelance attached to an LCD projector is a vast improvement over the old days of foils or 35mm slides, the human has given up a degree of control, freedom, and naturalness. The system described below provides for a more natural human interface.

Alice is to give a seminar about her $O_{2.5}$ project. As she walks into the seminar room, she allows herself to be be identified as the speaker. She does not need to carry a laptop with her slides on it – all of her files are globally accessible. Alice tells the system how to find her talk by simply supplying enough keywords to uniquely identify the file she wants. Her files are well indexed and so she merely describes the file in human terms and not with some bizarre syntax. The system knows where she is and marshals all the physical components that may be needed for her to control the display.

Alice wants to control the display so that it matches her current desires. A seminar is a live event and the dynamics depend on the audience and speaker. Although it is crucial that she control the presentation, this control should be of minimal distraction. The same is true for the audience – they should be able to see the visual content, hear her commentary, and take notes at the same time. Moreover, unexpected events should be handled in a natural way.

Even today, Oxygen technologies can make a computer-mediated presentation a more natural experience. In particular, three natural ways to control the slides are provided, as opposed to the traditionally way where Alice either clicks the mouse or hits the enter key. It is computer-centric to force the speaker to always walk over to the laptop in order to advance the slide. A wireless mouse is only a partial solution as it requires that something be held in a hand. For Bob this might be fine, but Alice likes to use a laser pointer to highlight objects on the screen and she finds holding two objects to be very awkward. An integrated pointer/mouse is no better since it now requires attention to find the correct button.

Alice can use her laser pointer to highlight words, draw lines and sketches, as well as to switch between slides. Holding the pointer in the bottom right corner means to advance to the next slide. A camera looking at the screen interprets Alice's laser pointer gestures. But not all humans like to use laser pointers. Some people, especially when they are continuously engaged in speaking, like to use verbal commands to control the presentation. This is done with a microphone and software that continuously tries to understand commands. All three modes of control will always be active, allowing the speaker to use whatever is convenient.

There is more to a presentation than just advancing slides. Alice may want to see her notes, see the next slide before the audience, skip a slide or present the slides in a different order. A laptop, handheld, or any other personal communication device can be used by the speaker. To skip to a different slide without anyone knowing it, is a task that is easily performed by simply clicking on a different slide image on her personal display. The personal display must remain consistent with the public display. So, whether Alice says "Next Slide," chooses a slide from her private computer (handheld or laptop), or uses the laser pointer, both displays are updated.

The audience should also have a choice of ways to observe the presentation. They can look at the large projection screen in the front of the room, as is usually the case, or they may choose to view the presentation on their own personal digital device. The output is simultaneously broadcast to these devices. Some people in the audience might like to take notes and have them correlate with the presentation itself. We propose broadcasting a URL or some other identification symbol for the current contents. This can be either used to display the slide on the laptop, or be inserted into their notes. Later on in the privacy of their own room, these notes can be merged with an archived version of the talk. The archived version will match the presentation rather than the original file. Alice may have many "emergency" slides prepared that will be shown only in response to a question.

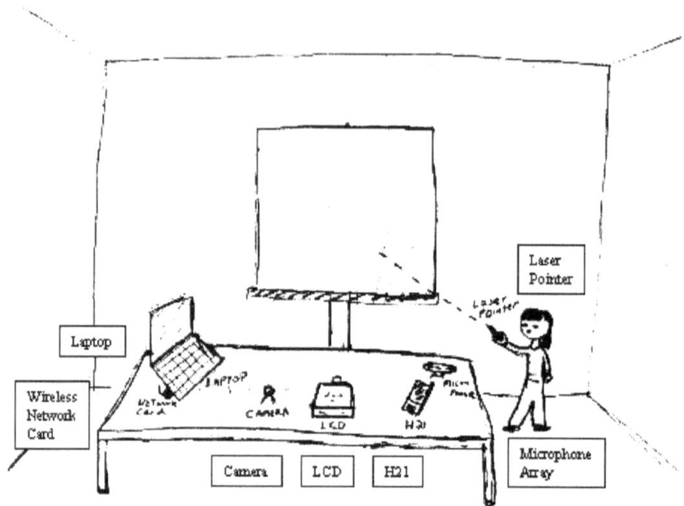

Fig. 2. The seminar room can be assembled from off-the-shelf components. The laptop controls the LCD, camera, microphone, and the networking parts of connecting to the file system, broadcasting, and archiving. The H21 is used by the speaker for personal notes and skipping slides. This application makes use of many of emerging technologies being pursued at the Lab. for CS and AI Lab at MIT.

To summarize, there are several output modalities: the LCD projection, a broadcast of the current content, an archival copy that can be accessed afterwards, and the ability to correlate the public presentation with her own personal view of the presentation.

Lastly, Alice also has "meta" operation control - e.g. switching to a different presentation package, such as a browser or Mathematica, or even to the contents of another presentation. She should also be able to control whether or not content is broadcast or archived.

3 Technology Overview

Research into many technologies that support the above scenario being pursued as part of Project Oxygen. Once again, we wish to emphasize that there are many competing technologies being developed elsewhere. We deliberately ignore them for several reasons[1]. First, to do justice to them all would make this article too large. Second, close physical proximity is usually required when making use of experimental, research systems. While it is possible to do this remotely for one component, it is nearly impossible do this for a number of research projects. We wish to provide feedback to these other research projects before they are ready

[1] The author wishes to apologize to all those who do not agree with these reasons. In a future, expanded version of this paper, many competing technologies will be cited. The author would be happy to learn about an relevant research.

for prime-time and we deliberately try to use them in some unintended way. While there are similar efforts in many of the intelligent or instrumented rooms, our example is simply geared towards exposing how components interact even with commodity hardware. As fun as it is, the particular demo of an oxygenated presentation is not the goal.

3.1 The Handy 21 (H21)

Although the commercial sector has been cranking out all kinds of hand-held devices, there is still much research to be done. The H21 should replace the plethora of communication gadgets with a single portable device. In particular, it should combine at least the functions of a cellular phone, wireless Internet connection, pager, radio, as well as a music and video player/recorder. Packing all this functionality into a single device appears to make it too heavy to be portable. So, industry strives to find the right set of combinations and to then sell add-ons to fill-in the missing pieces. The Oxygen approach is different: all that is needed is a minimal set of components built into the hardware with software and reconfigurable hardware used to provide whatever functionality is needed.

The SpectrumWare project [5] is developing a multipurpose communications system that can be programmed to receive and transmit many different types of signals in order to form a "communications chameleon." One can program the H21 to be a radio, cell phone, or television receiver. To fit in a small space it will need configurable hardware.

The RAW project [6] is developing a chip that will deliver unprecedented performance, energy efficiency and cost-effectiveness because of its flexible design, by exposing its wiring to the software system, the chip itself can be customized to suit the needs of whatever application is running on it. The Raw chip could be incorporated into a single device that could perform a wide variety of applications: encryption or speech recognition, games or communications.

The commercial sector also understands the need for low-power devices, especially handheld ones. However, to make substantial improvements, it is important to re-examine computer architecture from basic principles. The SCALE project [7] is aimed at doing just that.

Rather than waiting for this research to come to fruition, the Oxygen project will make use of commercial handheld computers. In fact, it is doubtful that we will ever build our own device. More likely, we will continue to modify and adapt commercial products that at a minimum, support Linux, audio and visual I/O, and multiple communication channels [8]. Although in an ideal world one will have the right devices for the job, in reality that is usually not the case. It is thus important to be able to make use of what is available. Users want to get the job done and so we expect to support a wide range of devices.

3.2 Networking, Naming and Location Management

One's personal data should be easily and universally accessible. Having a multitude of digital devices, each with some possibly inconsistent set of data, is neither natural nor geared towards the needs of the human. Having to remember a set of arbitrary names just to use a physical device sitting in plain sight is also demeaning to the human user.

The self-certifying file system, SFS [9], is a universal, secure filesystem that has a single global namespace but no centralized control. Other similar filesystems require the users to use a particular key management system to provide security and authentication. SFS separates key management from file system security, thereby allowing the world to share the filesystem no matter how individuals chose to manage their keys.

Within a building it is useful to know where things, including one's self, are physically located. The traditional approach is to have all things periodically broadcast their identity and to have sensors spread throughout the building that detect these things. To provide a degree of privacy, among other reasons, the Cricket [11] location-support system takes the opposite approach. Spread throughout the building are a set of beacons. The beacons are a combination of RF and infrared signals that broadcast physical and logical location information. Things in the environment sense these beacons. Thus, a handheld knows where it is located rather than the system knowing it. A person has the freedom to reveal his or her location – usually when some service or resource is required. All sorts of devices need to be integrated into the system having network connectivity and location awareness [12].

Knowing the location of things enables one to name digital devices by their intended use rather than by some specific URL or IP address. The Intensional Naming System [10] does just that by maintaining a database of currently active devices along with attributes describing their intended use. Devices periodically check-in to avoid having their entries time-out. INS supports early binding for efficiency and late binding for mobility. With INS, it is possible to route packets to the LCD projector that is located in the same room as the speaker or to route messages to whatever display device is near the intended recipient.

3.3 Security and Correctness

As evident by the central place of this subsection, the Oxygen Project considers security and privacy as a central component of a human-centric system. We are developing a personal identification device that has two interesting features. It has a very simple interface, perhaps only a single button to distinguish between identification and authorization [12,13]. The simpler the interface the easier it is to make the device secure. The second feature is that identification mechanisms provide privacy. A guiding philosophy is that privacy is the right to reveal information about oneself. When one chooses to make use of public system resources one is choosing to reveal information about one's self. Various schemes for secure, private group collaboration are being developed [13] as well.

As computers continue their infestation of human activities, their reliability becomes more important. Specifying the behaviors of interacting systems is particularly challenging. Research efforts, I/O Automaton (IOA) [14] and Term

Rewriting Systems (TRS) [15], aimed at proving the appropriateness of collective behaviors, have focused on precise and concise specifications.

3.4 Human Interfaces

There is no question that verbal and visual interfaces to computers are rapidly maturing and are already being successfully deployed. However, speech and natural language systems need to extend beyond simple dialog systems. The approach is to gather information from the speaker in a number of ways, to fuse this with information from other sources and to carry out tasks in an off-line fashion as well, so as to optimize the users time [16]. This effort is also trying to make it easy to develop domain-specific speech interfaces; moving the creation of a interface from an art to a science.

The focus on visual input system that recognize a range of gestures tries to leverage multiple input devices. Just like stereo vision makes it easier to differentiate objects, collaboration between multiple monitoring devices simplifies many recognition tasks. For example, a microphone array along with a array of cameras can be used to do speech processing of all the conversations going on in a meeting, even when several people talk at once [18]

On the output side, there is research aimed at building very large displays. The challenge is to overcome the bandwidth problem of getting all the pixels out of a computer. The approach is to embed processing, connectivity, and display all in one package so that the pixel generator is close to the pixel display, thereby creating a sufficiently rich environment to mimic the Holodeck of Star Trek fame [19]. A related effort is to develop an economical method for displaying 3D using an auto-stereoscopic display [20].

3.5 Collaboration

There is much to be done in the way of supporting computer-mediated human collaboration. Teleconferencing has made strives in allowing collaboration between people who are widely spatially disjoint, but it is still difficult to collaborate when people are temporally disjoint [21]. Much of this work is going on in the Artificial Intellegence Laboratory at MIT and unfortunately, I only know a little bit about it. The seminar presentation scenario described in this paper is just the beginning.

4 Implementing the Seminar Presentation System

We can now relate the technologies described in the previous section with the needs of our seminar presentation system. The explanation roughly follows the description in Section 2. We ignore traditional issues like authorization and allocation of resources and application code written in traditional ways.

When Alice enters the seminar room, she must be identified and the presentation manager must be initiated. A simple tag broadcasts her public key to her

H21 or, if she does not have one, then to the room computing E21 infrastructure. The H21, with Alice's permission, will initiate the seminar presentation manager application as an extension of Alice's computing environment. Rather than having applications run on machines with only a loose connection to the owner, they are all under direct control of the initiator, who has the ability to interact with them from any Oxygen supported I/O device..

Access to Alice's presentation files is via the secure, global filesystem, SFS. Advanced indexing systems, such as Haystack [22], will be used to find files or slides within a file. The Cricket location management system is being used to know where the presentation is occurring. It is possible for the speaker and the audience to seamlessly move to a larger seminar room without losing any content. The intentional naming system, INS, is used to route packets between the components of the system and provide for fault tolerance and mobility. If one component crashes, INS will help in finding an alternate or reconnect when the component comes back online.

For the input modes, speech and vision processing is used. The speech project, Galaxy, has been developed mostly for dialogs and is being adapted to an active monitoring mode. The vision system [18] is used for the initial laser pointer and later for human gesture recognition. A microphone array combined with a vision system that precisely located the position of the speakers mouth is being developed to allow the speaker more mobility. Recognition of drawing gestures, makes use of technology underlying the Rational Capture project [23].

The presentation itself will be controlled in a conventional manner. For powerpoint presentations running under windows, we use visual basic to connect to the rest of the application middleware as well as to control the presentation itself. For the speaker's note view, a stripped down web-browser is used with the application code written in Java.

The output side, at the moment, is the least sophisticated. We hope to make use of the Auto-stereoscopic Display work that will enable 3D image rendering and the Holodeck research that will support very large active displays. In addition, capturing the experience for later review will make use of the research in collaboration [21].

Finally, the presentation manager application is written in a special "communication oriented" language and middleware, described in the next section. Such communication oriented languages, along with IOA and TRS research will lead to the development of correct distributed systems that work first time out, and allow one to focus on performance as the system scales.

5 The Language Overview

This section highlights the core of a communication oriented language used to program some Oxygen applications. The work described in this section is preliminary and so the description is deliberately sketchy[2]. Although, Java could

[2] This work is so preliminary that the language has yet to be named.

be used, especially since most of the underlying technologies provide a Java interface, our language lets the application writer to focus on what is relevant, permits very aggressive optimizations and is more concise and precise.

At a high level of abstraction, there are only a few components that need to be manipulated: nodes, edges, messages, and actions. Nodes are just about anything that can be named and communicated via sockets. An edge is a directed connection between nodes. A message is an entity that flows along an edge. An *event* is the creation or destruction of one of these components; thus there are only six different types of events. Rules (or actions) make up the final component of the language. An *action* consists of a trigger and a consequence. A trigger is an event, such as the creation of a node or the destruction of an edge. A consequence is also an event. For example, the existence of a message on an edge can trigger a set of edges to be disconnected.

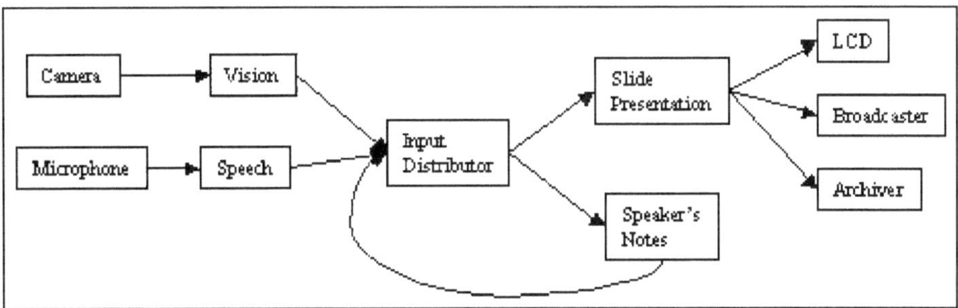

Fig. 3. A graphical view of the components and their connections.

Nodes are named in a way that is compatible with the intentional naming system [10] and consists of a collection of key/value pairs. When a node is named, these pairs are matched against a database of existing, functioning devices or services. For simplicity, any node can be created or destroyed. In actuality, it is only the connection that is created or destroyed when the node is a physical device or an enduring software service. Connections are IP/Port specifiers; it is assumed that all devices and third-party services have some kind of wrapper that converts input and output to the appropriate formats. Rather than have a special case to handle the case when a named node does not exist, it is assumed that such nodes are created and then immediately destroyed. The code to handle a non-existent node is exactly the same as the code to handle a node that was connected but becomes disconnected or destroyed.

Messages are self-describing and self-typed. They can be named, as with nodes, by a set of key/value pairs but must include a location, either a node or an edge. Very large messages or streams, such as audio, video, or screen images are conceptionally the same as text messages, but the implementation treats them different to ensure sufficient performance. As each message moves through the system, it is assumed to be created when it arrives at a location and destroyed when it leaves that location. This permits actions to treat message events and creation/destruction of node and edge events in the same uniform way.

All the action is, of course, with the actions. Actions are simple yet powerful rules. Actions can be created or destroyed, just like all other objects in the language, and are thus events. So, some event trigger can create new actions or remove current ones. Actions are needed to control what happens in a dynamic, sensor rich environment. When one enters a room from the hallway, two events happen: the link to the hallway is destroyed and the link to the room is created. Either of these events can serve as triggers for a whole slew of actions.

Figure 4 shows the specification of the presentation manager. The nodes are named using key/value pairs. The variable *owner* is a parameter of the system and is passed-in when the application begins. The location specifier could be done in the same way, but in the code in the Figure it is hardwired. Nodes, messages, and edges are all named to make it easier to read the code. Two sets of actions are specified. One disconnects the I/O devices. This, presumably is useful for situations in which the speaker wishes to temporarily pause the current presentation and to switch to a different one[3]. The first action is invoked whenever there is a "pause" message on the dialog-in port. The consequence of this action is to destroy the four edges that connect to the camera, microphone, LCD, and broadcast process. These will be used by the other application. The archiver is dedicated to this application and so can remain connected. A second set of actions show another example of disconnecting only the edges to the broadcaster and archiver nodes.

There is a middleware system that executes the language [4]. Initially it executes on a single machine, but soon will be made fault tolerant and decentralized. Nearly all actions performed by the middleware corresponds to an event and events can be triggers for other events.

Although, we expect that the language will be compiled for optimum performance and reliability, it is also possible to interpret commands during run time. A user can modify an application during run-time to adapt to changing needs. The simple structure makes this easy to do provided there is a way for a user to easily name nodes, edges, and messages.

6 Conclusion

Scientific endeavors have always alternated between periods of deep and narrowly focused research activities and periods of synthesis across many fields. I believe we are in the midst of a new computer revolution. The relentless doubling of performance every 18 months, the even faster exponential growth of the web and its communication infrastructure, and the maturing of many human-computer interface technologies means that things will not stay the same. While industry is doing some of this work, the emphasis is on producing products that are good at one thing. Oxygen is not producing products, rather it is exploring what is possible when one synthesizes the fruits of research across many fields.

[3] Hopefully, the speaker is not checking her mail during the presentation itself!

Application Name:
Seminar Presentation Manager (owner)

Nodes:
microphone : ["Device", "microphone", "Location","NE43-518"]
camera : ["Device", "camera", "Location","NE43-518"]
input : ["Process", "input-collector", "OS","Unix","Owner", owner]
ppt : ["Process", "powerpoint-displayer", "OS","Windows","Owner", owner]
ppt' : ["Process", "speaker-notes", "Platform","H21","Owner", owner]
lcd : ["Device", "lcd", "Location","NE43-518"]
broadcaster : ["Process", "broadcast-slides","OS","Unix","Owner", owner]
archiver : ["Process", "archive-slides", "OS","Unix","Owner", owner]

Edges:
m_{pause} : ["Message", "pause", "Location", dialog-in]
m_{resume} : ["Message", "resume", "Location", dialog-in]
$m_{confidential}$: ["Message", "confidential", "Location", dialog-in]
m_{public} : ["Message", "public", "Location", dialog-in]

Messages:
e_{ms}: (microphone , speech) , e_{si}: (speech , input)
e_{cv}: (camera , vision) , e_{vi}: (vision , input)

e_{ip}: (input , ppt) , $e_{ip'}$: (input , ppt')
e_{pl}: (ppt , lcd) , e_{pb}: (ppt , broadcaster)
e_{pa}: (ppt , archiver) ,

$e_{p'i}$: (ppt' , input)

Actions:
(m_{pause} , ($!e_{cv}$, $!e_{ms}$,$!e_{pl}$, $!e_{pb}$))
(m_{resume} , (e_{cv} , e_{ms} ,e_{pl} , e_{pb}))
($m_{confidential}$, ($!c_{pb}$, $!c_{pa}$))
(m_{public} , (e_{pb} , e_{pa}))

Fig. 4. Part of the communications program that expresses the connections. There are always two implicit nodes: dialog-in and dialog-out. The actions disconnect and reconnect the I/O devices on a pause or resume command. Presumably this is used to switch to another presentation. Similarly, the speaker may want to go "off-the-record" and show slides that are not archived nor broadcast. Going "public" reestablishes these links.

This paper describes a work in progress. Not only is this system still under development, but many of the technologies that it exploits are also under development. The presentation manager is simply a data point. It gives insight into the tools that will be needed in the future, gives feedback to those researchers developing the components, and is just one of several parallel efforts. These efforts will create the infrastructure necessary for the next decade. There is much to be done but we must keep the goal in sight – computers must become easier and more natural to use.

Acknowledgements

The author would like to thank John Ankorn for his help with all aspects of the integration effort as well as with this text. Thanks are also due to Hilla Rogel for her drawings and to Anant Agrawal for involving me in the Oxygen Project. This work is supported in part by our industrial partners,: Acer, Delta, Hewlett Packard, NTT, Nokia, and Philips, as well as by DARPA through the Office of Naval Research contract number N66001-99-2-891702.

References

1. Dertouzous, M. *The Unfinished Revolution*, Harper-Collins Publishers, New York, 2001.
2. Scientific American, August 1999 Issue.
3. MIT, *http://www.oxygen.lcs.mit.edu*
4. Rudolph, L. Poxygen *http://www.org.lcs.mit.edu*
5. Garland, S. and J. Guttag, "SpectrumWare" *http://nms.lcs.mit.edu/projects/spectrumware*
6. Agarwal, A. and S. Amarasinghe, "RAW" *http://www.cag.lcs.mit.edu/raw*
7. Asanovic, K., "SCALE" *http://www.cag.lcs.mit.edu/scale*
8. Ankorn, J. "EMS" *http://www.oxygen.lcs.mit.edu/ems*
9. Kaashoek, F. "SFS" *http://www.fs.net*
10. Balakrishnan, H. "INS" *http://nms.lcs.mit.edu/projects/ins*
11. Balakrishnan, H. "Cricket" *http://nms.lcs.mit.edu/projects/cricket*
12. Devadas, S. "Aries" *http://caa.lcs.mit.edu/caa/projects.html*
13. Rivest, R. "CIS" *http://theory.lcs.mit.edu/~cis/cis-projects.html*
14. Lynch, N. "Input Output Automata" *http://theory.lcs.mit.edu/tds/i-o-automata.html*
15. Arvind and Rudolph, L. "TRS" *http://www.csg.lcs.mit.edu/synthesis*
16. Zue, V. "Galaxy" *http://www.sls.lcs.mit.edu/sls*
17. Glass, J. "SpeechBuilder" *http://www.sls.lcs.mit.edu/sls*
18. Darell, T. "Visual Interface" *http://www.ai.mit.edu/~trevor/vision-interface-projects.html*
19. McMillan, L. "Image Based Rendering" *http://www.graphics.lcs.mit.edu/~mcmillan/IBRwork/*
20. McMillan, L. "Auto-stereoscopic Display" *http://www.graphics.lcs.mit.edu/~mcmillan*
21. Shrobe, H. "Knowledge-Based Collaboration" *http://kbcw.ai.mit.edu*
22. Karger, D. "Haystack" *http://haystack.lcs.mit.edu*
23. Davis, R. "Rational Capture" *http://www.ai.mit.edu/people/davis/davis.htm*

Database Systems Architecture: A Study in Factor-Driven Software System Design

Peter C. Lockemann

Fakultät für Informatik, Universität Karlsruhe
Postfach 6980, 76128 Karlsruhe, Germany
lockeman@ira.uka.de

Abstract. The architecture of a software system is a high-level description of the major system components, their interconnections and their interactions. The main hypothesis underlying this paper is that architectural design plays *the* strategic role in identifying, articulating, and then reconciling the desirable features with the unavoidable constraints under which a system must be developed and will operate. The hypothesis results in a two-phase design philosophy and methodology. During the first phase, the desirable features as well as the constraints are identified. The second phase is a decision process with features and constraints as the driving factors, and tradeoffs contingent on a value system that will always include subjective elements. It is of course impossible to validate the hypothesis in full generality. Instead, we restrict ourselves to an analysis – much of it retrospective – of architectures of database management systems in networks. The analysis demonstrates that the most challenging part of architectural design is to identify – very much in the abstract – those features that promise to have *the* major impact on the architecture. Further, by separating the features into two classes, a primary class with all those features that dominate the design, and a second class with those features that can then be treated orthogonally, the complexity of the design task is reduced.

1 Hypothesis

The architecture of a software system is – much in the tradition of classical systems analysis – a description of the major system components, their interconnections and their interactions. The description is on a high-level: Major features are identified, but little attention is as yet given to the details of ultimate implementation. Or in other words, developing a system architecture is "programming-in-the-very-large".

The main hypothesis underlying this paper is that architectural design is a vital first step in the development of software systems. We claim that architectural design plays *the* strategic role in identifying, articulating, and then reconciling the desirable features with the unavoidable constraints – technical, financial and personnel – under which a system must be developed and will operate. In a nutshell, architectural design is the means for explicating the major conflicts, and

K.R. Dittrich, A. Geppert, M.C. Norrie (Eds.): CAiSE 2001, LNCS 2068, pp. 13–35, 2001.
© Springer-Verlag Berlin Heidelberg 2001

for deciding and documenting the necessary tradeoffs on a strategic level, and thus at a time when no major implementation efforts and expenses have as yet occurred. This may sound platitudinous to most, but in practice all too many flaws or failures in business systems can be traced back to the lack of an explicit architectural design and implicit mistakes when viewed from an architectural perspective.

Our hypothesis drives a two-phase design philosophy and methodology. During the first phase, the desirable features as well as the constraints are identified. During the second phase, tradeoffs are determined that preserve as many of the features as possible while minimizing the effects of the constraints. The second phase clearly is a decision process with features and constraints as the driving factors, and tradeoffs contingent on a value system that will always include subjective elements.

A scientifically rigorous approach to verifying the hypothesis would require us to set up two development teams, supply them with the same system specifications, have them follow different design strategies of which only one is strictly architecture-based, and compare the results. Moreover, such an approach would have to cover a sufficiently wide spectrum of software systems. Obviously, all this is entirely impractical. The approach to verification we take in this paper is more circumstantial, then. For one, we concentrate on a few types of database system architectures as they may appear in distributed information systems. Second, much of our analysis is retrospective. We examine system architectures that have found wide acceptance, and try to reinterpret them in the light of factors we consider particularly critical.

2 Resource Managers in Distributed Information Systems

2.1 Shared Resources and Services

Distributed information systems are a reflection of modern distributed organizations – business, administration or service industry. Today, business processes are viewed as the central concept for organizing the way business is done. Designing the business processes is, therefore, considered a major challenge. Requirements for distributed information systems should be a major outcome of the design.

Information exchange is a vital part of business processes. Business processes operate across geographical distances, often on a worldwide scale, and they utilize corporate memory in the form of huge data repositories. Consequently, we limit ourselves to a view of distributed information systems that concentrates on those features that support information exchange. Since information exchange has a *spatial* and a *temporal* aspect, distributed information systems are expected to overcome temporal and spatial distances. This gives a first – trivial – architectural criterion, the separation of the two aspects into data communications systems and database management systems.

More abstractly, from the perspective of the business process an information system can be viewed as a set of information *resources* together with a suitable (*resource*) *manager* each. Spatial exchange is provided by data communication

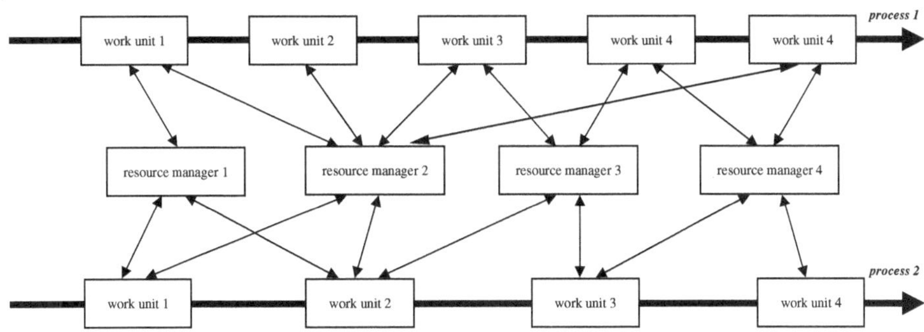

Fig. 1. Business processes and resource managers

managers, with local and global networks as their resources. Temporal exchange is supplied by database managers, with main and peripheral store as their physical resources and databases as their logical resources. The spectrum of assistance a resource manager offers to its customers is referred to as its *service*. Business processes, then, call upon the communications and database management services of a distributed information system.

Figure 1 illustrates the basic framework. A business process consists of a collection of work units. Each unit draws on the services of one or more resource managers, and different units may address the same manager. Assume for simplicity that each business process is totally ordered. Within a business process, then, resources are shared in temporal order. However, in general a large number of business processes – within the same enterprise or across different enterprises – take place in parallel.

From a service perspective, we refer to the resource managers as the *service providers* and the work units as the *service clients*.

2.2 Service Features

From an abstract perspective client and provider enter into a contract, a so-called *service agreement*. Crudely speaking, an agreement deals with two aspects: *What* is to be performed, and *how well* it is to be performed. For service features that make up the first aspect the provider is expected to give *absolute* guarantees, whereas for features within the second aspect *graded* guarantees can be negotiated. We refer to features of the first kind as *service functionality* and of the second kind as *service quality*. Ideally, service functionality would correspond to the desirable features and service quality to the constraints.

To start with the service functionality of distributed information systems, the features seem to fall into four broad categories.

1. *Utility*. This is the raison d'être for the agreement. It describes the collection of functions through which the client initiates spatial and temporal data exchange.

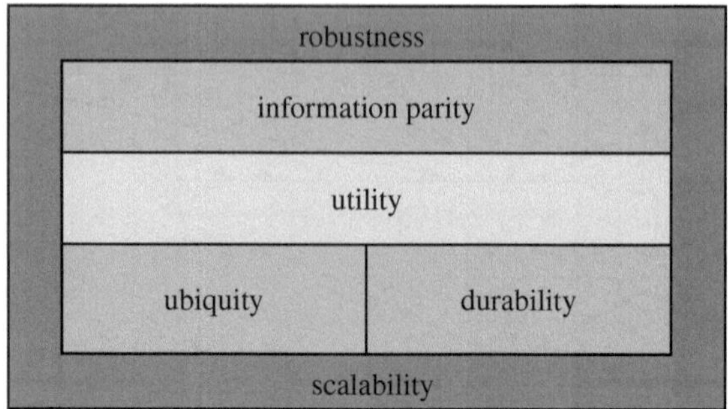

Fig. 2. Service functionality and qualities in distributed information systems

2. *Ubiquity.* Data exchange should be possible between any pair of clients at any time at any place. Data access should be possible for any client at any time from any place.
3. *Durability.* Access to stored data – unless explicitly overwritten – must remain possible at any time in an unlimited future.
4. *Information parity.* Data carries information, but it is not information by itself. To exchange information, the sender has to encode its information as data, and the receiver reconstructs the information by interpreting the data. Any exchange should ensure, to the extent possible, that the interpretations of sender and receiver agree, that is, that meaning is preserved. This requires some common conventions, e.g., a formal framework for interpretation. The requirement is more stringent for temporal exchange than for spatial exchange because the former does not offer an opportunity for direct communication to clear up any misunderstandings. Rather, in temporal exchange the conventions must be made known to the service provider to ensure that the interpretation remains the same on data generation and access.

For distributed information systems, service quality has two major aspects.

1. *Robustness.* The service must remain reliable under any circumstances, be they errors, disruptions, failures, incursions, interferences. Robustness must always be founded on a *failure model.*
2. *Scalability.* The service must tolerate a continuous growth of service requests, both for data transmission and data storage or retrieval.

Figure 2 illustrates how the service features interact. Ubiquity is primarily the responsibility of data communications, durability of database management. All other features are the shared responsibility of both.

As seen from the contract angle, architectural design is a parallel effort to contract negotiation. Given the service features, systems designers determine how they affect each other under further constraining factors such as limitations of physical resources. They decide how they may be traded against each other and, ultimately, whether the deal can be closed or terms must be renegotiated.

2.3 Service Dynamics

From Fig. 1 we observe three kinds of relationships in an information system.

1. *Client-provider.* Clients issue *service requests* to a provider by calling a service function. The provider autonomously fills the request and returns the result to the client. Client and provider run in separate processes, the communication may be synchronous or asynchronous.
2. *Client-client.* By sharing resources, business processes, through their work units, may interact. If the interaction is wanted because the processes pursue a common objective we have a case of *cooperation.* If it is unwanted we have a case of *conflict* between the processes.
3. *Provider-provider.* A business process may call upon the services of a number of providers. To ensure that the process reaches its final objective the providers involved must *coordinate* themselves.

2.4 Refining the Hypothesis

Chapter 2 provides us with a general framework for expressing the external factors that govern architectural design. Altogether we isolated six features, too many to be considered all at once. Hence, we refine our hypothesis to one that assumes that some features exert more influence than others. The major features are used to develop a gross architecture. A measure of correct choice would be that the other features affect only a single component of the gross architecture, or add a single component to it. We refer to this property as *design orthogonality.*

The ensuing four chapters will test the hypothesis in retrospective for various database system architectures. Some of them (Chaps. 3 and 6) are widely accepted, others (Chaps. 4 and 5) are less so that this paper could even be regarded as a – modest – original contribution.

3 Database Management Systems Reference Architecture

3.1 Service Features

Our focus is database management systems (DBMS). To indicate the focus, we use a specialized terminology for the features.

1. *Data model.* A data model expresses DBMS utility. The utility is generic: Due to the huge investment that goes into the development, DBMS must be capable of supporting a large and broad market of applications. As such a data model provides a collection of primitive state spaces and transition operators, and additional constructors that allow these to be grouped into more complex state spaces and transition procedures, respectively. More formally speaking, a data model can be compared to a polymorphic type system. Operators and procedures correspond to the service functions that may be called by clients. Scalability has a functional counterpart in a constructor for dynamic sets of record-structured elements.

2. *Consistency.* Given a state of the business world, the goal is to have the database reflect a suitable abstraction of this state (the miniworld) in an up-to-date version. Information parity is thus refined to a notion of restricting the content and evolution of the data store to a well-defined set of states and state transitions considered meaningful. Each state of the database is to be interpreted as a particular state of affairs in the miniworld. An update to the database intended to reflect a certain change in the miniworld is indeed interpreted as that same change by every observer of the database. However, since a DBMS cannot divine the true state of the miniworld, it would be too much to ask for preservation of meaning in this ideal sense. Instead, one settles for lesser guarantees. *Static consistency* means that the current state of the database is always drawn from a predefined set of consistent database states, which are considered to be valid and unambiguous descriptions of possible states of the miniworld. *Dynamic consistency* ensures that state transitions take consistent states to consistent states. To enforce the two, the appropriate sets of states and state transitions need to be defined in a *database schema.* Usually a database schema can be considered as a database type together with further state constraints. The polymorphic transition operators ensure consistency by observing the schema. Generally speaking, then, consistency refers to the degree of information parity between database and miniworld.

3. *Persistency.* Durability calls for the preservation of data on non-volatile storage, i.e. on a medium with an (at least conceptually) unlimited lifetime. Moreover, preservation should be restricted to those database states that are regarded consistent. Such states are called persistent. As a rule, only the outcome of executing a transactional procedure (called a (database) *transaction*) is considered to be persistent.

4. *Resilience.* A robust DBMS must be able to recover from a variety of failures, including loss of volatile system state, hardware malfunctions and media loss. All failure models to achieve robustness should be based on the notion of consistency. A distinction must be made, though, whether a transaction is affected in isolation or by conflict with others. In the first failure case the failure model causes the database to revert to an earlier persistent state or, if this proves impossible, to somehow reach an alternative consistent state. In the second case, the failure model is one of synchronizing the conflicting transactions so that the result is persistent both from the perspective of the individual transaction (internal consistency) and the totality of the transactions (external consistency).

5. *Performance.* DBMS functionality must scale up to extremely large volumes of data and large numbers of transactions. Scalability will manifest itself in the *response time* of a transaction. System administrators will instead stress *transaction throughput.* Collectively the two are referred to as DBMS performance.

In classical centralized database management systems, *ubiquity* is usually ignored. The topic becomes more important in networked database management systems.

3.2 Physical and Economical Bottlenecks

As noted in Sect. 2.2, utility is the raison d'être for DBMS, but so is durability. Moreover, both are closely intertwined: the former assumes the latter. Consequently, the first step is to analyze whether any of the remaining features directly affects one of them and thus indirectly the other. We claim that the feature that exerts a major influence is performance.

The effect of performance is due to constraints that originate with the physical resources. Even after decades durability is still served almost exclusively by magnetic disk storage. If we use processor speed as the yardstick, a physical bottleneck is one that slows down processing by at least an order of magnitude. The overwhelming bottleneck, by six orders of magnitude, is access latency, which is composed of the movement of the mechanical access mechanism for reaching a cylinder and the rotational delay until the desired data block appears under the read/write head. This bottleneck is followed by a second one of three orders of magnitude, transmission bandwidth.

We note that even main memory introduces one order of magnitude. Computers attempt to overcome it by staging data in a fast cache store. DBMS should follow a similar strategy of *data staging* by moving data early enough from peripheral to main storage. But now we observe a second, economical bottleneck: The price per bit for main memory is by two to three orders of magnitude higher than for magnetic disk. Data staging that finds a suitable balance between physical and economical bottlenecks will thus have to be one of the principles guiding the architectural design of DBMS.

3.3 Primary Tradeoff: Balancing Data Model and Performance

Section 3.2 suggests that utility and performance should dictate the design strategy, i.e., that they should have first priority. We thus introduce two diametrically opposed design directions: A top-down direction of mapping the data model to the physical resources, and a bottom-up direction of lessening the bottlenecks by data staging (Fig. 3). This is a complicated undertaking that by tradition requires a stepwise approach in order to break up the decision space into smaller portions. The result is a multi-layered architecture.

Determining the decision space for each layer is the paramount challenge. Our conjecture is that we have to find suitable abstractions for both utility and performance such that both match for the purpose of balancing the needs. Utility is in terms of function mapping, and performance in terms of criteria for "early enough" data staging. Figure 4 illustrates the approach. Mapping utility downwards is equivalent to reducing the expressiveness of the data model. Moving performance upwards is equivalent to widening the context for determining what "early enough" means.

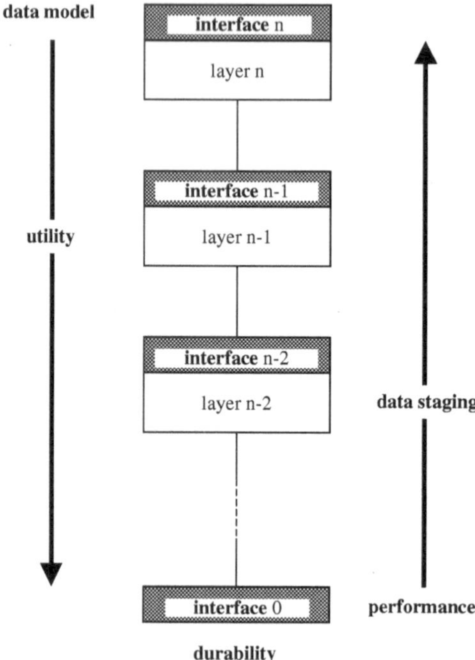

Fig. 3. Architectural design centered on data model and performance

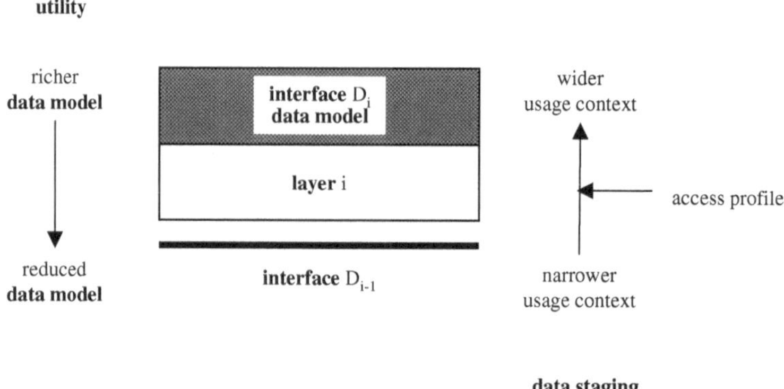

Fig. 4. Balancing functionality and performance within a layer

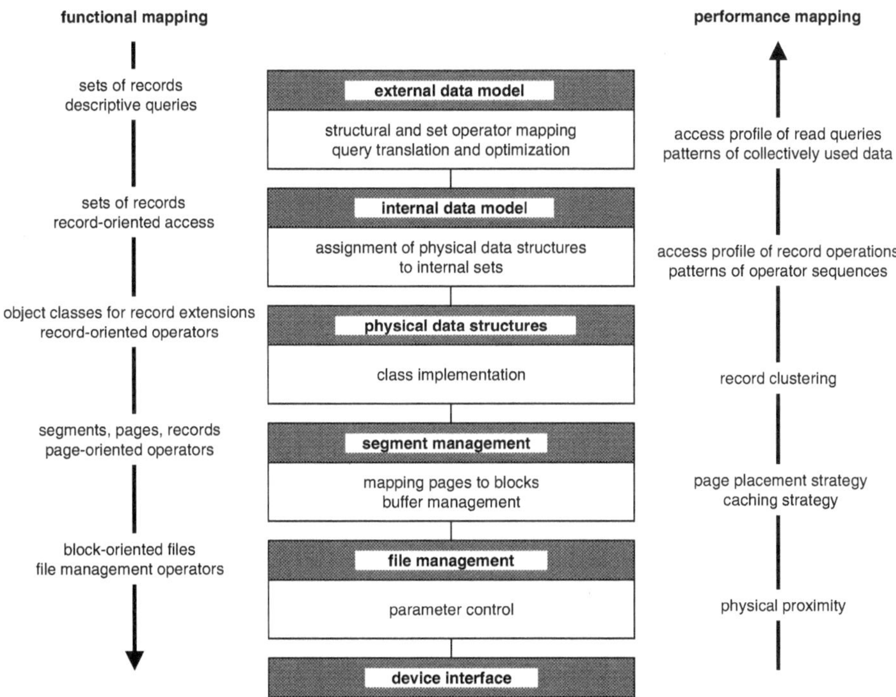

Fig. 5. Reference architecture for set/record-oriented database management systems

A well-known reference architecture for DBMS uses five layers. Figure 5 shows the architecture. We demonstrate that this architecture can be explained in terms of the two factors data model and performance, following the principal ideas just mentioned.

1. We assume a service-level data model that is set/record-structured and uses a set-oriented and, hence, descriptive query language such as SQL or OQL. For data staging on the topmost layer we assume a predominance of read queries. Consequently we examine the sequence of read queries and analyze the result for frequent patterns of collectively retrieved data (read profile). Data would then be staged by rearranging the database according to these patterns. This in turn determines the data model for the next lower layer (internal data model): It should again be set/record-structured.
2. We now turn to the functional mapping. Some reduction in expressiveness should take place. Since structurally there is little difference, the difference can only be in the operators. These will now be record-oriented, i.e., navigational. The mapping encompasses three aspects. One is the actual rearrangement of the external structures. The second concerns the translation of the queries into set-algebraic expressions over the rearranged database, the algebraic and non-algebraic optimization of the expressions with a cost function that minimizes estimated sizes of the intermediate results. The third pro-

vides implementations of the algebraic operators in terms of the next lower data model.

3. We alternate again with data staging. We analyze the entire query profile, now in terms of the record operators. Our hope is to find characteristic patterns of operator sequences for each internal set. Each pattern will determine a suitable data organization together with operator implementations. Consequently, the data model on the next lower layer offers a collection of something akin to classes (physical data structures).
4. Back to the functional mapping we mainly assign physical data structures to the internal sets.
5. At this point we change direction and start from the bottom. Given the storage devices we use physical file management as provided by operating systems. We choose a block-oriented file organization because it makes the least assumptions about subsequent use of the data and offers a homogeneous view on all devices. We use parameter settings to influence performance. The parameters concern, among others, file size and dynamic growth, block size, block placement, block addressing (virtual or physical). To lay the foundation for data staging we would like control over physical proximity: adjacent block numbering should be equivalent to minimal latency on sequential access. Unfortunately, not all operating systems offer this degree of control. The data model is defined by classical file management functions.
6. The next upper layer, segment management, is particularly critical from a functional perspective in that it forms a bridge between a world that is devoid of service content and just moves blocks of bytes around, and a world which has to prepare for the services. The bridge maintains the block structure, but makes the additional assumption that blocks contain records of much smaller size. This requires a more sophisticated space management within as well as across blocks and an efficient dynamic placement and addressing scheme for records. Because of all this added value the data model refers to blocks as pages and files as segments. The operators determine access to pages and records and placement of records.
7. The segment management will also determine the success of data staging on the upper layers. For one, if the physical proximity of blocks is to be exploited for pages, mapping of pages to blocks is critical (placement strategy). Even more critical is how the layer exploits main memory to improve page access by three to five orders of magnitude. Use is made of large page buffers. Aside from buffer size the crucial factor is the predictive model of "early enough" page loading (caching strategy).
8. This leaves the details of the physical data structures layer. Given a page, all records on the page can be accessed with main memory speed. Since each data structure reflects a particular pattern of record operations, we translate the pattern into a strategy for placing jointly used records on the same page (record clustering). The functional mapping is then concerned with the algorithmic solutions for the class methods.

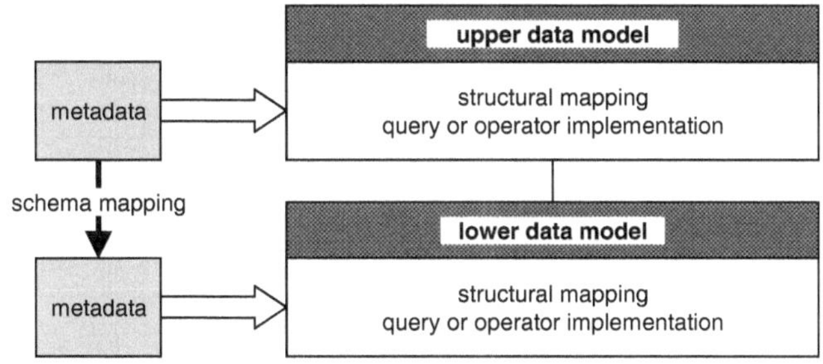

Fig. 6. Principle for adding schema consistency

3.4 Schema Consistency

If data model and performance are indeed the most critical design factors, then the remaining features should only have an orthogonal influence.

Consider the functional mapping in Sect. 3.3. It was entirely based on data models and, hence, generic. One of the features to be added at this point is schema consistency. Schema consistency is equivalent to type safety in programming. In generic solutions type checking is done at runtime. Consequently, type information – often referred to as *metadata* – must be maintained by the operations in each layer. The content of the metadata on each layer is derived from the metadata on the next upper level, by a mapping that is determined by the functional mapping for the data models. On the topmost layer the database schema constitutes the metadata. Figure 6 illustrates the principle, and demonstrates that consistency is indeed orthogonal to the data model.

Since from a data management perspective metadata is just data, the metadata of all layers are often collected into a separate repository called the *data dictionary*, thus perfecting the orthogonality (Fig. 7).

3.5 Consistency, Persistency and Resilience

Transactions define achievable consistency and persistency. They also incorporate the failure model for resilience – atomicity for failures and isolation for suppression of conflicts. Such transactions are said to have the ACID properties. Transaction management consists of three components: a transaction coordinator that does all the interaction with the clients and the necessary bookkeeping, a scheduler that orders the operations of a set of concurrent transactions to ensure serializability and recoverability, and a recovery manager that guarantees persistency and resilience.

Since buffer managers are in charge of a critical part of performance they operate fairly autonomously. On the other hand, atomicity requires close collaboration between recovery manager (including a log manager) and buffer manager. Therefore, recovery managers are made an integral part of segment management. If we wish to achieve orthogonality we should concentrate most or all other tasks of transaction management within this layer, i.e., within segment management.

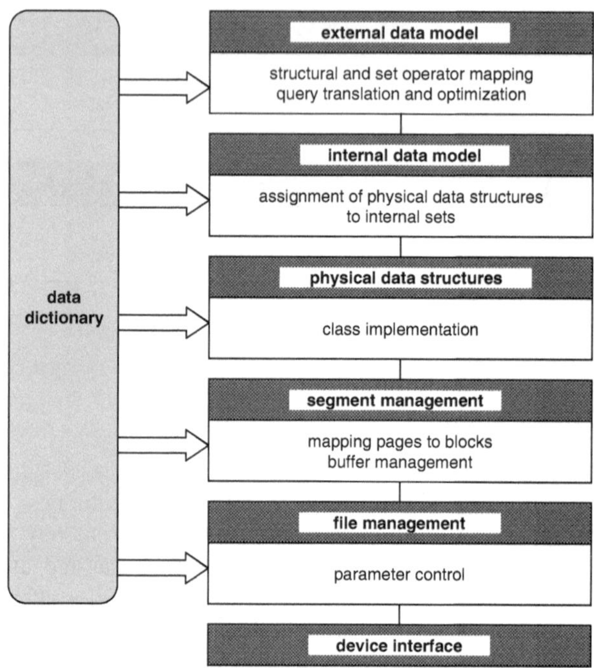

Fig. 7. Data dictionary as an orthogonal system component

As a consequence, we make schedulers a part of segment management as well, and synchronize on the basis of page accesses. Transaction coordinators remain outside the basic architecture because they are uncritical for performance. Figure 8 illustrates the solution.

4 Semistructured Database Management Systems

4.1 Service Features

In Chap. 3 we identified the data model as one of the predominant features. If we continue to verify our refined hypothesis, it makes sense to modify the assumptions for the data model. In this section we replace the data model of sets of small-size records and set operations by a data model for semistructured databases. The model reflects attempts to take a database approach to the more or less structured document databases of the Web or classical information retrieval.

1. *Data model.* The record of the reference model is replaced by a graph structure. A node is labeled by a non-unique tag, and even siblings may have identical tags. A node may include text of any length and optionally a traditional attribute/value list. From the system perspective the text is atomic.

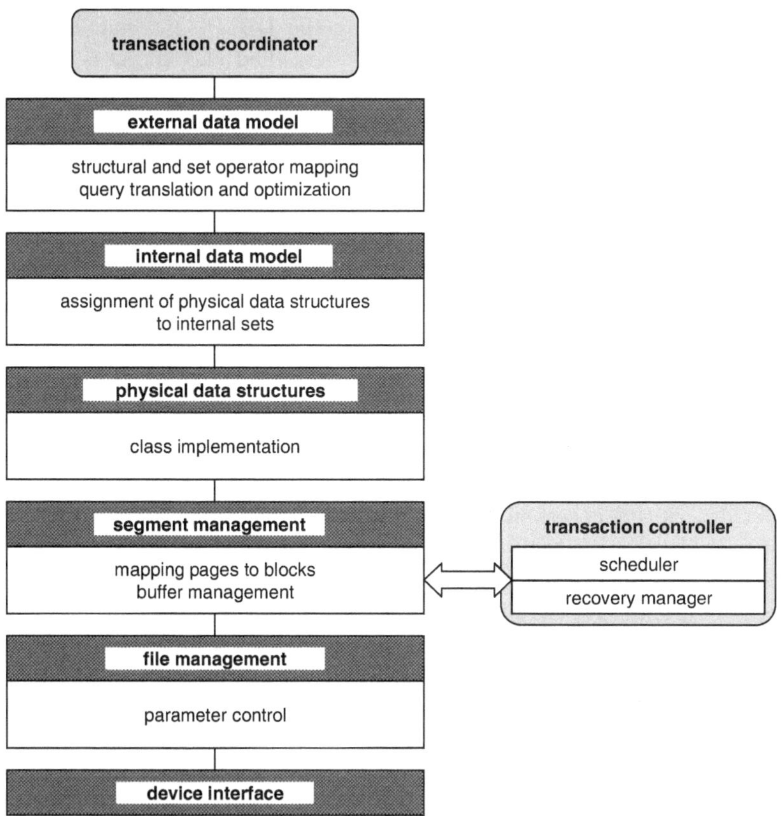

Fig. 8. Incorporating transaction management

Retrieval is considered the predominant service function and is, for all or parts of the graph structure and even across sets of graphs, by pattern matching. Changes to a graph structure, no matter how complicated, are treated as atomic.

The choice of data model has consequences for some of the other features.

2. *Consistency.* Regularity of structured databases is the basis for schema consistency. Whether the DBMS can guarantee information parity even in the limited form of schema consistency depends on the degree of regularity that one may observe for a semistructured database. Regularity will have to be defined in the less restrictive form of regular expressions. Graphs added to the database must then conform to the schema. Because consistency is defined on single graphs and updates to them are already atomic, transactional procedures seem to play a lesser role.
3. *Persistency.* With lesser need for transactions, durability usually implies persistency.
4. *Resilience.* Without transactions, a different failure model must be employed. Failures may occur as well, but there are no longer any fallback positions

that can automatically be recognized by the DBMS. Instead, clients must now explicitly identify persistent states, so-called checkpoints, to which to revert on failure. Since changes to a graph structure are atomic, the failure model does not have to take interferences into account.

5. *Performance.* The physical bottleneck remains the same. Consequently, data staging will have to remain the primary strategy for performance enhancement.

4.2 Architectures

Low Consistency. The notion of database schema makes sense only in an environment with some degree of regularity in database structures. Consequently, where there is no such regularity there is little to enforce in terms of consistency. There are further drawbacks. Little guidance can be given to the analysis of access profiles, and since there is no notion of type extension, little opportunity exists for descriptive access across sets and thus for query optimization. Hence, the upper two layers in terms of the reference architecture of Chap. 3 can exert only very little influence on performance, and there are no major challenges to functional mapping. We may thus collapse the layers into a single one.

There is a strong notion of physical proximity, though. Proximity is defined by the topology of a graph structure, and because reading access is by pattern matching, the topological information should be clustered on as few file blocks as possible. Since long text fields in the nodes are the biggest obstacle to clustering, these should be separated out and kept on other blocks, again in a clustered fashion. Contrary to the reference architecture of Chap. 3 where, due to small record sizes, the individual page is the primary object of concern, performance now dictates that clusters of blocks are considered.

Two tasks remain, then. For performance, clusters of adjacent blocks of data must be staged in main memory. For utility, the graph topology and node fields must be mapped to block clusters. The tasks become part of a layer that replaces the layers of physical data structures and segment management in the reference architecture. With a transaction concept lacking, no further factors exert an influence. Figure 9 shows the result of the discussion.

Schema Consistency. The solution of the previous section treats semistructured databases in isolation. This is against recent trends of combining the traditional structured databases of Chap. 3 with semistructured databases. Suppose that for reasons of *transparency* we maintain a single data model, and we agree on semistructured data as the more general of the two data models. Clearly though, incorporating structured databases forces us to include consistency as an important design factor. In turn we need graph structures that are similar in terms of the underlying regular expressions. These will have to be classified into types. Consequently, even though one will observe less regularity across their instances as in the traditional case there is a notion of database schema possible. To gain set orientation, access should cover complete type extensions so that

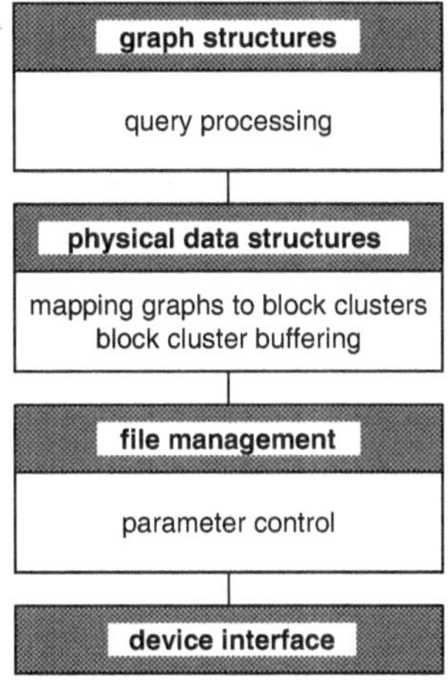

Fig. 9. Architecture for low-consistency semistructured database systems

query languages may again be descriptive, though their concern is retrieval only. Analyses of retrieval profiles may be done along the lines of the graph types. These are all arguments that favor an upper layer similar to the one in the reference architecture of Chap. 3. In the details of functional mapping this layer will in all probability be much more powerful.

From a performance standpoint the retrieval patterns should have an influence. Our conjecture is that a notion of proximity can again be associated with these patterns. If we assume that reading access is limited to extensions of single type, proximity can be translated to subgraphs that correspond to query patterns. Performance mapping, then, consists of dividing the original graph into (possibly overlapping) subgraphs. As above, the graphs should be separated into topology and long text fields or, more generally, arbitrary media data. Functional mapping now includes query optimization.

Subgraphs and long fields require different underlying implementations. In fact, by closer scrutiny of the subgraphs we may detect that some exhibit the strong regularity of traditional structured databases. Hence, there is a need for assigning different physical data structures and, consequently, a need for the second layer of our reference architecture. However, whereas the uppermost layer seems to be more complicated than the one in Chap. 3, the reverse seems to be true for the second layer.

For the implementation we take a two-track approach. There is a good chance that subgraphs have a size that can be limited to single blocks. Hence, their implementation may follow the reference architecture of Chap. 3 from the physical data structures layer on downwards, though in detail the techniques may differ. Fields with unstructured media data should follow the architecture of Fig. 9. Figure 10 summarizes the discussion.

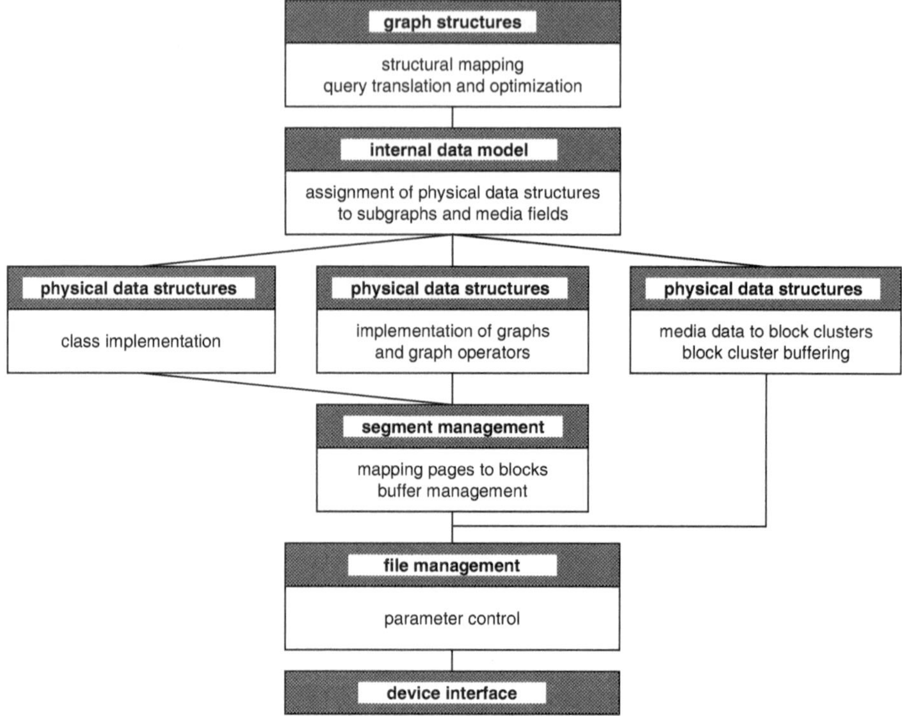

Fig. 10. Architecture for schema-consistency semistructured database systems

5 Multimedia Databases

5.1 Service Features

Even though one may attach any type of media data to a graph node, and the types may vary, semistructured databases are by no means multimedia databases. They become multimedia databases only if we offer services that take note of the interrelationships between different media and, hence, are capable of combining two or more media. The new service functionality, *cross synchronicity*, ensures that the contents of the related media match along a time scale. Take as an example the continuous playback of combined video and audio.

Cross synchronicity relies on *playback synchronicity*, the ability of a system to present media data under temporal conditions identical to those during recording. As a service feature, playback synchronicity straddles the line between functionality and quality. Poor playback synchronicity limits utility, but there is a certain tolerance as to speed, resolution and continuity of playback within which utility remains preserved. Hence, playback synchronicity could also be regarded as a service quality.

5.2 Architecture

To study the architectural effects we presume that cross synchronicity is a service functionality and playback synchronicity a service quality. Since cross synchronicity is a feature over and above those that gave rise to our reference architecture, we should try to treat it as an orthogonal feature, and encapsulate it in a separate layer or component somewhere on the upper levels.

Playback synchronicity seems closely related to performance, because it has to deal with the same kind of bottleneck – peripheral storage. However, conditions are even more stringent. Playback must remain fairly continuous over minutes up to an hour or so. Longer breaks or jitter due to relocation of access mechanisms or contention by other processes must stay within guaranteed limits. Hence, depending on speed special storage devices (stream devices) or dedicated disk storage are employed. Special buffering techniques will have to even out the remaining breaks and jitter. The layers further up should intervene as little as possible. Hence, for continuous media the right-hand branch in Fig. 10 must become even more specialized. The physical data structures layer should now restrict itself to storing an incoming stream of media data and managing it. Discontinuous media data and graph structures could follow the architecture of Fig. 10.

Now, given playback synchronicity of each individual media data, the upper layers of Fig. 10 may be preserved to deal with the structural aspects of the interrelationships between the media, again intervening only during storing of the data and during the setup phase of playback. On top we add another layer that controls cross synchronicity, usually along the lines of a given script. Figure 11 gives an overall impression of the architecture.

6 Networked Databases

6.1 Service Features

We return to Fig. 1. In distributed information systems the single business process and even a work unit faces a multiplicity of resource managers. All these may be placed at geographically disparate locations. Consequently, *ubiquity* enters the picture as an additional service functionality.

In Sects. 2 and 3.1 the service functionalities of consistency and persistency and the service quality of resilience where tied to the notion of database transaction and, hence, to a single resource manager. In general, these features ought to be tied to the business process as a whole. Technically then, we refer to a business process as an *application transaction*. Usually, an application transaction is distributed. Since each component deals with transactional properties on an individual basis, *coordination* comes into play as a further service functionality.

Service clients as well as service providers differ in a large number of characteristics. Some of these are entirely technical in nature, such as differences of hardware and operating system platforms or the transmission protocols they understand. These discrepancies are referred to as technical heterogeneity. Other

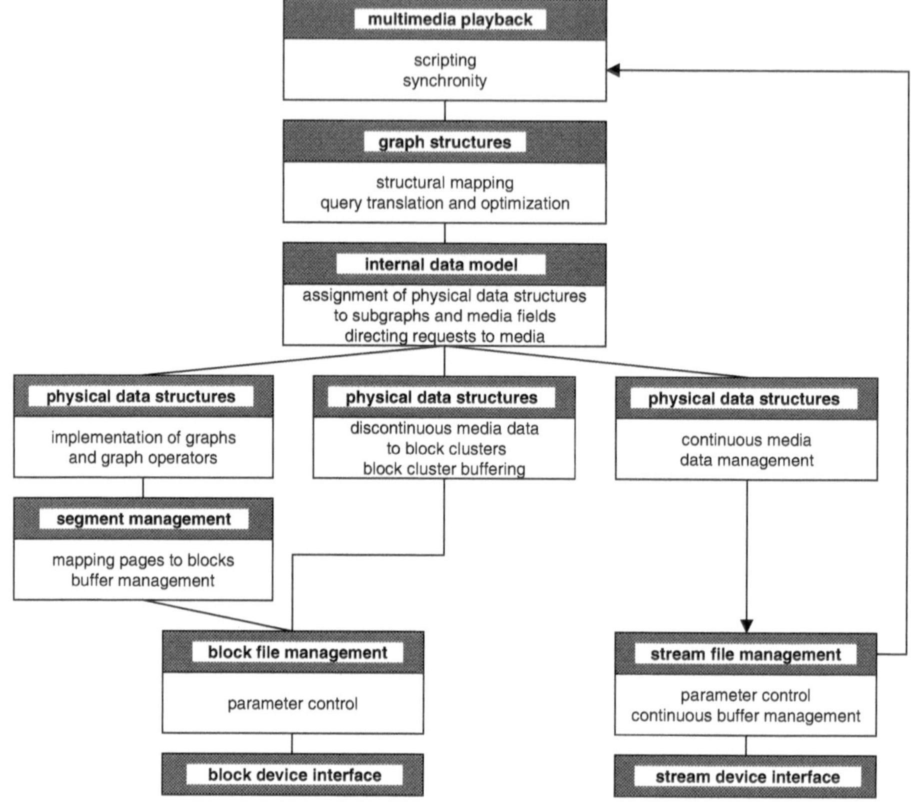

Fig. 11. Architecture for multimedia database systems

differences – noted as semantic heterogeneity – have to do with functionality, e.g., different data models, or consistency, e.g., different schemas. How much heterogeneity is acceptable seems more of a gradual decision. Consequently, we add a new service quality: *homogeneity of services.*

6.2 Middleware

If we wish to prove our hypothesis, we should follow our previous architectural strategy and realize the new features orthogonally whenever possible. In the case of distributed information systems, the strategy translates into touching the resource managers only lightly, and adding infrastructure that deals with the new features. This infrastructure goes under the name of *middleware.*

The first issue to deal with is ubiquity. The issue is resolved by utilizing the data communications infrastructure and employing a common high-level protocol, e.g. TCP/IP. The second is technical homogeneity. Middleware enforces the homogeneity by establishing an internal standard (take the IIOP protocol of CORBA) and requiring site-local adapters. Practically all modern middle-

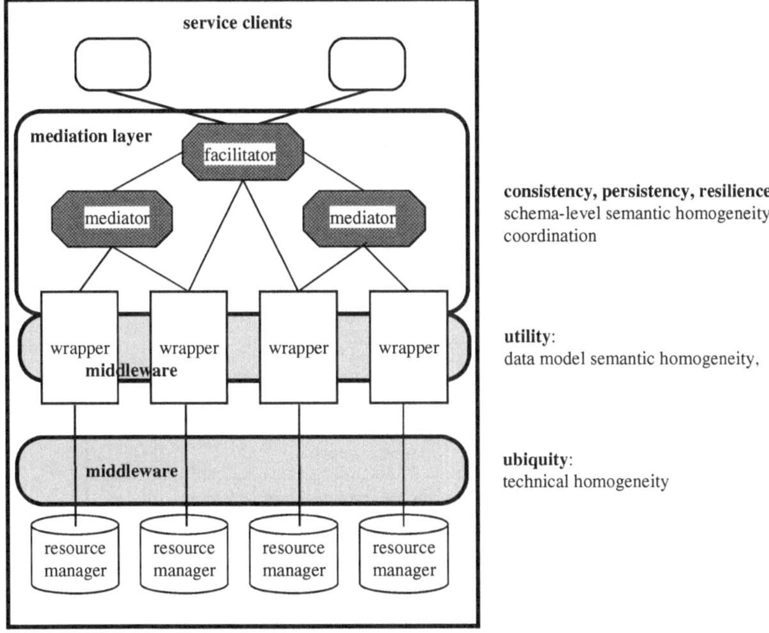

Fig. 12. Reference architecture for middleware

ware approach semantic homogeneity by setting their own data model standard (e.g., HTML documents for the WWW, remote objects for CORBA, DCOM and RMI).

This leaves as tasks those that are much more difficult to standardize, essentially all those that have to do with content. Content affects consistency and derived features such as persistency or resilience. If one cannot come up with standards, meta-standards may help. The architectural representation of meta-standards is by frameworks. A framework that is oriented towards semantic homogeneity on the schema level, and coordination issues, is the I^3 (Intelligent Integration of Information) architecture. Figure 12 combines the middleware and I^3 approaches into a single architecture.

Wrappers, mediators and facilitators are part of the I^3 framework. Wrappers adapt their resource managers on the semantic level, by mapping the local data model to the common, standardized data model and the schema to one expressed in terms of the common data model. Mediators homogenize the schemas by overcoming discrepancies in terminology and structure. Increasingly they rely on ontologies that are represented by thesauri, dictionaries, catalogues, or knowledge bases. Other mediators accept queries that span several resources, translate them according to the homogenized schemas, send them off to the resource managers, and collect the homogenized results. Facilitators ease the orientation of service clients in the network. Examples of facilitators are Web search engines or catalogues of data sources.

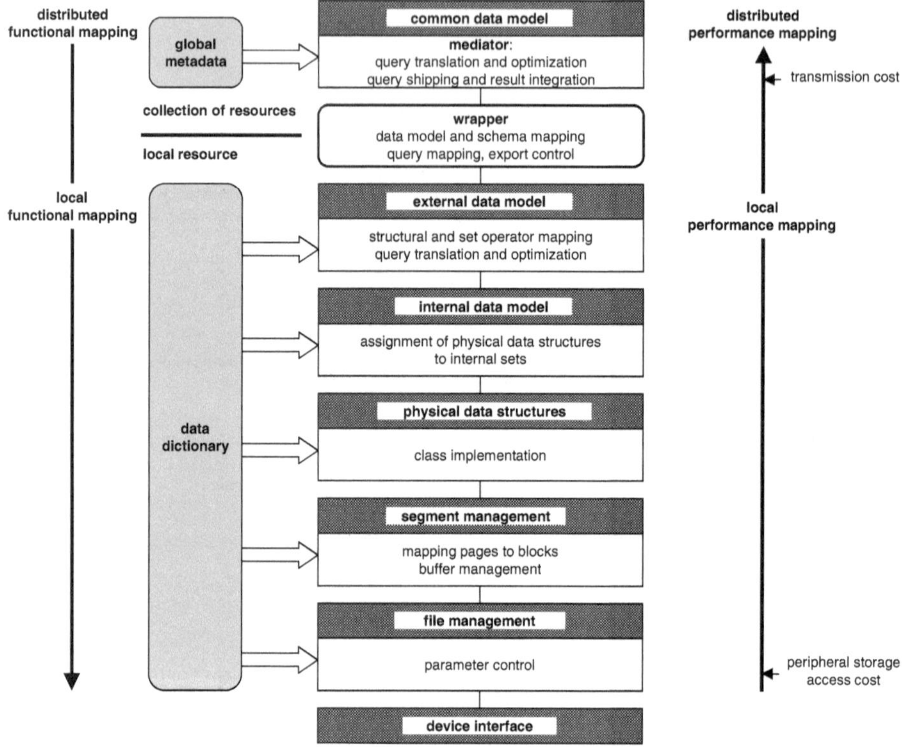

Fig. 13. Database management system in a network

6.3 Data Model, Consistency and Performance

Local utility and consistency are the responsibility of the individual resource manager. Collective utility and consistency are the realm of the mediator. From the viewpoint of the individual database management system all aspects of global utility are an add-on service functionality and, hence, relegated to a new top layer. Figure 13 illustrates this logical view. The view leaves open whether the mediator is a centralized component or distributed across the resource managers.

Mediators achieve on a network-wide basis what the external data model layer does locally: query translation, query optimization, query shipping and result collection and integration. These are tremendously complicated issues, but nonetheless orthogonal to all the local tasks. The mediator is supported by metadata structures such as the catalogues and ontologies mentioned before, or a global schema (the so-called federated schema).

The wrapper, besides mapping the schema, queries and results to the common data model, acts as a kind of filter in order to make visible only those parts of the database which are globally accessible (the so-called export schema).

Because the local database system remains unaffected, all new performance bottlenecks arise from distribution. In the past it was transmission time that dominated the mediation strategies, and it remains so still today to the extent that bandwidths have not kept pace with processor speed. Independent of speed, latency, i.e. the time delay between request and start of transmission, remains a serious handicap. Therefore, query optimization is largely governed by transmission cost.

6.4 Consistency, Persistency and Resilience

Globally, the service functionalities of consistency and persistency and the service qualities of concurrency and resilience are tied to an application transaction. Locally, the functionalities and qualities are guaranteed by ACID database transactions. We limit our considerations to application transactions that also are ACID. Consequently, unless it suffices for global isolation and atomicity to rely solely on local isolation and atomicity, some global – possibly distributed – coordination mechanisms must be introduced.

It is well known from transaction theory that local isolation is the basis for any kind of global isolation. Consequently, conflicts among business processes (see Fig. 1) are handled purely locally. Likewise it is well known that global atomicity – global persistency and resilience – require coordination in the form of global commit protocols. These protocols are centralized in the sense that they distinguish between a coordinator and its agents, where the agents reflect those resource managers that updated their databases. The most widely used protocol, Two-Phase-Commit (2PC), requires that even after local commit the agents are capable, during the so-called uncertainty phase, of rolling back the transaction. In purely local transaction management an uncertainty phase does not exist. As a consequence, the recovery manager must suitably be adapted.

This leaves the question of where to place coordination and, incidentally, transaction monitoring. Both are generic tasks. If we assume that application transactions pass through the mediator, it seems only natural to place the tasks with the middleware (Fig. 14). And indeed, this is what middleware such as CORBA and DCOM try to do. In fact, today's understanding of many of the earlier transaction processing systems is that of middleware.

In summary, accounting for distributed transaction management is still close to orthogonal: A global component is added, and a single local component is adapted.

7 Conclusions

Does the paper support our hypothesis that architectural design plays *the* strategic role in identifying, articulating, and then reconciling the desirable features with the unavoidable constraints under which a system must be developed and will operate? Is the design philosophy and methodology with a first phase for identifying the desirable features and the constraints and a second phase for determining the tradeoffs the correct consequence? We claim the paper does.

Equally important – and this seems the novel aspect of our approach and thus the contribution of the paper – is a refinement of the strategy that challenges

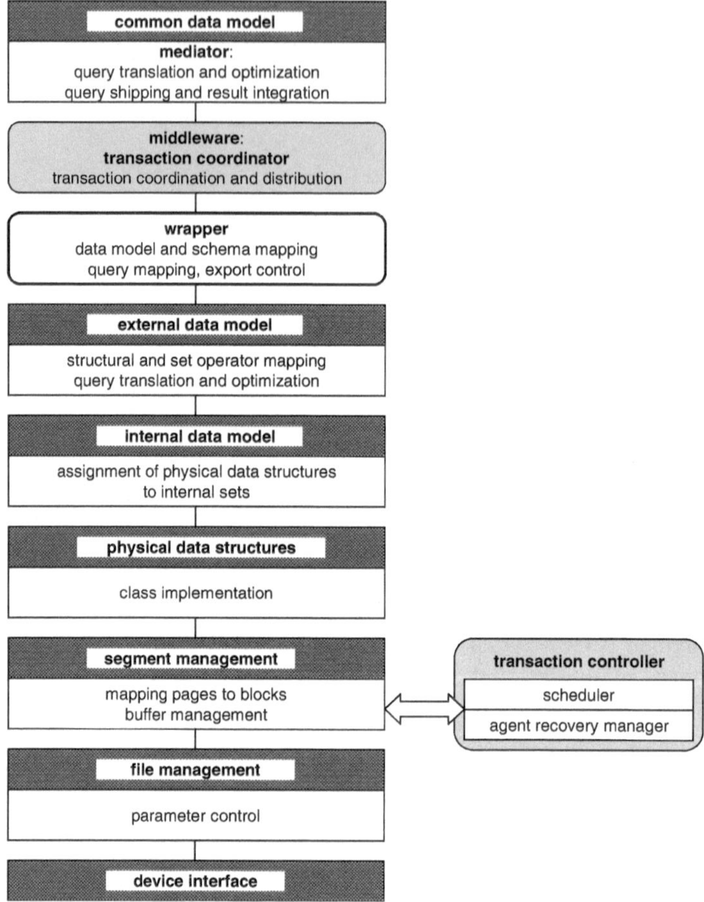

Fig. 14. Incorporating global transaction management

designers to identify – very much in the abstract – those features that promise to dominate the design of the architecture. If successful, the remaining features in the second class can be treated orthogonally. Orthogonality may either be perfect so that the features give rise to additional components, or at least sufficiently strong so that the features can be taken care in a single component within the architecture developed so far.

The proof was by circumstance and limited to one kind of system – database management systems in networks. Architectures that have proven their worth in the past were evaluated in retrospective. Others – DBMS architectures for semistructured and multimedia databases – still vary widely, and the ones we developed in this paper reflect those in the literature that sounded most convincing to the author. Also, there is by no means universal agreement on how to divide the responsibilities for the service features between resource managers and middleware. Our hope is that this paper will contribute to future design decisions.

One may argue that our base was way too small and too specialized to render statistically significant evidence. Clearly, the study should continue to cover other kinds of systems. Our next candidate for attention is data communications systems. Nonetheless, it should have become clear that a design strategy based on our hypothesis is little more than a conceptual framework. Architectural design of software systems remains a creative task, albeit one that should follow sound principles of engineering.

8 Bibliographic Notes

The reference architecture of Chap. 3 is due to T. Härder and A. Reuter and is itself based on the early System R architecture. A modern interpretation can be found in Härder, T.; Rahm, E.: *Datenbanksysteme: Konzepte und Techniken der Implementierung.* Springer, 1999 (in German). Many of the numerous and excellent textbooks on database systems that have appeared in the more recent past use similar architectures as a reference. Where publications on commercial database products present system architectures (unfortunately not too many do) they seem to indicate that overall the same principles were applied. A careful analysis of peripheral storage as the performance bottleneck is given in Gray, J., Graefe, G.: *The Five-Minute Rule Ten Years Later, and Other Computer Storage Rules of Thumb.* ACM SIGMOD RECORD, 1998. Among the textbooks and publications on semistructured database systems hardly any deal with issues of architecture to any detail. Of those on multimedia databases the situation is only slightly better. Architectures can be found in Apers, P.M.G.; Blanken, H.M.; Houtsma, M.A.W. (eds.): *Multimedia Databases in Perspective.* Springer, 1997 and Chung, S.M. (ed.): *Multimedia Information Storage and Management.* Kluwer Academic Publ., 1996.

The classical textbook on transaction management, which starts from an architectural view is Gray, J., Reuter, A.: *Transaction Processing: Concepts and Techniques.* Morgan Kaufmann Publ., 1993. An architectural approach is also taken by Bernstein, P.A., Newcomer, E.: *Principles of Transaction Processing.* Morgan Kaufmann, 1997. For middleware the reader is referred to the numerous literature over the past few years. A good pointer to the I^3 architecture is Wiederhold, G.: *Intelligent Integration of Information.* Kluwer Academic Publ., 1996, whereas the details have remained in draft form: Arens, Y. et al.: *Reference Architecture for the Intelligent Integration of Information.* Version 2, ARPA Tech. Report.

Acknowledgements

The author is grateful to Gerd Hillebrand for his thoughtful comments and discussion on an earlier version of the paper. Klaus Dittrich through his comments helped sharpen the focus of the paper.

Evolution not Revolution: The Data Warehousing Strategy at Credit Suisse Financial Services

Markus Tresch and Dirk Jonscher

CREDIT SUISSE FINANCIAL SERVICES
Technology & Services
8070 Zurich, Switzerland
{markus.tresch, dirk.jonscher}@csfs.com

Abstract. Data Warehousing is not new to Credit Suisse Financial Services. Over the past twenty years, a large number of warehouse-flavored applications was built, ranging from simple data pools to classical management information systems, up to novel customer relationship management applications using state-of-the-art data mining technologies.
However, these warehouse projects were neither coordinated nor are they based on the same infrastructure. Moreover, dramatic changes of the business design had a huge impact on information analysis requirements. Both together resulted in a nearly unmanageable complexity.
Therefore, Credit Suisse Financial Services started a 3-year enterprise-wide data warehouse re-engineering initiative at the beginning of 1999. This paper presents the motivation, experiences, and open issues of this strategic IT project.

1 Introduction

Driven by urgent business needs, various data warehouse-like systems were built in Credit Suisse Financial Services (CSFS) over the past twenty years. The early systems are mere data pools serving batch reporting purposes for financial and management accounting. The second generation of data warehouses supports preliminary ad-hoc analysis capabilities and serves credit, payment, and securities businesses. Customer relationship management (CRM) started ten years ago with a data pool dedicated to customer profitability analysis. The system was recently replaced with a state-of-the-art warehouse now supporting data mining and on-line analytical processing (OLAP).

Recent trends are nearly real-time data warehouses (based on operational data stores) and hybrid systems, where analysis results are fed back into online transaction processing systems (e.g. reclassification of banking customers in accordance with analysis results). Moreover, ERP packages, like SAP or Peoplesoft, are advertising their own interpretation of data warehousing.

K.R. Dittrich, A. Geppert, M.C. Norrie (Eds.): CAiSE 2001, LNCS 2068, pp. 36–45, 2001.
© Springer-Verlag Berlin Heidelberg 2001

Moreover, a large number of applications implemented warehousing and analysis functionality, like for example, economic research, credit early warning, comprehensive risk analysis, click stream analysis in e-business, or market positioning.

Summing up, data warehouses in CSFS have very complex application and data dependencies. Data sharing is not coordinated, data sources generate multiple (overlapping) feeder files for different warehouses (resulting in unnecessary load at the feeder systems), and there are cascading warehouses, etc. (cf. Fig. 1). Even worse, there is no global view which data are stored where and are used for which purpose.

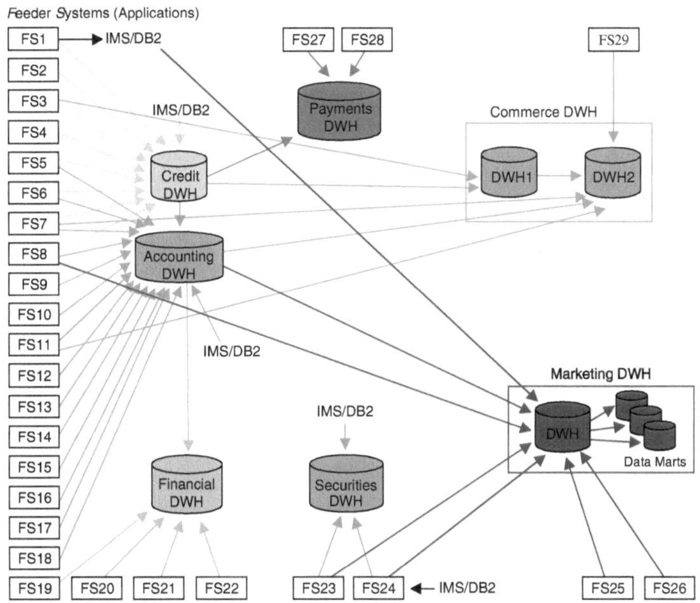

Fig. 1. Historically evolved data warehouse environment

From the technical point of view, many different platforms, tools, and architectures are in place. Platforms include Mainframe / DB2, IBM RS 6000 SP2 / DB2, Sun E10000 / Oracle and Mainframe / SAS. Tools like SAS, ETI, PowerCenter, Trillium, Darwin, Clementine, Microstrategy and Brio are heavily used. However, most of the old systems use their own home-made analysis tools. The systems are in different states of the product life cycle (from antic to state-of-the-art). Consequently, each warehouse was maintained by a dedicated team of IT people resulting in a very poor overall productivity.

From a business perspective, there is an end-to-end ownership and sponsorship of each data warehouse system.

2 Strategy of Managed Evolution

An in-depth analysis of the existing data warehousing systems revealed the following:

- Data warehouse development has become expensive and slow because of the high complexity of the historically evolved environments and the very different warehouse platforms being used. Even worse, a stable production environment could not be provided any more.
- A high percentage of resources were required to *maintain* old systems instead of developing new solutions for business users.
- The integration of off-the-shelf software was very difficult because interfaces were either not compliant with standards or were not clearly defined and documented.
- The implementation of new business requirements, like alternative distribution channels, was very hard.
- An enterprise-wide data quality process could hardly be established, since each warehouse implemented its own data semantics.

The following major alternatives to improve the situation were investigated:

- *Green field approach*

  Starting from scratch, a completely new huge (multi-terabyte) enterprise data warehouse is designed, which replaces all existing systems. This approach was not considered a real option due to very high costs, long duration, unpredictable risks, and a general lack of skills and resources.
- *Migration to an off-the-shelf package*

  A data warehouse solution, for example from an ERP vendor, is purchased that fulfils the requirements to a large degree. This was also no feasible option because of the huge data volume and the specific needs of CSFS.
- *Take over a proven data warehouse solution from another bank*

  A warehouse system is bought from a competitor. However, such a solution is only possible in case of a merger with or acquisition of another bank.
- *Opportunistic approach*

  Existing warehouses are improved without any fundamental change. In principle, this means to follow the same road down as we did in the past. As mentioned above, this is no long-term solution, especially since time-to-market of new projects gradually increases.
- *Managed evolution*

  Existing warehouses are improved, including *fundamental* changes with respect to the warehouse platform and architecture. Managed evolution requires a continuous balance between a purely IT-driven approach, opting for very fast solutions which are maximum coherent to a target data warehouse architecture, and a purely business driven approach, based on opportunistic implementations of business functionality for maximizing quick wins (cf. Fig. 2). For CSFS, managed evolution was the only realistic option.

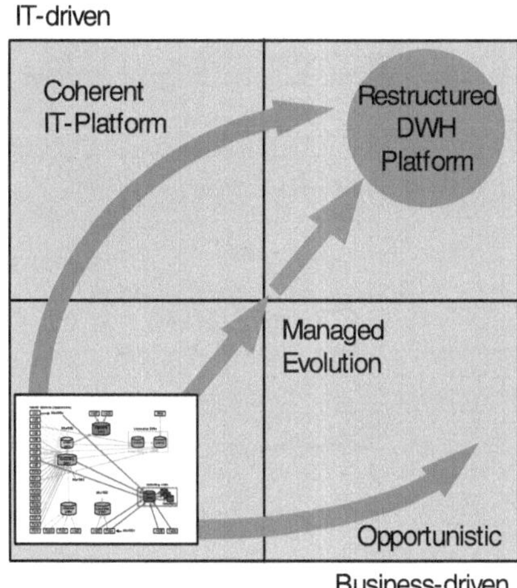

Fig. 2. Managed evolution data warehouse strategy

3 Data Warehouse Re-architecture Program

In 1999, the data warehouse re-architecture program was started to migrate all existing warehouses to an enterprise-wide target architecture in a step-wise process (managed evolution). The main task was to design and maintain an overall DWH architecture (target architecture), including system, application, data and metadata architecture. Furthermore, a migration path for all existing DWH to the target architecture was designed to ensure that all warehouse projects eventually conform to the target architecture. A standard tool suite for data warehouse design, implementation and maintenance was set up. A dedicated technical data warehousing platform was defined, built, and maintained. The required business organizations and processes/responsibilities to manage DWH were also implemented.

The migration is being executed in two major steps:

- migration phase: all warehouses are moved to the dedicated data warehousing *platform*
- consolidation phase: all warehouses conform to the standard warehousing *architecture*

The first phase was successfully completed at the end of 2000, except for one of the legacy systems.

The following picture shows a high level view of the data warehouse target architecture. Data warehouses are strictly separated from operational systems (in Fig. 3, the latter is referred to as application domains).

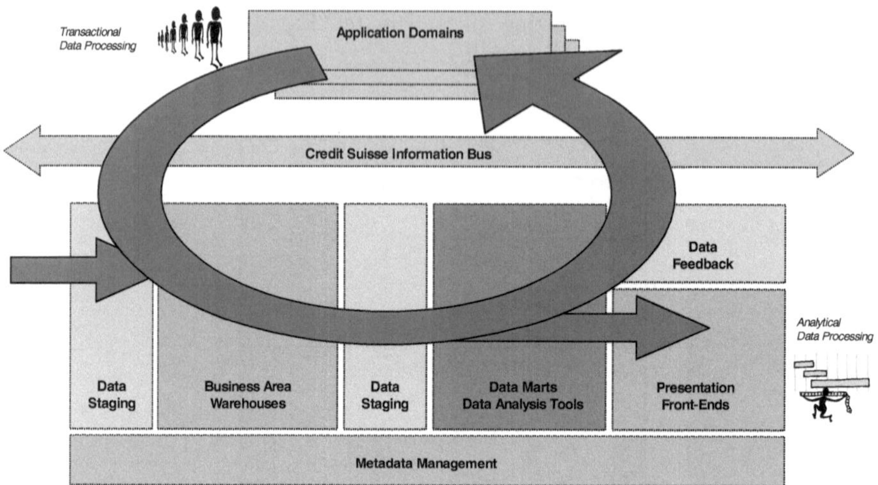

Fig. 3. Data warehouse target architecture

The Credit Suisse information bus (which is based on CORBA technology) is used to periodically transfer data from operational systems to the warehouse platform, and to feed analysis results back into operational systems. Note that data are never "improved" within the warehouse processing chain (errors have to be fixed in the source systems) and that no on-line access from DWH to OLTP systems for data extraction is allowed.

The warehouse domain consists of the following building blocks:

- The *data staging area* stores the feeder files which are delivered by feeder systems. An ETL-tool (PowerCenter from Informatica) is used to *e*xtract and *t*ransform these data and *l*oad them into business area warehouses.

  The data staging area ensures that each data item is extracted only once from the source systems and is shared between all warehouse applications. External data (like demographic information purchased from data providers) are also loaded into the data staging area.

- *Business area warehouses* are the large strategic warehouses which serve several analysis and decision support systems. They are centrally managed by IT experts.

  In contrast to an enterprise data warehouse, we are implementing multiple systems each storing data of one particular business area (e.g. financial data, credit data, marketing data, ...). These warehouses are fully normalized and store data at entity level (i.e. non-aggregated). They can be understood as the shared stock of data for all warehousing applications.

  Only a few power users are allowed to access business area warehouses directly.

- In a second staging step (where the same ETL-tool is used) data are extracted from business area warehouses and loaded into *data marts* which are optimized in accordance with well-defined business cases. This means, data marts only contain de-normalized subsets of all data based on multi-dimensional data models. Data marts have a single business owner and serve a rather homogeneous set of end users.

Several *data analysis* tools (application servers) can be used to access data marts (reporting, OLAP and data mining tools).

- *Presentation front-ends* – the warehouse clients – are strictly separated from analysis tools since we aim at a thin client strategy. End users only need a standard PC and a web browser like Netscape (which connects to application servers) for analyzing data.
- The *data feedback area* is like a staging area but prepares analysis results for being returned into operational systems.

  We support pull mechanisms, where operational systems fetch data from the feedback area, as well as push mechanisms, where the feedback area passes data to operational systems using standard application interfaces.
- *Metadata management* has an IT and a business view. The IT view allows IT professionals to centrally manage the warehouse (operational metadata). It also allows for an efficient implementation of new reports or even data marts. The business view allows end users to learn which data are where available in the warehouse system (data model and data location) and to get additional explanations on the data in the warehouse (data semantics and data lineage).

4 Experiences and Open Issues

After two years of the re-architecture program, the platform – including scalable hardware and all software tools – has been implemented to a large degree and is pretty stable. The platform consists of 2 Sun E10000 Starfires (2 x 64 SPARC II CPU), approx. 15 TB storage managed by 3 Hitachi disk subsystems (1 HDS 9900 and 2 HDS 7700) using Veritas files systems, and several Oracle 8 databases. The platform provides for separated development, test, and production environments. It is designed to support about 10'000 end users (mostly for reporting purposes).

A novel data staging process was implemented based on Informatica's PowerCenter. Migration of the existing warehouses (except of one) to the new platform is completed and the first business area warehouse is operational. The other business area warehouses will be implemented during the rest of this year.

The following experiences have been made (this list is not complete):

- The platform proved to be very powerful and reliable. The staging process is on average 4 times faster than on the old platforms, and the database performance is satisfactory[1].

[1] The remaining performance problems are mainly caused by poor physical designs of some data marts or poor application designs.

An efficient load process becomes increasingly important, since new business requirements demand for a daily load of data while the batch window decreases at the same time due to on-line banking (which does not care about "office hours")[2].

However, one should not spend too much time on evaluating platforms. Other high-end Unix servers and enterprise storage systems would do as well. Platform selection should mainly depend on the skills being available in a company, and whether the tools business users are most interested in run on this platform.

- The new architecture (based on business area warehouses) allows for a very quick implementation of new data marts (if the required data are already available in business area warehouses and the project team is familiar with the environment). Now a completely new data mart can be implemented within 3 months (from design to production).
- Interface management between the warehouse platform and the feeder systems is a very important issue. In the past, the warehouse projects were responsible for implementing the data extraction programs (usually in PL/I) in order to generate the feeder files. This approach (pull principle) leads to various problems:

  When a feeder system is changed, the extraction programs have to be adapted, too. However, if these changes have to be implemented by two different project teams, the release schedules of both projects must be synchronized. In a large-scale environment with hundreds of feeder systems and dozens of warehouses, such a synchronization is not easy.

  Moreover, if the warehouse platform and the platform of the feeder system are different, the warehouse project team has to maintain additional skills which often results in a waste of resources.

  Therefore, we switched over to a *push principle*. The warehousing and the feeder system projects sign *service level agreements* that define which data must be delivered in which format at which time to the warehouse platform. If a feeder system is changed, the same project team will also modify the extraction program.

  The push principle also simplifies a smooth transition of feeder systems to new banking paradigms. The distinction between on-line and batch processing (where data extraction is done as part of the end of day processing) will soon no longer exist (7 x 24 hours on-line processing). When the push principle is used, warehouse systems need not care whether the feeder files have been generated by a data extraction program during the end of day processing, or have been generated continuously using, e.g., message-oriented middleware.
- Reengineering existing systems “on-the-fly” is a tedious process which can only be successfully completed if this process is strongly supported by the top management.

---

[2] The next release of the reference architecture will include an operational data store (ODS) as part of the staging area to support nearly real-time warehousing. An ODS is also required when the batch window is completely closed (7 x 24 hours operation of banking applications). ODS data will be transferred to the warehouse platform using MQ Series middleware.

A business steering committee must be established that decides strategic issues of the whole warehousing platform and keeps all business projects on track. Since the re-architecture program is a long-term project that binds lots of resources, it is always tempting for smaller projects to opt for quick win solutions that (once again) do not comply with the defined standards.

A general problem is that it is easy to find business owners for data marts (each data mart serves a particular business need so that business people have a natural interest in these systems), but much more difficult to find business owners for business area warehouses (the common stock of data).

However, clearly defined responsibilities (including data security decisions and data quality measures) are even more important for business area warehouses than for data marts, since these databases are the common foundation of all data marts.

- The organization of data warehouse developers needs to be adapted in accordance with the new architecture. It is no longer meaningful that project teams care for the whole processing chain of a warehouse application (end-to-end). A separation between data staging/development of business area warehouses and development of data marts is required. The "classical" project teams then develop solutions for given business requests by defining data marts and developing reports, etc. A close cooperation of both groups is of course required since new business projects may need additional data that are not yet available in any business area warehouse.

In the following areas we did not make the expected progress[3]:

- Metadata management is still a big problem. On the one hand, metadata management tools are still immature, and on the other hand the warehousing tools (in particular staging and reporting tools) have poor interfaces to upload metadata from these tools into the global metadata repository.

  However, an automated management of metadata (bidirectional) is essential for end user acceptance. If IT users (developers of staging processes, warehouses and data marts) and power users (report developers) have to maintain metadata explicitly, the metadata repository and the warehousing system will diverge over time, rendering metadata management useless.

  Even if metadata can be captured automatically, user-friendly interfaces are required to present these data to the end users. Interfaces need to include search engines where casual users can look for warehousing data based on ontologies. Reports must be classified automatically in accordance with these (customizable) ontologies.

  A fundamental problem is data lineage at data item level where script-based transformation rules (staging tool) and reporting *programs* make it very questionable whether it is possible to achieve any progress at all.

  At least a case study would be helpful that describes examples of successful metadata management implementations, highlights success/failure factors (from

[3] In these areas additional (applied) research seems to be meaningful.

the technological to the business point of view) and gives a deeper understanding of realistic expectations.

- Systems management (in particular performance management) is also difficult in our warehousing environment. It is almost impossible to guarantee a defined minimum of system resources (in particular CPU) for critical applications if database instances are shared. So far performance management is only possible either at operating system level (Sun resource manager) or database level (Oracle resource manager), since both are not compatible with each other.

  This does certainly not require basic research, since the required technologies are known from the mainframe world for ages. In this respect, even high-end Unix servers are still in their infancy (not to mention NT).

  Another issue is performance monitoring in n-tier environments. It is easy to check whether the warehousing components are up and running (including network connections, web servers, application servers and database servers). However, it is hard to find out where the bottleneck actually is if end users experience a poor end-to-end performance.

- Today's warehousing tools each have their own security manager. Implementing security – in particular access control – over the whole warehousing chain such that data are consistently protected irrespective of their actual location and representation is almost impossible. In principle, a comprehensive end-to-end security model would be required that allows for an integrated privilege management over heterogeneous tools (including ad-hoc reporting). Otherwise, security management is not only an error-prone task, but also hardly understandable for auditors.

  So far, the problem has only partially been solved with integrated user administration tools like Control-SA from BMC, SAM from Systor/Schumann or Tivoli (User Administration and Security Management) from Tivoli/IBM.

  Another open issue is the provision of a *complete* audit trail over all warehousing tools. In environments with high security requirements like ours, auditors sometimes need to know which end user has read which data. Therefore, all tools must provide for the required log data and these log data must be integrated into a global log database.

  For integrating audit data, warehouse technologies can be applied (e.g. an audit warehouse), but today most tools do not make the required audit data available. Even the database system does not deliver these data, since permanently running Oracle in trace mode is not really an option.

  What we need is a publish/subscribe mechanism for all tools so that warehouse administrators can collect all data needed according to *their* requirements without introducing too much overhead on the warehouse platform.

- Performance can always be improved, in particular for on-line analytical processing (OLAP). Currently, OLAP based on very high data volumes (terabyte range) is only possible using relational OLAP (ROLAP). Since response times of ROLAP tools are rather high, we would prefer to use a multidimensional or hybrid OLAP system. Unfortunately, these tools are not yet able to handle such high data volumes, in particular if the underlying data change every day (creating a huge data cube can take several days).

- Another open issue is archiving in the context of evolving data models (schemas). Today it is easy to generate backups (enterprise storage systems even allow for serverless backups, and storage area networks can easily handle the data volume), but systems evolve over time. In most cases, it is not feasible to migrate all existing backups when new system releases are introduced.
- One of the major problems is data quality management. More than 500 feeder systems deliver data to our warehouse platform. The data quality differs from system to system considerably.

  Warehouse engineers can only measure the quality of data to some degree (using heuristics), but they cannot improve the quality. Data quality management has to be a business-driven process where the correction of errors is initiated by the business users. We have made the experience that it is very hard to convince business users in the early phases of a warehousing project to spend some time on data quality management. However, the more the warehouse platform is being used the higher gets the interest of business users to fix errors that they have found in their reports. A data quality management process must be in place which ensures that errors that have been found are really fixed.

  Unfortunately, business users often argue that these corrections should be made in the data warehouse (where it is often much easier than in the source systems to access these data, and the interfaces are much more user-friendly), since it "should be no problem" to feed these corrections back into the source system. We strongly argue *not* to apply this "solution". The resulting system will tend to be unmanageable, since cyclic dependencies between systems at the data instance level are never fully understood in large-scale environments.
- Finally, we envision that our warehouses will have to manage semistructured and multimedia data. There are plenty of examples of such data in e-commerce applications, e.g. contracts, insurance policies, geographical data for mortgages, and for risk management in the insurance business.

  Today's analysis tools (reporting, OLAP and data mining) are not ready yet to cope with such kinds of data.

Summing up, the basic technologies for implementing complex data warehousing and analysis systems are available. Though tool providers (database vendors and analysis tool vendors) may benefit from new and improved algorithms (in particular to improve the performance of their tools), only evolutionary improvements are to be expected.

According to our experiences, warehouse projects usually do not fail for technological reasons. They fail due to organizational problems, insufficient involvement of business users, missing top management attention or poor skills management.

A Comparison and Evaluation of Data Requirement Specification Techniques in SSADM and the Unified Process

Peter Bielkowicz and Thein T. Tun

Dept. of Computing, Information Systems and Mathematics
London Guildhall University
100 Minories, London EC3N 1JY, UK
{bielkowi, tthe08}@lgu.ac.uk

Abstract. During the analysis stage of a typical information system development process, user requirements concerning system functionality and data are captured and specified using requirement specification techniques. Most of these specification techniques are graphical (semi-formal), i.e. they involve modelling. This paper presents a comparison of data requirement specification techniques in SSADM, which is a strong data-centred method, and the Unified Process, an object-oriented method. In particular, we investigated how data groups (entities and classes), their attributes and relationships are identified, specified and validated in both methods. Data requirement specification techniques used in both methods are then evaluated against a set of detailed criteria based on five requirement specification quality attributes. Both methods seem to have a similar informal approach to producing initial data requirement specifications, but they differ when these initial models are refined. The refinement in SSADM is more rigorous. Therefore, this paper makes a few recommendations for the Unified Process.

Introduction

This paper has the following structure. The quality attributes of a good data requirement specification and criteria for evaluation used in this paper are discussed in the next section. It is followed by a brief overview of data requirement specification techniques in SSADM and an assessment of them. A summary of data requirement specification techniques in the Unified Process and an assessment of them are to be found in the subsequent section. The last two sections are devoted to the summary of our findings and, conclusions and recommendations.

Properties of a Good Data Requirements Specification

Authors such as [6], [7], [8], [10] and [11] have discussed quality attributes that general requirements specifications should posses. Based on some of the common criteria considered by those authors and others, we propose a set of characteristics that are important to a data requirement specification. They are *completeness*, *minimality*, *correctness*, *non-redundancy* and *consistency*. These characteristics apply to different

K.R. Dittrich, A. Geppert, M.C. Norrie (Eds.): CAiSE 2001, LNCS 2068, pp. 46–59, 2001.
© Springer-Verlag Berlin Heidelberg 2001

levels of comprehension of the user's requirements, namely, *description level*, *semantic level*, and *contextual level.*

Description Level

This is a superficial level at which one can check whether a data requirement specification contains all requirements wished by the user. Mere mentions of reasonably detailed requirements wanted by the user in the specification would satisfy the completeness of the specification.

1. **Completeness** – A data requirement specification is complete if the specification captures adequate information about the problem domain, i.e. the specification contains all elements desired by the user.

One will also have to make sure that the specification does not contain requirements that are not wanted by the user.

2. **Minimality** – A data specification is minimal if the specification does not contain elements that are not desired by the user.

The first characteristic necessitates that important data requirements are not missing in the specification. The second characteristic demands that the specification does not have any requirements from outside the problem domain. Combining the two characteristics, in effect, will amount to the (set of) requirements in the specification being roughly equal to the (set of) real user's requirements. However, it is not clear at this level whether the analyst's understanding of requirements is in line with the user's. Those misunderstandings can only be detected at a deeper semantic level.

Semantic Level

At this level, the analyst should strive to demonstrate that the semantics of the specification is the same as the user's real intentions.

3. **Correctness** - A data specification is correct if there are no differences in the user's understanding of the requirements specification and the semantics of the requirements specification itself.

By now, one should have a fairly reliable specification. To enforce a good structure onto the specification, one will need to remove superfluous requirements.

4. **Non-redundancy** – A data specification contains no redundant elements if no two elements represent the same information.

Non-redundancy characteristic will compel the requirement analyst to expose unnecessary overlaps in the specification. Please note that non-redundancy is a semantic level criterion, while minimality is a description level one, and see Table 1 for illustrations.

Finally, the data requirement specification should show consistency with other specification such as specification of system functionality.

Contextual Level

A data requirement specification will not exist in isolation. It will have to share information with other specifications. Therefore, it is important that requirements are defined consistently.

5. **Consistency** - A data specification is consistent with other specifications if the specification does not present any conflicting requirements. Some authors such as [7] and [6] distinguish between two kinds of consistencies: internal consistency and external consistency. Internal consistency is the consistency of various elements within a specification or model and external consistency is the consistency of specifications of elements across various models. In the context of data requirement specification technique, the external consistency is more plausible than the internal one. Since consistency is applied to various specifications, it is a very important criterion. Without it, it will be very difficult to ensure that other criteria, namely, completeness, minimality, correctness and non-redundancy are met.

The five characteristics discussed above apply to various elements in the specification.

Three Major Elements of Data Requirements and Their Characteristics

What exactly do we mean by 'data requirements'? In the context of system development methodologies, data requirements include the following elements, or in other words, data requirements are described using the following notions/concepts:

- Data groups
- Data attributes and
- Relationships between data groups

A data group is a collection of data attributes, such as entities in Relational Data Models and classes in a Class Model. Data attributes are pieces of information such as attributes of entities, and properties/attributes of a class. Relationship between data groups means a static connection between the data groups such as entity relationships, and class relationships, e.g. association, aggregation.

In a ‘good’ data requirement specification, all three elements (notions/concepts) should possess all the characteristics described above, except consistency that is applicable to specifications as a whole only. It gives rise to thirteen detailed criteria that can be used for evaluation of data requirement specification techniques. These criteria with illustrative examples are summarised in Table 1 that uses a small student library system as a case study scenario.

Implications for Methodologies

In the light of the discussion so far, it seems natural to expect a meticulous data requirement specification technique to provide mechanisms to guarantee that data requirement specifications meet some or all of the above-mentioned criteria. In fact, there are two different but interrelated issues: first, the strength of the technique itself to deliver a specification that meets the criteria above; second, checking mechanisms to validate that the criteria have been met. But we will not generally attempt to separate the two issues in this paper; rather, we will focus on overall rigour of the entire technique(s) to produce a good data requirement specification.

Table 1. Levels, characteristics, criteria and examples

Level	Character-istic	Element	Criterion	Meaning	Example
Descrip-tion	Complete-ness	Data groups	Complete-ness of data groups	No missing entity/class in the specification	Reader class/entity missing from the library system
Descrip-tion	Complete-ness	Data elements	Complete-ness of data elements	No missing attributes in entities	Reader name missing from the Reader entity/class
Descrip-tion	Complete-ness	Relation-ships	Complete-ness of relation-ships	No missing relationships in between classes/ entities	Reader-Loan relationship missing between Reader and Loan entity/class
Descrip-tion	Mini-mality	Data groups	Mini-mality of data groups	No extra entity/class in the specification	Publisher entity/class (where it is not deemed required)
Descrip-tion	Mini-mality	Data elements	Mini-mality of data elements	No extra attributes in classes/ entities	Reader's weight attribute in Reader entity
Descrip-tion	Mini-mality	Relation-ships	Mini-mality of relation-ships	No extra relationships between entities/ classes	Relationship between Book Title and Publisher entities/classes where Publisher is not required
Semantic	Correct-ness	Data groups	Correct-ness of data groups	No entity/class with invalid meaning	Assuming Reader the same as Student entity/class, when the business rule is that only those students who have borrowed at least one item is considered as Reader
Semantic	Correct-ness	Data elements	Correct-ness of data elements	No attributes with invalid meaning in an entity/class	Assuming Address attribute holds home address when it should be current address
Semantic	Correct-ness	Relation-ships	Correct-ness of relation-ships	No relationship with invalid meaning between entities/ classes	Wrong cardinalities etc...

Level	Character-istic	Element	Criterion	Meaning	Example
Semantic	Non-redundan-cy	Data groups	Non-redun-dancy of data groups	No two entities/ classes with the same meaning in the specification	Student and Reader entities when they are the same
Semantic	Non-redundan-cy	Data elements	Non-redun-dancy of data elements	No derivable attributes and attributes that share same information	Having Date of Birth and Age attributes, or Address and Post Code attributes when Address also contains Post Code
Semantic	Non-redundan-cy	Relation-ships	Non-redun-dancy of relation-ships	No entity/class relationships which are to provide alternative navigation routes	Circular relationships between entities/classes where some of them are solely for alternative route for navigation
Contextu-al	Consis-tency	Related specifi-cations	Consis-tency of the specifi-cations	No two specifications of a requirement in different models should be conflicting	Data specification suggests that Book Reservation is not needed but the functional requirement specification indicates that students should be able to reserve books.

Table 2. Scales for different degrees of rigour

Rigour	Scale
Not present	**0**
Present but weak or not explicit	**1**
Explicit and rigorous	**2**

We would also like to stress that it is not enough if a methodology just mentions something like 'Look for missing classes and attributes in this step' or 'Review the class model to identify hidden attributes'. It does not tell methodology-users much. A good methodology should say, and methodology-users need to know, how to correctly identify and specify all three elements and how to check whether the specification meets the above criteria. It means one not only needs to know 'what to do', but perhaps more importantly, one needs to know 'how to do it'.

Thus, when assessing various methodologies and related techniques, we will try to gauge their rigour using various measurement scales as shown in Table 2.

Structured Methods

Method: SSADM Version 4/4+

Structured System Analysis and Design Method (SSADM) was first launched in 1981, and has been used by UK Governmental Departments ever since. It has gone through many improvements over the years and the major version SSADM Version 4 was launched in 1990. More minor changes were made later and SSADM 4+ was then released in 1996. See [12].

A Summary of Data Requirement Specification Techniques in SSADM

In SSADM 4 and 4+ the major data analysis model is Logical Data Model (LDM) that includes Logical Data Structure (LDS), otherwise known as Entity Relationship Diagram (ERD). See [2] and [12].

In ***Step 140 Investigate Current Data***, a draft LDM is produced after gathering candidates for entities coming from various sources such as: physical data stores in the current environment, requirement specification, specialist knowledge of the application domain and so on. Together with entities, their main attributes, e.g. key attributes, and some other obvious attributes are discovered. Entity relationships and their degrees are also identified. But there is no urgent need for an entirely valid LDM at this point, since the model will be refined progressively. Towards the end of ***Step 140***, the LDM is revised first by removing redundant relationships such as 'indirect relationships'. Then the LDM is validated against the Data Flow Model (DFM, which consists of Data Flow Diagrams) that is being developed at the same time. For an entity, SSADM requires that, there must be at least one DFD process that creates the entity, at least one process that sets attribute values, and at least one process that deletes the entity. The only exception is a case where given entities are used for references only. The checking is done by drawing up the Process/Entity Matrix. ***Step 140*** concludes with the examination of 'Access Paths' to demonstrate that elementary processes that need to access a number of entities can do so by navigating through the defined relationships.

In ***Step 320 Develop Required Data Model***, the LDM is adjusted according to the user's choice of system functionality. That generally involves removing entities that are not required by the new system and adding new entities that will be required. Entity relationships are then revised and improved in the light of changes in entities. LDM documentation is completed by filling up Entity Description forms, which include information such as names of all known attributes, primary and foreign keys etc... When documenting the entities, cross-references are made to the data stores in the DFM. The Store/Entity Cross-Reference ultimately determines the content of data stores for the DFM.

In ***Step 330 Derive System Functions***, various functions are identified from data flow diagrams, and are documented using Function Definition forms. A function in SSADM is 'a set of processing that users wish to carry out at the same time' [12]. Functions are triggered by events. User interface of each of these functions is defined using I/O Structure diagram(s) that use Jackson Structure Diagram-like notations to show detailed exchange of data items between the user and the system. These diagrams are then validated by users.

In ***Step 340 Enhance Required Data Model***, the LDM, which has been developed using a top-down approach, is 'enhanced and confirmed' by Normalised Tables that are developed using the bottom-up Relational Data Analysis (RDA) technique. RDA is performed on some complex I/O (Input/Output) Structures, which were produced as part of Function Definitions. I/O Structures contain rich information about data coming in and going out of the system. Using the RDA technique, the I/O Structures are successively transformed into the Third Normal Form (3NF). After that, 3NF tables are converted into LDS fragments using a mechanistic process that involves four steps. Subsequently, the LDS fragments are compared with the original LDS to resolve differences. When doing so, SSADM suggests that, one should look for a) differences in attributes, such as uncovered attributes, attributes in different entities etc… b) additional entity types and c) additional relationships.

In ***Step 360 Develop Processing Specification*** an Entity Access Matrix is developed which demonstrates how events affect entities during their lifetime. These effects are detailed by operations in Entity Life History modelling; they are later incorporated into operations on Effect Correspondence Diagrams. ECDs show how an event affects various entities, and how these entity effects correspond to each other [12].

Assessment

The main data requirement specification in SSADM is mainly represented by LDM. The LDM is supported by Logical Data Structure, Entity Specifications, Attribute Specifications and Relationship Specifications. Table 3 summarises the assessment of data requirement specification techniques in SSADM.

Table 3. Summary of the assessment of data requirement specification techniques in SSADM

Criterion	How it is met (if at all)	Rigour
Completeness of entities	**Task 30 of Step 140 Check that the LDM will support the Elementary Processes from the DFM.** **Task 40 of Step 140 Record any new data requirements or problems with current data in the Requirements Catalogue:** Missing entities can be uncovered when comparing LDS fragments from Normalisation with the original LDS.	2
Completeness of entity attributes	**Task 40 of Step 340 Compare the mini LDM with the Required System LDM, and resolve any difference:** Comparing fragments of normalised LDS with the original LDS. Normalisation is performed on I/O Structures that is a set of data attributes that the users can provide the system with and expect from the system.	2
Completeness of entity relationships	**Task 40 of Step 340 Compare the mini LDM with the Required System LDM, and resolve any difference:** Comparing fragments of normalised LDS with the original LDS. Normalisation establishes entity relationships via foreign keys.	2

Criterion	How it is met (if at all)	Rigour
Minimality of entities	**Task 30 of Step 140 Check that the LDM will support the Elementary Processes from the DFM.** **Task 10 of Step 320 … Exclude elements of the LDM which are outside the new system…** **Task 10 of Step 360 Develop the Entity Access Matrix, and document each event and enquiry in the Event and Enquiry Catalogue:** In Entity Access Matrix modelling, entities that are affected by no events or too few events are to be questioned.	2
Minimality of entity attributes	**Step 340 Enhance Required Data Model:** Since RDA is the analysis of data items in I/O Structures that are confirmed by the user, certain attributes that were initially thought to be necessary when in fact they are not, could be discovered when merging RDA results with the initial LDS. **Step 360 Develop Processing Specification:** Entity Life History also provides an opportunity to check whether all attributes are affected by event operations.	2
Minimality of entity relationships	**Step 340 Enhance Required Data Model:** RDA. One can generally assume that all relationships defined in RDA in terms of foreign keys are necessary and others should be examined further.	2
Correctness of entities	SSADM provides very detailed documentation standard for describing entities. User's participation in validating the documents can be valuable.	1
Correctness of entity attributes	**Task 30 of Step 330 Create I/O Structures or [*sic*] for each off-line function. Create I/O Structures for on-line functions as appropriate. Study more involved functions using User Interface Design techniques:** User involvement in the development of I/O Structures, and related User Interface Design Prototyping is important.	1
Correctness of entity relationships	**Step 340 Enhance Required Data Model:** RDA. All entity relationships defined via foreign keys are generally correct.	1
Non-redundancy of entities	**Step 340 Enhance Required Data Model:** RDA. Normalisation necessitate the removal or combination of entities with similar of same attributes.	2
Non-redundancy of entity attributes	**Step 340 Enhance Required Data Model:** RDA. Normalisation specifically requires the removal of derivable attributes etc…	2
Non-redundancy of entity relationships	**Step 140 Investigate Current Data:** It clearly asks to remove redundant relationships such as 'indirect relationships'. **Step 340 Enhance Required Data Model:** RDA. Relationships derived from the Normalisation are always optimal.	2

Criterion	How it is met (if at all)	Rigour
Consistency of the specifications	There are various stages where cross-checking between models is done. In **Step 140**, for instance, one checks whether all functional requirements can be supported by the data requirement specification by drawing up a small Entity/Process Matrix. The comparison of the Normalisation results with the original LDS is a very powerful inter-model validation.	2
	Total	23

Object-Oriented Methods

Method: The Unified Process

The Unified Process published in 1999 is a development process model to accompany the Unified Modelling Language. It describes how to make an effective use of the UML models. The Unified Process evolved from the earlier method, the Objectory that later became the Rational Unified Process, and is a product of a collaborative work of three well-known methodologists with contributions from other experts [4].

A Summary of Data Requirement Specification Techniques in the Unified Process
One of the early activities within the Requirements Workflow involves identification of Domain/Business classes in the form of a Business Model or a Domain Model alongside the Use Case Model and user-interface descriptions [1], [4], and [9]. A Business Model or a Domain Model, which is a less elaborate kind of business model, can be used to establish the system context.
A Domain Model, that basically is a UML class model, shows the most important domain classes together with some obvious attributes and relationships but no or very few operations in the system that are discovered by analysing the requirements specification [4]. Candidates for domain classes are a) Business objects b) Real-world objects and c) Events.
Alternatively, a Business Model can be produced, which encompasses a Business Use Case Model and a Business Class Model. The business use case model details business processes used by business actors. Therefore, use case model of the system will contain a sub-set of business use cases plus some extra use cases if required. The business class model describes the realisation of each of the business use cases. It should be noted that candidates for classes in business class modelling come from the business use case model, while in domain modelling, candidates are derived from the real world. Also, the business class model will have more complete information about the classes, their attributes and operations, since realisation of business use cases have already been demonstrated. A Logical User Interface Design, that shows how 'attributes of the use cases' are utilised by the users, is produced for each use case. The Unified Process suggests looking for information the actors supply to the system and information the actors require from the system etc.
When the Requirements Workflow is completed we will have a tentative class model of the system that is enhanced in the following analysis stage.

When performing analysis of use case realisation, the Unified Process suggests the use of collaboration diagrams over sequence diagrams [4]. First for each use case, one has to identify participating analysis classes such as boundary classes, control classes and entity classes. Boundary classes deal with the system's interaction with the outside world, and control classes deal with the synchronisation of contributions made by various classes. (In this paper, we will largely concentrate on entity classes). Entity classes handle stored information. These entity classes are discovered using a combination of input from the early business class model, and techniques such as grammatical analysis of use case descriptions and so on.

Identifying Analysis Classes [4] step involves identification of analysis classes together with their responsibilities, attributes, and relationships. The following guidelines are provided (extracted for entity classes only).

- Entity classes are discovered by analysing use-case descriptions, existing domain model etc

After discovering object interactions and class responsibilities, ***Identifying Attributes*** step gives us the following guidelines.

- Attribute names are nouns and attribute data types should be conceptual. Try to reuse existing data types.
- Attributes may be separated into classes if there are too many of them.
- Entity class attributes are easy to find. Domain/Business classes provide a good starting point.
- Attributes sometimes need just a definition of them, not the actual attributes, etc…

Identifying Associations and Aggregations requires one to investigate links between objects in collaboration diagrams. These links indicate that the associated classes have a kind of static relationship such as association or aggregations. These relationships 'must exist in response to demands from the various use-case realisations'. So, one should not be concerned with most efficient navigation routes at this stage [4].
Sequence diagrams can be used instead of collaboration diagrams during use case realisation step if one wants to emphasise the timing of messages passed between objects. However, the Unified Process recommends the use sequence diagram in the subsequent design stage. See [4] and also [5].

Assessment

The principal data requirement specification in the Unified Process is represented by the class model, in particular, by entity classes, their attributes and relationships. The use case model describes the functional requirements of the system and the two models are fed into the use case realisation models. A summary of the assessment of data requirement specification techniques in the Unified Process is shown in Table 4.

Table 4. Summary of the assessment of data requirement specification techniques in the Unified Process

Criterion	How it is met (if at all)	Rigour
Completeness of classes	**Use Case Realisation**. When we demonstrate how various objects contribute to the execution of a use case, we are bound to discover some missing classes. But the Unified Process does not say how to discover	1

Criterion	How it is met (if at all)	Rigour
	such missing classes.	
Completeness of class attributes	The **Logical User Interface Design**. It does present a chance for users to verify some attributes from classes. Again it is not explicit.	1
Completeness of class relationships	**Use Case Realisation**. The Unified Process clearly states that all classes that interact must have a relationship.	2
Minimality of classes	**Use Case Realisation**. The Unified Process states that 'an entity class that does not participate in any use-case realisation is unnecessary'.	2
Minimality of class attributes	The **Logical User Interface Design**. It also can be used to negotiate what information the users do not need from the system.	1
Minimality of class relationships	**Use Case Realisation**. The Unified Process clearly states that all classes that interact must have a relationship. Other relationships are to be questioned.	2
Correctness of classes	The **Logical User Interface Design**. The users involved in the development can indirectly confirm it.	1
Correctness of class attributes	The **Logical User Interface Design**. Users may be able to verify what these attributes represent by involving in this process.	1
Correctness of class relationships	The Unified Process clearly says when to add a relationship, but it does not say what kind of relationship, in terms of cardinality etc, we should add. So it is not present.	0
Non-redundancy of classes	The Unified Process does not say how to identify semantically redundant classes in the specification. It is not present either.	0
Non-redundancy of class attributes	Not present.	0
Non-redundancy of class relationships	The Unified Process also specifically asks for the removal of class relationships that are solely for easy navigation.	2
Consistency of the specifications	In the Unified Process, the use case realisation is the only major modelling that brings together the use cases and classes. Other models are a little fragmented. For instance, Logical User Interface Design is unrelated to boundary classes. It also mentions use case attributes but it never suggests where they come from and how they are connected with attributes of classes.	1
	Total	14

Summary of Findings

From the descriptions of each method in the previous two sections, one can see that the way initial LDS and Class Model are produced is very similar. Both methods use a combination of rather informal techniques such as grammatical analysis, investigation of paper documents, consultation with users etc.

When it comes to refinement of the initial model, one can see a gulf of differences.

How to Identify Classes/Entities and Check Them

The class modelling in the Unified Process completely relies on a top-down analysis approach. It identifies major classes first, then attributes, followed by relationships.
SSADM, on the other hand, uses a top-down technique to produce an initial Logical Data Model. In addition to that, SSADM also uses a bottom-up technique, RDA. The two techniques in SSADM start from different points, one from the abstract requirements and the other, from concrete information about data in I/O Structures. Then RDA rules are applied to some of the complex I/O Structures, which results in entities whose attributes and relationships are clearly defined. The two Logical Data Structures are then compared to sort out differences.
The Unified Process does not have many checking mechanisms to perform cross-checking between models. Use case realisation is the only major modelling activity that brings together use cases and classes, hence there are some opportunities to reveal missing classes.
SSADM has plenty of steps where the checking is done explicitly and rigorously. For instance, **Step 140** clearly states that all entities must be able to support the processing units. That statement is materialised by a small but important matrix, the Entity Process Matrix that summarises how entities are affected by processes.
One can also observe some similarities between SSADM's Effect Correspondence Diagram and Unified Process's Collaboration Diagram. They both show how entities/classes are affected by/contribute to a process/the realisation of a use case. But how do we know which classes/entities contribute to a particular use case/process?
Prior to Effect Correspondence Diagramming in SSADM, one has to produce an Entity Access Matrix that shows which entities are affected by each event. To determine if an event (assuming that an event is a process) affects an entity, one has to see if any instances of entity are created or updated or deleted. One also applies simple rules such as having at least one event to create the entity, another event to modify and yet another one to delete each entity. Only then ECDs are produced. The matrix gives us a good opportunity to check if the specification contains all necessary entities, does not contain extra entities, etc.
The Unified Process attempts to jump straight to show the dynamics of object interactions for each use case without first establishing which and how classes will contribute to various use cases.
Therefore, EAM-like matrix and logical analysis of data items that go in and out of the system are important activities that ensure the data requirement specification in SSADM meet many quality criteria, and such important activities have no counterparts in the Unified Process.

How to Identify Attributes and Check Them

As far as attributes are concerned, the Unified Process entirely depends on the analyst's insight into the problem. Hardly any concrete guidelines are provided as to how one can identify attributes and validate them.
SSADM also depends on the analyst's insight into the problem at the initial stages. Once the logical user interface (I/O Structures) for the system is designed and agreed by the user, Normalisation is carried out on some complex I/O Structures, resulting in

LDS fragments that are then compared with the original LDS. There is little room for errors and omissions.
Again, one can see the importance of logical analysis of data items that cross the system's boundaries.

How to Identify Relationships and Check Them

The Unified Process obviously states that classes that communicate must have a kind of relationship.
SSADM uses Normalisation to validate entity relationships that were created by abstract thinking.
The differences lead us to the following conclusions and recommendations.

Conclusions and Recommendations

The data requirement specification techniques of SSADM demonstrate a high degree of thoroughness, which in some cases might prove to be a little 'bureaucratic' and over-enthusiastic [3]. But the data requirement specification techniques of the Unified Process lack a kind of rigour that one can find in SSADM. In particular, there are not many crosschecks between models. We believe that this can be rectified as follows.

1. Generally models like I/O Structures are very important. Because, if and only if we can establish what information users want from the system (outputs) and what they can provide to the system (inputs), we can have a reliable basis from which we can derive important attributes of entity classes (and classes themselves in some situations). We are also of opinion that there is a scope in the Unified Process to incorporate I/O Structure-like diagrams to Logical User Interface Designs of use cases or more specifically boundary classes[1]. The primary focus of the diagrams will be in describing data attributes coming in and going out of the system in a logical way. SSADM uses Relational Data Analysis technique to group these data items. One may also use a similar, but less formal, technique, such as a simple mapping technique, to ensure that entity classes can support the attributes specified in the boundary classes.
2. I/O Structure-like diagrams can also be useful when looking for classes that contribute to a use case realisation. It means that if data items defined in an I/O Structure-like diagram match attributes of certain classes, these classes would somehow have to contribute to the use case realisation.
3. Finally, we believe that a tool such as Entity Access Matrix is required to describe clearly which classes contribute to which use cases. See Fig.1. Such matrix gives us an overall picture of various classes' participation in use cases and the matrix can be used to detect invalid classes and use cases as one can do in SSADM.

---

[1] In this paper, we did not discuss Boundary Class in length. The Unified Process does mention that boundary classes should be used to capture incoming and outgoing data items, rather than getting overwhelmed with GUI objects. But the Unified Process treats boundary classes as rather informal, in a sense that information from these classes are never used by models. Therefore, we would like to promote importance of these boundary classes by using systematic ways to capture information and utilise them appropriately.

	Class 1	Class 2	Class ...	Class N
Use Case 1	X	X		
Use Case 2		X	...	
Use Case ...	X		...	
Use Case N				

Fig. 1. An EAM-like matrix to show allocation of classes to use cases

Future Work

We are also investigating some other popular object-oriented methods such as OMT, in order to critically assess the rigour of their data requirement specification techniques using the same criteria proposed in this paper. Further to the third conclusion, our next research will concentrate on a comparison and evaluation of requirement specification techniques concerning system functionality used by Structured and Object-oriented methods.

References

1. Booch G., Rumbaugh J., and Jacobson I. (1999) *The Unified Modeling Language User Guide*, Addison Wesley Longman, Massachusetts
2. Goodland M. and Slater C. (1995) *SSADM Version 4: A Practical Approach.* McGraw-Hill Europe, Berkshire, England
3. Hares, J. S. (1994) *SSADM version 4: The Advanced Practitioner's Guide*, John Wiley & Sons Ltd, West Sussex, England
4. Jacobson I., Booch G., Rumbaugh J. (1999) *The Unified Software Development Process: The complete guide to the Unified Process from the original designers*, Addison Wesley
5. Jacobson I., Christerson M., Jonsson P., Overgaard G. (1992) *Object-oriented Software Engineering: A use case driven approach*, Addison-Wesley Longman, Essex, England
6. Kotonya G. and Sommerville I. (1998) *Requirements Engineering: process and techniques*, John Wiley & Sons, West Sussex, England
7. Loucopoulos P. & Karakostas V. (1995) *System Requirements Engineering*, McGraw-Hill
8. Pfleeger S. L. (1998) *Software Engineering: Theory and practice*, Prentice Hall
9. Rumbaugh J., Jacobson I., and Booch G. (1999) *The Unified Modeling Language Reference Manual*, Addison-Wesley
10. Sommerville I.(1995) *Software Engineering* (Fifth Edition), Addison Wesley Longman
11. Vliet J. C. van (1993) *Software Engineering: Principles and Practice* (July 1997 reprint), John Wiley & Sons, West Sussex, England
12. Weaver P. L., Lambrou N. and Walkley M. (1998) *Practical SSADM 4+: A complete tutorial guide (second edition)*, Financial Times Professional Limited (Pitman Publishing)

From User Requirements to User Interfaces: A Methodological Approach*

Juan Sánchez Díaz, Oscar Pastor López, and Juan J. Fons

Department of Information Systems and Computation
Valencia University of Technology. Valencia, Spain
{jsanchez;opastor;jjfons}@dsic.upv.es

Abstract: When a software product is designed and implemented, it is very important to assure that the user requirements have been properly represented. To achieve this objective, a guided software production process is needed, starting from the initial requirements engineering activities and through to the resultant software product. In this paper, a methodological approach for generating user interfaces corresponding to the user requirements is introduced. By doing this, we go a step further in the process of properly embedding requirements engineering in to the software production process, because users can validate their requirements as early as possible, through the validation of the user interfaces generated as a software representation of their requirements. Also, these interfaces can be reused for further refinement as a useful starting point in the software development process.

1 Introduction

When conceiving a computer system, the first stage consists of understanding and showing the user´s needs in a suitable manner. This process is generically called requirements engineering and has been acknowledged as a crucial task within the development process ([2],[16]). It is widely known that errors which originate in this requirement stage can go undetected up to the operating step, thereby causing faults which have serious consequences, especially in critical systems.

Errors during the stage of elicitation of requirements are mainly caused by the gap which exists between the users and the development process. The users are presented with an abstract specification of the system, which is generally incomprehensible to them. This often makes the value of performing requirements engineering techniques unclear and originates well-known comments as "requirements engineering is fine, but we do not have time to deal with it!". To face this problem, techniques for obtaining user interfaces which correspond to user requirements within a precise methodological approach are needed.

* The research reported in this paper has been partially supported by the Spanish Ministries of Education and Industry, CICYT- FEDER project number TIC1FD97-1102

K.R. Dittrich, A. Geppert, M.C. Norrie (Eds.): CAiSE 2001, LNCS 2068, pp. 60–75, 2001.
© Springer-Verlag Berlin Heidelberg 2001

A system's behaviour may be intuitively described through the use of scenarios. A scenario is defined as a partial description of a system's behaviour in a particular situation ([2]). This description's partial nature allows for a scenario to cover parts of a system's behaviour. This is an interesting characteristic, as different users can perceive the system in different ways.

Scenarios are a valuable tool for capturing and understanding requirements and for analysing interactions between man and machine ([17]). A standard requirements engineering process, one based on scenarios ([25]), has two main tasks. The first one generates specifications using the scenarios to describe the system's behaviour. The second one consists of validating the scenarios with the user by means of simulation or prototyping. Both tasks are tedious if not backed up by an automatic or semi–automatic tool.

The validation stage is tackled in some cases by using RAD tools (Rapid Application Development) ([15]). The process of defining and generating the user interface prototype is, in any case, a manual process, since each object must be explicitly created.

In order to overcome all these problems, we consider that it is necessary to define a methodological approach to derive user interfaces from user requirements. This is why we are interested in the validation of scenarios by means of automatically generating user interface application prototypes and their symbolic implementation.

In this paper, such a software production environment is introduced. As opposed to other proposals, we defend the idea of having a process with a high degree of automation where the generation of user interfaces corresponding to precise user requirements has a methodological guidance. Furthermore, a corresponding visual tool which allows us to almost completely automate the entire process has been implemented. An important contribution of the method is that it automatically generates an inter–form model of navigation which is based on the relationships *include* and *extend* specified among the use cases. The introduction of this navigation feature makes possible to use the generated interfaces in web environments.

In short, this paper presents both a methodological proposal and the associated support tool which backs it up, within the field of requirements engineering. They are based on the Unified Modelling Language (UML), extended by the introduction of Message Sequence Charts (MSC) ([12]). As we view MSCs as extended UML Sequence Diagrams by adding the needed stereotypes, the approach can be considered UML-compliant from the notational point of view.

A clear, precise iterative process allows us to derive user interface prototypes in a semi-automatic way from scenarios. Furthermore, a formal specification of the system is generated and represented through state transition diagrams. These diagrams describe the dynamic behaviour of both the interface and control objects associated to each use case or each MSC. The method has four main steps: the first two steps require analyst assistance to some degree, whereas the last two steps make the process of scenario validation fully automated by means of prototyping

The work is structured as follows: the second section introduces the UML models which we use in our proposal, together with the example that will illustrate the process of prototype generation. Section 3 contains a detailed discussion of the activities and processes which our approach entailed. Related work is discussed in section 4. Lastly, future lines of work and conclusions will be presented. Images captured on the CASE tool implemented for the method are include throughout the paper, to make the process easier to understand.

2 The Unified Modelling Language

The Unified Modelling language (UML) ([4]) is widely accepted as a standard for the modelling of object oriented systems. Among the different views or models proposed in UML, our approach is based on the use case model, the state transition diagrams and a variant of the sequence diagrams: the MSC. We assume the existence of an initial class diagram model that is used in the building stage of the MSC to show classes which emit or receive events. To illustrate the proposal, we shall use a simplified order management system as described in [20].

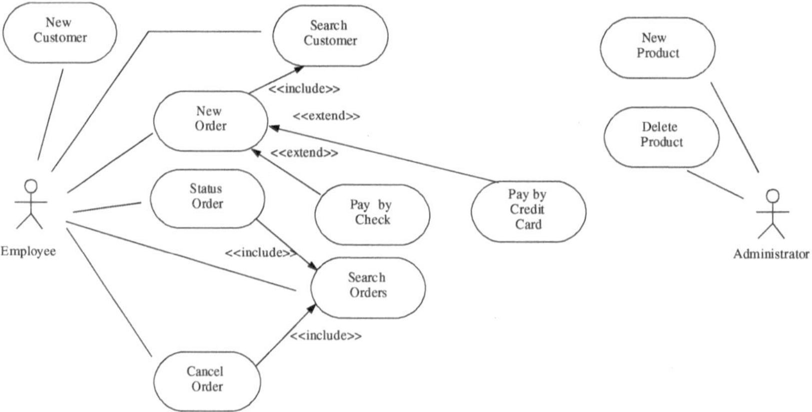

Fig. 1. Symplified use case model, order entry system.

Next, we are going to briefly introduce the main features of the UML models that we use, focusing on how we use them and which expressiveness has been added in order to understand why they are needed in our proposal.

2.1 The Use Case Model

The use case model contains the agents or actors which can interact with the system, represented as a use case. A use case, as introduced by I. Jacobson [13], constitutes a complete course of interaction that takes place between an actor and the system. Figure 1 shows a use case model (simplified due to shortage of space) for an order entry system. There are two actors: *employee* and *administrator*. The employee is in charge of adding new customers, introducing orders, finding out the order status (admitted, ready to send, sent) and lastly canceling them. The administrator is in charge of: adding new products and deleting products.

Two relationships can be defined: include and extend. The "include" relationship is employed when a flow of events can be textually inserted within another use case. On the other hand, the "extend" relationship shows an alternative execution flow. This can, under certain conditions, be considered as an inclusion.

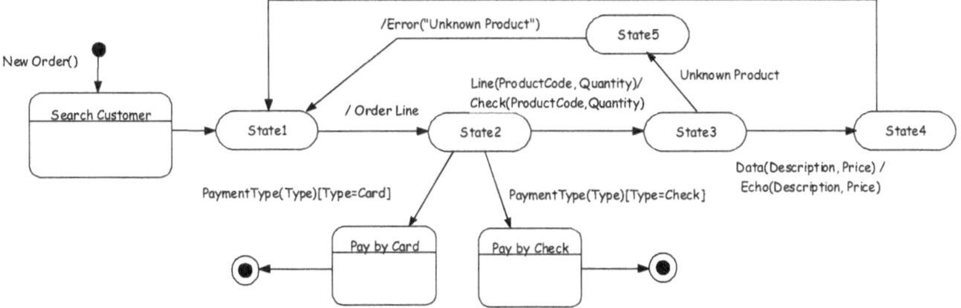

Fig. 2. State transition diagram, user interface object in *New Order* MSC.

2.2 The MSC

The MSCs are widely used in the field of telecommunications to describe the exchange of information among system classes. In their simplest form, with no information on control or conditions, they are considered to be equivalent to the UML interaction diagrams. Figure 5 contains the notation used in our tool.

The vertical lines represent system classes or actors, whereas the sending of events or the exchange of information is shown as horizontal lines. Each event can have a number of arguments associated with it, and these can be basic types (integer, string, boolean, enumerated, real etc.) or class types. There is an additional notation that reflects the repetition of events, alternatives, exceptions, interruptions and conditions which can show what state the system is in. To keep the proposal within the UML umbrella, this extra-information can be seen as a UML sequence diagram extension based on the introduction of the corresponding stereotypes. The MSC can be broken down into levels, in such a way that the analyst can employ the desired degree of abstraction in each diagram. ITU ([12]) or Telelogic ([26]) may be consulted for further details.

2.3. State Transition Diagrams

The state transition diagrams ([9]) are normally used within the object oriented methodologies to describe the behaviour of the system classes. They are useful for representing the objects life-cycles of the objects, and they can also be used to describe the behaviour of the application user interface ([10]). In our approach, we shall employ them to describe the object of the user interface and the object of control that appears in each MSC diagram. The STD will be used in the process of user interface animation.

In Figure 2, a standard STD expressiveness (UML compliant) is used to describe the behaviour of the user interface object used in the *New Order* MSC. This diagram will be explained in detail in the next section. At the moment, we introduce it only to demostrate that we use the expressiveness provided by the STD of the UML.

3 Description of the Proposal

In this section we present precise method to guide the generation of user interfaces corresponding to the user requirements, according to the ideas introduced in section 1 and using the UML diagrams commented on the previous section. Figure 3 shows a schematic representation of the activities contained in the proposed method. As we have commented above, the first two activities, namely scenario representation and synthesis of use cases, are manual activities which the analyst must carry out. The last two, specification generation and generation of prototypes, are totally automatic.

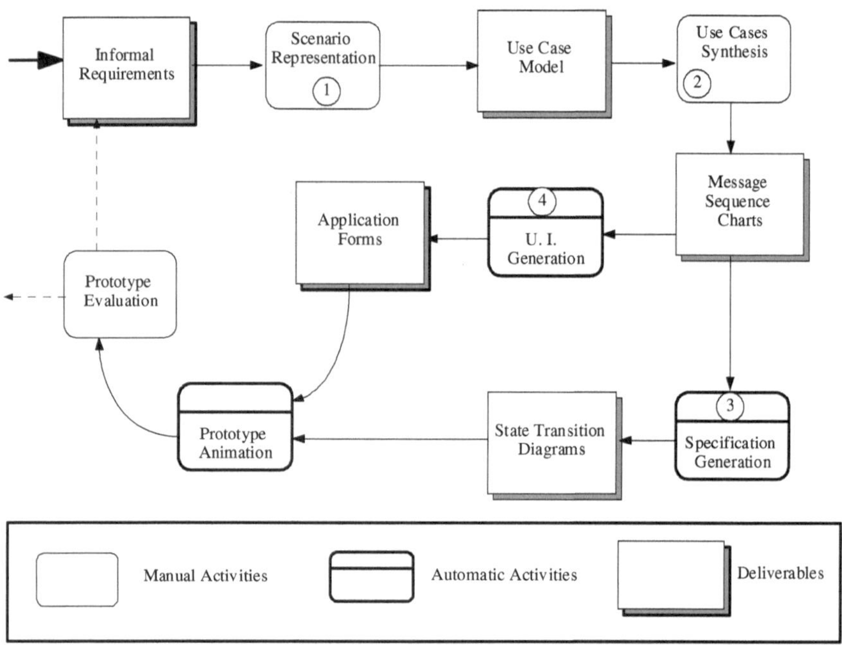

Fig. 3. Schematic representation of the method.

Figure 3 also fixes the order in which these activities should be performed. The process begins at the scenario representation stage where a use case model is created. The next stage consists of describing use cases by means of MSC. During the stage of specification generation, a state transition diagram (STD) for the class User Interface and another STD for the Control Object are automatically obtained from a given MSC. Lastly, the final stage consists of automatically generating the user interface prototypes as well. The method is iterative; in consequence, the prototyping is used to validate and enrich the initial requirements. We shall now proceed to explain each stage in detail.

3.1 Scenario Representation

During the scenario representation stage, the analyst has to construct the system's use case model. This process is structured in three layers or stages: *the initial model*, *the description model* and the *structured model* ([22],[23]).

The initial model shows the actors and use cases grouped together according to whether they carry out similar functions. Figure 4 shows the initial model of the order entry system.

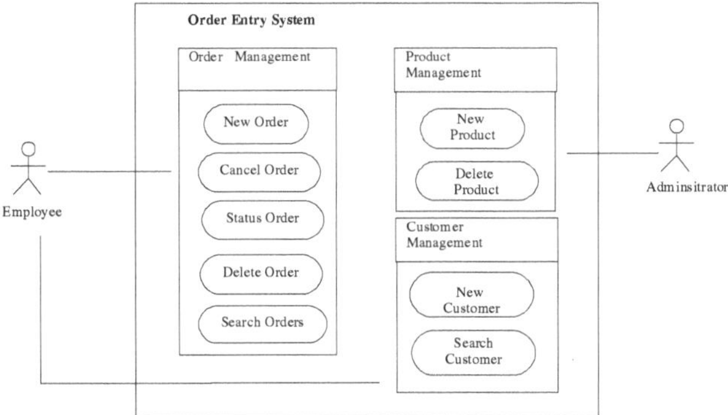

Fig. 4. Use case initial model.

We use the initial model for describing actors, non-structured uses cases and system services. For instance, the *order management* includes the use cases: new order, cancel order, status order, delete order and search orders. Once this is done, the description model contains the use case descriptions, using a natural language template. We shall use a variant on the template proposed by L. Constantine et al ([7]), which is composed of the following elements:

- Name: Identifies the use cases and should reflect its function or immediate purpose.
- Preconditions: the state of the system required for the use cases to be executed
- Post-conditions: the state the execution of the use cases leads the system to.
- Primary actor: This is the initiator actor who generates the starting stimulus to the system.
- Secondary actors: Other participating actors who communicate with the use case.
- Event flow description: it shows the events generated by the actors (user intentions) and the system commitments (system responsibilities).
- Extension points: Reflect alternatives, conditional o exceptional courses of interaction that can alter the main flow of events. We use two kinds of extension points: synchronous extensions and asynchronous extensions.
- Relationships: Identifies other use cases which are related in some way (include, extended) to the use case being described.

The template in Table 1 shows the description of the use case "search customer". The event flow is divided into two columns: user intentions and system responsibilities. This division allows us to identify when the actors request services, and when the system acts as information supplier. This is important for constructing the MSC in the next stage of the method as we will be seen below.

Table 1. Use case template, *Search Customer*.

Name: Search Customer	
Relations	
Include:	None
Extend:	None
Description	
Preconditions	None
Postconditions	None
Primary Actor	Employee
Secondary Actors	None
Event Flow	
User intentions	**System responsibilities**
1. The employee selects *Search Customer*	
2. The employee introduces Customer Name or Customer Code	
3. The employee selects "*Apply*"	
	4. The system searches customers by Name or Code
	5. The system displays the result of search
Asynchronous extensions	
6. The employee can select *Stop* at any point	
Synchronous extensions	
7. If there is no *Name* or *Code* then system displays a message error at point 3.	

Finally, the structured model graphically shows the use cases relationships *include* and *extend* (see Figure 1), producing the corresponding final graphical representation. With this step, the process of capturing user requirements is considered to be finished.

3.2 Synthesis of Use Cases

Once we have obtained the Use Case Model, we need to work with the involved use case information to undertake the process of designing a software system. To do this, use cases must be formally described: the formal definition of a use case is achieved by using a graphic, non-ambiguous language, such as MSC. In this phase, which is a manual one, the use case templates are used as help so that the analyst can detect the events sent by the actors and by the classes of problem domain In each MSC, besides the participating actors (initiator, supporting actors), one class for the user interface and one class that acts as control object are introduced, according to the initial Objectory proposal ([13]) and according to the UML approaches for a software production process ([14]).

The formalism that we use, as opposed to other well-known methods (for example, the trace diagrams of OMT [18]), allows us to show alternatives in the sending of events, iterations, exceptions, etc. Furthermore, the notation allows for a graphic representation of the dependence that exists among use cases. If a case A uses a case B, a rectangle with the name B is placed in A's MSC. This indicates that the diagram can extend to diagram B. If the given extension depends on a given condition, this extension relationship can be treated in the graphic level following the MSC semantics by properly representing the condition that must hold and calling the diagram produced by the extension.

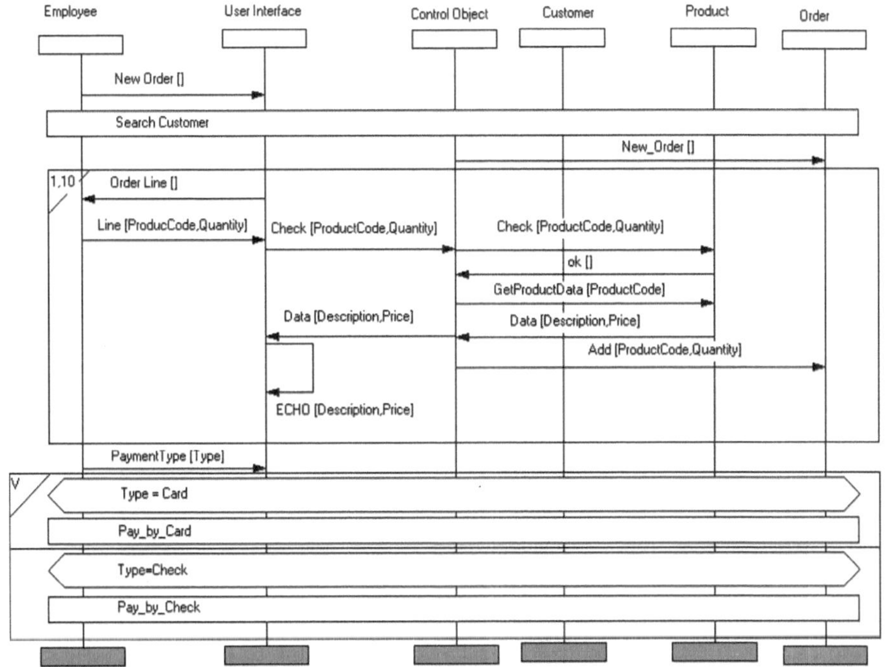

Fig. 5. MSC *New Order.*

Figure 5 show an "new order" MSC corresponding to the "new order" use case. This diagram shows the events which occur when an order is introduced into the system. The "include" relationship corresponding to the search customer use case appears in the diagram as a rectangle that is labeled with the name of the associated MSC diagram that contains the expansion (*search customer* in this case). The extension point attached to the payment by credit card use case is represented through the corresponding MSC condition (*type=card*) and the rectangle that refers to the MSC representation of the use case extension (*payment* by *credit card* in this case).

In the above figure, the actor *employee* initiates the MSC sending the event New Order. At this point the MSC search customer is "executed", and an empty order is created. The rectangle labelled with 1,10 shows the repetition of events: the system asks for an order line and the employee enters product codes and quantities. The user interface object checks the product code sending the event *CheckProduc(ProductCode, Quantity)* to the control object. The control object verifies the information with *Product*, returns the event *Data(Description, Price)* and adds a new order line to the current order. The employee enters the payment information (*PaymentType(Type)*) and finally, if the condition *Type=Card* holds, the MSC *Pay_by_Card* is "executed". It is important to note that this MSC shows a primary scenario of interaction without error conditions.

An important piece of data that must be introduced in this step is constituted by the labels that will appear in the user interface to identify relevant pieces of information. When following the flow of events specified in the MSC, a given piece of information enters or exits the user interface object, the analyst must specify the corresponding label, that will play a basic role in the process of generating the user interface.

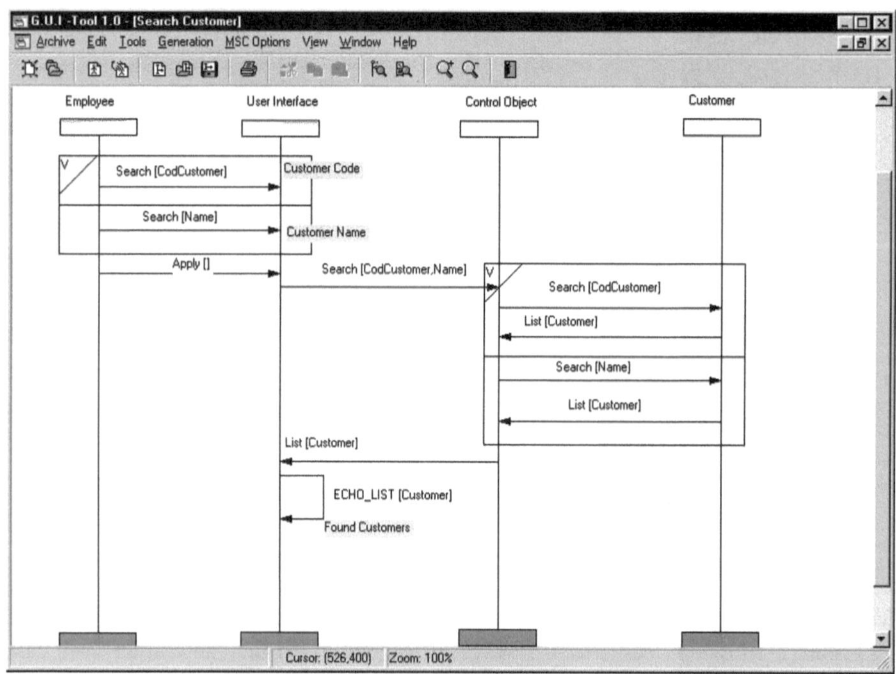

Fig. 6. Labelled MSC chart, *Search Customer*

Figure 6 shows the MSC corresponding to the use case "*search customer*". The analyst has placed 3 labels: "customer code", "customer name" and "founds customers". These correspond to two areas/fields of entry and one echo or area/field of exit. This information will literally appear in the form associated with the use case.

Apart from the labels, information about the type of the arguments of each event specified in the diagram must be introduced. The allowed types are the basic data-valued types (number, boolean, character, strings, enumerated) and the object-valued types corresponding to the system classes. The type of the event arguments together with the class attributes are used in the process of the interface generation, as will be seen to in the following sections.

3.3 Generation of the Specification

The specification is formed by a group of state transition diagrams that describe the system behaviour. One diagram for each user interface object and another for each control object that appears in a selected MSC are generated. To put this idea to work, in the tool, the analyst selects the MSCs for which he wants to generate the specification, and the resulting specification is stored in the tool´s repository. To make the implementation easier, we do not employ a graphic representation for the generated STDs; they are stored and viewed using a transition table.

In the process of generating transition tables, we use a variant of the algorithm proposed by Systa ([24]), the details of which can be seen in Sánchez ([21]). The process consists of associating pathways within a STD found in an MSC. Messages directed towards a particular object O, are considered events in the state transition diagram for O. Messages directed away from O are considered actions. References to another diagram in an MSC are transformed into super-states.

The figure 2 shows the STD for the user interface object in the use case *new order*. The states *search customer*, *pay by card* and *pay by check* are super-states. Each super-state has a flat state transition diagram associate to it. The path:

Search_Customer.(State1.State2.State3.State4.State1)$^{1..10}$*.Pay_By_Card*

corresponds with the MSC or normal scenario of figure 5. The alternate paths (ie. *State3.State5*) correspond to an exceptional scenario.

3.4 User Interface Generation

The process of generating the user interface prototype to develop the final software product starts with the analysis of the use case structured model, or the equivalent, resultant MSCs obtained from them. For each actor that plays a role of service activator, a view model is obtained by generating a form that contains the set of forms that can be called up directly from the application menu. Each use case that an actor can execute is converted into a form. The model describing the way of navigating among these forms is obtained by analyzing the include and extend relationships.

Referring back to the example introduced in Figure 1, the following forms would be generated: new customer, search customer, order status, search order, cancel order, and new order. Notice that no forms for Pay by Check or Pay by Credit Card are generated in the view model: as they are the result of using "*include*" or "*extend*" relationships, they follow the following rules:

- If A (New Order) is one case that uses a case B (Search Customer), then there is a button placed in the form associated with A which allows navigation towards the form associated with B.
- If B (Payment by Check) is a case which extends a case A (New Order) and B cannot be directly executed, a button is placed in A which allows navigation towards B.

Let's see how MSC services are converted into user interface components. For such a service, a form is generated. This form will contain the widgets needed to have the corresponding software representation at the interface level. To make this process of conversion clearer, in Table 2 we present the set of transformations that are done depending on which kind of information input/output do we have.

For instance, given a service $E(a_1,..a_n)$, if a_1 is an enumerated type (as specified in the corresponding MSC), depending on the size (greater or less than 4), a set of radio buttons or a list selection widget will be generated in the resulting user interface.

Table 2. User interfaces widgets.

Information Input: $E(a_1,...,a_n)$		
Argument type (a_i)	**Condition**	**Form Component**
Enumerated	Size =<4	Enabled radio buttons widget
	Size >4	Enabled List Selection widget
Boolean	-	Enabled Radio buttons widget
Character, Number, String	-	Enabled Text Edit widget
$a_1,..,a_n$ class dependent	-	Grid, an input entry for each attribute class
Information Output:		
Basic	-	Disabled Text Edit widget
Class type/ $a_1,..,a_n$ class dependant	-	Disabled Grid widget
Class type List	-	Disabled Grid widget with scroll bars
Error message	-	Message Box widget with confirmation button

The scheme set out in Table 2 is used by the tool to create the set of software components that constitutes the user interface. In fact, this is a closed group of rules that guides the generation process. The above table is based on heuristics and results found in the literature ([11]).

For the MSC shown in Figure 6, and in accordance with the previous table, the tool automatically generates a form with two fields of entry and with the labels "Customer Code" and "Customer Name". As the system will produce an echo, and as this is made up of a list of objects, a grille/grid will be generated, where each column corresponds to Customer class attributes. The buttons *Apply* and *Exit* are always generated for every form, to represent the two basic options of committing an action or cancelling it in a standard way.

As the "*search customer*" use case is neither used nor extended by any other use case, there are no connecting buttons to allow navigation. What is possible is to reach "*search customer*" from the form associated with "*new order*". This is the consequence of the "*include*" relationship declared between "*new order*" and "*search customer*" (see Fig.1).

Once a form has been generated according to our method, it can be modified using the target visual programming environment. It can also be animated by using symbolic execution, simulating the execution of the associated state transition diagram of the user interface object.

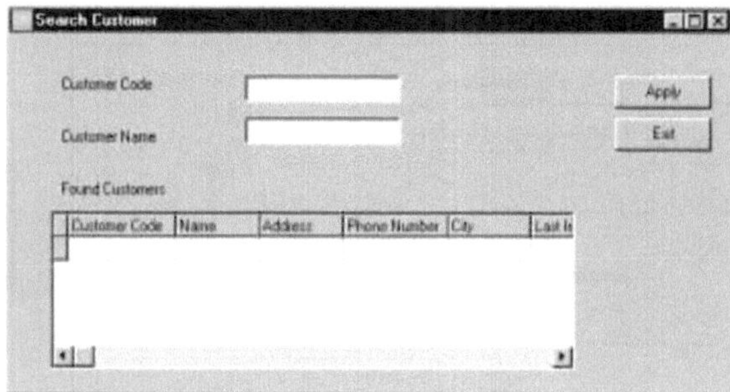

Fig. 7. Form associated to use cases Search Customer.

3.4.1 Actor View Model

The tool integrates the forms generated for the various use cases (MSCs) in a single application (see Fig. 8). The application comprises a menu where the options are the use cases that each actor can execute. For the example under study, this model shows the forms that the actors, employee and administrator, can activate.

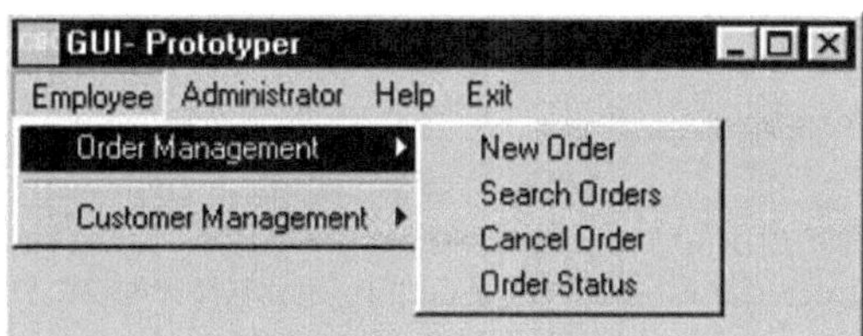

Fig. 8. Actor View Model.

For instance, the actor employee can execute the following services: *new order*, *search orders*, *cancel order* and *order status,* from the menu option *order management* and *search customer*, *new customer* from the menu option *customer management*. The actor administrator, on the other hand, can execute: *new product* and *cancel product*. We use the primary use case model to group the application menu options.

3.4.2 Interform Navigational Model

The inter-form navigation model represents the set of forms that an actor needs in order to complete all the required tasks, as well as the actor's potential options to go from one form to another.

In Fig. 9 the bold line represents the unconditional activation of a form, and the implicit return to the activation point after execution. This unconditional activation corresponds to the use case relationship "include". On the other hand, the activation conditions represents the extension points of a use case. By exploiting these relationships among the use cases or rather among the different MSCs, we automatically obtain the interform navigation model, from the structured use cases model.

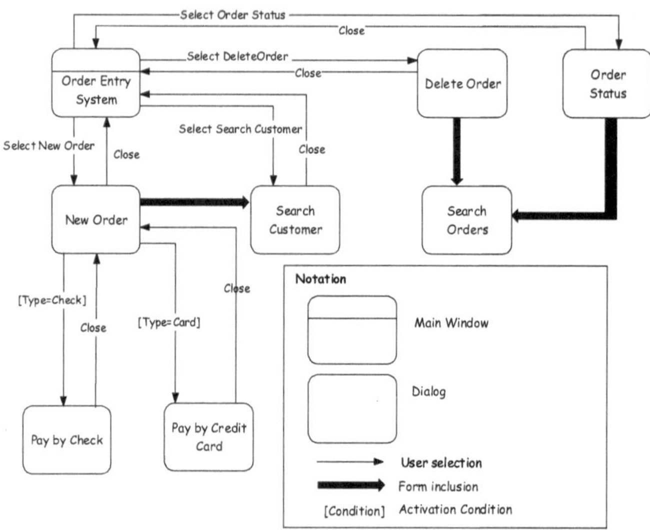

Fig. 9. Interform navigation model, actor employee.

4 Related Work

Within the field of user interface generation, there is a varied range of proposals, that go from those that use data models and generate only the static part of the interface to those that are scenario-based and support different degrees of automation (from partial to total).

If we take into account the proposals based on data models, it is normal to use a variant of the entity-relationship model, or rather an object model. However, if we consider those scenario-based models, they tend to use Petri nets or state transition diagrams as specification techniques. We shall now discuss some of the more relevant proposals in these two categories.

Worth citing from among the approaches based on data models which only generate the static part of the interface, are Janus and Trident. In Janus ([1]), an object model is built and a window is associated to each non-abstract class of the model. The analyst manually selects the methods and attributes which are relevant to that interface. The setting then generates a static view. Even if this approach is interesting, if we compare it with our method, it can be seen that the treatment is merely static, which is an important weak point that we have overcome.

Trident ([27]) uses a more sophisticated approach. Working from an entity – relationship model, an activity graph is drawn which connects interactive tasks with the system's functions and data. The graph is used as a means of entry to a process that selects the different units of presentation. As in the previous case, only the static part is generated, what again marks the main difference with our work.

In TADEUS ([19]), a stage generically called dialogue design is used, and this covers two tasks which must be carried out manually: the design of the navigational dialogue that describes the sequence of the different views, and the design of the processing dialogue which in turn describes the changes undergone by the objects within a dialogue view. The formalism used is based on the Petri nets. During the generation process, scripts are obtained which can be included in an interface management system. The manual Petri Net building process is, in our oppinion, the main drawback of the proposal: it is very hard to use Petri Nets for specifying the user interface properties.

Lastly, the most similar proposal to our own is SUIP ([8]) from Montreal University. They use collaboration diagrams, which they define by means of flat files which are enriched by information from the user interface. They focus on behavioural aspects as we do, but at a lower level of abstraction. We think that starting with use cases is more appropriate than using collaboration diagrams. Furthermore, in contrast to our proposal, it does not obtain a navigational model automatically. It is interesting for us to note that are currently modifying their proposal to use Petri nets instead of STD.

5 Conclusions and Future Work

A methodological approach to guide the process of generating user interfaces from the user requirements elicitation has been presented in this paper. As a main contribution, the approach uses a functional style to capture requirements using a Use-Case background, instead of starting from a merely static system view as other similar approaches do. This is an important point, because analysts can focus on the behavioural system aspects, which are represented by a Use Case Model in a natural way. This Use Case Model is the starting point for the user interface generation process.

Furthermore, the proposal has a high degree of automation, which allows us to provide a CASE tool to support the method. We have built a metamodel for the MSCs, this metamodel has been used to design a relational data base which acts as a repository for information. The tool is programmed in Delphi, and the data base in Interbase 5.1, but the underlying ideas could be implemented in any other conventional client-server software development environment.

The proposal has been successfully put to work to validate user requirements, by executing the user interface prototypes generated in an automated way following the four-step method presented in the paper.

Another aspect which is worth highlighting is that the method allows us to obtain the interform navigational model automatically. This is being used at the moment to generate web clients where the navigational capabilities are derived from the requirements specification, which is an interesting line of research. As a main objective which is underway at this moment, we want to focus on the appearance of the generated forms, with the final objective of being able to use them not only for prototyping purposes, but also as a ready-to-use final software product.

To reach this objective, we are presently trying to incorporate our scheme for the generation and definition of interfaces in the RSI (Requirements/Service/Interface) of M. Collins ([6]). This approach uses 3 categories of use cases: requirements, interface and services. The requirement use cases document business processes which can be automated. The interface model statically describes the user interfaces of the application, by only using drawings with their different components. Lastly, the service model describes what functions the system has to provide, regardless of the needs of one particular user interface. Our proposal would fit perfectly into the interface model, that is, the interface level can be described by means of an automatically generated prototype.

Finally, if some of the figures containing images caught by the tool are observed, a file labelled as "data" can be seen. Our idea is to define real data and to try to animate the prototype with this data, following the same line as those who validate object models based on STD ([5])

References

[1] Balzert, H.; "From OOA to GUIs: The Janus System". IEEE Software, 8(9), Febraury 1996, pp 43-47.

[2] Benner K.M.; Feather M.S.; Johnson W.L. "Utilizing Scenarios in the Software Development Process. En Information System Development Process, pp 117-134. Elsevier Science Publisher, 1993. Editores: N. Prakash, C. Rolland, y B. Percini.

[3] Bennett, D.W. "Designing Hard Software: The Essential Tasks". Manning Publications co, 1997.

[4] Booch G; Rumbaugh J; Jacobson I. "The unified modelling language". Addison-Wesley. 1999.

[5] Bridge Point. CASE TOOL, Project Technology. http://www.projtech.com

[6] Collins-Cope, M; "RSI- A Structured Approach to Use Cases and HCI Design". Personal Communication, Ratio Group Ltd. 1999.

[7] Constantine L.L; Lockwood L.A.D. "Software for Use: A practical Guide to the Models and Methods of Usage-Centered Design". Addison Wesley 1999.

[8] Elkoutbi M; Khriss I; Keller R; "Generating User Interface Prototypes from Scenarios". Proceedings of the Fourt IEEE International Symposium on Requeriments Engineering (RE'99). Limerick Ireland 1999.

[9] Harel D; "State Charts: a visual formalism for complex systems". Science of Computer Programming, 8(3), 231-274. 1987.

[10] Horrocks I. "Constructing the User Interface with Statecharts". Addison-Wesley, 1998.

[11] IBM, Systems Application Architecture: Common User Access- Guide to User Interface Design- Advanced Interface Design Reference, IBM, 1991.

[12] ITU: Recommendation Z. 120: Message Sequence Chart (MSC). ITU, Geneva, 1996.

[13] Jacobson I et al. "Object-Oriented Software Engineering: A use case driven approach". New-York ACM Press, 1992.

[14] Jacobson I; Booch G; Rumbaugh J. "The Unified Software Development Process". Addison-Wesley,1999.

[15] Kerr J; Hunter R; "Inside RAD". McGraw-Hill 1994.

[16] Kotonya, G; Sonmmerville, I. "Requirements Engineering: Process and Techniques". John Wiley & Sons. 1998.

[17] Nielsen J. "Scenarios in Discount Usability Engineering". Scenario-Based Design: Envisioning Work and Technology in System Development. John Wiley & Sons, 1995. pp 59-85.

[18] Rumbaugh J; et al. "Object Oriented Modeling". Prentice Hall 1997.

[19] Schlungbaum E; Elwert T; "Modeling a Netscape-like browser using TADEUS". CHI'96. Vancuver 1996. Pp 19-24

[20] Schneider G.; Winters J.P. "Applying Use Cases: A practical Guide". Addison Wesley 1998.

[21] Sánchez Díaz J; "Generating State Transition Diagrams from Message Sequence Charts" Technical Report -DSIC,-1999. Valencia University of Technology (in Spanish).

[22] Sánchez J; Pelechano V; Pastor O; "Un entorno de generación de interfaces de usuario a partir de casos de uso". Workshop On Requeriments Engineering' 99. pp 106-116. Buenos Aires. Septiembre 1999.

[23] Sánchez J; Pelechano V; Insfrán E; "Un entorno de generación de prototipos de interfaces de usuario a partir de diagramas de interacción". IDEAS 2000. Pp 145-155. Cancún (Mexico). Abril 2000.

[24] Systa T. "Automated Support for Constructing OMT Scenarios and State Diagrams". Department of Computer Science. University of Tampere. Technical Report A-1997-8

[25] Somé S.S.; Dssouli R; Vaucher, J. "Toward an Automation of Requirements Engineering using Scenarios". Journal of Computing and Information, vol 2,1, 1996, pp 1110-1132,

[26] Telelogic. CASE TOOL Telelogic Tau 3.6.2. http://www.telelogic.com. 1999.

[27] Vanderdonckt J. "Knoledge-Based Systems for Automated User Interface Generation: the TRIDENT Experience". RP-95-010. March 1995.

The ADORA Approach to Object-Oriented Modeling of Software

Martin Glinz[1], Stefan Berner[2], Stefan Joos[3], Johannes Ryser[1], Nancy Schett[1], and Yong Xia[1]

[1] Institut für Informatik, Universität Zürich, Winterthurerstrasse 190, CH-8057 Zurich, Switzerland
{glinz, ryser, schett, xia}@ifi.unizh.ch

[2] FJA, Zollikerstrasse 183, CH-8008 Zurich, Switzerland
stefan.berner@fja.com

[3] Robert Bosch GmbH, Postfach 30 02 20, D-70469 Stuttgart, Germany
stefan.joos@de.bosch.com

Abstract. In this paper, we present the ADORA approach to object-oriented modeling of software (ADORA stands for Analysis and Description of Requirements and Architecture). The main features of ADORA that distinguish it from other approaches like UML are the *use of abstract objects* (instead of classes) as the basis of the model, a *systematic hierarchical decomposition* of the modeled system and the *integration* of all aspects of the system *in one coherent model*. The paper introduces the concepts of ADORA and the rationale behind them, gives an overview of the language, and reports the results of a validation experiment for the ADORA language.

1 Introduction

When we started our work on object-oriented specification some years ago, we were motivated by the severe weaknesses of the then existing methods, e.g. [3][5][15]. In the meantime, the advent of UML [16] (and to a minor extent, OML [6]) has radically changed the landscape of object-oriented specification languages. However, also with UML and OML, several major problems remain.

There is still no true integration of the aspects of data, functionality, behavior and user interaction. Neither do we have a systematic hierarchical decomposition of models (for example, UML packages are a simple container construct with nearly no semantics). Models of system context and of user-oriented external behavior are weak and badly integrated with the class/object model [9].

So there is still enough motivation not to join simply the UML mainstream and to pursue alternatives instead. We are developing an object-oriented modeling method for software that we call ADORA (Analysis and Description of Requirements and Architecture) [1][13]. ADORA is intended to be used primarily for requirements specification and also for logical-level architectural design. Currently, ADORA has no language elements for expressing physical design models (distribution, deployment) and imple-

K.R. Dittrich, A. Geppert, M.C. Norrie (Eds.): CAiSE 2001, LNCS 2068, pp. 76-92, 2001.
© Springer-Verlag Berlin Heidelberg 2001

mentation models. However, as ADORA models are object-oriented, we can implement a smooth transition from an ADORA architecture model to detailed design and code written in an object-oriented programming language.

In this paper, we present the ADORA language. We discuss the general concepts and give an overview of the language. The main contributions of the ADORA language are

- a concept for systematic hierarchical decomposition of models which is particularly useful when modeling distributed systems,
- the integration of different aspects into one coherent model,
- the ability to visualize a model in its context,
- language elements for tailoring the formality of ADORA models.

Throughout this paper, we will use a distributed heating control system as an example. The goal of this system is to provide a comfortable control for the heating system of a building with several rooms. An operator can control the complete system, setting default temperatures for the rooms. Additionally, for every room individual temperature control can be enabled by the operator. Users then can set the desired temperature using a control panel in the room. The system shall be distributed into one master module serving the operator and many room modules.

However, ADORA is not only applicable for the specification of industrial control systems. For the validation of the usefulness of the language, we have modeled a distributed information system with ADORA (see section 4).

The rest of the paper is organized as follows. In section 2 we discuss the basic concepts of ADORA and their rationale. In section 3 we give an overview of the language. In section 4 we present the results of a first validation of the ADORA language. Finally, we compare the concepts of ADORA with those of UML and conclude with a discussion of results, state of work and future directions.

2 Key Concepts and Rationale of the ADORA Approach

In this section, we briefly describe the five principles that ADORA is based on and give our rationale for choosing them.

2.1 Abstract Objects instead of Classes

When we started the ADORA project, all existing object-oriented modeling methods used class diagrams as their model cornerstone. However, class models are inappropriate when more than one object of the same class and/or collaboration between objects have to be modeled [9][12]. Both situations frequently occur in practice. For an example, see the buttons in Fig. 1. Moreover, class models are difficult to decompose. As soon as different objects of a class belong to different parts of a system (which often is the case), hierarchical decomposition does no longer work for class models [12]. Wirfs-Brock [19] tries to overcome the problems of class modeling by using classes in different *roles*. However, decomposition remains a problem: what does it mean to decompose a role?

We therefore decided to use abstract, prototypical objects as the core of an ADORA model (Fig. 1). An equivalent to classes (which we call types) is only used to model common characteristics of objects: types define the properties of the objects and can be organized in subtype hierarchies. In order to make models more precise, we distinguish between *objects* (representing a single instance) and *object sets* that represent a set of instances. Modeling of collaboration and of hierarchical decomposition (see below) becomes easy and straightforward with abstract objects.

In the meantime, others have also discovered the benefits of modeling with abstract objects. UML, for example, uses abstract objects for modeling collaboration in collaboration diagrams and in sequence diagrams. However, without a notion of abstraction and decomposition, only local views can be modeled. Moreover, class diagrams still form the core of a UML specification.

2.2 Hierarchical Decomposition

Every large specification must be decomposed in some way in order to make it manageable and comprehensible. A good decomposition (one that follows the basic software engineering principles of information hiding and separation of concerns) decomposes a system recursively into parts such that

- every part is logically coherent, shares information with other parts only through narrow interfaces and can be understood in detail without detailed knowledge of other parts,
- every composite gives an abstract overview of its parts and their interrelationships.

The current object-oriented modeling methods typically approach the decomposition problem in two ways: (a) by modeling systems as collections of models where each model represents a different aspect or gives a partial view of the system, and (b) by providing a container construct in the language that allows the modeler to partition a model into chunks of related information (e.g. packages in UML). However, both ways do not satisfy the criteria of a good decomposition. Aspect and view decompositions are coherent only as far as the particular aspect or view is concerned. The infor-

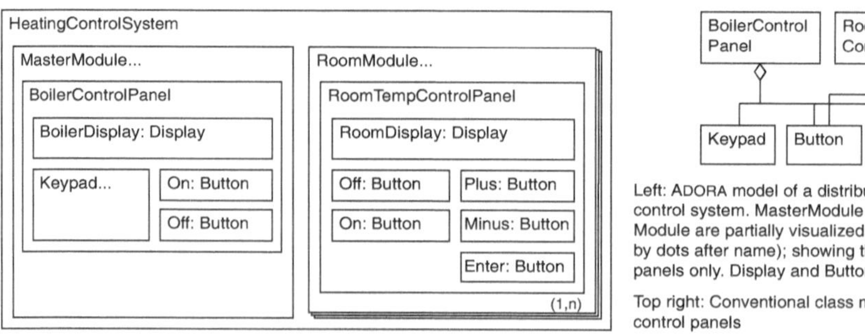

Left: ADORA model of a distributed heating control system. MasterModule and RoomModule are partially visualized (indicated by dots after name); showing the control panels only. Display and Button are types.

Top right: Conventional class model of the control panels

Fig. 1. An ADORA object model (left) vs. a conventional class model (top right)

mation required for comprehending some part of a system in detail is *not* coherently provided. Container constructs such as UML packages have semantics that are too weak for serving as composites in the sense that the composite is an abstract overview of its parts and their interrelationships. This is particularly true for multi-level decompositions. Only the ROOM method [18] can decompose a system in a systematic way. However, as ROOM is also based on classes, the components are not classes, but class references. This asymmetry makes it impossible to define multi-level decompositions in a straightforward, easily understandable way.

In ADORA, the decomposition mechanism was deliberately chosen such that good decompositions in the sense of the definition given above become possible. We recursively decompose objects into objects (or elements that may be part of an object, like states). So we have the full power of object modeling on all levels of the hierarchy and only vary the degree of abstractness: objects on lower levels of the decomposition model small parts of a system in detail, whereas objects on higher levels model large parts or the whole system on an abstract level.

2.3 Integrated Model

With existing modeling languages, one creates models that consist of a set of more or less loosely coupled diagrams of different types. UML is the most prominent example of this style. This seems to be a good way to achieve separation of concerns. However, while making separation easy, loosely coupled collections of models make the equally important issues of integration and abstraction of concerns quite difficult.

In contrast to the approach of UML and others, an ADORA model integrates all modeling aspects (structure, data, behavior, user interaction...) in one coherent model. This allows us to develop a strong notion of consistency and provides the necessary basis for developing powerful consistency checking mechanisms in tools. Moreover, an integrated model makes model construction more systematic, reduces redundancy and simplifies completeness checking.

Using an integrated model does of course not mean that everything is drawn in one single diagram. Doing so would drown the user in a flood of information. We achieve separation of concerns in two ways: (1) We *decompose* the model *hierarchically*, thus allowing the user to select the focus and the level of abstraction. (2) We use a *view concept* that is based on *aspects*, not on various diagram types. The *base view* consists of the objects and their hierarchical structure only. The base view *is combined with* one or more *aspect views*, depending on what the user wishes to see. These two concepts – hierarchy and combination of views – constitute the *essence* of organizing an ADORA model.

So the complete model is basically an abstract one – it is almost never drawn in a diagram. The concrete diagrams typically illustrate certain aspects of certain parts of a model in their hierarchical context. However, as every concrete diagram is a view of an integrated model of the complete system, we can build strong consistency and completeness rules into the language and build powerful tools for checking and maintaining them. Readability of diagrams is achieved by selecting the right level of abstraction, by restricting the number of aspects being viewed together, and by splitting complex diagrams into an abstract overview diagram and some part diagrams. For example, if Fig. 2 is perceived to be too complex, it can be split into an overview diagram (Fig. 8) and two part diagrams, one for MasterModule and one for RoomModule.

2.4 Adaptable Degree of Formality

An industrial-scale modeling language should allow its users to adapt the degree of formalism in a specification to the difficulty and the risk of the problem in hand. Therefore, they need a language with a broad spectrum of formality in its constructs, ranging from natural language to completely formal elements.

In ADORA, we satisfy this requirement by giving the modeler a choice between informal, textual specifications and formal specifications (or a mixture of both). For example, an object may be specified with an informal text only. Alternatively, it can be formally decomposed into components. These in turn can be specified formally or informally. As another example, state transitions can be specified in a formal notation or informally with text or with a combination of both.

The syntax of the ADORA language provides a consistent framework for the use of constructs with different degrees of formality.

2.5 Contextual Visualization

Current modeling languages either lack capabilities for system decomposition or they visualize decompositions in an explosive zoom style: the composites and their parts are drawn as separate diagrams. Thus, a diagram gives no information about the context that the presented model elements are embedded in. In ADORA, we use a fisheye view concept for visualizing a component in its surrounding hierarchical context. This simplifies browsing through a set of diagrams and improves comprehensibility [1].

3 An Overview of the ADORA Language

An ADORA model consists of a basic hierarchical object structure (the base view, as we call it) and a set of aspect views that are combined with the base view. In this section we describe these views and their interaction.

3.1 Basic Hierarchical Object Structure

The object hierarchy forms the basic structure of an ADORA model.

Objects and object sets. As already mentioned above, we distinguish between objects and object sets. An *ADORA object* is an abstract representation of a single instance in the system being modeled. For example, in our heating control system, there is a single boiler control panel, so we model this entity as an object. Abstract means that the object is a placeholder for a concrete object instance. While every object instance must have an object identifier and concrete values for its attributes, an ADORA object has neither of these. An *ADORA object set* is an abstract representation of a set of object instances. The number of instances allowed can be constrained with a cardinality. For example, in an order processing system we would model suppliers, parts, orders, etc. as object sets. In the heating control system, we have a control panel in every room and we control at least one room. Thus we model this panel as an object set with cardinality (1,n); see Fig. 2.

Structure of an ADORA object. An object or object set has a twofold inner structure: it consists of a set of properties and (optionally) a set of parts.

The *properties* are attributes (both public and private ones), directed relationships to other objects/object sets, operations and so called standardized properties. The latter are user-definable structures for stating goals, constraints, configuration information, notes, etc.; see below.

The *parts* can be objects, object sets, states and scenarios. Every part again can consist of parts: objects and object sets can be decomposed recursively as defined above, states can be refined into statecharts, scenarios into scenariocharts (as we call them, see below). Thus, we get a hierarchical whole-part structure that allows modeling a hierarchical decomposition of a system. The decomposition is strict: an element neither can contain itself nor can it be a part of more than one composite. We stick to a strict decomposition due to its inherent simplicity and elegance. Commonalities between objects in different positions of a decomposition hierarchy can be modeled by assigning them the same type (see the paragraph on types below and Fig. 1).

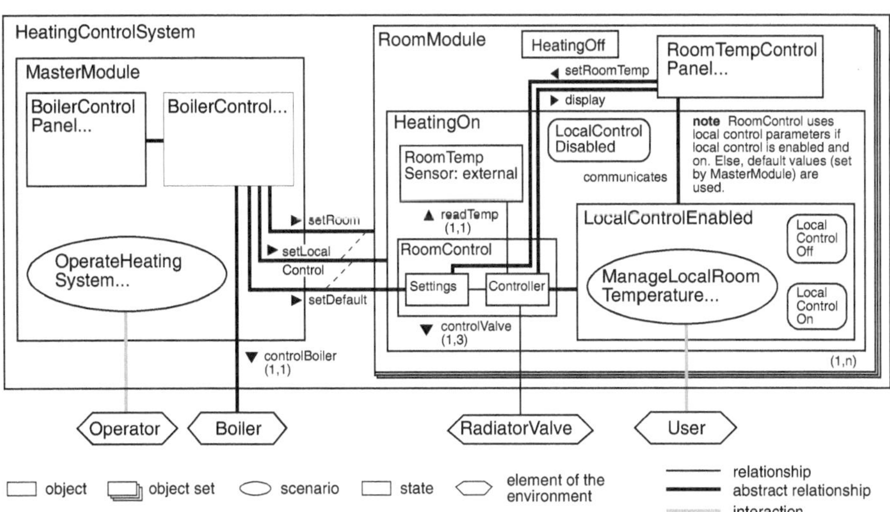

Fig. 2. An ADORA view of the heating control system: base view combined with structural view and context view

Graphic representation. In order to exploit the power of hierarchical decomposition, we allow the modelers to represent an ADORA model on any level of abstraction, from a very high-level view of the complete system down to very detailed views of its parts.

We achieve this property by representing ADORA objects, object sets, scenarios and states by nested boxes (see Fig. 1 and 2). The modeler can freely choose between drawing few diagrams with deep nesting and more diagrams with little nesting. In order to distinguish expanded and non-expanded elements in a diagram, we append three dots to the name of every element having parts that are not or only partially drawn on that diagram.

Types. Frequently, different objects have the same inner structure, but are embedded in different parts of a system. In the heating system for example, the boiler control panel and the room control panels both might have a display with the same properties. In these situations, it would be cumbersome to define the properties individually for every object. Instead, ADORA offers a type construct. An ADORA type defines

- the attributes and operations of all objects/object sets of this type
- a structural interface, that means, information required from or provided to the environment of any object/object set of this type. This facility can be used to express *contracts*.

A type defines neither the relationships to other objects/object sets nor the embedding of the objects of that type. Types can be organized in a subtype hierarchy.

An object can have a type name appended to its name (for example, RoomDisplay: Display in Fig. 1). In this case, the object is of that type and the type is separately defined in textual form. Otherwise, there is no other object of the same type in the model and the type information is included in the definition of the object.

```
propertydef goal numbered Hyperstring constraints unique;
propertydef created Date;
propertydef note Hyperstring;
object HeatingControlSystem...
goal 1 "Provide a comfortable control for the heating of a building with several rooms."
created 2000-11-04
note "Constraints have yet to be discussed and added."
end HeatingControlSystem.
```

Fig. 3. Definition and use of standardized properties

Standardized properties. In order to adapt ADORA in a flexible, yet controlled way to the needs of different projects, application domains or persons, we provide so called *standardized properties*. An ADORA standardized property is a typed construct consisting of a header and a body. Fig. 3 shows the type definitions for the properties goal, created and note and the application of these properties in the specification of the object HeatingControlSystem. As name and structure of the properties are user-definable, we get the required flexibility. On the other hand, typing ensures that a tool nevertheless can check the properties and support searching, hyperlinking and cross-referencing.

3.2 The Structural View

The structural view combines the base view with directed relationships between objects. Whenever an object A references an information in another object B (and B is not a part of A or vice-versa) then there must be a relationship from A to B. Referencing an information means that A

- accesses a public attribute of B,
- invokes an operation of B,
- sends an event to B or receives one from B.

Every relationship has a name and a cardinality (in the direction of the relationship). Bidirectional relationships are modeled by two names and cardinalities. Relationships are graphically represented by lines between the linked objects/object sets. An arrow preceding the name indicates the direction of the relationship (Fig. 2).

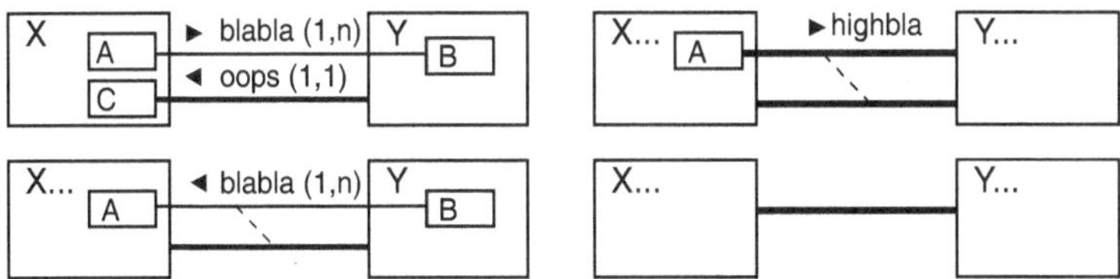

Fig. 4. Four static views of the same model on different levels of abstraction

Static relationships must reflect the hierarchical structure of the model. Let objects A and B be linked by a relationship. If A is contained in another object X and B in an object Y, then the relationship A → B implies abstract relationships X → Y, A → Y and X → B. Whenever we draw a diagram that hides A, B or both, the next higher abstract relationship must be drawn. Abstract relationships are drawn as thick lines. They can, but need not be named. In case of partially expanded objects, we sometimes have to draw both a concrete and a corresponding abstract relationship. In this case, we indicate the correspondence by a dashed hairline (Fig. 4). In the view shown in Fig. 2, we have some examples. All relationships from BoilerControl to other objects are abstract ones because their origins within BoilerControl are hidden in this view. The relationships readTemp from Controller to RoomTempSensor and controlValve from Controller to RadiatorValve are elementary relationships. Hence they are drawn with thin lines. If we had chosen a view that hides the contents of RoomControl, we had drawn two abstract relationships from RoomControl to RoomTempSensor and to RadiatorValve, respectively.

3.3 The Behavioral View

Combining objects and states. For modeling behavior, ADORA combines the object hierarchy with a statechart-like state machine hierarchy [7][8][12]. Every object represents an abstract state that can be refined by the objects and/or the states that an object contains. This is completely analogous to the refinement hierarchy in statecharts [10] and can be given analogous semantics for state transitions. We distinguish pure states (represented graphically by rounded rectangles) and objects with state (see Fig. 5). Pure states are either elementary or are refined by a pure statechart. Objects with state additionally have properties and/or parts other than states.

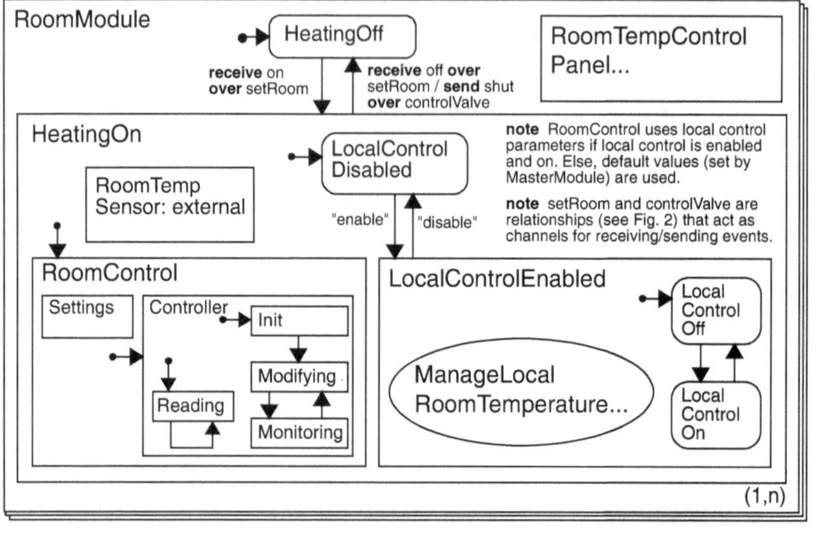

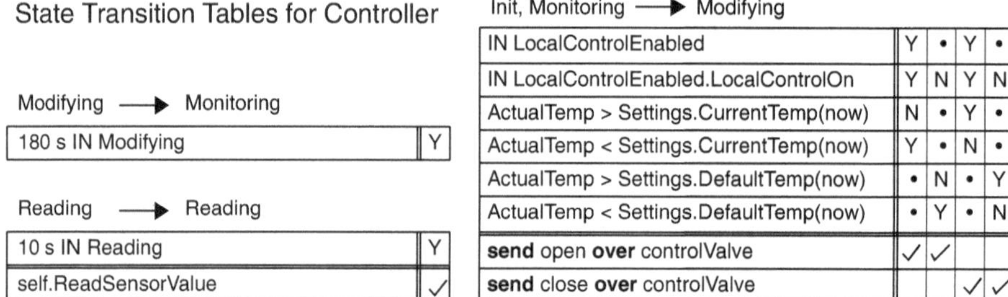

State Transition Tables for Controller

Modifying → Monitoring

180 s IN Modifying	Y

Reading → Reading

10 s IN Reading	Y
self.ReadSensorValue	✓

Init, Monitoring → Modifying

IN LocalControlEnabled	Y	•	Y	•
IN LocalControlEnabled.LocalControlOn	Y	N	Y	N
ActualTemp > Settings.CurrentTemp(now)	N	•	Y	•
ActualTemp < Settings.CurrentTemp(now)	Y	•	N	•
ActualTemp > Settings.DefaultTemp(now)	•	N	•	Y
ActualTemp < Settings.DefaultTemp(now)	•	Y	•	N
send open **over** controlValve	✓	✓		
send close **over** controlValve			✓	✓

Fig. 5. A partial ADORA model of the heating control system; base view and behavior view

We do not explicitly separate parallel states/state machines as it is done in statecharts. Instead, objects and states that are part of the same object and have no state transitions between each other are considered to be parallel states. Objects that neither are the destination of a state transition nor are designated as initial abstract states are considered to have no explicitly modeled state.

By embedding the behavior model into the object decomposition hierarchy, we can easily model behavior on all levels of abstraction. On a high level, objects and states may represent abstract concepts like operational modes (off, startup, operating...). On the level of elementary objects, states and transitions model object life cycles.

State transitions. Triggering events and triggered actions or events can either be written in the traditional way as an adornment of the state transition arrows in the diagrams, or they can be expressed with transition tables [14]. For large systems with complex transition conditions the latter notation is more or less mandatory in order to keep the model readable. Depending on the degree of required precision, state transition expressions can be formulated textually, formally, or with a combination of both.

Fig. 5 shows the graphic representation of a behavior view with some of the variants described above. When the system is started, then for all members of the object set

RoomModule the initial state HeatingOff is entered. The transition to the object HeatingOn is specified formally. It is taken when the event on is received over the relationship set-Room (cf. Fig. 2). If this transition is taken, the state LocalControlDisabled and the object RoomControl are entered concurrently. Within RoomControl, the object Controller is entered and within Controller the parallel states Init and Reading. This is equivalent to the rules that we have for statecharts. The state transitions between LocalControlDisabled and LocalControlEnabled are specified informally with a text only. This makes sense in situations where we do not yet precisely know in which situation an event has to be triggered by which component. The transitions within Controller are specified in tabular form, because they are quite complex.

Timing and event propagation. In ADORA we use the quasi-synchronous timing and event propagation semantics defined in [8]. In contrast to usual statecharts and other than in [8] we do not broadcast events in ADORA. Instead, events have to be explicitly sent and received. Thus we avoid global propagation of local events.

3.4 The Functional View

The functional view defines the properties of an object or object set (attributes, operations...) that have not already been defined by the object's type. When there is only one object of a certain type, the complete type information is embedded in the object definition. The functional view is not combined with other views; it is always represented separately in textual form.

Joos [13] has defined a formal notation for specifying functions in ADORA, building upon existing notations. As there is nothing conceptually new with function definitions, we do not go into further detail here.

3.5 The User View

In the last few years, the importance of modeling systems from a user's viewpoint, using scenarios or use cases, was recognized (for example, see [4][8][11] and many others). In ADORA, we take the idea of hierarchically structured scenarios from [8] a step further and integrate the scenarios into the overall hierarchical structure of the system. In our terminology, a scenario is an ordered set of interactions between partners, usually between a system and a set of actors external to the system. It may comprise a concrete sequence of interaction steps (instance scenario) or a set of possible interaction steps (type scenario). Hence, a use case is a type scenario in our terminology.

We view scenarios and objects to be complementary in a specification. The scenarios specify the stimuli that actors send to the system and the system's reactions to these stimuli. However, when these reactions depend on the history of previous stimuli and reactions, that means on stored information, a precise specification of reactions with scenarios alone becomes infeasible. The objects specify the structure, functions and behavior that are needed to specify the reactions in the scenarios properly.

In the base view of an ADORA model, scenarios are represented with ovals. In the user view, we combine the base view with grey lines that link the scenario with all

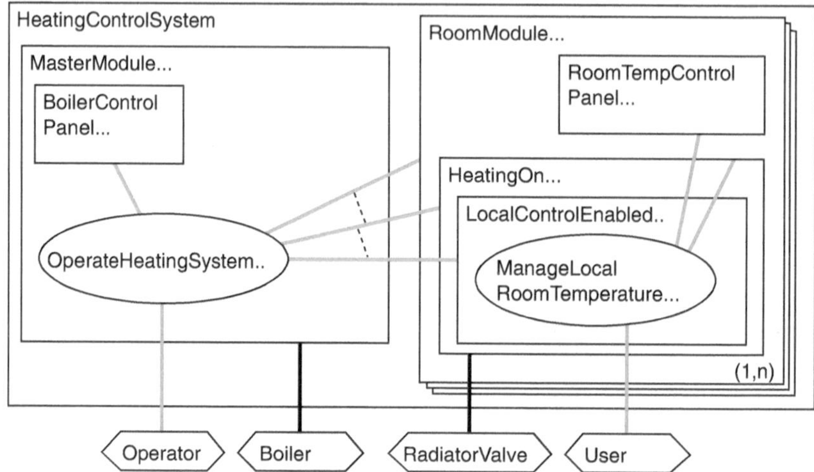

Fig. 6. A user view of the heating control system

objects that it interacts with (Fig. 6). For example, the scenario ManageLocalRoomTemperature specifying the interaction between the actor User and the system is localized within the object LocalControlEnabled. Internally, the scenario interacts with RoomTempControlPanel and with an object in HeatingOn which is hidden in this view.

An individual scenario can be specified textually or with a statechart. In both cases, ADORA requires scenarios to have one starting and one exit point. Thus, complex scenarios can be easily built from elementary ones, using the well-known sequence, alternative, iteration and concurrency constructors. In [8] we have demonstrated statechart-based integration of scenarios using these constructors. However, when integrating many scenarios, the resulting statechart becomes difficult to read. We therefore use Jackson-style diagrams (with a straightforward extension to include concurrency) as an additional means for visualizing scenario composition. We call these diagrams scenariocharts (Fig. 7).

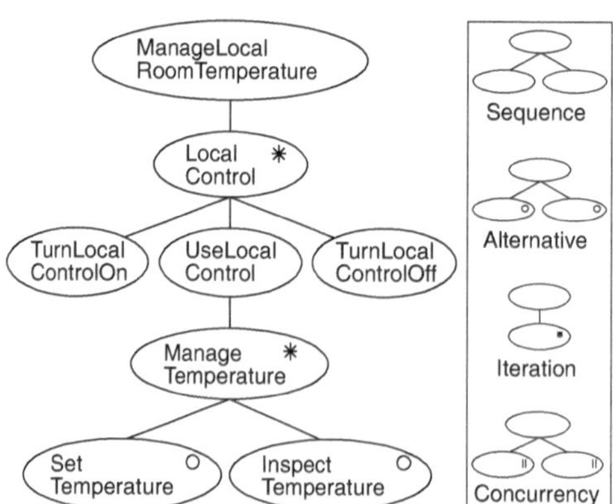

Fig. 7. A scenariochart modeling the structure of the ManageLocalRoomTemperature scenario

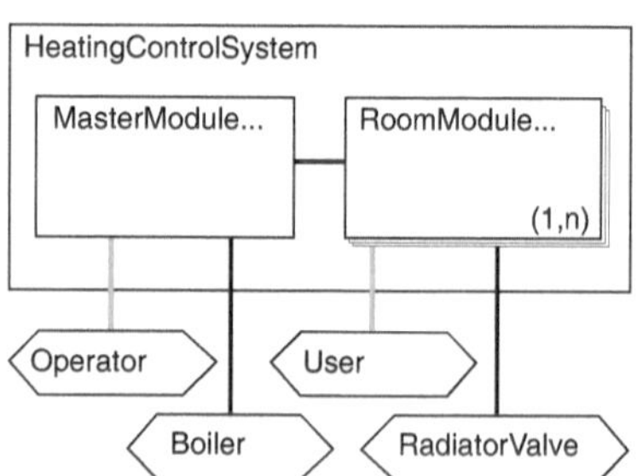

Fig. 8. A context diagram of the heating control system

Thus, we have a hierarchical decomposition in the user view, too. The object hierarchy of the base view allocates high-level scenarios (like ManageLocalRoomTemperature in our heating system) to that part of the system where they take effect. The scenario hierarchy decomposes high-level scenarios into more elementary ones. As a large system has a large number of scenarios (we mean type scenarios/use cases here, not instance scenarios), this facility is very important for grouping and structuring the scenarios.

Note that the ADORA user view is a logical view of user-system interaction only; it does not include the design of the user interface.

3.6 The Context View

The context view shows all actors and objects in the environment of the modeled system and their relationships with the system. Depending on the degree of abstraction selected for the system, we get a context diagram (Fig. 8) or the external context for a more detailed view of the system (Fig. 2).

In addition to external elements that are not a part of the system being specified, an ADORA model can also contain so called *external objects*. We use these to model preexisting components that are part of the system, but not part of the specification (because they already exist). External objects are treated as black boxes having a name only. In the notation, such objects are marked with the keyword external (for example, the object RoomTempSensor in Fig. 2). In any specification where COTS components will be part of the system or where existing components will be reused, modeling the embedding of these components into the system requires external objects.

3.7 Modeling Constraints and Qualities

Constraints and quality requirements are typically expressed with text, even in specifications that employ graphical models for functional specifications. In traditional specifications and with UML-style graphic models we have the problem of interrelating functional and non-functional specifications and of expressing the non-functional specifications on the right level of abstraction.

In ADORA, we use two ADORA-specific features to solve this problem. (1) The decomposition hierarchy in ADORA models is used to put every non-functional requirement into its right place. It is positioned in the hierarchy according to the scope of the requirement. The requirements themselves are expressed as ADORA standardized properties. Every kind of non-functional requirement can be expressed by its own property kind, for example performance constraint, accuracy constraint, quality...).

3.8 Consistency Checking and Model Verification

Having an integrated model allows us to define stringent rules for consistency between views, for example "When an object A references information in another object B in any view and B is not a part of A or vice-versa, then there must be a relationship from A to B in the static view." A language for the formulation of consistency constraints and a compiler that translates these constraints into Java have been developed [17]. By executing this code in the ADORA tool, the tool is enabled to check or enforce these constraints. The capabilities for formal analysis and verification of an ADORA model depend on the chosen degree of formality. In the behavior view, for example, a sufficiently formal specification of state transitions allows to apply all analyses that are available for hierarchical state machines.

4 Validation of ADORA

In our opinion, there are two fundamental qualities that a specification language should have: the language must be easy to comprehend (a specification has more readers than writers) and the users must like it.

Therefore, we experimentally validated the ADORA language with respect to these two qualities. We conducted an experiment with the following goals.

- Determine the comprehensibility of an ADORA specification both on its own and in comparison with an equivalent specification written in UML – today's standard modeling language – from the viewpoint of a reader of the specification.
- Determine the acceptance of the fundamental concepts of ADORA (using abstract objects, hierarchical decomposition, integrated model...) both on its own and in comparison with UML from the viewpoint of a reader/writer of models.

4.1 Setup of the Experiment

In order to measure these goals, we set up the following experiment [2]. We wrote a partial specification of a distributed ticketing system both in ADORA and in UML. The system consists of geographically distributed vending stations where users can buy tickets for events (concerts, musicals...) that are being offered on several event servers. Vending stations and event servers shall be connected by an existing network that needs not to be specified.

Then we prepared a questionnaire consisting of two parts. In the first part, the "objective" one, we aimed at measuring the comprehensibility of an ADORA model. We created 30 questions about the contents of the specification, for example "Can a user at a point of sale terminal purchase an arbitrary number of tickets for an event in a single transaction?" 25 questions were yes/no questions; the rest were open questions. For every question, we additionally asked, whether the answering person was sure or unsure about her or his answer and how difficult it was to answer the question.

In the second part, the "subjective" one, we tested the acceptance of ADORA vs. UML. We asked 14 questions about the personal opinion of the answering person concerning distinctive features of both ADORA and UML, for example "Does it make sense to use an integrated model (like ADORA) for describing all aspects of a system"?

We ran the experiment with fifteen graduate and Ph.D. students in Computer Science who were not members of our research group. The participants were first given an introduction both to ADORA and to UML. Then we divided the participants into two groups. The members of group A answered the objective part of the questionnaire first for the ADORA specification and then for the UML specification; group B members did it vice-versa. Finally, both groups answered the subjective part of the questionnaire. In order to avoid answers being biased towards ADORA, we ensured the anonymity of the filled questionnaires.

Two participants did not finish the experiment; another person's answers could not be scored because his answers revealed insufficient base knowledge of object technology. So we finally had twelve complete sets of answers.

4.2 Some Results

Due to space limitations, we only can present some key results here. The complete results are given in [2]. As the differences between groups A and B are marginal, we consolidate the results for both groups in the results given below.

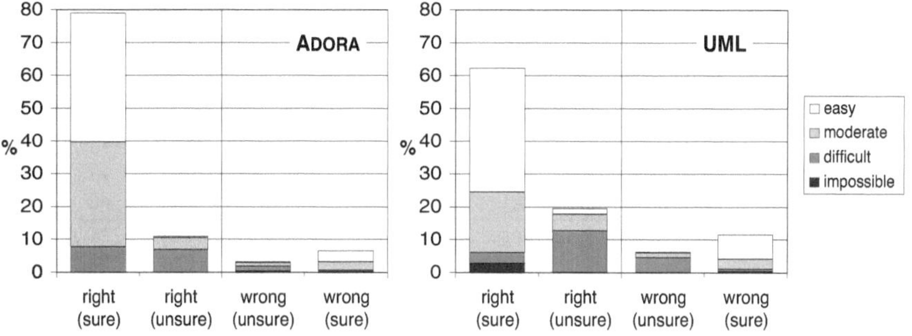

Fig. 9. Comprehensibility of models. Right and wrong answers to the questions in the objective part of the questionnaire for ADORA vs. UML models. The graphics also shows how certain the participants were about their answers and how they rated the difficulty of answering.

Fig. 9 shows the overall results of the first part of the questionnaire. For each model, we had a total of 360 answers (30 questions times 12 participants). For every answer, we determined whether the answer was objectively right or wrong. The answers were further subdivided into those where the answering person was sure about her or his answer and those where she or he was not. The subdivision of the columns indicates how difficult it was to answer the questions in the participants' opinion. (For example, about 79% of the questions about the ADORA model were answered correctly and the participants were sure about their answer. For about half of these answers, the participants judged the answer to be easy to give.)

Despite the fact that the number of participants was fairly small, these results strongly support the comprehensibility hypothesis and also show a clear trend that an

ADORA specification is easier to comprehend than an UML specification. Moreover, we have two important results that are statistically significant at a level of 0.5% [2]:

- The percentage of correctly answered questions is higher for the ADORA model than for the UML model. That means, reading the ADORA model is less error-prone than reading the UML model.
- When answering a question correctly, the readers of the ADORA model are more confident of themselves than the readers of the UML model.

Table 1 summarizes the results of the subjective part of the questionnaire. Again, the results strongly support our hypothesis that users like the fundamental concepts of ADORA and that they prefer them to those of UML.

Table 1. Acceptance of distinct features; ADORA vs. UML

Statement		strongly agree	mostly agree	mostly disagree	strongly disagree
The specification gives the reader a precise idea about the system components and relationships	ADORA	23%	62%	8%	8%
	UML	8%	46%	31%	15%
The structure of the system can be determined easily	ADORA	54%	31%	8%	8%
	UML	8%	38%	23%	31%
The specification is an appropriate basis for design and implementation	ADORA	25%	75%	0%	0%
	UML	0%	50%	33%	17%
Using an integrated model (ADORA) makes sense		42%	25%	33%	0%
Using a set of loosely coupled diagrams (UML) makes sense		8%	17%	67%	8%
Hierarchical decomposition eases description of large systems		15%	69%	15%	0%
ADORA eases focusing on parts without losing context		38%	46%	15%	0%
Decomposition in ADORA eases finding information		46%	38%	15%	0%
Integrating information from different diagrams is easy in UML		15%	15%	46%	23%
Specifying objects with their roles and context is adequate		31%	54%	15%	0%
Describing classes is sufficient		0%	15%	62%	23%

Even if we subtract some potential bias (maybe some of the participating students did not want to hurt us), we can conclude from this experiment that the ADORA language is clearly a step into the right direction.

5 Yet Another Language? ADORA vs. UML

The goal of the ADORA project is not to bless mankind with another fancy modeling language. When UML became a standard, we of course investigated the option of making ADORA a variant of UML. The reason why we didn't is because ADORA and UML differ too much in their basic concepts (Table 2). The most fundamental difference is the concept of an integrated, hierarchically decomposable model in ADORA vs. a flat, mostly non-decomposable collection of models in UML. Using packages for an ADORA-like decomposition fails, because UML packages are mere containers. Using stereotypes for integrating ADORA into UML would be an abuse of this concept, as the resulting language would no longer behave like UML. A real integration of ADORA-concepts into UML would require major changes in the UML metamodel [9].

Table 2. Comparison of basic concepts of ADORA vs. UML

ADORA	UML
Specification is based on a model of abstract objects, types are supplementary	Specification is based on a class model, object models are partial and supplementary
Specifies all aspects in one integrated model; separation of concerns achieved by decomposition and views	Uses different models for each aspect. Separates concerns by having a loosely coupled collection of models
Hierarchical decomposition of objects is the principal means for structuring and comprehending a specification	Class and object models are flat. Only packages can be decomposed hierarchically
Scenarios are tightly integrated into the specification; they can be structured and decomposed systematically	Use cases (=type scenarios) are loosely integrated with class and object models. Structuring capabilities are weak, decomposition is not possible.
Precise rules for consistency between aspect views	Nearly no consistency rules between aspect models
Conceptual visualization eases orientation and navigation in the specification and improves comprehensibility	UML tools provide traditional scrolling and explosive zooming only

6 Conclusions

Summary. We have presented ADORA, an approach to object-oriented modeling that is based on object modeling and hierarchical decomposition, using an integrated model. The ADORA language is intended to be used for requirements specifications and high-level, logical views of software architectures.

Code generation. ADORA is not a visual programming language. Therefore, we have not done any work towards code generation up to now. However, in principle the generation of prototypes from an ADORA model is possible. ADORA has both the structure and the language elements that are required for this task.

State of work. We have finished a first definition of the ADORA language in 1999 [13]. In the meantime we have evolved some language concepts and have conducted an experimental validation. The ADORA tool is still in the proof-of-concept phase. We have a prototype demonstrating that the zooming algorithm, which is the basis of our visualization concept, works.

Future plans. The work on ADORA goes on. In the next years, we will develop a real tool prototype and investigate the use of ADORA for partial and incrementally evolving specifications. Parallel to that, we want to apply ADORA in projects and evolve the language according to the experience gained.

References

1. Berner, S., Joos, S., Glinz, M., Arnold, M. (1998). A Visualization Concept for Hierarchical Object Models. *Proceedings 13th IEEE International Conference on Automated Software Engineering (ASE-98)*. 225-228.
2. Berner, S., Schett, N., Xia, Y., Glinz, M. (1999). *An Experimental Validation of the ADORA Language*. Technical Report 1999.05, University of Zurich.
3. Booch, G. (1994). *Object-Oriented Analysis and Design with Applications*, 2nd ed. Redwood City, Ca.: Benjamin/Cummings.

4. Carroll, J.M. (ed.)(1995). *Scenario-Based Design*. New York: John Wiley & Sons.
5. Coad, P., Yourdon E. (1991). *Object-Oriented Analysis*. Englewood Cliffs, N. J.: Prentice Hall.
6. Firesmith, D., Henderson-Sellers, B. H., Graham, I., Page-Jones, M. (1998). *Open Modeling Language (OML) – Reference Manual*. SIGS reference library series. Cambridge: Cambridge University Press.
7. Glinz, M. (1993). Hierarchische Verhaltensbeschreibung in objektorientierten Systemmodellen – eine Grundlage für modellbasiertes Prototyping. [Hierarchical Description of Behavior in Object-Oriented System Models – A Foundation for Model-Based Prototyping (in German)] In Züllighoven, H. et al. (eds.): *Requirements Engineering '93: Prototyping*. Stuttgart: Teubner. 175-192.
8. Glinz, M. (1995). An Integrated Formal Model of Scenarios Based on Statecharts. In Schäfer, W. and Botella, P. (eds.): *Software Engineering – ESEC'95*. Berlin: Springer. 254-271.
9. Glinz, M. (2000). Problems and Deficiencies of UML as a Requirements Specification Language. *Proceedings of the Tenth International Workshop on Software Specification and Design*. San Diego. 11-22.
10. Harel, D. (1987). Statecharts: A Visual Formalism for Complex Systems. *Sci. Computer Program.* **8** (1987). 231-274.
11. Jacobson, I., Christerson, M., Jonsson, P., Övergaard, G. (1992). *Object-Oriented Software Engineering – A Use Case Driven Approach*. Reading, Mass.: Addison-Wesley.
12. Joos, S., Berner, S., Arnold, M., Glinz, M. (1997). Hierarchische Zerlegung in objektorientierten Spezifikationsmodellen [Hierarchical Decomposition in Object-Oriented Specification Models (in German)]. *Softwaretechnik-Trends*, **17**, 1 (Feb. 1997), 29-37.
13. Joos, S. (1999). *ADORA-L – Eine Modellierungssprache zur Spezifikation von Software-Anforderungen* [ADORA-L – A modeling language for specifying software requirements. (in German)]. PhD Thesis, University of Zurich.
14. Leveson, N.G., Heimdahl, M.P.E., Reese, J.D. (1999). Designing Specification Languages for Process Control Systems: Lessons Learned and Steps to the Future. In Nierstrasz, O. and Lemoine, M. (eds.): *Software Engineering – ESEC/FSE'99*. Berlin: Springer. 127-145.
15. Rumbaugh, J., Blaha, M., Premerlani, W., Eddy, F., Lorensen, W. (1991). *Object-Oriented Modeling and Design*. Englewood Cliffs, N. J.: Prentice Hall.
16. Rumbaugh, J., Jacobson, I., Booch, G. (1999). *The Unified Modeling Language Reference Manual*. Reading, Mass.: Addison-Wesley.
17. Schett, N. (1998). *Konzeption und Realisierung einer Notation zur Formulierung von Integritätsbedingungen für ADORA.Modelle*. [A notation for integrity constraints in ADORA models – Concept and implementation (in German)]. Diploma Thesis, Univ. of Zurich.
18. Selic, B., Gullekson, G., Ward, P. T. (1994). *Real-Time Object-Oriented Modeling*. New York: John Wiley & Sons.
19. Wirfs-Brock, R., Wilkerson, B., Wiener, L. (1993). *Designing Object-Oriented Software*. Englewood Cliffs, N. J.: Prentice Hall.

Techniques for Reactive System Design: The Tools in TRADE

Roel J. Wieringa and David N. Jansen*

Department of Computer Science, University of Twente, P.O. Box 217, 7500 AE, The Netherlands
{roelw, dnjansen}@cs.utwente.nl

Abstract. Reactive systems are systems whose purpose is to maintain a certain desirable state of affairs in their environment, and include information systems, groupware, workflow systems, and control software. The current generation of information system design methods cannot cope with the high demands that originate from mission-critical application, geographic distribution, and a mix of data-intensive, behavior-intensive and communication-intensive properties of many modern reactive systems. We define an approach to designing reactive software systems that deals with these dimensions by incorporating elements from various information system and software design techniques and extending this with formal specification techniques, in particular with model checking. We illustrate our approach with a smart card application and show how informal techniques can be combined with model checking.

1 Introduction

The past ten years have shown an explosion of different types of information technology in which the classical distinction between information systems and control software has disappeared. In addition to the data-intensive applications like order administrations, and control-intensive applications like production control, there is now widespread use of email, office agendas, shared workspaces, workflow management, enterprise resource planning (ERP), electronic data interchange (EDI), and internet-based ecommerce applications. In these applications, we see a varying mix of complexity along the three major dimensions of functional software properties: data, behavior, and communication. (That these are the three important dimensions of functional software properties is argued elsewhere [13].)

A fourth dimension has emerged as important as well: geographical distribution. For example, classical information systems may be distributed over several sites, and they may be connected to classical production control systems in a complex network of applications that may even cross organizational boundaries.

A fifth dimension relevant to software is the degree to which it is mission critical. This is not a property of the software as such but of the way it is used by an organization. We now have many companies large parts of which basically *are*

* Partially supported by NWO-SION, project number 612-323-419.

K.R. Dittrich, A. Geppert, M.C. Norrie (Eds.): CAiSE 2001, LNCS 2068, pp. 93–107, 2001.
© Springer-Verlag Berlin Heidelberg 2001

software systems, operated by a few employees. This has long been the case in the finance business but it is now also the case for application service providers, supply chain integration, and business-to-consumer electronic commerce. When these applications fail, their businesses lose money by the minute. As a consequence, there must be ample attention to reliability, security, safety and other mission-critical attributes. We claim that formal techniques have a role to play in this area and later in this paper we argue how this can be done.

The current generation of functional and object-oriented methods do not suffice to deal with all of these dimensions. Structured techniques [11,15] tend to spaghetti-like data flow diagrams with a sloppy semantics, and object-oriented techniques have evolved into a Unified Modeling Language (UML) whose complexity is not motivated by the complexity of the systems to be designed, but by the number of stakeholders involved in defining the notation [12,10]. In addition, the complexity of the UML, as well as the complexity of the process in which the UML is defined, leads to a continuous stream of revisions and an incomplete and ambiguous semantics.

In this paper we propose a simple approach that picks the elements from structured and object-oriented approaches that have turned out in practice to be very useful, and extends this with a formal specification approach to deal the increasing need for reliability and precision. We claim that the resulting approach, called TRADE (Techniques for Requirements and Architecture DEsign) is useful bag of tools to use when designing complex information systems.

The unifying view that we present starts from the concept of a reactive system, introduced by Harel and Pnueli [4] 15 years ago. This is explained in section 2. Section 3 defines our mix of structured and object-oriented techniques and discusses how formal techniques can be combined with informal techniques. Appendix 4 illustrates out claim by an example specification of a smart card application, and of two properties that we model checked in our specification. Details about the techniques and guidelines are given elsewhere [3,6,14].

2 Reactive Systems

A reactive system is a system that, when switched on, is able to respond to stimuli as and when they occur, in a way that enforces or enables desired behavior in the environment [4,7]. Stimuli may arrive at arbitrary times and the response must be produced soon enough for the desirable effect to take place. Somewhere in the environment, *events* occur, that cause *stimuli* at the system interface. The system *responds* based on an internal model that it maintains of its environment, and the response leads to, or enables, a desired *action* in the environment. Examples of reactive systems are information systems, workflow management systems, email systems, systems for video conferencing, shared office agenda systems, chat boxes, group decision support systems, process control software, embedded software and game software.

Reactive systems are to be distinguished from *transformational systems*, which are systems that, when switched on, accept input, perform a terminating computation, produce output, and then turn themselves off. Examples of transformational systems are compilers, assemblers, expert systems, numerical routines, statistical procedures, etc. A transformational system can be viewed as computing single, isolated stimulus-response pairs. Transformational systems have no relevant internal state that survives a single stimulus-response pair. To specify a transformational system, you must specify a terminating algorithm, whereas to specify a reactive system, you must specify stimulus-response behavior. Both kinds of system must be switched on before used, but when a reactive system switches itself off this is because something wrong has happened. For a transformational system the opposite is true: when it does *not* switch itself off, something is wrong.

3 Requirements Specification for Reactive Systems

3.1 Functions

Reactive systems exist to provide services to their environment. They provide these services by responding to stimuli. We define a *function* of a system as any service delivered by the system to its environment. A function delivers value to its environment. It is the ability to deliver value to its environment that motivates someone to pay money for the system.

To specify a function, we must at least specify what value is delivered by the function and when it is delivered, i.e. which event triggers it. Figure 6 gives an example. In our view it is not a good idea to specify detailed scenarios for functions. These scenarios obscure the view of what the function is for (which value is delivered). They also mix requirements (what do we need the system for) with architecture (what are the high-level components that will deliver these services). As we illustrate in the appendix, a precise architecture description of the system will include a detailed specification of the behavior of the system.

3.2 The Environment

The interactions of a software system always consist of exchanges of symbol occurrences (e.g. data items, event occurrences) with the environment. A truly implementation-independent description of the data manipulated by the system restricts itself to the symbol occurrences that cross the external interface shown in the context diagram. These symbol occurrences have a *meaning,* which must be understood by the designers and users in order to understand the behavior of the system. The subject of these exchanges is called the *subject domain* of the interactions. For example, the subject domain of our ticket selling example system in the appendix consists of tickets, routes, etc. These are the entities about which the system communicates with its environment. The subject domain is often modeled by an entity model, such as illustrated in Fig. 9.

In addition to the subject domain, the environment of the system consists of a *connection domain*, which consists of entities that can observe the subject

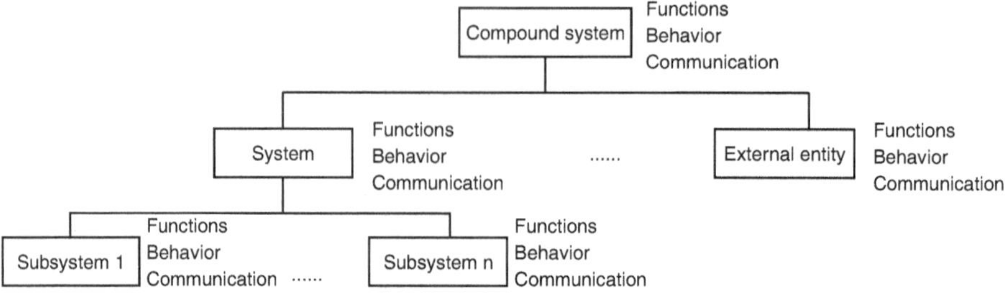

Fig. 1. Aggregation hierarchy and system aspects.

domain and provide stimuli to the system, and entities that can act on the subject domain based on the responses of the system. In control software, these entities are called sensors and actuators, but people can play these roles too.

A third element of the environment is the *implementation platform,* which is the collection of programmable entities that will contain the software. When software engineers talk about the software environment, they often mean the implementation platform. When information engineers talk about the environment, they usually mean the subject and connection domains.

3.3 Requirements and Architecture

We define a *requirement* as a desired property of the system and an *architecture* as the way components are put together in order to realize these desired properties. The architecture dimension introduces an aggregation hierarchy, in which components at a lower level jointly realize properties of a component at a higher level.

In our view, requirements are not restricted to external properties. A requirement of a system is just any desired property of the system, be it a desired function or a desire that the software be executable on a certain implementation platform. An important kind of requirement is of course the *functional* requirement, which is basically a description of the desired system functions. In addition to the value delivered to the environment, there are at least two important aspects of software system functionality that usually have to be specified, namely *behavior,* which is the ordering of stimulus-response pairs in time, and *communication,* which is the ordering of stimulus-response pairs in "space", i.e. the communication links that connect the stimuli and responses with events and actions in the subject domain. These aspects are repeated at every aggregation level. So for each part of the system, we can ask what its desired functions, behavior and communication properties are, all the way down to individual software objects (Fig. 1).

A software system has many other properties, including safety, security and reliability properties etc. For some of these properties, formal techniques are needed to show their presence or absence. We give an example in the appendix.

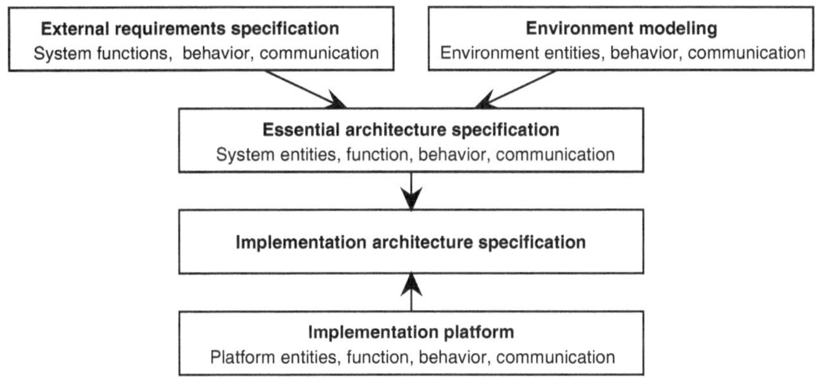

Fig. 2. The TRADE approach.

3.4 Design Approach

In this paper, by *design* we mean an activity in which the desired properties of a product are decided. Design is making decisions, and it results in a specification that documents these decisions. This use of the term agrees with that in other branches of manufacturing, where design is followed by planning the production process, manufacturing the components, assembling the components and distribution and sales.

This means that specifying requirements in our terminology is a design activity! Requirements specification is solution specification; the problem to be solved exists in the environment of the system that should solve it (it would be embarrassing if it would reside *in* the system) [5]. Figure 2 summarizes the TRADE approach. The *environment model* is an outcome of problem analysis. It represents the environment as consisting of entities, their behavior and their communication. We assume the problem to be solved has been clearly stated and analyzed. The *external requirements* are desired external system properties and include desired system functions, behavior and communication with external entities. The *essential architecture* of a software system is the architecture it would have if we had ideal implementation technology. It is motivated by the structure of the external environment and the system requirements and not by any implementation considerations [8]. Other terms that have been used for this are *logical* and *conceptual* architecture. The *implementation platform* is the collection of programmable entities that will contain the software. These too have functions, behavior and communication. The *implementation architecture* is the mapping of the essential architecture to the programmable entities in the available implementation platform.

3.5 Design Techniques

There are very few techniques needed to specify the aspects listed in Fig. 1. The second column of table 3 lists a few simple techniques that are sufficient.

	Simple	Complex
Decomposition	ERD	SSD
Functions	Mission statement, refined to function descriptions	
Behavior	Event list, possibly including state transition diagram	Statechart
Communication	Communication diagram	

Fig. 3. The techniques in TRADE.

In the following explanation, the term "entity" is used to indicate whatever a description technique is applied to: the entities in the environment, the system itself, or parts of its architecture or of its implementation platform.

– An *entity-relationship diagram* (ERD) represents entity types and their cardinality properties. We restrict the meaning of ERDs to classification and identification (counting) properties of entities. ERDs can be used to represent the decomposition of the environment or of an architecture into entities. It is often used to represent a decomposition of the subject domain into entities. A complex extension of ERDs is the UML *static structure diagram* (SSD), which allows the declaration of services (interfaces, operations, signal receptions etc.) offered by entities, which are now called "objects" in object-oriented methods. Usually there is too much interface detail in a system to be all represented in diagram form.
– *Mission statement* and *function descriptions* describe the things that an entity will do for its environment in natural language. They should emphasize the value delivered for the environment. They can be used to specify the mission and functions of the system and of the entities in the system's architecture.
– An *event list* of an entity is a list of all events that the entity should respond to, and the desired response of the entity. It can be a list of informal natural language descriptions, but this can be refined into state transition tables or diagrams. Statecharts are complex state transition diagrams that can represent information in an event list in diagram form.
– A *communication diagram* consists of boxes and arrows that represent entities and their communication channels. The boxes may represent individual entities or entity types. In the second case, the arrows represent communication channels between instances of the types. A communication diagram may be used to represent communication in the environment, between the system and its environment, between entities in the system architecture, and between entities in the implementation platform.

The appendix contains examples of the use of these techniques, discusses their meaning informally and indicates their use.

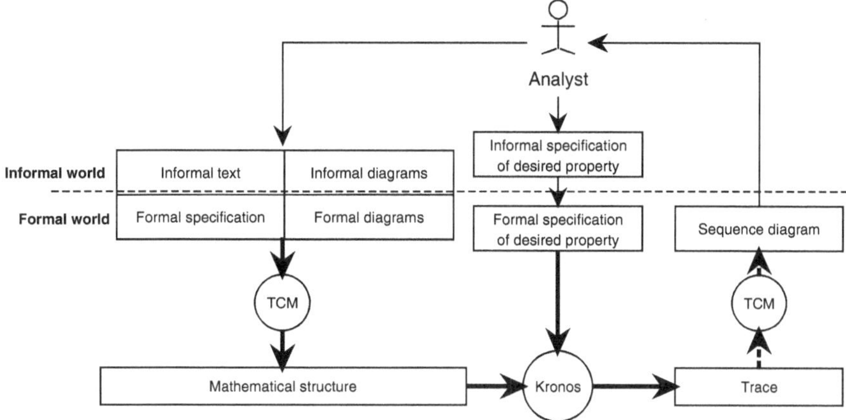

Fig. 4. Combining informal techniques with model checking.

3.6 Formal Techniques

In our view the relationship between the formal and informal parts of a specification (Fig. 4) is that the formal part rephrases fragments of the informal part. The formal part can consist of text or diagrams with a formal semantics. To illustrate the viability of this approach, we have defined a formal execution semantics for object-oriented statecharts that is suitable to represent essential-level behavior [3]. Barring unrestricted object creation, the semantics of a collection of object-oriented statecharts is a finite-state labelled transition system (LTS), which is the mathematical structure in Fig. 4. We have also defined an extension of computation tree logic with actions and real time (ATCTL) to be used as a property language for reactive systems, and defined a translation of ATCTL into the property language of the Kronos model checker [1]. We implemented these using the diagram editing tool TCM [2] as a front-end to Kronos. Space restrictions prevent us from giving more details about this.

The fat arrows in Fig. 4 represent manipulations done by machine. The solid fat arrows have been implemented in TCM, and we used this on our example specification in the appendix to check some desirable properties. The analyst does not have to know or understand the translations behind the fat arrows. The combination of TCM and Kronos thus helps the analyst to understand the design in an early stage without implementing it and without having to learn a complex formal language. It also helps making the informal parts of the model precise.

4 Summary and Further Work

In this paper we have only discussed techniques and showed their place in an approach to reactive system specification and design. We have not discussed precise or formal semantics, or guidelines for using these techniques. A compendium of these guidelines has been prepared elsewhere [14]. Current work includes implementing the dashed fat arrows in Fig. 4 and applying the resulting tool to a number of case studies.

Acknowledgement. The paper benefited from many discussions with and detailed comments of Michael Jackson, and from comments by Jaap-Henk Hoepman on security mechanisms.

References

1. M. Bozga, C. Daws, O. Maler, A. Olivero, S. Tripakis, and S. Yovine. Kronos: a model-checking tool for real-time systems. In A.J. Hu and M.J. Vardi, editors, *Computer aided verification (CAV)*, pages 546–550. Springer, 1998. LNCS 1427.
2. F. Dehne and R.J. Wieringa. Toolkit for Conceptual Modeling (TCM): User's Guide. Technical Report IR-401, Faculty of Mathematics and Computer Science, *Vrije Universiteit*, De Boelelaan 1081a, 1081 HV Amsterdam, 1996. http://www.cs.utwente.nl/~tcm.
3. H. Eshuis and R. Wieringa. Requirements level semantics for UML statecharts. In S.F. Smith and C.L. Talcott, editors, *Formal Methods for Open Object-Based Distributed Systems (FMOODS) IV*, pages 121–140, 2000.
4. D. Harel and A. Pnueli. On the development of reactive systems. In K. Apt, editor, *Logics and Models of Concurrent Systems*, pages 477–498. Springer, 1985. NATO ASI Series.
5. M.A. Jackson. *Problem Frames: Analysing and Structuring Software Development Problems.* Addison-Wesley, 2000.
6. D.N. Jansen and R.J. Wieringa. Extending CTL with actions and real time. In *International Conference on Temporal Logic 2000*, pages 105–114, October 4–7 2000.
7. Z. Manna and A. Pnueli. *The Temporal Logic of Reactive and Concurrent System Specification.* Springer, 1992.
8. S. M. McMenamin and J. F. Palmer. *Essential Systems Analysis.* Yourdon Press/Prentice Hall, 1984.
9. W. Reif and K. Stenzel. Formal methods for the secure application of java smartcards. Technical report, University of Ulm, December 1999. Presentation slides. http://www.informatik.uni-augsburg.de/swt/fmg/projects/javacard_presentation.ps.gz.
10. J. Rumbaugh, I. Jacobson, and G. Booch. *The Unified Modeling Language Reference Manual.* Addison-Wesley, 1999.
11. K. Shumate and M. Keller. *Software Specification and Design: A Disciplined Approach for Real-Time Systems.* Wiley, 1992.
12. UML Revision Task Force. *OMG UML Specification.* Object Management Group, march 1999. http://uml.shl.com.
13. R.J. Wieringa. A survey of structured and object-oriented software specification methods and techniques. *ACM Computing Surveys*, 30(4):459–527, December 1998.
14. R.J. Wieringa. *Design Methods for Reactive Systems: Yourdon, Statemate and the UML.* Morgan Kaufmann, To be published.
15. E. Yourdon. *Modern Structured Analysis.* Prentice-Hall, 1989.

- **Name of the system:** Electronic Ticket System (ETS).
- **Purpose of the system:** Provide capability to buy and use tickets for a railway company using a Personal Digital Assistant and a smart card.
- **Responsibilities of the system:**
  - Sell a ticket
  - Show a ticket on request
  - Stamp a ticket for use
  - Refund a ticket
- **Exclusions:**
  - The system will not provide travel planning facilities.
  - Only tickets for making a trip by one person will be considered.

Fig. 5. Mission of the ETS.

The Electronic Ticket System (ETS)

The Electronic Ticket System (ETS) is a software system by which travelers can buy a railway ticket with a smart card when put in their Personal Digital Assistant (PDA). The ticket is a digital entity than can be created on the smart card itself. Payment is done through a wireless connection with the computer system of the railway company, that itself is connected to the computer systems of a clearing house [9].

External Requirements: Functions and Other Properties

The *mission statement* of a system states the purpose of the system and should be written down for all reactive systems. It lists its major responsibilities and, possibly, things that the system will not do. Responsibilities are things the system should do, exclusions are things the system will not do. There are infinitely many things the system will not do, but writing down a few of these is an important tool in expectation management. Figure 5 shows the mission statement of the ETS.

Functions can be presented in the form of an indented list called a *function refinement tree*. This is useful for all reactive systems, and can be used to bound the discussion about desired functionality with the stakeholders. The current system is so simple that all its functions have already been listed in the mission statement, and a separate function refinement tree is not needed. Each function should be specified from the standpoint of the system, emphasizing the value delivered to the environment. See Fig. 6 for an example.

One of the required ETS properties is that it should offer the functionality described above. Other required properties include the ones listed in Fig. 7. We discuss the formal specification and model checking of these properties at the end of the appendix.

- **Name:** Sell a ticket.
- **Triggering event:** Traveler requests to buy a ticket.
- **Delivered value:** To sell a railway ticket to a traveler at any time and place chosen by the traveler.

Fig. 6. Description of an ETS function.

P1 A traveler cannot get a ticket without paying for it.
P2 A traveler who has paid for a ticket gets it.
P3 A refunded ticket cannot be used any more.
P4 A fully used ticket cannot be refunded.
P5 It is not possible to use a ticket twice.

Fig. 7. Other required ETS properties.

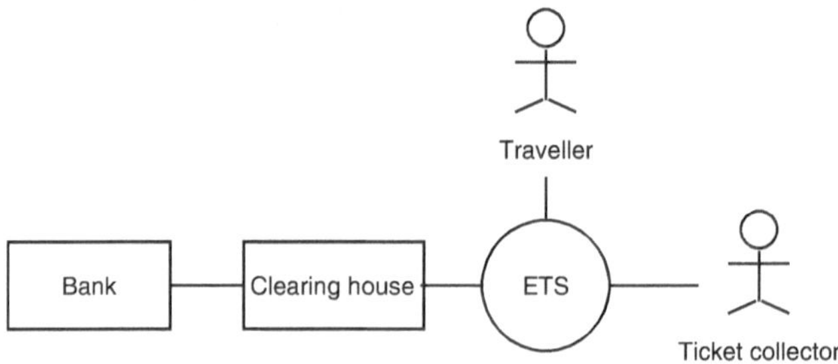

Fig. 8. Communication context of ETS.

Environment: Entities and Communication

Desired properties, including system functions, are provided by interacting with the environment in stimulus-response behavior. Figure 8 shows the communication architecture of the environment, including the external interfaces of the system. The diagram abstracts from the internal distribution of the software system over physical entities of the implementation platform, and from the way communication channels are realized. That distribution will be part of the architecture specification shown later. The context diagram views the system as a black box offering certain functionality and shows which communication channels with the environment exist. It is always useful to draw a communication diagram of a reactive system. It separates the part of the world to be designed (the system) from the part of the world that is given (the environment).

Figure 9 shows an entity model of the subject domain of the ETS. It should be supplemented by a number of constraints, such as that all segments of a route

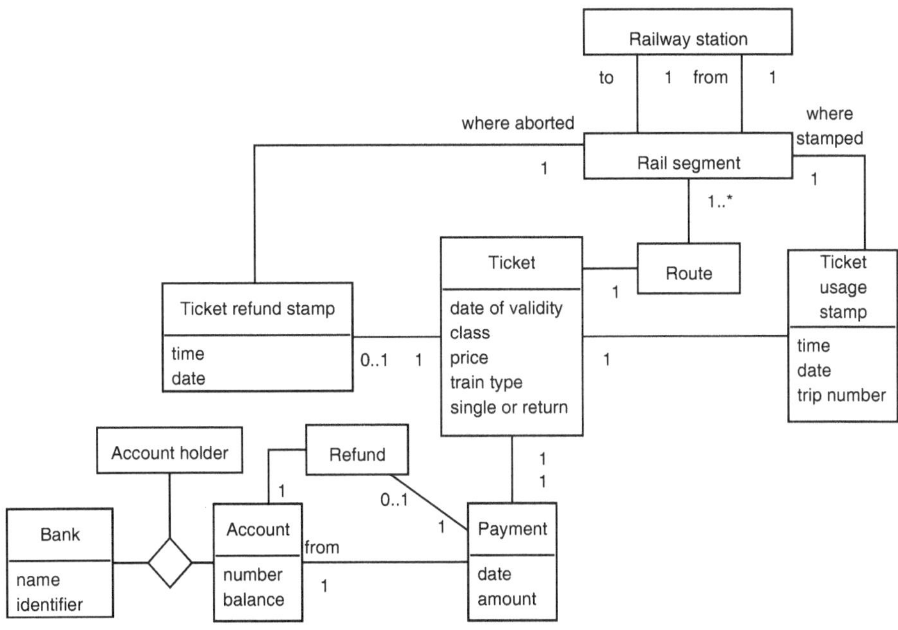

Fig. 9. Subject domain of the ETS.

should be connected. Note that the ERD is a model of the environment, not of the data stored by the system. Note also that the traveler may be different from the account holder.

Whenever there is potential misunderstanding about the meaning of the data that crosses the external system interface, as in data-intensive reactive systems, it is useful to make a subject domain ERD that represents the types of entities to which the external interactions of the system refer.

External Requirements: System Event List

Each function is triggered by an external event, a condition change, or a temporal event. Each function can be refined into a set of possible transactions, where a transaction is an atomic interaction of the system. Each transaction in turn can be represented by a tuple (event, stimulus, current state, next state, response, action), which tells us which stimulus triggers the transaction, which event is supposed to have caused this stimulus, what the response should be, given the current state of the system, and what action is assumed to be caused by the response. The current and next state in the tuple are really states of the dialog between the system and its environment. To perform its function, the system must maintain a representation of this state. In general, there may be a many-many relationship between transactions and functions: One transaction may occur in several functions and one function may contain several transactions.

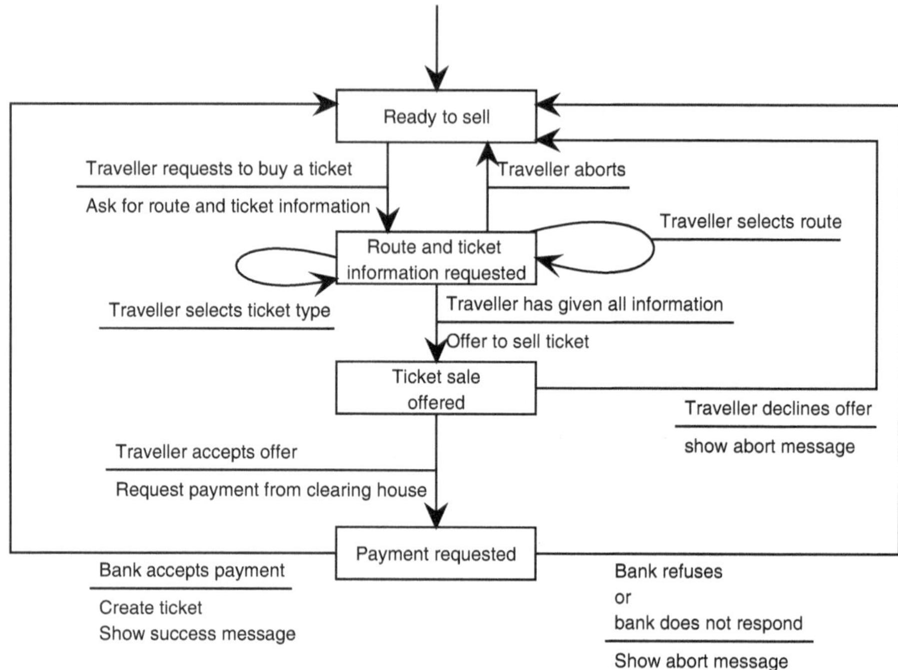

Fig. 10. Selling dialog.

Often, we can represent parts of the system event list by a state transition diagram, e.g. a statechart or its simpler ancestors Mealy and Moore diagrams. Figure 10 gives the event list for the selling function in the form of a Mealy diagram. Rectangles represent states, arrows state transitions, and arrow labels list events above the line and actions below the line.

An event list is always useful to make, but different reactive systems require different levels of detail. Some information systems have merely two types of external events to respond to, namely `query` and `update`, but in many other cases there are also state change events and temporal events to respond to. Refining the system function descriptions into an event list uncovers these desired responses.

Architectures and Implementation Platform

Essential Architecture. The essential, distribution-independent architecture of the system is shown in Fig. 11. It uses a hybrid notation in which parallel bars represent data stores and rectangles represent stateful objects.

- The selling dialog (Fig. 10) has been allocated to a single object class, each instance of which can execute this dialog.
- The other functions are simple transformations that produce output or perform updates on request.

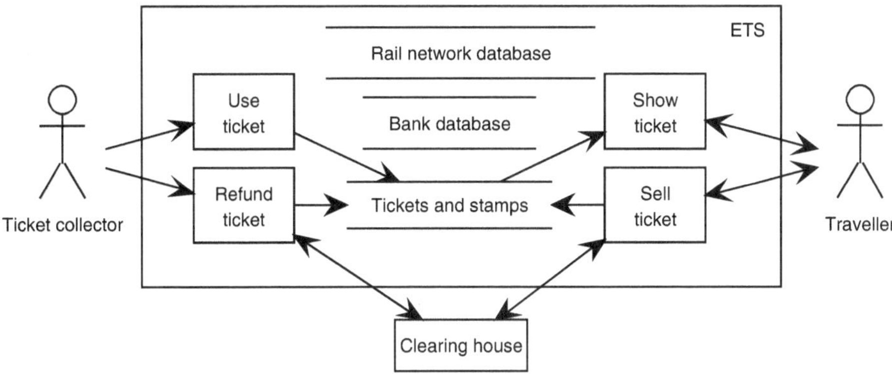

Fig. 11. Essential communication architecture of the system in a hybrid notation

– Data about the subject domain has been partitioned into three data stores.

Figure 12 represents the structure of the data stores. Now we reuse the subject domain ERD to represent the structure of the data and allocate the data to stores. Note that the system does not contain data about all subject domain relationships.

A communication diagram such as 11 is very useful in reactive system design to show simple communication architectures. They can be used to show the high-level communication architecture of a system and simple lower-level architectures, as in the ETS. However, for most systems, at lower aggregation levels they quickly become too complicated to be useful.

Checking Desired Properties. To check whether the essential architecture satisfies all desired properties, we created our model with the TCM editor and generated Kronos input for it. We also formalize the desirable properties in ATCTL, and checked whether they hold in our essential architecture, using Kronos. As an illustration, consider these two properties:

P1 You cannot get a valid ticket without paying. $\neg\exists(\top\ _{\mathbf{-bank\_pay\_yes}}\mathcal{U}\mathsf{valid})$.
P3 A refunded ticket will never become valid again. $\mathsf{refunded} \rightarrow \neg\exists\Diamond\mathsf{valid}$

P1 was shown to be true using Kronos. The negation of P3 was checked with Kronos and proved to be unreachable. Because the negation is unreachable, the property holds.

The meaning of these proofs is that the system, if implemented this way, will be secure. It does not mean that the environment will be secure. For example, it does not mean that an insecure connection to the railway company or the clearing house is impossible. And with such an insecure connection, third parties could masquerade as bank, and cause the railway company to give a ticket to the passenger without valid payment. If that should be excluded, then the system boundary should be extended to include the connections and we should formulate properties of these connections.

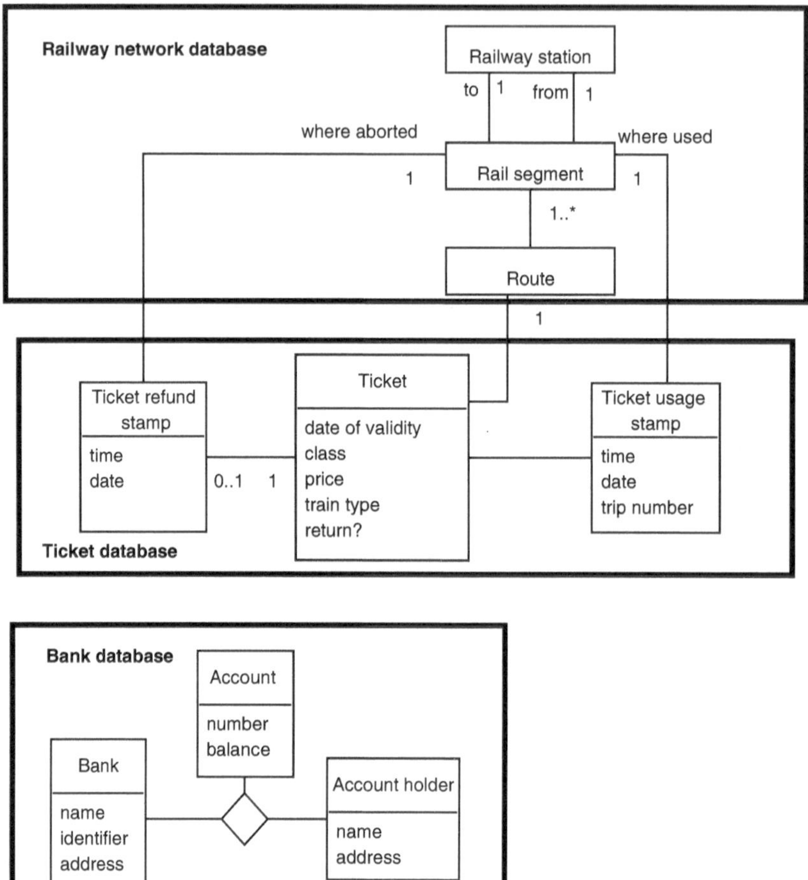

Fig. 12. Data about the subject domain.

Implementation Architecture. The physical distribution architecture of the system is represented by the UML deployment diagram of Fig. 13. This is the implementation platform given to the designer.

Figure 14 shows the allocation of the essential architecture elements of Fig. 11 to the nodes in the deployment network. The following design decisions have been made:

- Functionality is allocated to nodes where it is needed.
- Duplication of data stores is added to avoid frequent communications. The price to pay is that duplicate data stores may be mutually inconsistent. Because the rail network is not likely to change very frequently, and changes are planned far in advance, there is a small risk that this will ever happen.
- Communication interfaces are added.

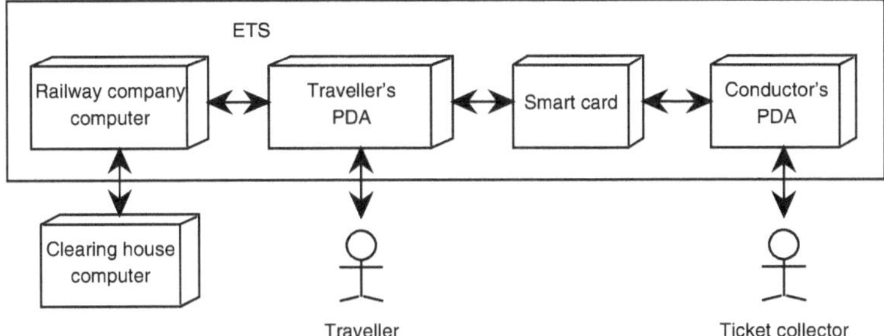

Fig. 13. Physical implementation platform and context.

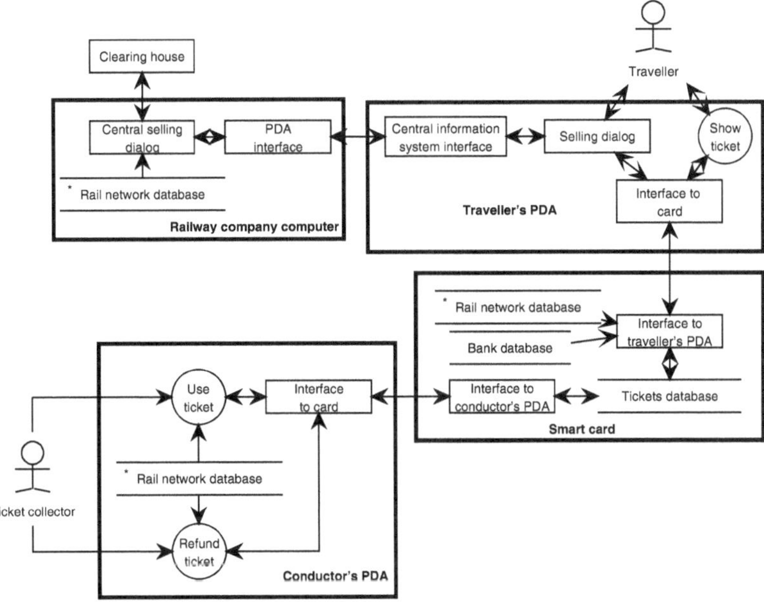

Fig. 14. Allocation of components of the essential architecture to the physical deployment network.

The next stage in creating a secure design would be to check that the implementation is correct with respect to the higher-level design. The implementation is represented by Fig. 14, the data model of Fig. 12 and the textual specifications of all elements of these diagrams (these are not given here). The specification is given by the high-level diagram of Fig. 11 and the textual specifications of the data stores, transformations and object classes contained in it. Both models have a formal semantics in terms of labeled transition systems, which makes classical bisimulation equivalence checking techniques applicable, at least in principle. Working this out for practical examples is subject of future research.

A Requirements-Driven Development Methodology*

Jaelson Castro[1], Manuel Kolp[2],and John Mylopoulos[2]

[1] Centro de Informática, Universidade Federal de Pernambuco, Av. Prof. Luiz Freire S/N, Recife PE, Brazil 50732-970**
jbc@cin.ufpe.br

[2] Department of Computer Science, University of Toronto, 10 King's College Road, Toronto M5S3G4, Canada
{mkolp,jm}@cs.toronto.edu

Abstract. Information systems of the future will have to better match their operational organizational environment. Unfortunately, development methodologies have traditionally been inspired by programming concepts, not organizational ones, leading to a semantic gap between the system and its environment. To reduce as much as possible this gap, this paper proposes a development methodology named *Tropos* which is founded on concepts used to model early requirements. Our proposal adopts the *i** organizational modeling framework [18], which offers the notions of *actor*, *goal* and (actor) *dependency*, and uses these as a foundation to model early and late requirements, architectural and detailed design. The paper outlines *Tropos* phases through an e-business example. The methodology seems to complement well proposals for agent-oriented programming platforms.

1 Introduction

Information systems have traditionally suffered from an impedance mismatch. Their operational environment is understood in terms of actors, responsibilities, objectives, tasks and resources, while the information system itself is conceived as a collection of (software) modules, entities (e.g., objects, agents), data structures and interfaces. This mismatch is one of the main factors for the poor quality of information systems, also the frequent failure of system development projects.

One cause of this mismatch is that development methodologies have traditionally been inspired and driven by the programming paradigm of the day. This means that the concepts, methods and tools used during all phases of development were based on those offered by the pre-eminent programming paradigm. So, during the era of structured programming, structured analysis and design techniques were proposed [9,17], while object-oriented programming has given rise more recently to object-

* The Tropos project has been partially funded by the Natural Sciences and Engineering Research Council (NSERC) of Canada, and Communications and Information Technology Ontario (CITO), a centre of excellence, funded by the province of Ontario.

** This work was carried out during a visit to the Department of Computer Science, University of Toronto. Partially supported by the CNPq – Brazil under grant 203262/86-7.

K.R. Dittrich, A. Geppert, M.C. Norrie (Eds.): CaiSE 2001, LNCS 2068, pp. 108–123, 2001
© Springer-Verlag Berlin Heidelberg 2001

oriented design and analysis [1,15]. For structured development techniques this meant that throughout software development, the developer can conceptualize the system in terms of functions and processes, inputs and outputs. For object-oriented development, on the other hand, the developer thinks throughout in terms of objects, classes, methods, inheritance and the like.

Using the same concepts to align requirements analysis with system design and implementation makes perfect sense. For one thing, such an alignment reduces impedance mismatches between different development phases. Moreover, such an alignment can lead to coherent toolsets and techniques for developing system (and it has!) as well, it can streamline the development process itself.

But, why base such an alignment on implementation concepts? Requirements analysis is arguably the most important stage of information system development. This is the phase where technical considerations have to be balanced against social and organizational ones and where the operational environment of the system is modeled. Not surprisingly, this is also the phase where the most and costliest errors are introduced to a system. Even if (or rather, when) the importance of design and implementation phases wanes sometime in the future, requirements analysis will remain a critical phase for the development of any information system, answering the most fundamental of all design questions: "what is the system intended for?"

Information systems of the future like ERP, Knowledge Management or e-business systems should be designed to match their operational environment. For instance, ERP systems have to implement a process view of the enterprise to meet business goals, tightly integrating all functions from the operational environment. To reduce as much as possible this impedance mismatch between the system and its environment, we outline in this paper a development framework, named *Tropos*, which is requirements-driven in the sense that it is based on concepts used during early requirements analysis. To this end, we adopt the concepts offered by *i** [18], a modeling framework offering concepts such as *actor* (actors can be *agents*, *positions* or *roles*), as well as social dependencies among actors, including *goal*, *softgoal*, *task* and *resource* dependencies. These concepts are used for an e-commerce example[1] to model not just early requirements, but also late requirements, as well as architectural and detailed design. The proposed methodology spans four phases:

- Early requirements, concerned with the understanding of a problem by studying an organizational setting; the output of this phase is an organizational model which includes relevant actors, their respective goals and their inter-dependencies.
- Late requirements, where the system-to-be is described within its operational environment, along with relevant functions and qualities.
- Architectural design, where the system's global architecture is defined in terms of subsystems, interconnected through data, control and other dependencies.
- Detailed design, where behaviour of each architectural component is defined in further detail.

The proposed methodology includes techniques for generating an implementation from a *Tropos* detailed design. Using an agent-oriented programming platform for the implementation seems natural, given that the detailed design is defined in terms of

[1] Although, we could have included a simpler (toy) example, we decided to present a realistic e-commerce system development exercise of moderate complexity [6].

(system) actors, goals and inter-dependencies among them. For this paper, we have adopted JACK as programming platform to study the generation of an implementation from a detailed design. JACK is a commercial product based on the BDI (Beliefs-Desires-Intentions) agent architecture. Early previews of the *Tropos* methodology appeared in [2, 13].

Section 2 of the paper describes a case study for a B2C (business to consumer) e-commerce application. Section 3 introduces the primitive concepts offered by *i** and illustrates their use with an example. Sections 4, 5, and 6 illustrate how the technique works for late requirements, architectural design and detailed design respectively. Section 7 sketches the implementation of the case study using the JACK agent development environment. Finally, Section 8 summarizes the contributions of the paper, and relates it to the literature.

2 A Case Study

Media Shop is a store selling and shipping different kinds of media items such as books, newspapers, magazines, audio CDs, videotapes, and the like. *Media Shop* customers (on-site or remote) can use a periodically updated catalogue describing available media items to specify their order. *Media Shop* is supplied with the latest releases and in-catalogue items by *Media Supplier*. To increase market share, *Media Shop* has decided to open up a B2C retail sales front on the internet. With the new setup, a customer can order *Media Shop* items in person, by phone, or through the internet. The system has been named *Medi@* and is available on the world-wide-web using communication facilities provided by *Telecom Cpy*. It also uses financial services supplied by *Bank Cpy*, which specializes on on-line transactions.

The basic objective for the new system is to allow an on-line customer to examine the items in the *Medi@* internet catalogue, and place orders.

There are no registration restrictions, or identification procedures for *Medi@* users. Potential customers can search the on-line store by either browsing the catalogue or querying the item database. The catalogue groups media items of the same type into (sub)hierarchies and genres (e.g., audio CDs are classified into pop, rock, jazz, opera, world, classical music, soundtrack, ...) so that customers can browse only (sub)categories of interest.

An on-line search engine allows customers with particular items in mind to search title, author/artist and description fields through keywords or full-text search. If the item is not available in the catalogue, the customer has the option of asking *Media Shop* to order it, provided the customer has editor/publisher references (e.g., ISBN, ISSN), and identifies herself (in terms of name and credit card number).

3 Early Requirements with *i**

During early requirements analysis, the requirements engineer captures and analyzes the intentions of stakeholders. These are modeled as goals which, through some form of a goal-oriented analysis, eventually lead to the functional and non-functional requirements of the system-to-be [7]. In *i** (which stands for "distributed intentionality''), early requirements are assumed to involve social actors who depend on each other for goals to be achieved, tasks to be performed, and resources to be furnished. The *i** framework includes the *strategic dependency model* for describing the network of relationships among actors, as well as the *strategic rationale model* for describing and supporting the reasoning that each actor goes through concerning its relationships with other actors. These models have been formalized using intentional concepts from AI, such as goal, belief, ability, and commitment (e.g., [5]). The framework has been presented in detail in [18] and has been related to different application areas, including requirements engineering, software processes and business process reengineering.

A strategic dependency model is a graph, where each node represents an *actor*, and each link between two actors indicates that one actor depends on another for something in order that the former may attain some goal. We call the depending actor the *depender* and the actor who is depended upon the *dependee*. The object around which the dependency centers is called the *dependum*. Figure 1 shows the beginning of an *i** model.

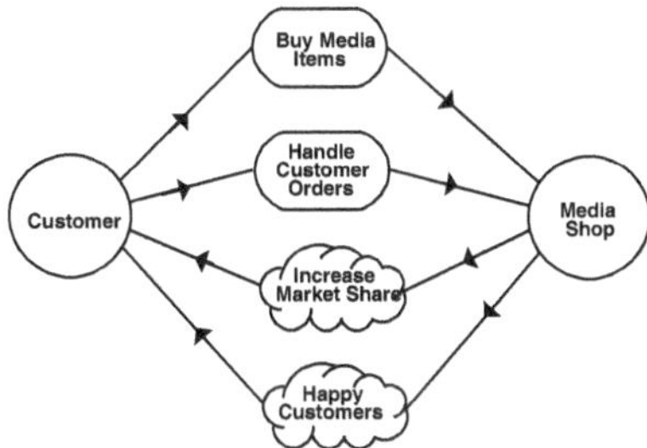

Fig. 1. "*Customers* want to buy media items, while the *Media Shop* wants to increase market share, handle orders and keep customers happy"

The two main stakeholders for a B2C application are the consumer and the business actors named respectively in our case *Customer* and *Media Shop*. The customer has one relevant goal *Buy Media Items* (represented as an oval-shaped icon), while the media store has goals *Handle Customer Orders, Happy Customers,* and *Increase Market Share*. Since the last two goals are not well-defined, they are represented as softgoals (shown as cloudy shapes).

Once the relevant stakeholders and their goals have been identified, a strategic rationale model determines through a means-ends analysis how these goals (including softgoals) can actually be fulfilled through the contributions of other actors. A strategic rationale model is a graph with four types of nodes – *goal*, *task*, *resource*, and *softgoal* – and two types of links – means-ends links and process decomposition links. A strategic rationale graph captures the relationship between the goals of each actor and the dependencies through which the actor expects these dependencies to be fulfilled.

Figure 2 focuses on one of the (soft)goal identified for *Media Shop,* namely *Increase Market Share*. The analysis postulates a task *Run Shop* (represented in terms of a hexagonal icon) through which *Increase Market Share* can be fulfilled. Tasks are partially ordered sequences of steps intended to accomplish some (soft)goal. Tasks can be decomposed into goals and/or subtasks, whose collective fulfillment completes the task. In the figure, *Run Shop* is decomposed into goals *Handle Billing* and *Handle Customer Orders*, tasks *Manage Staff* and *Manage Inventory,* and softgoal *Improve Service* which together accomplish the top-level task. Sub-goals and subtasks can be specified more precisely through refinement. For instance, the goal *Handle Customer Orders* is fulfilled either through tasks *OrderByPhone, OrderInPerson* or *OrderByInternet* while the task *Manage Inventory* would be collectively accomplished by tasks *Sell Stock* and *Enhance Catalogue*.

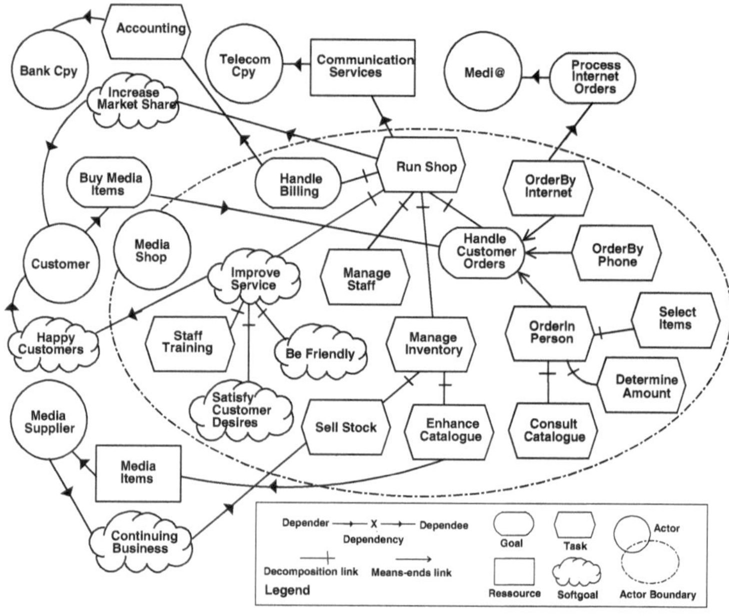

Fig. 2. Means-ends analysis for the softgoal *Increase Market Share*

4 Late Requirements Analysis

Late requirements analysis results in a requirements specification which describes all functional and non-functional requirements for the system-to-be. In *Tropos*, the information system is represented as one or more actors which participate in a strategic dependency model, along with other actors from the system's operational environment. In other words, the system comes into the picture as one or more actors who contribute to the fulfillment of stakeholder goals. For our example, the *Medi@* system is introduced as an actor in the strategic dependency model depicted in Figure 3.

With respect to the actors identified in Figure 2, *Customer* depends on *Media Shop* to buy media items while *Media Shop* depends on *Customer* to increase market share and remain happy (with *Media Shop* service). *Media Supplier* is expected to provide

Media Shop with media items while depending on the latter for continuing long-term business. He can also use *Medi@* to determine new needs from customers, such as media items not available in the catalogue. As indicated earlier, *Media Shop* depends on *Medi@* for processing internet orders and on *Bank Cpy* to process business transactions. *Customer*, in turn, depends on *Medi@* to place orders through the internet, to search the database for keywords, or simply to browse the on-line catalogue. With respect to relevant qualities, *Customer* requires that transaction services be secure and usable, while *Media Shop* expects *Medi@* to be easily maintainable (e.g., catalogue enhancing, item database evolution, user interface update, ...). The other dependencies have already been described in Figure 2.

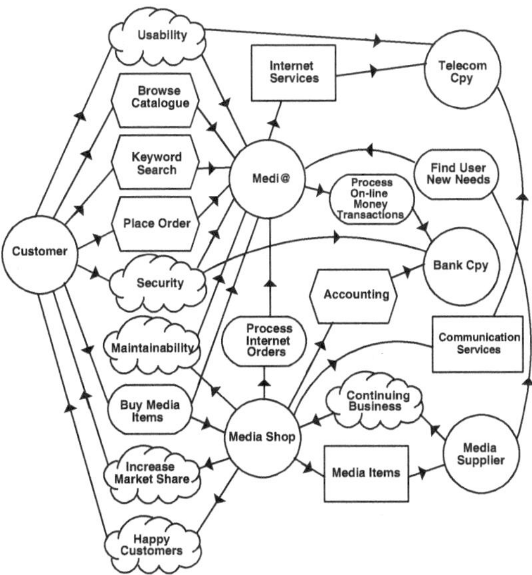

Fig. 3. Strategic dependency model for a media shop

As late requirements analysis proceeds, *Medi@* is given additional responsibilities, and ends up as the depender of several dependencies. Moreover, the system is decomposed into several sub-actors which take on some of these responsibilities. This decomposition and responsibility assignment is realized using the same kind of means-ends analysis along with the strategic rationale analysis illustrated in Figure 2. Hence, the analysis in Figure 4 focuses on the system itself, instead of a external stakeholder.

The figure postulates a root task *Internet Shop Managed* providing sufficient support (++) [3] to the softgoal *Increase Market Share*. That task is firstly refined into goals *Internet Order Handled* and *Item Searching Handled*, softgoals *Attract New Customer*, *Secure* and *Usable* and tasks *Produce Statistics* and *Maintenance*. To manage internet orders, *Internet Order Handled* is achieved through the task *Shopping Cart* which is decomposed into subtasks *Select Item*, *Add Item*, Check *Out,* and *Get Identification Detail*. These are the main process activities required to design an operational on-line shopping cart [6]. The latter (goal) is achieved either through sub-goal *Classic Communication Handled* dealing with phone and fax orders or *Internet Handled* managing secure or standard form orderings. To allow for the ordering of new items not listed in the catalogue, *Select Item* is also further refined into two alternative subtasks, one dedicated to select catalogued items, the other to preorder unavailable products.

To provide sufficient support (++) to the *Maintainable* softgoal, *Maintenance* is refined into four subtasks dealing with catalogue updates, system evolution, interface updates and system monitoring.

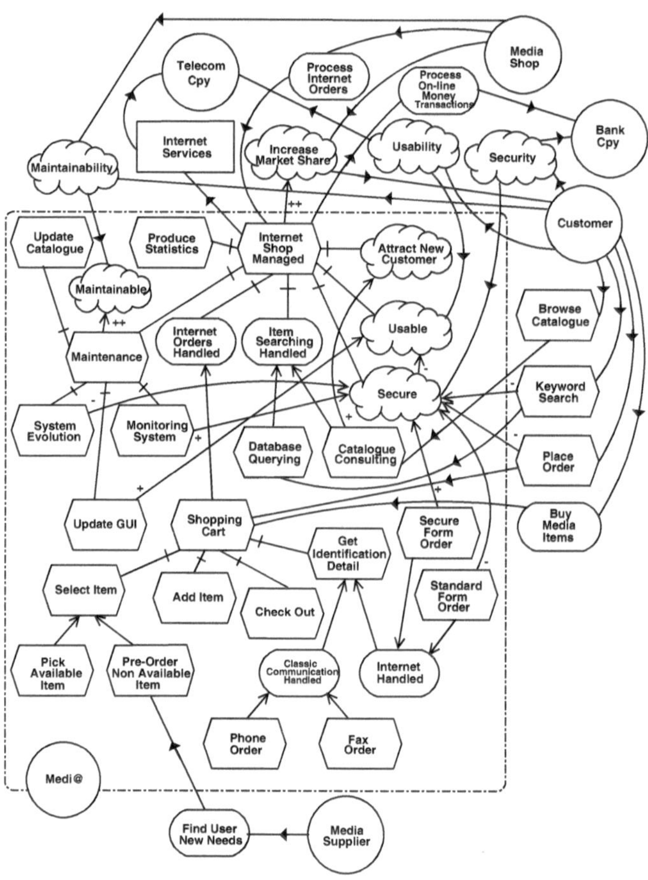

Fig. 4. Strategic rationale model for *Medi@*

The goal *Item Searching Handled* might alternatively be fulfilled through tasks *Database Querying* or *Catalogue Consulting* with respect to customers' navigating desiderata, i.e., searching with particular items in mind by using search functions or simply browsing the catalogued products.

In addition, as already pointed, Figure 4 introduces softgoal contributions to model sufficient/partial positive (respectively ++ and +) or negative (respectively - - and -) support to softgoals *Secure*, *Usable*, *Maintainable*, *Attract New Customers* and

Increase Market Share. The result of this means-ends analysis is a set of (system and human) actors who are dependees for some of the dependencies that have been postulated.

Figure 5 suggests one possible assignment of responsibilities identified for *Medi@*. The *Medi@* system is decomposed into four sub-actors: *Store Front*, *Billing Processor*, *Service Quality Manager* and *Back Store*.

Store Front interacts primarily with *Customer* and provides her with a usable front-end web application. *Back Store* keeps track of all web information about customers, products, sales, bills and other data of strategic importance to *Media Shop*. *Billing Processor* is in charge of the secure management of orders and bills, and other financial data; also of interactions to *Bank Cpy*. *Service Quality Manager* is introduced in order to look for *security* gaps, *usability* bottlenecks and *maintainability* issues.

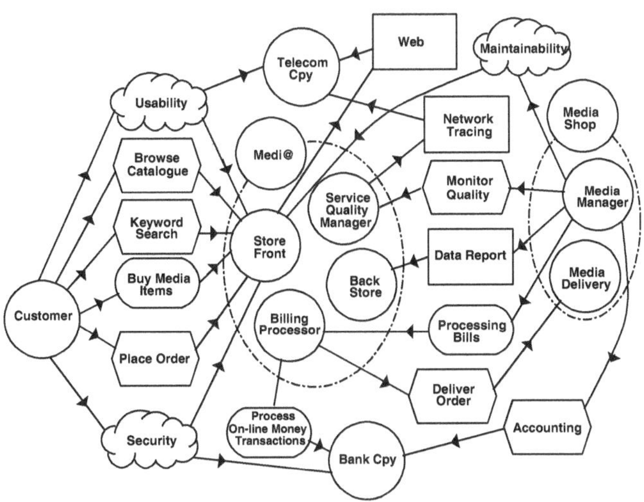

Fig. 5. The web system consists of four inside actors, each with external dependencies

All four sub-actors need to communicate and collaborate. For instance, *Store Front* communicates to *Billing Processor* relevant customer information required to process bills. For the rest of the section, we focus on *Store Front*. This actor is in charge of catalogue browsing and item database searching, also provides on-line customers with detailed information about media items. We assume that different media shops working with *Medi@* may want to provide their customers with various forms of information retrieval (Boolean, keyword, thesaurus, lexicon, full text, indexed list, simple browsing, hypertext browsing, SQL queries, etc.).

Store Front is also responsible for supplying a customer with a web shopping cart to keep track of selected items. We assume that different media shops using the *Medi@* system may want to provide customers with different kinds of shopping carts with respect to their internet browser, plug-ins configuration or platform or simply personal wishes (e.g., Java mode, simple browser, frame-based, CGI shopping cart,...)

Finally, *Store Front* initializes the kind of processing that will be done (by *Billing Processor*) for a given order (phone/fax, internet standard form or secure encrypted form). We assume that different media shop managers using *Medi@* may be processing various types of orders differently, and that customers may be selecting the kind of delivery system they would like to use (UPS, FedEx, ...).

Resource, task and softgoal dependencies correspond naturally to functional and non-functional requirements. Leaving (some) goal dependencies between system actors and other actors is a novelty. Traditionally, functional goals are "operationalized" during late requirements [7], while quality softgoals are either operationalized or "metricized" [8]. For example, *Billing Processor* may be operationalized during late requirements analysis into particular business processes for processing bills and orders. Likewise, a security softgoal might be operationalized by defining interfaces which minimize input/output between the system and its environment, or by limiting access to sensitive information. Alternatively, the security requirement may be metricized into something like "No more than X unauthorized operations in the system-to-be per year".

Leaving goal dependencies with system actors as dependees makes sense whenever there is a foreseeable need for flexibility in the performance of a task on the part of the system. For example, consider a communication goal "communicate X to Y". According to conventional development techniques, such a goal needs to be operationalized before the end of late requirements analysis, perhaps into some sort of a user interface through which user Y will receive message X from the system. The problem with this approach is that the steps through which this goal is to be fulfilled (along with a host of background assumptions) are frozen into the requirements of the system-to-be. This early translation of goals into concrete plans for their fulfillment makes systems fragile and less reusable.

In our example, we have left three goals in the late requirements model. The first goal is *Usability* because we propose to implement *Store Front* and *Service Quality Manager* as agents able to automatically decide at run-time which catalogue browser, shopping cart and order processor architecture fit best customer needs or navigator/platform specifications. Moreover, we would like to include different search engines, reflecting different search techniques, and let the system dynamically choose the most appropriate. The second key softgoal in the late requirements specification is *Security*. To fulfil it, we propose to support in the system's architecture a number of security strategies and let the system decide at run-time which one is the most appropriate, taking into account environment configurations, web browser specifications and network protocols used. The third goal is *Maintainability*, meaning that catalogue content, database schema, and architectural model can be dynamically extended to integrate new and future web-related technologies.

5 Architectural Design

A system architecture constitutes a relatively small, intellectually manageable model of system structure, which describes how system components work together. For our case study, the task is to define (or choose) a web-based application architecture. The canonical web architecture consists of a web server, a network connection, HTML/XML documents on one or more clients communicating with a Web server via HTTP, and an application server which enables the system to manage business logic and state. This architecture is not intended to preclude the use of distributed objects or Java applets; nor does it imply that the web server and application server cannot be located on the same machine.

By now, software architects have developed catalogues of web architectural styles (e.g., [6]). The three most common styles are the *Thin Web Client*, *Thick Web Client* and *Web Delivery. Thin Web Client* is most appropriate for applications where the client has minimal computing power, or no control over its configuration. The client requires only a standard forms-capable web browser. *Thick Web Client* extends the *Thin Web Client* style with the use of client-side scripting and custom objects, such as ActiveX controls and Java applets. Finally, *Web Delivery* offers a traditional client/server system with a web-based delivery mechanism. Here the client communicates directly with object servers, bypassing HTTP. This style is appropriate when there is significant control over client and network configuration.

The first task during architectural design is to select among alternative architectural styles using as criteria the desired qualities identified earlier. The analysis involves refining these qualities, represented as softgoals, to sub-goals that are more specific and more precise and then evaluating alternative architectural styles against them, as shown in Figure 6. The styles are represented as operationalized softgoals (saying, roughly, "make the architecture of the new system *Web Delivery-/Thin Web-/Thick Web*-based") and are evaluated with respect to the alternative non-functional softgoals as shown in Figure 6. Design rationale is represented by claim softgoals drawn as dashed clouds. These can represent contextual information (such as priorities) to be considered and properly reflected into the decision making process. Exclamation marks (! and !!) are used to mark priority softgoals while a check-mark "✔" indicates a fulfilled softgoal, while a cross "✖" labels a unfulfillable one.

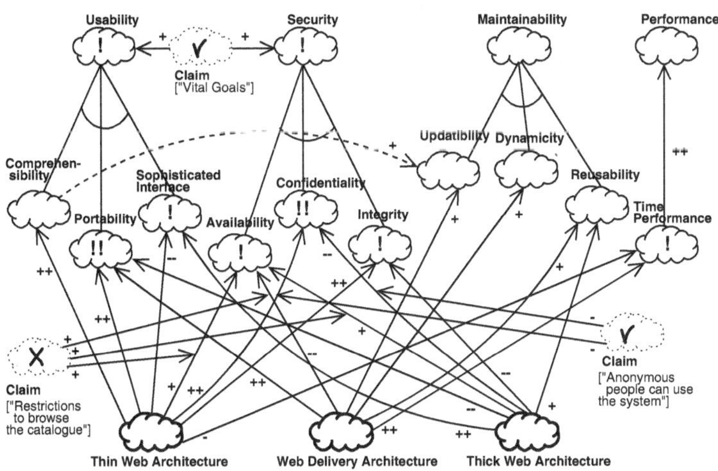

Fig. 6. Refining softgoals in architectural design

The *Usability* softgoal has been AND-decomposed into sub-goals *Comprehensibility*, *Portability* and *Sophisticated Interface.* From a customer perspective, it is important for *Medi@* to be intuitive and ergonomic. The look-and-feel of the interface must naturally guides customer actions with minimal computer knowledge. Equally strategic is the portability of the application across browser implementations and the quality of the interface. Note that not all HTML browsers support scripting, applets, controls and plug-ins. These technologies make the client itself more dynamic, and capable of animation, fly-over help, and sophisticated input controls. When only minimal business logic needs to be run on the client, scripting is often an easy and powerful mechanism to use. When truly sophisticated logic needs to run on the client, building Java applets, Java beans, or ActiveX controls is probably a better approach. A comparable analysis is carried out for *Security* and *Maintainability.*

As shown in Figure 6, each of the three web architectural styles contributes positively or negatively to the qualities of interest. For instance, *Thin Web Client* is useful for applications where only the most basic client configuration can be guaranteed. Hence, this architecture does well with respect to *Portability*. However, it has a limited capacity to support *Sophisticated User Interfaces*. Moreover, this architecture relies on a connectionless protocol such as HTTP, which contributes positively to system availability.

On the other hand, *Thick Web Client* is generally not portable across browser implementations, but can more readily support sophisticated interfaces. As with *Thin Web Client*, all communication between client and server is done with HTTP, hence its positive contribution to *Availability*. On the negative side, client-side scripting and custom objects, such as ActiveX controls and Java applets, may pose risks to client confidentiality. Last but not least, *Web Delivery* is highly portable, since the browser has some built-in capabilities to automatically download the needed components from the server. However, this architecture requires a reliable network.

This phase also involves the introduction of new system actors and dependencies, as well as the decomposition of existing actors and dependencies into sub-actors and sub-dependencies which are delegated some of the responsibilities of the key system actors introduced earlier.

Figure 7 focuses on the latter kind of refinement. To accommodate the responsibilities of *Store Front*, we introduce *Item Browser* to manage catalogue navigation, *Shopping Cart* to select and custom items, *Customer Profiler* to track customer data and produce client profiles, and *On-line Catalogue* to deal with digital library obligations. To cope with the non-functional requirement decomposition proposed in Figure 6, *Service Quality Manager* is further refined into four new system sub-actors *Usability Manager*, *Security Checker*, *Maintainability Manager* and *Performance Monitor*, each of them assuming one of the top main softgoals explained previously. Further refinements are shown on Figure 7.

An interesting decision that comes up during architectural design is whether fulfillment of an actor's obligations will be accomplished through assistance from other actors, through delegation ("outsourcing"), or through decomposition of the actor into component actors. Going back to our running example, the introduction of other actors described in the previous paragraph amounts to a form of delegation in the sense that *Store Front* retains its obligations, but delegates subtasks, sub-goals etc. to other actors. An alternative architectural design would have *Store Front* outsourcing some of its responsibilities to some other actors, so that *Store Front* removes itself from the critical path of obligation fulfilment. Lastly, *StoreFront* may be refined into an aggregate of actors which, by design work together to fulfil *Store Front*'s obligations. This is analogous to a committee being refined into a collection of members who collectively fulfil the committee's mandate. It is not clear, at this point, how the three alternatives compare, nor what are their respective strengths and weaknesses.

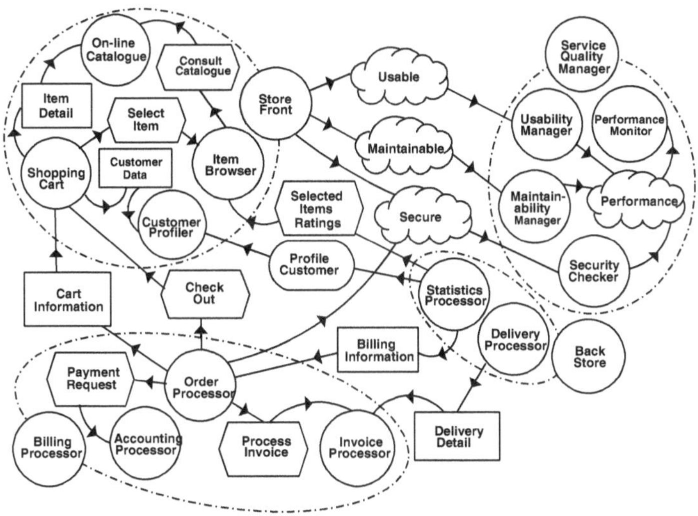

Fig. 7. Strategic Dependency Model for *Medi@* actors

6 Detailed Design

The detailed design phase is intended to introduce additional detail for each architectural component of a system. In our case, this includes actor communication and actor behavior. To support this phase, we propose to adopt existing agent communication languages, message transportation mechanisms and other concepts and tools. One possibility, for example, is to adopt one of the extensions to UML proposed by the FIPA (Foundation for Intelligent Agents) and the OMG Agent Work group [14]. The rest of the section concentrates on the *Shopping cart* actor and the *check out* dependency.

To specify the *checkout* task, for instance, we use AUML - the Agent Unified Modeling Language [14], which supports templates and packages to represent *checkout* as an object, but also in terms of sequence and collaborations diagrams.

Figure 8 focuses on the protocol between *Customer* and *Shopping Cart* which consists of a customization of the FIPA Contract Net protocol [14]. Such a protocol describes a communication pattern among actors, as well as constraints on the contents of the messages they exchange.

When a *Customer* wants to check out, a request-for-proposal message is sent to *Shopping Cart*, which must respond before a given timeout (for network security and integrity reasons). The response may refuse to provide a proposal, submit a proposal, or express miscomprehension. The diamond symbol with an "**X**" indicates an "exclusive or" decision. If a proposal is offered, *Customer* has a choice of either accepting or canceling the proposal. The internal processing of *Shopping Cart*'s *checkout* plan is described in Figure 9.

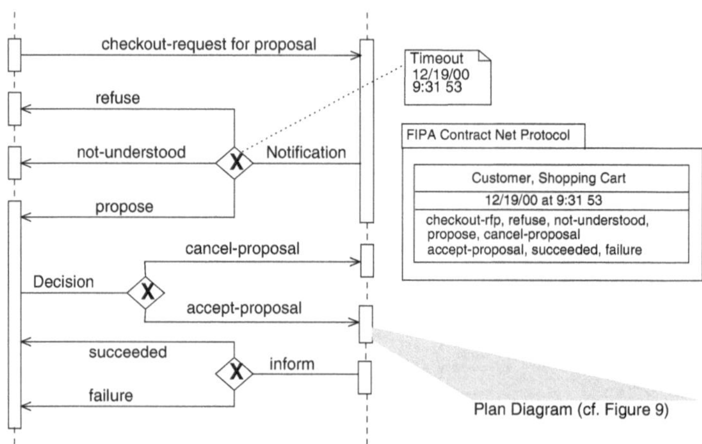

Fig. 8. Agent interaction protocol focusing on a *checkout* dialogue

At the lowest level, we use plan diagrams [12] (See Figure 9), to specify the internal processing of atomic actors. The initial transition of the plan diagram is labeled with an activation event (*Press checkout button*) and activation condition (*[checkout button activated]*) which determine when and in what context the plan should be activated. Transitions from a state automatically occur when exiting the state and no event is associated (e.g., when exiting *Fields Checking*) or when the associated event occurs (e.g., *Press cancel button*), provided in all cases that the associated condition is true (e.g., *[Mandatory fields filled]*). When the transition occurs any associated action is performed (e.g., *verifyCC()*).

An important feature of plan diagrams is their notion of failure. Failure can occur when an action upon a transition fails, when an explicit transition to a fail state (denoted by a small no entry sign) occurs, or when the activity of an active state terminates in failure and no outgoing transition is enabled.

Figure 9 depicts the plan diagram for *checkout*, triggered by pushing the checkout button. Mandatory fields are first checked. If any mandatory fields are not filled, an iteration allows the customer to update them. For security reasons, the loop exits after 5 tries ([i<5]) and causes the plan to fail. Credit Card validity is then checked. Again for security reasons, when not valid, the CC# can only be corrected 3 times. Otherwise, the plan terminates in failure. The customer is then asked to confirm the CC# to allow item registration. If the CC# is not confirmed, the plan fails. Otherwise, the plan continues: each item is iteratively registered, final amounts are calculated, stock records and customer profiles are updated and a report is displayed. When finally the whole plan succeeds, the *ShoppingCart* automatically logs out and asks the *Order Processor* to initialize the order. When, for any reason, the plan fails, the *ShoppingCart* automatically logs out. At anytime, if the cancel button is pressed, or the timeout is more than 90 seconds (e.g., due to a network bottleneck), the plan fails and the *Shopping Cart* is reinitialized.

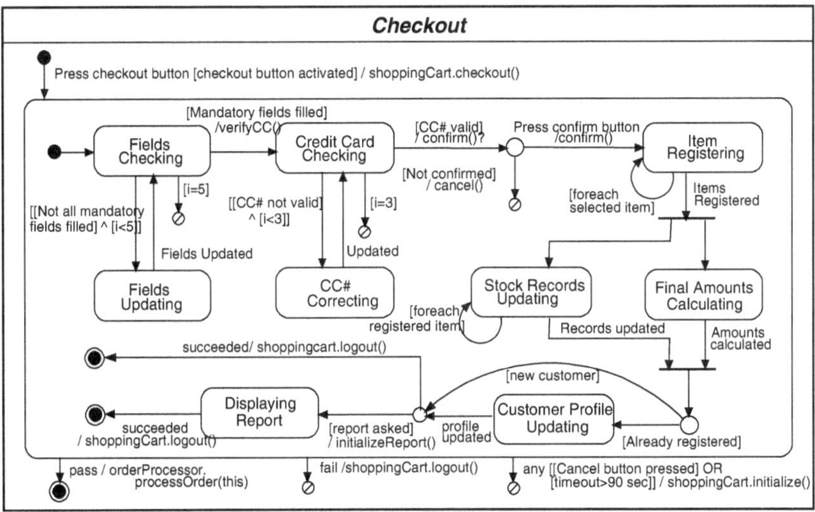

Fig. 9. A plan diagram for *checkout*

7 Generating an Implementation

JACK Intelligent Agents [4] is an agent-oriented development environment designed to provide agent-oriented extensions to Java.

JACK agents can be considered autonomous software components that have explicit *goals* to achieve, or *events* to cope with (desires). To describe how they should go about achieving these desires, agents are programmed with a set of *plans* (intentions).

Each plan describes how to achieve a goal under different circumstances. Set to work, the agent pursues its given goals (desires), adopting the appropriate plans (intentions) according to its current set of data (beliefs) about the state of the world. To support the programming of BDI agents, JACK offers five principal language constructs. These are *agents*, *capabilities*, *database relations*, *events*, and *plans*.

*I** actors, (informational/data) resources, softgoals, goals and tasks will be respectively mapped into BDI agents, beliefs, desires and intentions. In turn, a BDI agent will be mapped as a JACK agent, a belief will be asserted (or retracted) as a database relation, a desire will be posted (sent internally) as a BDIGoalEvent (representing an objective that an agent wishes to achieve) and handled as a plan, and an intention will be implemented as a plan. Finally, an *i** dependency will be directly realized as a BDIMessageEvent (received by agents from other agents).

Figure 10 depicts the JACK layout presenting each of the five JACK constructs as well as the implementation of the first part of the dialogue shown in Figure 8. The request for proposal *checkout-rfp* is a MessageEvent (*extends MessageEvent*) sent by

Customer and handled by the *Shopping Cart*'s *checkout* plan (*extends Plan*) as detailed in Figure 9. Finally, *Timeout* (which we consider a belief) is implemented as a closed world (i.e., true or false) database relation asserting for each *Shopping Cart* one or several timeout delays.

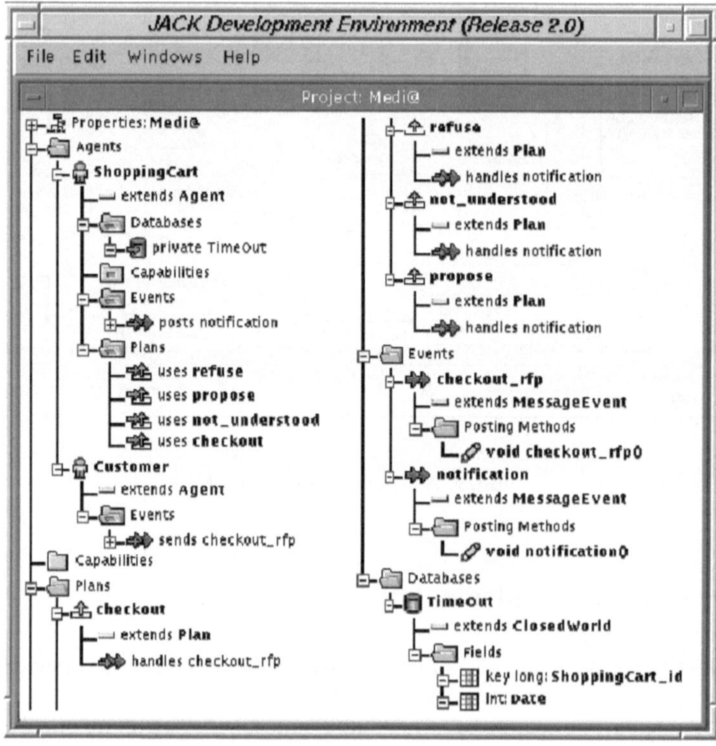

Fig. 10. Partial implementation of Figure 8 in JACK

8 Conclusion and Discussion

We have proposed a development methodology founded on intentional concepts, and inspired by early requirements modeling. We believe that the methodology is particularly appropriate for generic, componentized systems like e-business applications that can be downloaded and used in a variety of operating environments and computing platforms around the world. Preliminary results suggest that the methodology complements well proposals for agent-oriented programming environments.

There already exist some proposals for agent-oriented software development, most notably [10, 11, 14, 16]. Such proposals are mostly extensions to known object-oriented and/or knowledge engineering methodologies. Moreover, all these proposals focus on design – as opposed to requirements analysis – for agent-oriented software and are therefore considerably narrower in scope than *Tropos*.

Of course, much remains to be done to further refine the proposed methodology and validate its usefulness with real case studies. We are currently working on the development of formal analysis techniques for *Tropos*, also the development of tools which support different phases of the methodology.

References

[1] Booch, G., Rumbaugh, J. and Jacobson, I., *The Unified Modeling Language User Guide*, The Addison-Wesley Object Technology Series, Addison-Wesley, 1999.

[2] Castro, J., Kolp, M. and Mylopoulos, J., Developing Agent-Oriented Information Systems for the Enterprise, *Proceedings of the Second International Conference On Enterprise Information Systems* (ICEIS00), Stafford, UK, July 2000.

[3] Chung, L. K., Nixon, B. A., Yu, E. and Mylopoulos, J., *Non-Functional Requirements in Software Engineering*, Kluwer Publishing, 2000.

[4] Coburn, M., *Jack Intelligent Agents: User Guide version 2.0*, AOS Pty Ltd, 2000.

[5] Cohen, P. and Levesque, H., "Intention is Choice with Commitment", *Artificial Intelligence, 32(3)*, 1990, pp. 213-261.

[6] Conallen, J., *Building Web Applications with UML*, The Addison-Wesley Object Technology Series, Addison-Wesley, 2000.

[7] Dardenne, A., van Lamsweerde, A. and Fickas, S., "Goal–directed Requirements Acquisition", *Science of Computer Programming, 20*, 1993, pp. 3-50.

[8] Davis, A., *Software Requirements: Objects, Functions and States*, Prentice Hall, 1993.

[9] DeMarco, T., *Structured Analysis and System Specification*, Yourdon Press, 1978.

[10] Iglesias, C., Garrijo, M. and Gonzalez, J., "A Survey of Agent-Oriented Methodologies", *Proceedings of the 5th International Workshop on Intelligent Agents: Agent Theories, Architectures, and Languages* (ATAL-98), Paris, France, July 1998, pp. 317-330.

[11] Jennings, N. R., "On agent-based software engineering", *Artificial Intelligence, 117*, 2000, pp. 277-296.

[12] Kinny, D. and Georgeff, M., "Modelling and Design of Multi-Agent System", *Proceedings of the Third International Workshop on Agent Theories, Architectures, and Languages* (ATAL-96), Budapest, Hungary, August 1996, pp. 1-20.

[13] Mylopoulos, J. and Castro, J., "Tropos: A Framework for Requirements-Driven Software Development", Brinkkemper, J. and Solvberg, A. (eds.), *Information Systems Engineering: State of the Art and Research Themes*, Springer-Verlag, June 2000, pp. 261-273.

[14] Odell, J., Van Dyke Parunak, H. and Bauer, B., "Extending UML for Agents", *Proceedings of the Agent-Oriented Information System Workshop at the 17 National Conference on Artificial Intelligence*, pp. 3-17, Austin, USA, July 2000.

[15] Wirfs-Brock, R., Wilkerson, B. and Wiener, L., *Designing Object-Oriented Software*, Englewood Cliffs, Prentice-Hall, 1990.

[16] Wooldridge, M., Jennings, N. R. and Kinny D., "The Gaia Methodology for Agent-Oriented Analysis and Design", *Journal of Autonomous Agents and Multi-Agent Systems, 3(3)*, to appear, 2000.

[17] Yourdon, E. and Constantine, L., *Structured Design: Fundamentals of a Discipline of Computer Program and Systems Design*, Prentice-Hall, 1979.

[18] Yu, E., *Modelling Strategic Relationships for Process Reengineering*, Ph.D. thesis, Department of Computer Science, University of Toronto, Canada, 1995.

Constructing Libraries of Typical Plans

Antonio L. Furtado and Angelo E.M. Ciarlini*

Departamento de Informática - Pontifícia Universidade Católica do R.J.
22.453-900 Rio de Janeiro, Brazil
{furtado,angelo}@inf.puc-rio.br

Abstract. Databases able to represent, not only facts, but also *events* in the mini-world of the underlying information system can be seen as repositories of narratives about the agents and objects involved. The events treated in our approach are those attributed to executions of predefined application-oriented operations. This work addresses the identification of *typical plans* adopted by agents, by analysing a *Log* registering the occurrence of events, as represented by executions of such operations. The analysis is done by applying a previously formulated set of *goal-inference rules* to sequences of interrelated events, called *plots*, taken from the *Log*. The obtained *Library of Typical Plans*, together with the goal-inference rules, constitute the behavioural level of our proposed three-level conceptual schemas for the specification of information systems. A prototype Prolog implementation of the method for extracting typical plans is operational. A simple example is used to illustrate the discussion.

1. Introduction

The initial emphasis of the database approach to the conceptual specification of information systems was mostly on the static description of objects and their properties. At a later stage, however, attention was also given to functional characteristics [1]. Theoretical and practical work at these two complementary levels has led to entity-relationship schemes, object-oriented classes, and workflows, among other important contributions. The use of observed instances for building such specifications has been taken into consideration by recent research on workflow/process mining [10]. More recently, there has been a growing realization that the specification of an information system must also consider the *agents* [17,18,13] which will eventually put it to use. What agents do is not fortuitous; they organize *plans* in an attempt to reach specific *goals*. In turn, goals arise when certain *situations* occur. Besides adding to the definition of objects a characterization of their functional aspects, a third stage of specification is therefore needed, where agents and their expected interactions are modelled. Informally speaking, agents cause the occurrence of *events* affecting the existence and various properties of entities in the mini-world of a given information system. And what they make happen in this mini-world, which as a consequence of their actions traverses a series of intermediate states, can be viewed, borrowing from traditional literary terminology, as *narratives*.

* The work of the second author was supported by FAPERJ - Fundação de Amparo à Pesquisa do Estado do Rio de Janeiro, Brazil

K.R. Dittrich, A. Geppert, M.C. Norrie (Eds.): CAiSE 2001, LNCS 2068, pp. 124–139, 2001.
© Springer-Verlag Berlin Heidelberg 2001

Accordingly, we specify schemas at three successive levels. The first is the *static level*, where the types of facts about entities to be stored in the database are declared, according to the Entity-Relationship model extended with is-a hierarchies for entity types. Secondly, *application-oriented operations* are defined, in a STRIPS-like formalism [5], to provide the *dynamic level.* A third level, the *behavioural level*, is added, in order to model the reason why operations are executed and the way they are typically combined and performed. The behavioural level is composed of *goal-inference rules* and a *Library of Typical Plans*. Each goal-inference rule declares, for an agent or class of agents, that the occurrence of a certain situation tends to motivate the agent to pursue a given goal. Typical plans, consisting of partially ordered sets of one or more executions of the application-oriented operations, represent the expected patterns of database usage by the various agents, towards their goals.

The availability of a *Library of Typical Plans* enormously increases the understanding of strategies and policies habitually adopted by agents. The automatic construction of such a library is therefore useful for the specification of a system that models such agents. If a good plan is identified and generalized, it can be automated and reused in many similar situations. If a bad plan is identified, it means that wrong policies have probably been adopted and should then be corrected. At the final stage of system implementation, typical plans can be used by plan-recognition algorithms to simulate and evaluate the behaviour of the system. In [2], we describe a tool in which plan-recognition and plan-generation are combined in order to simulate both typical and non-typical interactions of database agents. Finally, when the system is fully operational, the *Library* can be used by plan-recognition algorithms to detect, by matching observed actions of an agent against the *Library*, whether such actions fit in some known plan or plans. As soon as a plan is detected, the system may be able to infer how to help or block such a plan, depending on the interest of the corporation.

Our three-level specification approach is especially useful if the database implemented provides the two following features: (a) updates can only be performed through the execution of the application-oriented operations introduced at the dynamic level; and (b) each execution of an operation triggers the insertion in a *Log* of a record containing the name and arguments of the operation executed, together with a time-stamp indicating the moment of execution.

A time-stamp-ordered sequence of records of executions of interrelated operations, extracted from the *Log*, clearly corresponds to a sequence of events, which justifies calling one such sequence a *plot* [2,21]. We mentioned above the notion of narratives happening in the context of an information system. Plots can then be interpreted as summaries of such narratives. As demonstrated in a companion paper [8], if a *text-generator* is available, plots can be interpreted to produce natural language answers to queries like: "What happened to Mary between time instants t1 and t2?". But the thrust of the present paper is how to use plots to help formulating the behavioural level of conceptual specifications. As a first step towards this objective, we assume that the goal-inference rules have been introduced at a preliminary stage. How to discover the rules is, of course, a difficult knowledge-discovery task [14,12], which we are investigating separately. Once the static and dynamic levels have been specified, and the goal-inference rules are available, we proceed to a trial phase, where the prospective agents are called to operate on a prototype implementation of the information system, thereby allowing the *Log* to grow to a size estimated large enough to constitute a sample. Our tool, called *BLIB*, analyses then a series of plots taken from the *Log*, on the basis of the goal-inference rules, in order to identify typical plans toward such goals and, with the designer's participation, to build a *Library of Typical Plans*.

The paper is organized as follows. Section 2 presents the three-level modelling concepts, emphasizing the visualization of plots as the result of plans of the various agents. Section 3 describes the method for obtaining typical plans from plots taken from the *Log*. Section 4 contains concluding remarks, pointing out aspects where certain tentative decisions adopted in the implemented prototype may be revised, by considering different alternatives. A small example is used throughout the paper to illustrate how the process works. For a more formal treatment of our modelling approach see [3]; other related formalisms can be found in [4,15,16].

2. Three-Level Specifications

The concepts used at each level will be introduced with the help of the very simple example of a *Company Alpha*'s database. Schemas are specified, at each level, in a notation compatible with logic programming.

2.1. The Static Level

At the static level, *facts* are classified according to the Entity-Relationship model. Thus, a fact may refer either to the existence of an entity instance, or to the values of its attributes, or to its relationships with other entity instances. Entity classes may form an *is-a* hierarchy. Entities must have one privileged attribute, which identifies each instance at all levels of the is-a hierarchy. Moreover, we shall restrict ourselves to single-valued attributes and binary relationships without attributes. All kinds of facts are denoted by *predicates*.

```
% COMPANY ALPHA EXAMPLE
dbowner('Company Alpha').
entity(person, name).
entity(employee).              is_a(employee, person).       attribute(employee, level).
entity(company, denomination).
entity(client).                is_a(client,company).         attribute(client, account_status).
entity(course, title).

relationship(serving, [employee, client]).
relationship(dissatisfied_with, [client, employee]).
relationship(taking, [employee, course]).
```

Fig. 1. Static sub-schema

The example static schema — given in Fig. 1 — includes, among the entity classes, *person*, *company*, and *course*; in addition, class *employee* is a specialization of *person*, and *client* a specialization of *company*. The identifying attributes are *name* (for *person*, and consequently also for *employee*), *denomination* (*company* and *client*) and *title* (*course*). For the attribute *level* (of *employee*) there are only two possible values: 1 and 2. *Account_status* is an attribute of *client*, referring to the status of the client's account, whose only value that will concern us here, because of its criticality, is *inactive*. Relationships *serving* and *dissatisfied_with* are defined between employees and clients; employees and courses are related by *taking*. With respect to onomastic criteria, notice that nouns are used to name entity classes (e.g. *person*) and attributes (e.g. *level*). For relationships, we favour past or present participles (e.g. *serving*). Examples of predicate instances representing facts are: (a) entity instance: *person('Mary')*; (b) attribute of entity: *level('Mary',1)*; (c) relationship: *serving('Mary','Beta')*. The set of all predicate instances of all types holding at a given instant constitutes a *state*. In temporal database environments [19], one can ask whether or not some fact F holds at a state S associated with a time instant t.

2.2. The Dynamic Level

The dynamic level covers the *events* happening in the mini-world of interest. A real world event is perceived in a temporal database environment as a *transition* between database states. Our dynamic level schemas (Fig. 2, for the current example) specify a fixed repertoire of *operations*, whose execution provides the only kind of admissible events, i.e. the only way to cause state transitions [9]. Accordingly, from now on we shall equate the notion of event with the execution of one of these operations.

As in the STRIPS formalism [5], each operation is defined through its signature, pre-conditions, and post-conditions or effects. Both pre-conditions and effects are expressed in terms of facts, thus establishing a connection with the static level. Preconditions are conjunctions of positive (or negated) facts, which should hold (or not hold) before the execution, whereas effects consist of facts added and/or deleted by the operation. When defining the signature of an operation, we declare the type of each parameter (which implicitly imposes a preliminary pre-condition to the execution of the operation) and its semantic role, borrowing from Fillmore's case grammars [6], a major contribution from the field of Linguistics. From the cases proposed by Fillmore, we retained *agent* (denoted by the letter "a") and *object* ("o"); we found convenient to denote the other cases (e.g. *beneficiary*, *instrument*, etc.) by some preposition able to suggest the role when used as prefix. The agent is, of course, whoever is in charge of executing the operation. In our example, operation *complain* is the only one whose definition indicates the agent explicitly. If none of the parameters is indicated as playing the role of agent, the database owner is assumed by default to have the initiative. Thus the clause

oper(replace(E1,E2,C), [employee/o, employee/by, client/for])

allows us to interpret the event *replace('Mary','Leonard','Beta')* unambiguously as *"Company Alpha replaces employee Mary by employee Leonard for client Beta"*.

The other clauses defining the operation (cf. Fig. 2) give its preconditions and effects. As a consequence of these clauses, as the reader can verify, this particular *replace* event will indeed produce the state transition below, whose net effect is that, in state *Sj*, *Leonard*, instead of *Mary*, is serving *Beta*:

Si		*Sj*
employee('Mary').		*employee('Mary')*
employee('Leonard')	⟶	*employee('Leonard')*
client('Beta').		*client('Beta').*
serving('Mary','Beta').		*serving('Leonard','Beta').*

The other operations make it possible for *Company Alpha* to *sign a contract* with a company (so as to make it one of its clients), to *hire* a person as employee with initial level 1, to *assign* an employee to the service of a client, to *enroll* an employee in a training course, to *promote* an employee by raising the level to 2, and to *fire* an employee. To clients it is allowed to formally *complain* about the service rendered by the assigned employee, with the contractual effect of suspending all business transactions (*account_status* = *inactive*).

```
% Operations
oper(sign_contract(C), [company/ with]).
added(sign_contract(C), client(C)).

oper(hire(E), [person/ o]).
added(hire(E), (employee(E), level(E, 1))).

oper(assign(E,C), [employee/ o, client/ to]).
added(assign(E,C), serving(E,C)).
precond(assign(E,C), ((not serving(E,C1)), (not serving(E1, C)))).

oper(enroll(E,T), [employee/ o, course/ in]).
added(enroll(E,T), taking(E,T)).
deleted(enroll(E,T), (dissatisfied_with(C,E), account_status(C,inactive))).
precond(enroll(E,T), (serving(E,C), not taking(E,T1))).

oper(promote(E), [employee/ o]).
added(promote(E), level(E,2)).
deleted(promote(E), level(E,1)).
precond(promote(E), (serving(E,C), not dissatisfied_with(C,E), level(E,1))).

oper(replace(E1,E2,C), [employee/ o, employee/ by, client/ for]).
added(replace(E1,E2,C), serving(E2,C)).
deleted(replace(E1,E2,C), serving(E1,C)).
precond(replace(E1,E2,C), (serving(E1,C), not serving(E2,C1))).

oper(fire(E), [employee/ o]).
deleted(fire(E), (employee(E), level(E,N), dissatisfied_with(C,E), account_status(C,inactive))).
precond(fire(E), (not serving(E,C))).

oper(complain(C,E), [client/ a, employee/ about]).
added(complain(C,E), (dissatisfied_with(C,E), account_status(C,inactive))).
precond(complain(C,E), serving(E,C)).
```

Fig. 2. Dynamic sub-schema

Pre-conditions and effects are usually tuned in a combined fashion, aiming at the enforcement of integrity constraints. It can be shown that the integrity constraints below, among others, will be preserved if, in consonance with the abstract data type discipline, the initial database is consistent and these pre-defined operations are the only way to cause database transitions: (a) an employee can serve at most one client and a client can be served by at most one employee(i.e. *serving* is a 1-1 relationship); (b) an employee can only be fired if currently not serving any client; and (c) to have a level raise, an employee must be serving a client whose account is not inactive.

Verbs are employed to name the operations, possibly with trailing prepositions or other words or particles, separated by underscore.

2.3. The Behavioural Level

Carefully designed application-oriented operations enable the various agents to handle the database in a consistent way. The question remains of whether they will coexist well with a system supporting such operations, and, if so, what actual usage patterns will emerge. Ideally, the designers of an information system should try to predict how agents will behave within the scope of the system, so as to ensure that the specification at the two preceding levels is adequate from a *pragmatic* viewpoint. The ability to make predictions about behaviour is also crucial for decision-making based on simulations of future events.

To model the reactions of prospective agents, our behavioural sub-schema for the *Company Alpha* example — given in Fig. 3 — contains a few illustrative *goal-inference rules*, plus some *typical plans* (represented as *complex operations*).

```
% Goal-inference rules and typical plans
gi_rule('company Alpha', (employee(E), not serving(E,C)), not employee(E)).
gi_rule('company Alpha', (serving(E,C), account_status(C,inactive)), not account_status(C,inactive)).
gi_rule(employee(E1), (level(E1,1),level(E2,2)), level(E1,2)).

op_complex(renovate_assistance(C,E2,E1),[client/ to, person/ with, employee/ 'in the position of']).
components(renovate_assistance(C,E2,E1), [f1: hire(E2), f2: replace(E1,E2,C), f3: fire(E1)], [f1-f2, f2-f3]).

op_complex(advance_the_career(E), [employee/ of]).
components(advance_the_career(E), [f1: enroll(E,C), f2: promote(E)], [f1-f2]).

op_complex(improve_service(C), [client/ for]).
is_a(enroll(E,T), improve_service(C): serving(E,C)).
is_a(renovate_assistance(C,E2,E1), improve_service(C)).
```

Fig. 3. Behavioural sub-schema

A goal-inference rule has, as antecedent, some *situation* which, if observed at a database state, will arouse in a given agent the impulse to act in order to reach some *goal*. Two rules refer to *Company Alpha*, the database owner. The first one indicates that, if employee *E* is not currently serving any client, *Alpha* will want that *E* cease to be an employee. The goal in the second rule is that *Alpha* will do an effort to placate any client *C* who, being dissatisfied with the employee assigned to its service, has assumed an inactive status. (Notice, incidentally, that "keeping a client happy" is, in the terminology of [17], a *soft goal*, i.e. an imprecisely defined objective; in our example, it assumes a more firm aspect through its dependence on the concrete consideration of the *account_status* attribute). A goal is indicated for employees: if *E1* has merely level 1, whilst some other employee *E2* has been raised to level 2, then, presumably moved by emulation, *E1* will want to reach this higher level.

The specification of behaviour is complemented by a *Library of Typical Plans*. A *typical plan* is a description of how an agent (or class of agents) usually proceeds towards some goal. It consists of either a set of partially ordered operations or plans, or of a set of specialized alternative plans able to achieve the goal. Plans of both kinds are expressed in the *Library* as complex operations. Let us call the operations introduced in the previous section *basic operations*. Then, a *complex operation* can be defined from the repertoire of basic operations (or from other complex operations, recursively) by either composition (part-of hierarchy), giving origin to *composite* operations, or by generalization (is-a hierarchy), yielding *generic* operations. In case of composition, the definition must specify the component operations and the ordering requirements, if any (noting that we allow plans to be *partially-ordered*). In case of generalization, the specialized operations must be specified.

In our example, complex operation *renovate_assistance* is composed of basic operations *hire*, *replace* and *fire*. In turn, complex operation *improve_service* generalizes basic operation *enroll* and complex operation *renovate_assistance*. (A minor technical detail: the fact *serving(E,C)*, introduced by ":" in the first is_a clause is needed to identify *E*, which is not in the parameter list of *improve_service*). Notice that the two (specialized) forms of *improve_service* have, among others, the effect of removing the undesired de-activation of a client's account. Both can be regarded as reflecting customary strategies (typical plans) of *Company Alpha* to placate a complaining client: it either trains the faulty employee or "renovates" the manpower offered to the client. And therefore both are adequate to achieve the goal expressed in the second rule of Fig. 3.

Complex operation *advance_the_career* has an apparent peculiarity, in that it deviates from the usual norm of plan-generation algorithms, whereby operations are chained together exclusively as needed for the satisfaction of pre-conditions. Here, however, the component operation *enroll* is not required for satisfying a pre-condition for *promote* (except in the special case where training is the chosen way to remove the effects of a pending complaint). Our notion of typical plans, similarly to scripts [20], allows however a looser interpretation. A plan is typical if it reflects the usages and policies, imposed or not by rational reasons, that are observed (or anticipated) in the real-world environment. Thus, we may imagine that the employer, *Company Alpha*, is sensed to be more favourable to promoting employees who, even in the absence of complaints against their service, seek the training program.

Through an analysis of the component or alternative operations of a given complex operation, it is possible to determine the pre-conditions for its execution. It is also possible to identify, among the facts that necessarily hold (or do not hold) after the execution, a goal to be achieved by the operation. Given, as input, observations concerning the execution of a few operations, the *Library of Typical Plans* can be used by plan-recognition algorithms to detect which possible plans the agents may be trying to perform. The recognized plan (or plans) can then be used in simulations of future events. Also, plan detection implies the detection of the respective goals and pre-conditions. Once the pre-conditions are obtained, they can be analysed to check whether the plan can be completed. In turn, the detected goals can serve as input to plan-generation algorithms to produce still other plans able to achieve them, which may also be worthy to be tried in simulation runs.

In a previous work [2] we have used the three-level schemata for *simulation* purposes, with the help of a *plan-recognition / plan-generation* method, combining algorithms introduced in [11] and [22], and supported by a Prolog prototype. In that context, a simulated process is enacted, whereby, at each state reached, the goal-inference rules are applied to propose goals by detecting situations affecting each agent. For attempting to fulfil such possibly collaborating or conflicting goals, *plans* are taken from a *Library of Typical Plans* or built by the plan-generator component. In turn, the execution of such plans leads to other states, where the goal-inference rules are again applied, and again plans are obtained and executed, so that the multistage process will continue until it reaches a state where no more goals arise, or until it is arbitrarily terminated.

3. Using Goal-Inference Rules to Extract Plans from Plots

More often than not it is difficult for the designers of real-life information systems to anticipate the usage patterns that will emerge after the system is delivered to operation. Hence, it may be necessary to postpone some design steps and interpose a trial phase, wherein agents are given access to a prototypical version of the system, with a *Log* of executed operations being recorded for later analysis. The missing design steps can then be undertaken with the benefit of the sample experimental evidence extracted from the *Log*.

In section 2.3. we were considering the formulation of: (a) goal-inference rules, and (b) typical plans (consisting of basic or complex operations). Now suppose instead that, assuming a more realistic scenario, it was possible, by interviewing the prospective agents of an information system being designed, to achieve step (a) to a reasonable extent, whereas (b) could not be completed, since the people consulted felt unable to predict beforehand how they would use the proposed basic operations. So, as suggested above, we resort to the trial phase strategy, allowing the agents to interact with a prototypical version until such time as the operation *Log* is sensed to contain enough data for a comprehensive analysis. We proceed by extracting from the *Log* a series of plots. Each plot is a sequence of events, ordered by their time-stamps, where the first event occurred at a time instant t1 and the last event at t2. In other words, if the entire *Log* is regarded as a sequence of events, then a plot PL is the subsequence of the *Log* circumscribed to a given time interval t1..t2. To avoid excessively long plots, involving many disparate events, it is possible (and often useful), as shown in [8], besides restricting plots to time intervals, to filter them so as to only retain events directly or indirectly related to certain specified objects. In the sequel, we shall assume that all plots to be processed have passed, whenever convenient, through this preliminary filtering step.

3.1. Interpreting Plots to Detect Plans

Having isolated a number of plots to be used as input, we analyse them by applying the various goal-inference rules supposed to have been determined at a preliminary phase. Our method relies on the assumption that plots generally reflect the interaction of diverse plans — not always totally executed and successful with respect to the intended goals — undertaken by the various agents.

When considering the plots, differently from the context of our previous work (cf. end of section 2.3), we are not looking at simulation runs, but rather at observed actions, which may not be entirely rational. Hence, our use of goal-inference rules falls into an *abductive* mode of reasoning, as explained in the sequel. Assume that a rule R indicates that agent A, confronted with situation S, will have the desire to achieve a goal G. Now suppose that in the plot being examined an operation (or sequence of operations) O is present, with the effect of achieving goal G for A, and suppose further that, in the state before the execution of O, the motivating situation S prevailed. We then formulate the *hypothesis* that the event can be explained by this rule R, i.e. that agent A executed (or was able to induce an authorized agent A' to execute) operation O because A previously observed the occurrence of S, being thereby motivated to achieve G. This kind of reasoning is no more than hypothetical, because there may exist other reasons (possibly expressed in other goal-inference rules) that may better explain why O was executed. So, each goal-inference rule helps us to suggest one *interpretation* for the events in a narrative.

Our Prolog prototype tool, *BLIB*, builds a *Library of Typical Plans* by examining a succession of plots taken from the *Log*. It begins by trying to extract from each such plot PL one or more plans, P, that can be *associated* with a goal-inference rule R = gi_rule(A,S,G), establishing that situation S motivates agent A to pursue goal G. A subsequence of events P' from PL is said to be associated with R if, prior to the execution of the first event in P', the situation S holds and, after the execution of the last event in P', a state is reached where G finally holds. Plan P is obtained from P' by a second more refined filtering process, which only keeps the events whose post-conditions contribute to G, plus, proceeding backwards, recursively, those that achieve pre-conditions of events already included in P. It should be stressed, in view of the preceding considerations, that a plurality of plans can be extracted from the same plot, and that the same plan can be associated with more than one goal-inference rule. Accordingly, all interpretations warranted by the existing rules are taken into account by the tool.

3.2. Overview of the Library Construction Process

The algorithm accumulates its output in a data structure, called the *ASG-Index* (from now on simply referred to as *Index*), which is organized as a table, whose entries correspond to each goal-inference rule gi_rule(A,S,G), defined on agent A, situation S, and goal G. Each entry of the table is consequently indexed by [A,S,G], and stores an (initially empty) list of operations, which can be either basic or complex. Each of these operations incorporates a specific plan associated with the rule. At any instant, the *Index* contains, in its essential elements, a representation of the current stage of the *Library of Typical Plans* being constructed. Consider, for instance, the plot:

PL = [s0, complain('Beta','Mary'), hire('Leonard'), hire('John'),
replace('Mary','Leonard','Beta'), fire('Mary')]

where s0 denotes the preceding database state. Now, consider the rule below, to be tentatively applied to PL:

gi_rule('company Alpha', account_status(C,inactive), not account_status(C,inactive)).

which says that, for agent A = *Alpha*, the situation S where the account of a client has become inactive induces the goal G of bringing it back to activity.

When analysing PL to check whether or not the rule is applicable, the chaining of pre-conditions and effects in PL is verified by a conventional *holds* meta-predicate, which, incidentally, is the basis for simple plan-generators following STRIPS formalisms. A fact F holds after an operation O is executed at a state s_Q reached by executing a previous sequence of operations Q if either: (1) O is the pseudo-operation s0 and F belongs to the corresponding database state; (2) F is among the facts declared to be added by O, and the pre-conditions of O hold at s_Q; or (3) F already held at s_Q and is not among the facts declared to be deleted by O.

Clearly, in case (1) the tool must have access to facts concerning the objects involved, holding at the temporal database state corresponding to s0, either directly retrieved by the tool itself or prefetched (as happens with the present version). To check facts at states reached along the execution of operations in PL, the *holds* meta-predicate simply resorts to the definitions of the operations. Notice that (2) and (3) make the process recursive (fixing the pre-conditions as sub-goals, or looking for F in the effects of the operations in Q), and that (3) is a standard solution for the frame problem (facts not affected by O continue to hold). By using *holds* we not only check coherence but also provide for the instantiation of some of the variables in the pre-conditions and effects that do not correspond to the parameters.

In our example, the rule is found to be applicable to PL, allowing the extraction of a sequence, which is readily refined to a plan P (by eliminating the irrelevant *hire('John')*): *P=hire('Leonard'),replace('Mary','Leonard','Beta'),fire('Mary').*

Notice that S occurs as a consequence of *complain('Beta','Mary')* and that G holds immediately after *fire('Mary').*

Having extracted a plan, *BLIB* must decide about its relevance to the construction of the *Index*. The plan may be simple, involving a single event, or compound. If the plan is compound, it is first put in a standard representation consisting of a set of tagged events and of a set of order dependencies, expressed as tag-pairs, where the order dependencies are determined exclusively on the basis of the satisfaction of pre-conditions by post-conditions, and where dependencies deducible by transitivity are omitted. In standard representation, the compound plan P exemplified above becomes:

– set of tagged events:

[f1:hire('Leonard'),f2:replace('Mary','Leonard','Beta'), f3:fire('Mary')]

– set of dependencies: *[f1-f2, f2-f3]*

In the *Index*, all operations involving a compound plan are kept in this format, which is convenient for testing if a candidate plan Pi brings a novel contribution. The inclusion or not of Pi at an entry [A,S,G] depends on a comparison with the operations already in that entry. One possibility is that an operation Oj defined on a (simple or compound) plan Pj identical to Pi is already there — in which case nothing is done. Another case is that of similar plans. We say that two plans Pi and Pj are *similar* if, even with different parameter values and executed in a different order, they involve the same: (a) number and type of events; (b) order dependencies; and (c) co-designation/ non-co-designation schemes.

Co-designation (or, respectively, non-co-designation) allows (forbids) the occurrence of the same value (constant or variable) in different parameter positions. Notice, for instance, that in the example above *'Mary'* occurs in the first position of *replace* and in *fire*; a plan with *'John'* in both places (or, say, *X* in both places) would meet the same co-designation requirement. To verify whether the order dependencies are the same, one looks for a renaming of the tags of one of the plans that can render the sets of order dependencies in the two plans identical.

Now, if we are considering a candidate plan Pi and there already exists an operation Oj with a similar plan Pj in the *Index*, Pj will be replaced by the *most specific generalization* P* of Pi and Pj, whenever P* is more general (contains a larger number of variables) than Pj; otherwise Pi is discarded. Reliance on most specific generalization [7] is a fundamental feature of our approach; we strive to stay as close as possible to the evidence supplied by the plots, and thus, in particular, keep constants that tend to repeat (e.g. one certain course taken by all employees who later succeeded to be promoted).

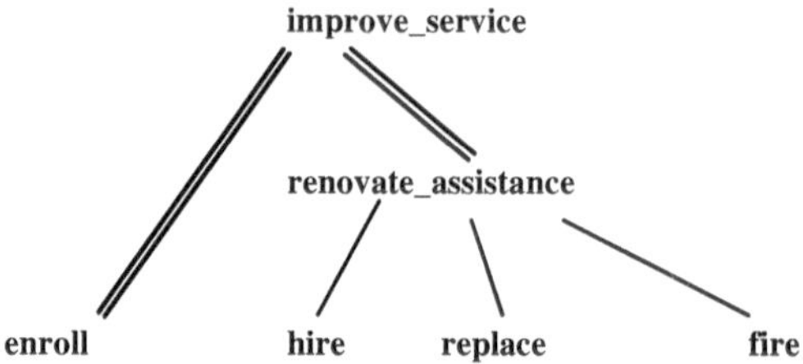

Fig. 4. Fragment of the library of typical plans

The third case is that of a Pi with no similar plan in the [A,S,G] entry. If Pi is simple, it is immediately added to the entry. If compound, *BLIB* asks the designer's help (noting that he can decline, in which case Pi is discarded) to indicate the signature and parameter roles of the new *composite* complex operation Oi. The generated clauses for the example, to be stored in the *Index* at the [A,S,G] entry are shown below:

op_complex(renovate_assistance('Beta','Leonard','Mary'),
[client/ to, employee/ with, employee/ 'in the position of']).
components(renovate_assistance('Beta','Leonard','Mary'),
[f1:hire('Leonard'),f2:replace('Mary','Leonard','Beta'),f3:fire('Mary')],[f1-f2,f2-f3]).

If the added basic or composite Oi is the first in the [A,S,G] entry, nothing else is done. Otherwise, *BLIB* regards the existence of more than one plan associated with the rule as an opportunity to create a *generic* complex operation. Accordingly, it asks the designer (who, again, can decline) to supply the signature of the complex operation. Suppose that, before introducing the new *renovate_assistance* operation, the respective entry already contained the basic operation *enroll*. With the designer's help, generic operation *improve_service* can be further added, with the following clauses being recorded in the entry (notice that all parameters are filled with variables, so that no propagated change will be needed in view of future detection of similar plans):

op_complex(improve_service(C), [client/for]).
is_a(enroll(E,T), improve_service(C): serving(E,C)).
is_a(renovate_assistance(C,P,E), improve_service(C)).

The current version of BLIB supports only one generic operation Og per entry. So, if Oi is introduced when there already exist two or more operations in the entry, and Og has already been introduced, then the definition of the existing Og is merely expanded through an additional is_a clause, relating Oi with Og.

Fig. 4 shows a fragment of the generated *Library of Typical Plans*. Following the conventions in Kautz's plan-recognition project [11], single arrows denote part-of links (composite operations) and double arrows are for is-a links (generic operations).

3.3. Reorganizing the Index

One software engineering requirement to be met by the organization of the *Library of Typical Plans* is that it should be conducive to a *modular architecture* for the later implementation of the operations. Multi-level composition naturally reflects in operational modules calling other modules on a shared basis, and multi-level generalization leads to the conditional choice of modules appropriate to each different case.

Yet, in its current version, the *first phase* of *BLIB* creates no more than one-level composition and generalization hierarchies, the only slightly more involved possibility being the presence of composite operations as alternatives of a generic operation, as exemplified in Fig. 4. Due to our concern to avoid complicated and time-consuming cases of propagation, we decided to keep this simple structure as long as new plots continue to be submitted as input. However, at any time after a batch of plots has been processed, *BLIB* can be called to execute a reorganization *second phase* over the *Index* (and, hence, over the represented structure of the *Library*), so as to produce certain improved multi-level hierarchies. If the acquisition of plans is resumed later, then the original restricted structure must be first reinstated (which is made possible by keeping a back-up copy prior to second phase runs).

The second phase of *BLIB* allows three kinds of restructuring, which are attempted in the indicated order: (1) multi-level generalizations; (2) generic operations as components of composite operations; and (3) multi-level compositions.

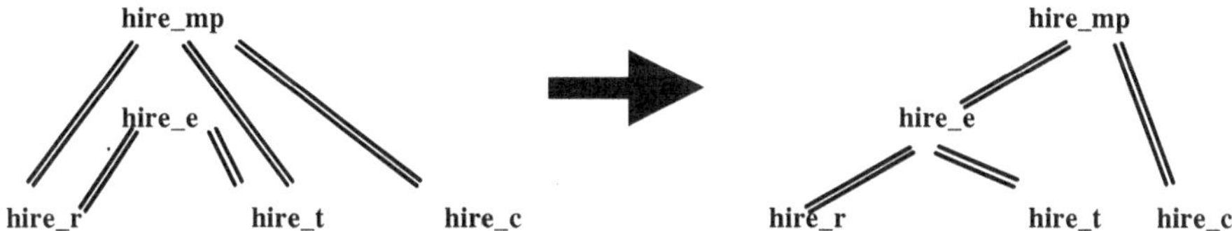

Fig. 5. Multi-level generalization

Reflecting what often happens in practice, we expanded our example application, which resulted in new larger versions of the three schemas (not reproduced here, for space considerations), and processed additional plots. The second phase transformations were then executed. One instance of each kind is displayed below in diagrammatic form.

1. Suppose that entry [A1,S1,G1] has a generic operation Og1, and entry [A2,S2,G2] has a generic operation Og2, and that the entire set of alternatives of Og2 is a proper subset {O1, O2, ..., Ok} of the alternatives of Og1. Then, transformation (1) can be applied to replace, at the definition of Og1, all the k alternatives by a single occurrence of Og2. Fig. 5 illustrates the transformation. In the extended application, besides hiring regular employees (operation *hire*, renamed to *hire_r*), *Company Alpha* was allowed to hire trainees (*hire_t*) and consultants (*hire_c*); notice that both regular employees and trainees are specializations of employee, whereas consultants are a separate category of manpower. At the first phase, operation *hire_mp* (hire manpower) was introduced as a generalization of *hire_r*, *hire_t* and *hire_c*, and operation *hire_e* (hire employee) as a generalization of *hire_r* and

hire_t. After the transformation, *hire_mp* becomes a generalization of *hire_e* and *hire_c*.

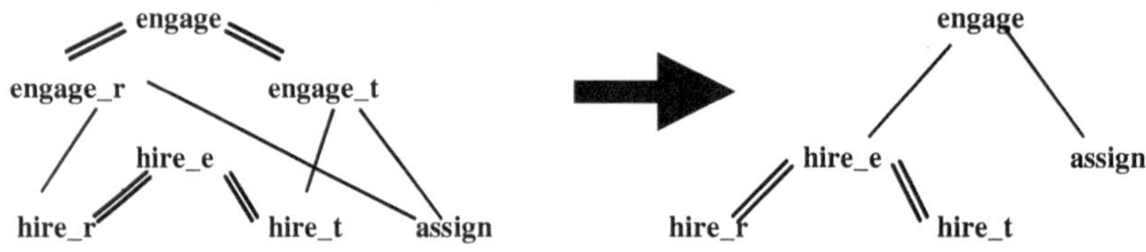

Fig. 6. Generic operation as component of composite operation

2. The second transformation replaces a generic operation, whose alternatives are composite operations differing by a single component, by one composite operation with a generic component. Assume that entry [A1,S1,G1] has a generic operation Og1, and that all the alternatives O1, O2, ... , On generalized by Og1 are composite operations with the same number m of components, of which m-1 involve the same operations, and that the order dependencies are analogous. So, each Oj alternative essentially differs from the others by only one component Oji. Let the set {O1i, O2i, ... , Oni} of the dissimilar components of the n alternatives correspond exactly to the set of alternatives under another generic operation Og2, contained in a separate entry [A2,S2,G2] of the *Index*. Then, transformation (2) can be applied to replace the entire contents of entry [A1,S1,G1] by one composite operation, keeping for convenience the same name as Og1. The components of the new Og1 will result from the most specific generalization of the components of O1, O2, ..., On, to be computed after replacing by Og2 the dissimilar component Oji of each Oj. Since employees (both regulars and trainees) cannot be fired as long as they stay associated with some client, they strive not simply to be hired but, as soon as possible, to become fully engaged through an assignment. The complex operation *engage* emerged at the first phase of the process with two specializations: *engage_r* and engage_t, for each kind of employee. As shown in Fig. 6, a considerably simpler structure results from adopting *hire_e* as a (generic) component of *engage*.
3. Finally, assume that entry [A1,S1,G1] has a composite operation Oc1, and entry [A2,S2,G2] has another composite operation Oc2, such that the entire set of components $\Sigma 2$ = {O1, O2, ... , Ok} of Oc2 is compatible, without imposing restrictions, with a proper subset $\sigma 1$ of the set $\Sigma 1$ of components of Oc1 (i.e. $\Sigma 2$ can be unified with $\sigma 1$, and the number of variables in $\sigma 1$ remains the same after unification). Suppose further that the order dependencies between the components of Oc2 are exactly those holding for those of subset $\sigma 1$ in the Oc1 definition. Transformation (3) then tries to replace, at [A1,S1,G1], all components of Oc1 in the $\sigma 1$ subset by a single component Oc2, and adjusts each order dependency [Oi-Oj] as follows, letting $\sigma 1^-$ be the complement subset $\Sigma 1$ - $\sigma 1$: (a) if both Oi and Oj are in $\sigma 1^-$, keep [Oi-Oj]; (b) if Oi is in $\sigma 1$ and Oj in $\sigma 1^-$, replace [Oi-Oj] by [Oc2-Oj]; (c) if Oi is in $\sigma 1^-$ and Oj in $\sigma 1$, replace [Oi-Oj] by [Oi-Oc2]; and (d) if both Oi and Oj are in $\sigma 1$, simply drop [Oi-Oj]. Cases (b) and (c) may obviously yield duplicates, which should be eliminated. A situation that is treated as an inconsistency, and

causes the transformation to fail, is the simultaneous presence of [Oc2-Op] and [Op-Oc2] among the resulting order dependencies, which will arise whenever there existed dependencies [Oq-Op] and [Op-Or], with both Oq and Or in σ1. For *Company Alpha*, to obtain a new client involves signing a contract and then providing assistance through the appointment of an employee. The chosen employee could be anyone previously hired or freed (by replace) from a previous assignment. But one possibly typical plan is to perform a new hiring specifically for immediate assignment to the new client, as indicated in composite operation obtain. On the other hand, the combination of hiring and assigning can be compactly expressed by composite operation *engage*. Fig. 7 shows a restructuring over the part-of hierarchy, which parallels what is done with the is-a hierarchy in Fig. 5.

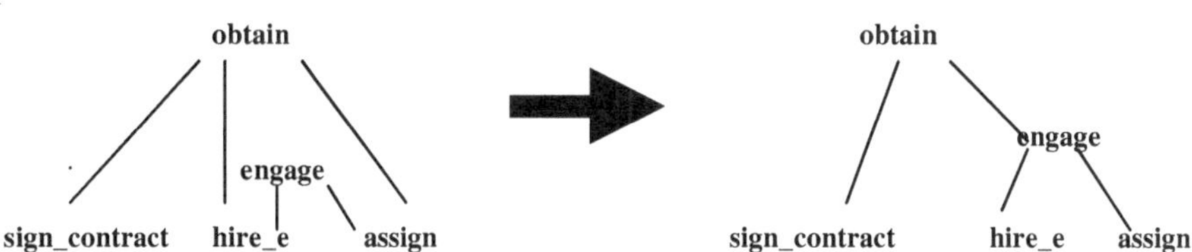

Fig. 7. Multi-level composition

4. Concluding Remarks

The present work is part of a larger research project, centred on the use of three-level schema specifications for a variety of purposes. For the simulation of possible futures in the mini-world of information systems — as well as for the interactive generation of plots of narratives, belonging to real-life or literary genres —, we developed a framework, formally described in [3]. In conformity with the framework, a prototype tool was implemented [2], called *Interactive Plot Generator (IPG)*, which utilizes logic programming and constraint programming features to conduct simulation experiments, supporting goal-inference rules of greater generality than those exemplified here. The architecture of *IPG* is displayed in Fig. 8. Rectangular boxes represent modules, all of which, except the *Rule Formulation* module, have been implemented. The *BLIB* tool introduced in this paper corresponds to the main constituent of the (shaded) *Library Construction* module.

Library construction terminates by extracting, from the *Index* built by *BLIB*, the clauses defining the complex operations and, after adding complementary information (especially the derived pre-conditions and the associated goals, which constitute the main effects of the operations), composing the *Library* in the exact format required for access by the other *IPG* modules.

Even though *BLIB* is fully operational, some of our decisions concerning its implementation were only tentative, and may be revised as we experiment with an ampler variety of examples, ideally adapting the tool to work directly with databases of a realistic size. The following points deserve special mention.

All extracted plans formed by more than one operation are discarded if the user does not choose to give to it the status of a named composite operation. One may, instead, prefer to keep record of such non-used plans for future analysis.

At the restructuring phase we limited ourselves to a few transformations which we found relatively safe. Other transformations that are both safe and useful may be identified in the future, leading, for instance, to complex operations that are generic and, at the same time, contain components (as found in [11], and even in our full version of *IPG*). Also, the preferred order of application of the transformations was chosen in view of how they are currently implemented, but a more systematic study of their mutual interactions is needed (and should be redone if new types of transformations are added).

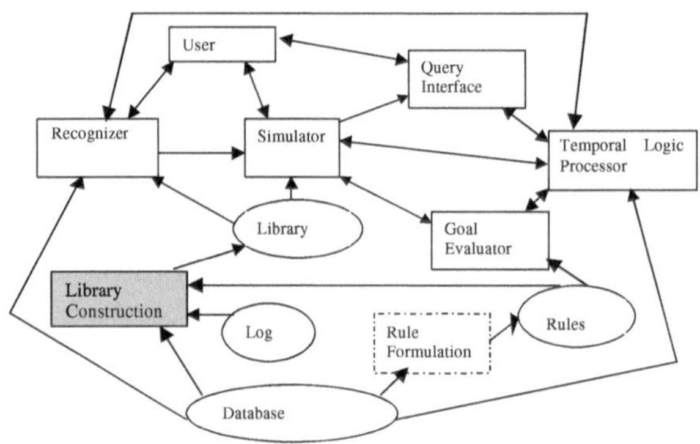

Fig. 8. General architecture of IPG

The present approach exhibits an *extensional*, rather than an *intensional* bent. For instance, for the first transformation we check whether the set of specialized operations of an *Index* entry [A1,S1,G1] currently contains the set of specialized operations of another entry [A2,S2,G2]. But, in principle, this may follow from an implication R2 → R1 relating the corresponding goal-inference rules, with the consequence that R1 = gi_rule(A1,S1,G1) would be more widely applicable than R2 = gi_rule(A2,S2,G2). This optional approach would involve a logical comparison of the two situation-goal pairs.

Turning to a broader issue, recall that we assumed the goal-inference rules available before the typical plans. One may find more convenient to adopt a different strategy, possibly trying to identify the typical plans on the basis of criteria not depending on the rules. Moreover, being "typical" carries a notion of frequent occurrence, which indicates that statistic measures should be incorporated, at least as a confirmation criterion. With the continuation of the project we intend, regardless of the preferred strategy for detecting typical plans, to focus our attention on goal analysis and on methods and tools for the discovery of goal-inference rules.

References

1. Batini, C., Ceri, S., Navathe, S.B.: Conceptual Database Design: an Entity-Relationship Approach. Benjamin-Cummings (1992)
2. Ciarlini, A.E.M., Furtado, A.L.: Simulating the Interaction of Database Agents. In: Proc. DEXA'99 Database and Expert Systems Applications Conference. Florence, Italy (1999)
3. Ciarlini, A.E.M., Veloso, P.A.S., Furtado, A.L.: A Formal Framework for Modelling at the Behavioural Level. In: Proc. The Tenth European-Japanese Conference on Information Modelling and Knowledge Bases. Saariselkä, Finland (2000)
4. Cohen, P.R., Levesque, H. J.: Intention is Choice with Commitment. Artificial Intelligence, 42. (1990) 213-261
5. Fikes, R.E., Nilsson, N. J.: STRIPS: A New Approach to the Application of Theorem Proving to Problem Solving. Artificial Intelligence , 2(3-4) (1971)
6. Fillmore, C.: The Case for Case. In: Bach, E., Harms, R. (eds.): Universals in Linguistic Theory. Holt, Rinehart and Winston (1968)
7. Furtado, A.L.: Analogy by Generalization and the Quest of the Grail. ACM/SIGPLAN Notices, 27, 1 (1992)
8. Furtado, A.L., Ciarlini, A.E.M.: Generating Narratives from Plots using Schema Information. In: Proc. NLDB'00 Applications of Natural Language to Information Systems. Versailles (2000)
9. Furtado, A. L., Neuhold, E. J.: Formal Techniques for Data Base Design. Springer-Verlag, Berlin (1986)
10. Herbst, J.: A Machine Learning Approach to Workflow Management. In: Proc. EMCL'2000 European Conference on Machine Learning. Barcelona, Spain (2000)
11. Kautz, H. A.: A Formal Theory of Plan Recognition and its Implementation. In: . Allen, J. F. et al (eds.): Reasoning about Plans.Morgan Kaufmann, San Mateo (1991)
12. Kolodner, J. L.: Case-Based Reasoning. Morgan Kaufmann, San Mateo (1993)
13. Kowalski, R., Sadri, F.: From Logic Programming towards Multi-Agent Systems. In: Annals of Mathematics and Artificial Intelligence, 25, 1-2. (1999) 391-419
14. Matheus, C.J., Chan, P.K., Piatesky-Shapiro, G.: Systems for Knowledge Discovery in Databases. IEEE Transactions on Knowledge and Data Engineering, 5, 6. (1993).
15. Meyer, J. J., Hoek, W., Linder, B.: A Logical Approach to the Dynamics of Commitments. Artificial Intelligence, 113 (1-2). (1999) 1-41
16. Miller, R., Shanahan, M.: Narratives in the Situation Calculus. Journal of Logic & Computation, 4, 5. (1994)
17. Mylopoulos, J., Castro, J.: Tropos: a Framework for Requirements-driven Software Development. In: Brinkkemper, S., Lindencrona, E., Solvberg, A. (eds.): Information Systems Engineering. Springer-Verlag, London. (2000)
18. Mylopoulos, J., Chung, L., Yu, E.: From Object-oriented to Goal-oriented Requirements Analysis. Communications of the ACM, 42, 1. (1999) 31-37
19. Ozsoyoglu, G., Snodgrass, R.T.: Temporal and Real-time Databases: a Survey. IEEE Transaction on Knowledge and Data Engineering, 7, 4. (1995).
20. Schank, R.C., Abelson, R.P.: Scripts, Plans, Goals and Understanding. Lawrence Erlbaum Associates, Hillsdale, NJ. (1977)
21. Sgouros, N. M.: Dynamic Generation, Management and Resolution of Interactive Plots. Artificial Intelligence, 107. (1999) 29-62
22. Yang, Q., Tenenberg, J., Woods, S.: On the Implementation and Evaluation of Abtweak. Computational Intelligence Journal, 12, 2. (1996) 295-318

The P2P Approach to Interorganizational Workflows

Wil M.P. van der Aalst and Mathias Weske

Department of Technology Management, Eindhoven University of Technology
P.O. Box 513, NL-5600 MB, Eindhoven, The Netherlands
{w.m.p.v.d.aalst, m.weske}@tm.tue.nl

Abstract. This paper describes in an informal way the Public-To-Private (P2P) approach to interorganizational workflows, which is based on a notion of inheritance. The approach consists of three steps: (1) create a common understanding of the interorganizational workflow by specifying a shared public workflow, (2) partition the public workflow over the organizations involved, and (3) for each organization, create a private workflow which is a subclass of the respective part of the public workflow. Using an example, we explain that the P2P approach yields an interorganizational workflow which is guaranteed to realize the behavior specified in the public workflow.

1 Introduction

In today's corporations, products and services are typically created by business processes, and workflow technology can be used for enhancing the flexibility and efficiency of these processes [14,19]. Corporations often operate across organizational boundaries, for example in E-commerce and extended enterprises [11,20,27]. Consequently, workflows between organizations – interorganizational workflows – are becoming increasingly important [21,12]. Interorganizational workflows are typically subject to conflicting constraints of the organizations involved. On the one hand, there is a strong need for coordination to optimize the flow of work in and between organizations. On the other hand, the organizations involved are essentially autonomous and have the freedom to create or modify workflows at any point in time. Some of the issues resulting from these conflicting goals will be tackled in this paper: We introduce the Public-To-Private (P2P) approach to interorganizational workflows which provides the means to specify a common public workflow, to partition it according to the organizations involved and to allow for private refinement of the parts by the organizations, based on a notion of inheritance. The P2P approach guarantees that the private workflows of the participating organizations (or, as we prefer to say, the domains) satisfy the public workflow as agreed upon; it consists of the following steps:

- *Step 1:* The organizations involved agree on a common public workflow, which serves as a contract between these organizations.

K.R. Dittrich, A. Geppert, M.C. Norrie (Eds.): CAiSE 2001, LNCS 2068, pp. 140–156, 2001.
© Springer-Verlag Berlin Heidelberg 2001

- *Step 2:* Each task of the public workflow is mapped onto one of the domains. Each domain is responsible for a part of the public workflow, referred to as its public part.
- *Step 3:* Each domain can now make use of its autonomy to create a private workflow. To satisfy the correctness of the overall interorganizational workflow, however, each domain may only choose a private workflow which is a subclass of its public part.

This paper introduces the P2P approach in an informal way, guided by an example of an electronic bookstore. The paper is structured according to the steps mentioned, and for each step concepts and notations are introduced when required; the complete definitions and the technical details of the proofs can be found in [4]. Sections 2 through 4 present the phases of the P2P approach, and Section 5 summarizes the main results. A discussion of related work and concluding remarks complete this paper.

2 Designing the Public Workflow (Step 1)

The example used throughout this paper is inspired by electronic bookstores such as Amazon [8] and Barnes and Noble [9]. In this section, we design the public workflow for ordering books. The scope of the workflow process includes the ordering, billing and shipping of books, involving the customer, the bookstore, the publisher, and the shipper.

The P2P approach uses *workflow nets* (WF-nets) [2] for modeling workflows, which are a specific form of Petri nets. In WF-nets, tasks are modeled by transitions, and causal dependencies are modeled by places and arcs. In fact, a place corresponds to a condition which can be used as pre- and/or post-condition for tasks. An AND-split corresponds to a transition with two or more output places, and an AND-join corresponds to a transition with two or more input places. OR splits/OR joins correspond to places with multiple outgoing/ingoing arcs. A WF-net has one source place and one sink place because any case (i.e., workflow instance) represented by the WF-net is created when it enters the workflow management system and is deleted once it is completely handled. An additional requirement is that there should be no dangling tasks or conditions, i.e., tasks and conditions which do not contribute to the processing of cases. Therefore, all the nodes of the workflow should be on some path from source to sink. WF-nets with these properties are called *sound* [1,2].

Figure 1 shows the public workflow N^{publ} of the electronic bookstore. This workflow can be regarded as a contract between the domains, i.e., the customer, the bookstore, the publisher, and the shipper. We stress that the public workflow does not necessarily show the way the tasks are actually executed; the real process may be much more detailed, and it may involve much more tasks. The public workflow only contains the tasks which are of interest to all parties. The public workflow shown in Fig. 1 is defined as a WF-net. While the mapping of the tasks to domains is only done in the next step, one can think of the tasks in the left column as performed by the customer, for instance the *place_c_order* task.

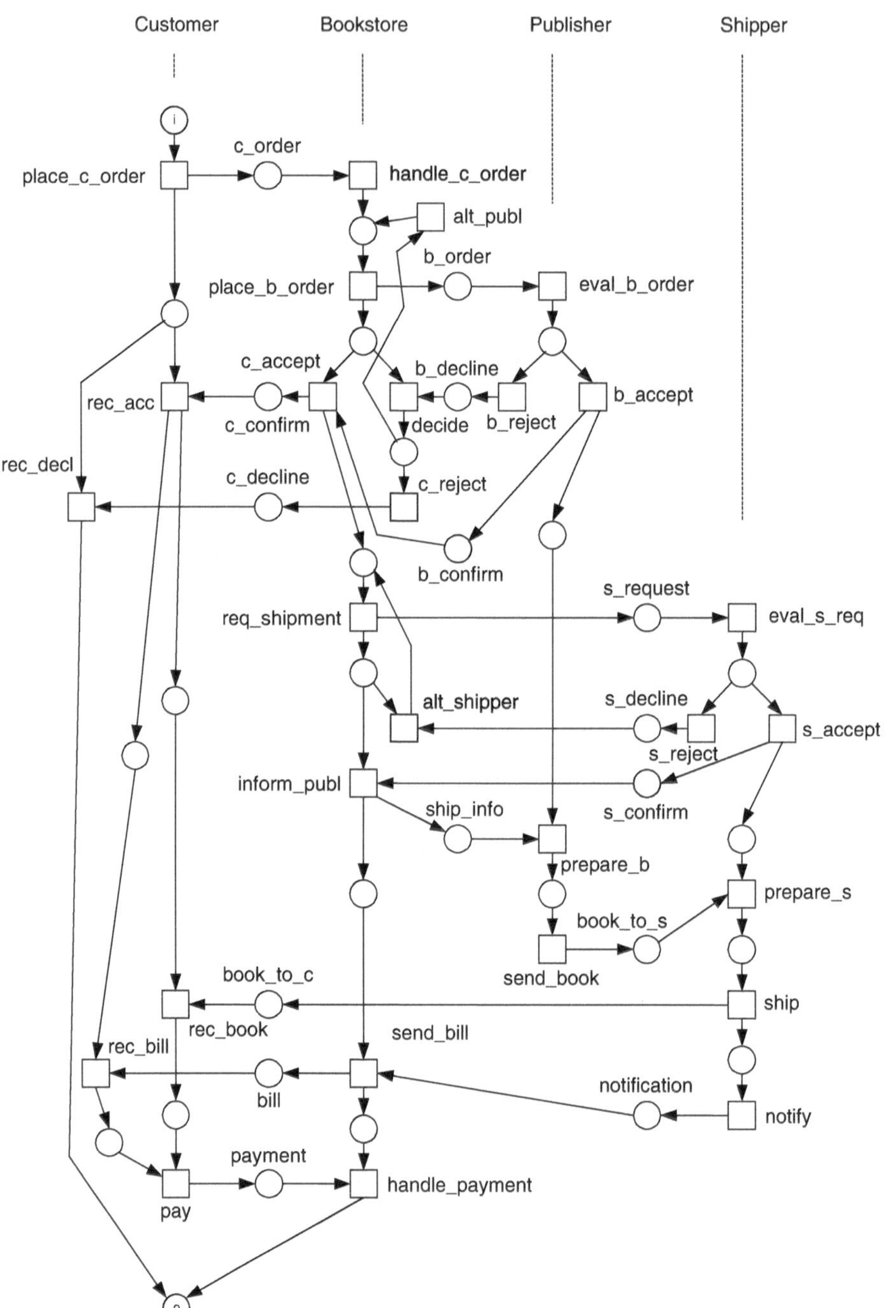

Fig. 1. The public workflow N^{publ}.

The next columns to the right belong to the bookstore (containing, e.g., the *handle_c_order* task to handle the customer order), the publisher (e.g., *eval_b_order*), and the shipper (e.g., *eval_s_req*), respectively.

The workflow process is initiated by a customer placing an order (represented by the task *place_c_order*). This customer order is sent to and is handled by the bookstore (*handle_c_order*). The electronic bookstore is a virtual company which has no books in stock. Therefore, the bookstore transfers the order of the desired book to a publisher (*place_b_order*). The bookstore order is evaluated by the publisher (*eval_b_order*) and either accepted (*b_accept*) or rejected (*b_reject*). In both cases an appropriate signal is sent to the bookstore. If the bookstore receives a negative answer, it decides (*decide*) to either search for an alternative publisher (*alt_publ*) or to reject the customer order (*c_reject*). If the bookstore searches for an alternative publisher, a new bookstore order is sent to another publisher, etc. If the customer receives a negative answer (*rec_decl*), then the workflow terminates. If the bookstore receives a positive answer (*c_accept*), the customer is informed (*rec_acc*), and the bookstore continues processing the customer order.

Once the order is confirmed, the bookstore sends a request to a shipper (*req_shipment*), the shipper evaluates the request (*eval_s_req*) and either accepts (*s_accept*) or rejects (*b_reject*) the shipping request. If the bookstore receives a negative answer, it searches for another shipper. This process is repeated until a shipper accepts. Note that, unlike the unavailability of the book, the unavailability of a shipper can not lead to a cancellation of the order. After a shipper is found, the publisher is informed (*inform_publ*), the publisher prepares the book for shipment (*prepare_b*), and the book is sent from the publisher to the shipper (*send_book*). The shipper prepares the shipment to the customer (*prepare_s*) and actually ships the book to the customer (*ship*). The customer receives the book (*rec_book*) and the shipper notifies the bookstore (*notify*). The bookstore sends the bill to the customer (*send_bill*). After receiving both the book and the bill (*rec_bill*), the customer makes a payment (*pay*). Then the bookstore processes the payment (*handle_payment*) and the interorganizational workflow terminates.

The public workflow shown in Fig. 1 is indeed a sound WF-net, since it has exactly one input place and one output place, at the moment when the workflow reaches the output place, all tasks have completed, and there are no dead transitions, i.e., all tasks of the WF-net are in fact reachable during workflow executions.

3 Partitioning the Public Workflow (Step 2)

In the second step of the P2P approach, the public workflow is partitioned according to the domains, and the public parts are related to each other, making up an interorganizational workflow. An interorganizational workflows is defined by an *interorganizational workflow net* (IOWF-net). An IOWF-net consists of a set of WF-nets, a set of channels, a set of methods, and a channel flow relation.

In our example, the public workflow is partitioned over four domains: the customer domain, the bookstore domain, the publisher domain, and the shipper domain, as shown in Fig. 2. Methods of the domains are represented by shaded boxes, and they are linked to channels by the channel flow relation, which is represented by arrows. In Fig. 2, the public parts of the customer, the bookstore, the publisher and the shipper are represented by boxes N_C^{part}, N_B^{part}, N_P^{part}, and N_S^{part}, respectively. Channels are represented by icons, and the channel flow relation represented by arrows specifies the linkage of the domains. For example, the *c_order* channel and the attached arrows represent the fact that customer order information flows from the customer domain to the bookstore domain, while the confirmation of the order flows in opposite direction, making use of channel *c_confirm*.

Based on this description it is clear how the public workflow needs to be partitioned. The public part of the customer domain is quite simple (cf. Fig. 3): The customer first places an order, using the method *place_c_order*. Then either the order is accepted, the book and the bill are received and the bill is paid, or the order is declined. Notice that for each transition in the WF-net, there is a method linked to it by a dotted line, representing the actual function which is invoked when the task is executed.

The public part of the bookstore workflow is slightly more complex (cf. Fig. 4): After the order arrives, the bookstore checks for a publisher ready for providing the ordered book. If no publisher can be found, the order is rejected. Otherwise, shipment is requested from a shipper, and payment is handled. The public parts of the publisher and shipper workflow are shown in Fig. 5.

The IOWF-net is a high-level representation of the domains and their dependencies; its semantics are given in terms of a labeled P/T net. A IOWF-net is transformed into a labeled P/T net by taking the union of all WF-nets, adding a place for each channel, connecting transitions to these newly added places, and removing superfluous source and sink places. We call this the flattening of the interorganizational workflow. As shown in [4], we can easily make sure that the partitioning is valid, i.e., all public parts are sound WF-nets and there is no multiple activation. We mention that the flattened IOWF-net equals the public workflow. Hence, flattening the interorganizational workflow shown in Fig. 2 results in the public workflow shown in Fig. 1.

4 Designing the Private Workflows (Step 3)

After partitioning the public workflow, each domain can realize the corresponding public part of the interorganizational workflow in any way they want, as long as they make sure that their private workflow is a subclass of their public part.

The subclass relationship between WF-nets is based on a specific notion of inheritance, called *projection inheritance*. Projection inheritance has been defined in [6,10] and uses encapsulation as a mechanism to establish subclass-superclass relationships. The basic idea of projection inheritance can be characterized as follows:

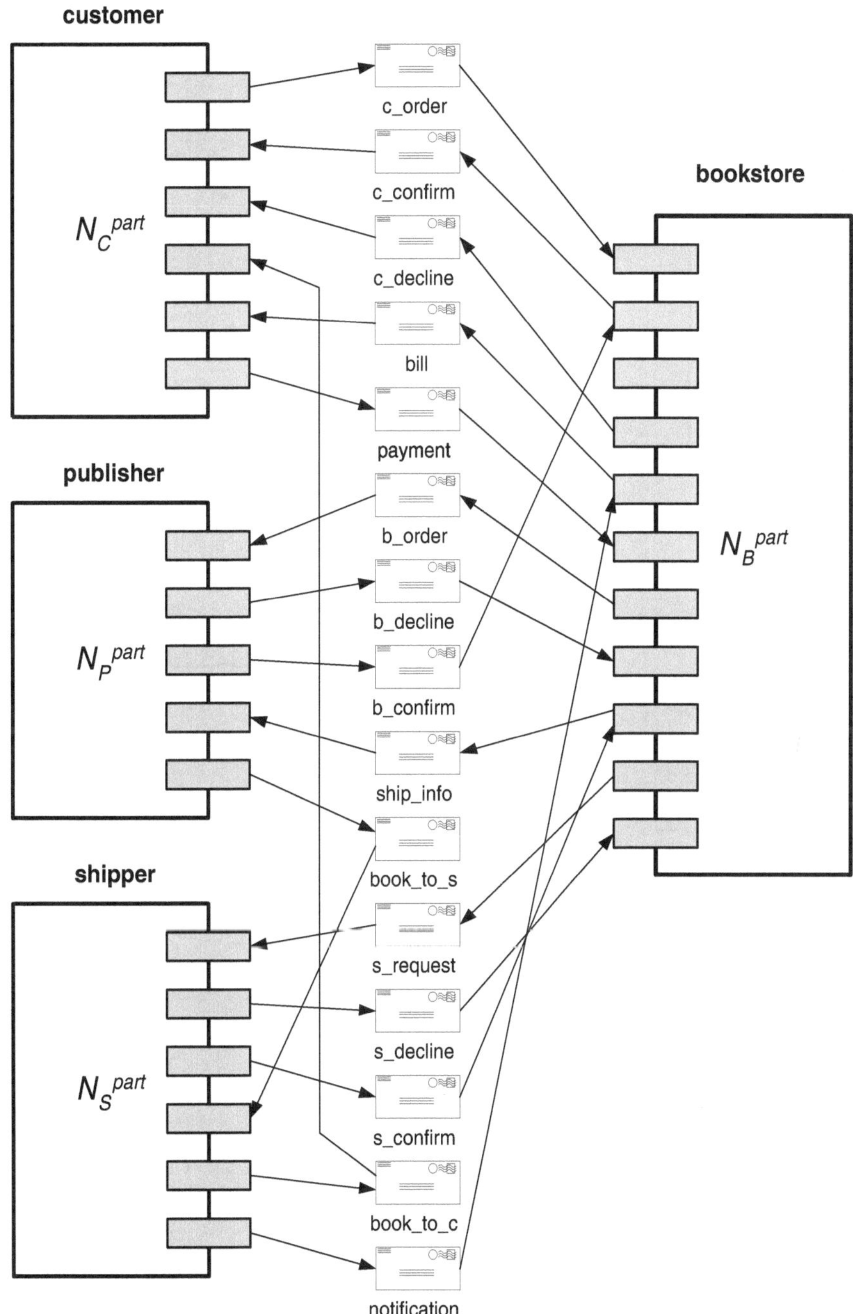

Fig. 2. The interorganizational workflow Q^{part}

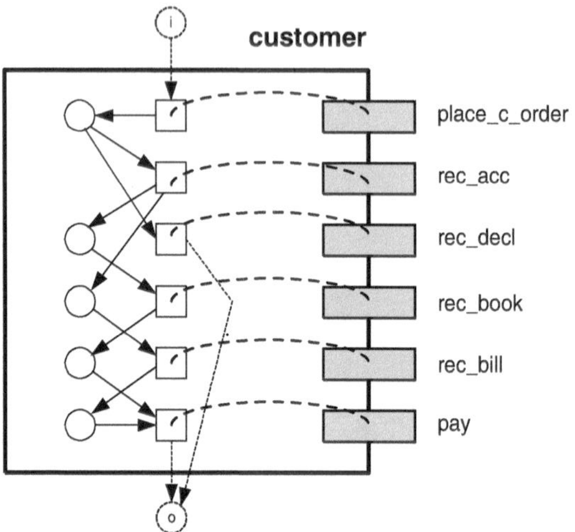

Fig. 3. The WF-net N_C^{part} (public part of the customer domain).

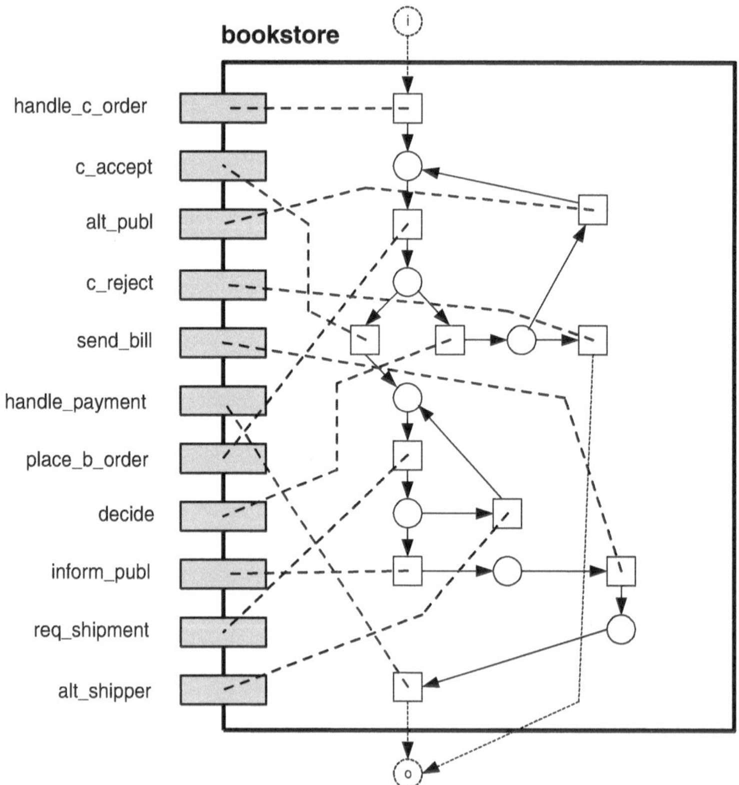

Fig. 4. The WF-net N_B^{part} (public part of the bookstore domain).

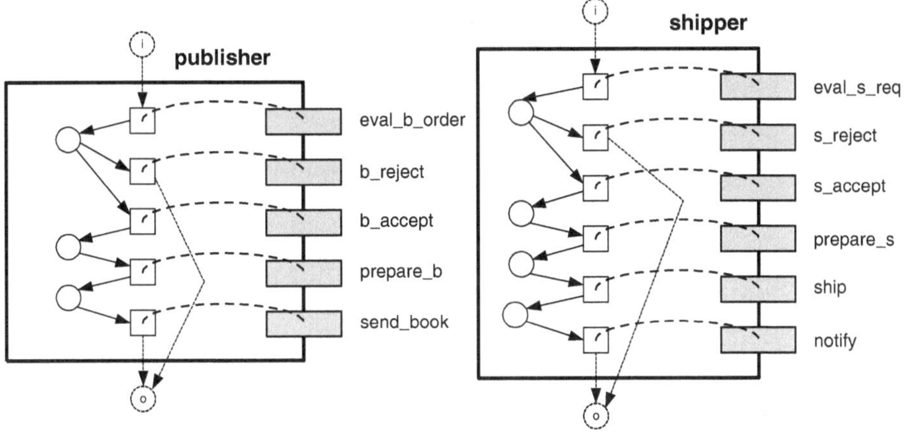

Fig. 5. The WF-net N_P^{part} (public parts of the publisher and shipper domains).

If it is not possible to distinguish the behaviors of x and y when arbitrary methods of x are executed, but when only the effects of methods that are also present in y are considered, then x is a subclass of y.

Projection inheritance is based on branching bisimilarity as the standard equivalence relation on marked, labeled P/T-nets [15]. For projection inheritance, all new methods (i.e., methods added in the subclass) are hidden; an abstraction operator τ is used to hide methods.

For any two sound WF-nets N and N', N' is a subclass of N under projection inheritance if and only if the externally visible behavior of N' is branching bisimilar to N. Let us consider the five WF-nets shown in Fig. 6. N_1 is not a subclass of N_0 because hiding of the new task d results in a potential execution where a is followed by c without executing b, i.e., the WF-net where d is hidden is not branching bisimilar. N_2 is a subclass of N_0 because hiding e in N_2 results in a behavior equivalent to the behavior of N_0, i.e., the addition of e only postpones the execution of b and does not allow for a bypass such as the one in N_1. N_3 is also a subclass of N_0: Hiding the parallel branch containing f yields the original behavior. Finally, N_4 is also a subclass of N_0.

Based on the notion of projection inheritance we have defined three *inheritance-preserving transformation rules.* These rules correspond to design patterns when extending a superclass to incorporate new behavior: (1) adding a loop (rule *PPS*), (2) inserting methods in-between existing methods (rule *PJS*), and (3) putting new methods in parallel with existing methods (rule *PJ3S*). The formal definitions of these transformation rules, their preconditions, and the proofs that these rules actually preserve projection inheritance are given in [6,10].

In the P2P approach, projection inheritance is used as a formal link between the public parts of the domains and the private workflows which are actually executed. Transformation rules are the key mechanism to create specializations of a given WF-net, making use of the fact that applying these rules to a given

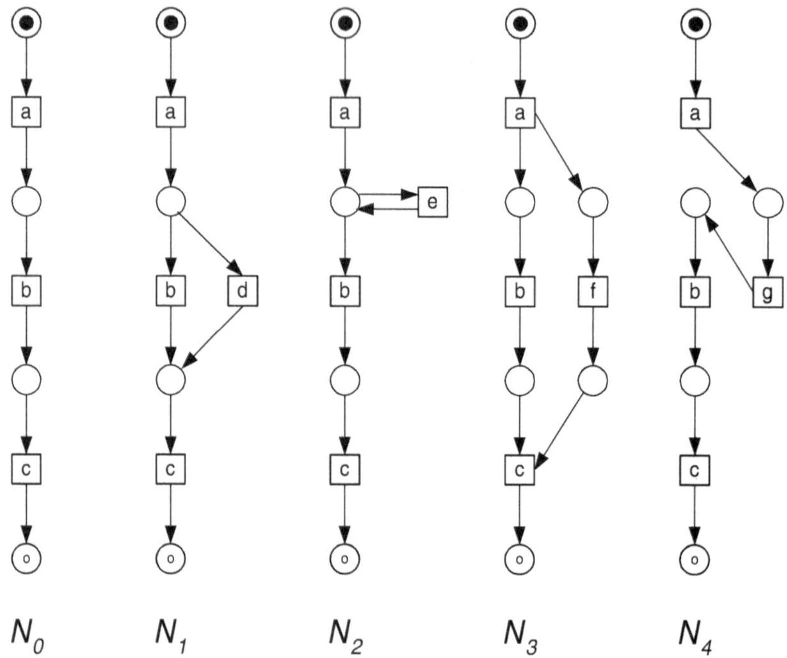

Fig. 6. N_2, N_3, and N_4 are subclasses of N_0 under projection inheritance.

WF-net is guaranteed to create a subclass of that WF-net. Hence, the P2P approach is constructive in the sense that any modification applied to a WF-net via transformation rules *PPS*, *PJS*, and *PJ3S* yields a subclass of the WF-net.

Figure 7 shows the private workflow of the bookstore. Five new tasks, i.e., tasks not present in the public workflow, have been added. After the customer order is handled, the customer profile (information about the interests of the customer) is updated (*update_customer_profile*). This task is executed in parallel with the placement of the bookstore order. After both tasks have been executed, the marketing department is informed (*inform_marketing*). The tasks *monitor_order*, *monitor_shipment*, and *monitor_payment* have been added to monitor the behavior of the publisher, shipper, and customer. The task *monitor_order* can be executed as long as the bookstore is waiting for a response of the publisher. The task *monitor_shipment* can be executed between the moment the publisher is informed and the moment the shipper sends a notification. The task *monitor_payment* can be executed after the bill is sent to the customer. Note that each of the monitor tasks can be executed multiple times. For example, the bookstore checks every week whether the customer has paid and if needed takes action, e.g., sending a bailiff.

We now show by construction that the private workflow N_B^{priv} (Fig. 7) is indeed a subclass of N_B^{part} (Fig. 4): tasks *monitor_order*, *monitor_shipment*, and *monitor_payment* can be added by applying transformation rule *PPS* three times;

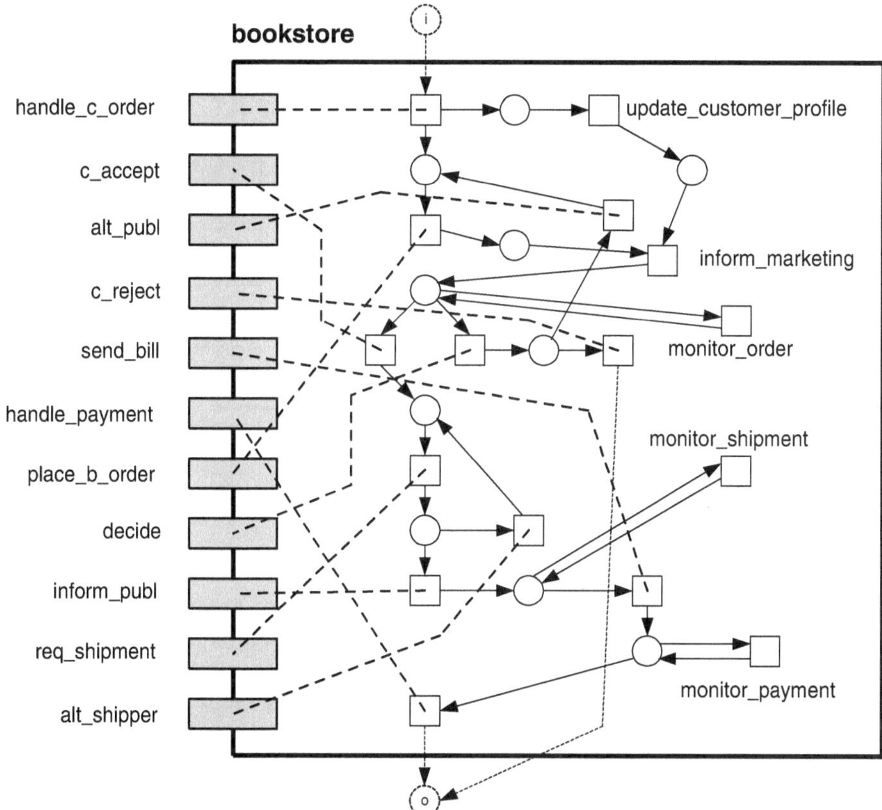

Fig. 7. The WF-net N_B^{priv} (private workflow of the bookstore domain).

task *inform_marketing* can be added using transformation rule *PJS*. To complete the construction, the task *update_customer_profile* can be added using transformation rule *PJ3S*.

Similarly, the private workflow of the publisher (Fig. 8) has been created by applying transformation rule *PJS* to the public part: The task *check_warehouse* has been added in-between the receipt of the order and the decision. In fact, the decision is based on the result of *check_warehouse*. After accepting the order of the bookstore, the corresponding inventory item is locked (*lock_inventory*), the stock is replenished (if possible) (*replenish*), and the book is moved to the part of the warehouse reserved for books which are waiting for shipment (*move_book_to_release_buffer*). It is easy to verify that the private workflow N_P^{priv} is a subclass of N_P^{part}, using the transformation rule *PJS*.

Figure 9 shows the private workflow of the shipper. Using the transformation rules, six new tasks have been added: Task *check_availability_trucks* is executed after the request by the bookstore is received. Based on this task the request is accepted or rejected. Tasks *update_file* and *quality_control* are executed in

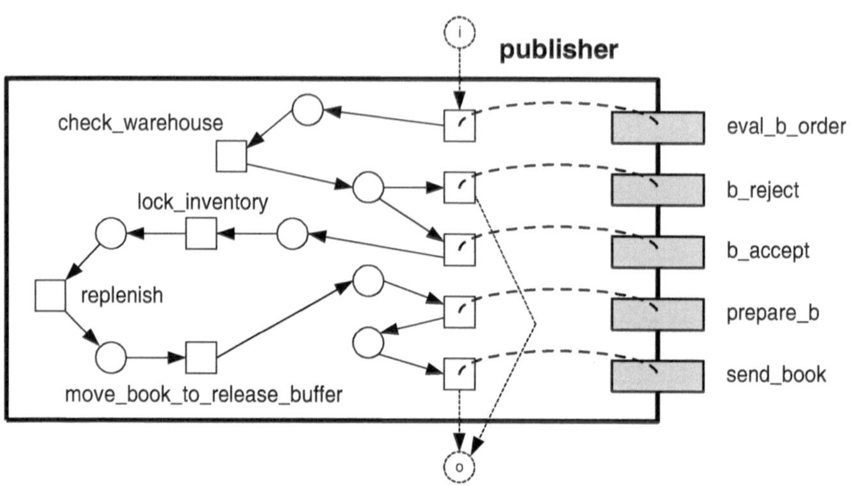

Fig. 8. The WF-net N_P^{priv} (private workflow of the publisher domain).

parallel with the preparation and shipment tasks. After preparation, shipments are assigned to trucks (*assignment*). Based on the assignment, the routing of the truck is determined (*routing*). In-between tasks *assignment* and *routing* the task *re-assignment* can be executed multiple times. Again it is easy to verify that the WF-net shown in Fig. 9 is indeed a subclass of the one shown in Fig. 5. Task *check_availability_trucks* can be added using transformation rule PJS. Tasks *update_file* and *quality_control* can be added using transformation rule $PJ3S$. Tasks *assignment*, *re-assignment*, and *routing* can be added using transformation rule PJS. Note that it is also possible to first add tasks *assignment* and *routing* using PJS, and then add task *re-assignment* using transformation rule PPS.

The design of the interorganizational workflow involving a customer, bookstore, publisher, and shipper presented in this paper is a simplification of the real process. In the real process customers can order multiple books at the same time, the customer can return books, the customer can refuse to pay, etc. One can imagine that for realistic interorganizational workflows where the public part consists of more than fifty tasks and the overall workflow consists of hundreds of tasks, a structured approach is needed to avoid all kinds of anomalies. In our opinion, the P2P approach could be used as a starting point for a more comprehensive approach which also deals with other aspects such as data and security.

5 Summary and Main Results

To summarize the P2P approach, in the first step the public workflow is specified in terms of a sound WF-net; it serves as a contract between the business partners involved. In the second step, the public workflow is partitioned over the set of domains. Note that each domain corresponds to an organizational entity. As a

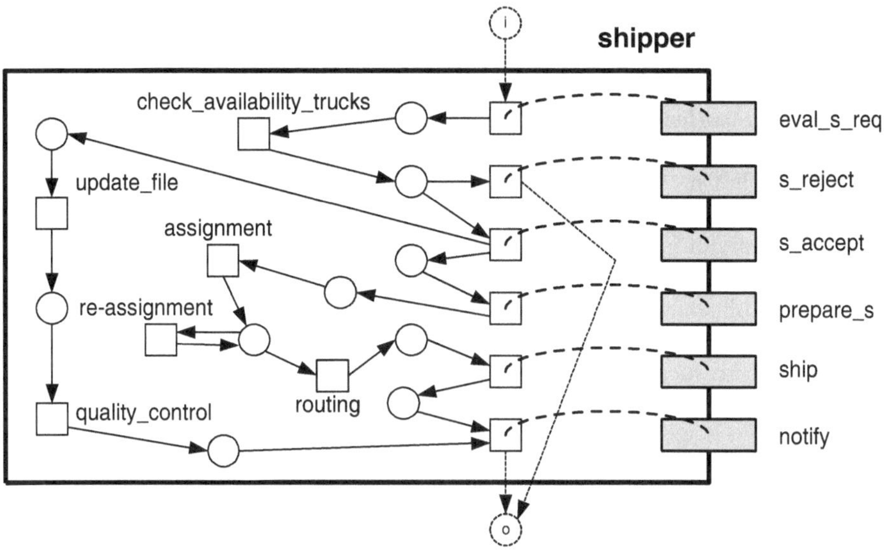

Fig. 9. The WF-net N_S^{priv} (private workflow of the shipper domain).

result of the partitioning, each fragment of the partitioned workflow corresponds to one of the domains and is represented by a sound WF-net, called public part. In the final step, the public parts are replaced by private workflows. Each private workflow corresponds to an actual workflow as it is executed in one of the domains. The P2P approach guarantees that each private workflow is a subclass of the corresponding public part under projection inheritance. It is important to note that the P2P approach is constructive: By applying the three transformation rules introduced above, the design is guaranteed to be correct without the need to check whether each private workflow is actually a subclass of the corresponding public part.

Following the general tone of this paper, we explain the main results informally and introduce concepts if and when required. Please refer to [4] for a detailed theoretical discussion. The first result concerns the *overall workflow*, which consists of all private workflows of the participating domains.

Result 1: The overall workflow is a sound WF-net.

This property is based on the observation that a part of a WF-net (called subflow) can be replaced by a specialization (i.e., a subclass subflow) without endangering soundness of the overall workflow. This result is proven in [4], based on a theorem which shows the compositionality of projection inheritance. From an application point of view, Result 1 makes sure that the P2P approach guarantees that the overall workflow is free of deadlocks and other anomalies.

Result 2: The overall workflow is a subclass of the public workflow.

This result shows that the dynamic behavior of the interorganizational workflow which the business partners agreed upon in the public workflow is in fact guaranteed to be satisfied by the execution of the interorganizational workflow, i.e., the overall workflow. From an application point of view, this is an important result, since it provides the business partners with the ability to perform any private modifications to their public workflow part, as long as the subclass relationship holds. Transformation rules are used for this purpose. Hence, an organization can be sure that its private workflow indeed satisfies the requirements specified in the contract, i.e., the public workflow.

The next result is based on the notion of *local views* of the domains. To introduce local views, we mention that each domain is aware of its private workflow and of the public parts of the other domains. The information which each domain has with respect to the overall workflow is called the local view of that domain. With respect to local views, the following interesting result can be obtained, which stresses the soundness of the P2P approach.

> *Result 3:* The overall workflow is a subclass of the local views of all domains, which in turn are subclasses of the public workflow.

For the final two properties we have to introduce some notation. Since projection inheritance is a partial ordering on the set of WF-nets, the Greatest Common Denominator (GCD) and the Least Common Multiple (LCM) can be defined. GCD and LCM are general concepts that apply to any ordering, and there are different applications of these concepts in the context of WF-nets, as described in more detail in [6]. In essence, the GCD of a set of WF-nets is a WF-net that captures the part these nets have in common, i.e., the part where they agree on. The LCM captures all possible behaviors. Note that projection inheritance is a partial order but not a lattice. Therefore, suitable definitions of GCD and LCM are far from trivial but can be defined as is shown in [6].

For an illustration of these concepts, consider the WF-nets N_0, N_2, N_3, and N_4 shown in Fig. 6. The GCD of these four nets is N_0, i.e., each of the four WF-nets is a subclass of this net and it is not possible to find a different WF-net which is also a superclass of N_2, N_3, and N_4 and at the same time a subclass of N_0. Figure 10 shows $N_{GCD} = N_0$ as the GCD of N_0, N_2, N_3, and N_4. Figure 10 also shows the WF-net N_{LCM}. N_{LCM} is a subclass of each of the four nets considered. Moreover, it is not possible to find a different WF-net which is also a subclass of N_0, N_2, N_3, and N_4 and at the same time a superclass of N_{LCM}. Any execution sequence generated by one of the four nets can also be generated by N_{LCM} after the appropriate abstraction. Based on the characterization of GCD and LCM we are now ready to present the following result:

> *Result 4:* The GCD of all local views is the public workflow.

The application specific interpretation of this result is as follows: The public workflow is the superclass of the local views of all domains, and it is minimal in the sense that no different WF-net can be found, which is a superclass of the local views and at the same time a subclass of the public workflow. This is an interesting, yet not surprising result. It shows that the local views of the domains have exactly the public workflow in common.

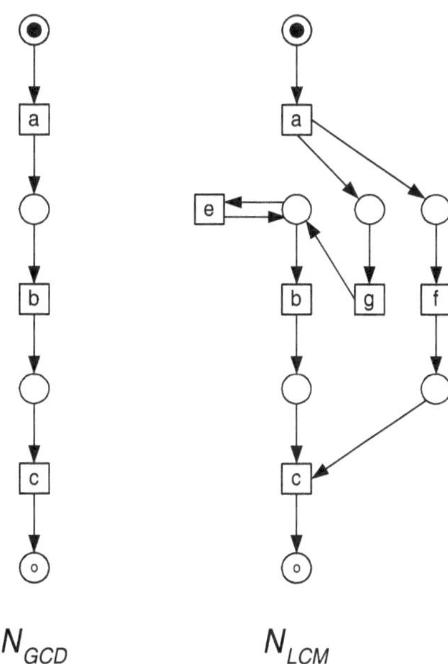

Fig. 10. The greatest common divisor N_{GCD} and least common multiple N_{LCM} of N_0, N_2, N_3, and N_4 shown in Fig. 6.

Analogously to the discussion of Result 4, the final result states a relationship between the local views of the domains and the overall workflow, as it is executed:

Result 5: The LCM of all local views is the overall workflow.

We interpret Result 5 as follows: The overall workflow is a specialization of all local views; conversely, the local views are superclasses of the overall workflow. The overall workflow is minimal in the sense that it is not possible to find a different WF-net which is also a subclass of all local views and which is a superclass of the overall workflow.

6 Related Work and Conclusions

Petri nets have been proposed for modeling workflow process definitions long before the term "workflow management" was coined and workflow management systems became readily available. Consider for example the work on Information Control Nets, a variant of the classical Petri nets, in the late seventies [13].

Only a few papers in the literature focus on the verification of workflow process definitions. In [16] some verification issues have been examined and the complexity of selected correctness issues has been identified, but no concrete verification procedures have been suggested. In [1] and [7] concrete verification procedures based on Petri nets have been proposed. This paper builds upon the work presented in [1] where the concept of a sound WF-net was introduced. The technique presented in [7] has been developed for checking the consistency of transactional workflows including temporal constraints. However, the technique is restricted to acyclic workflows and only gives necessary conditions (i.e., not sufficient conditions) for consistency. In [23] a reduction technique has been proposed. This reduction technique uses a correctness criterion which corresponds to soundness and the class of workflow processes considered are in essence acyclic free-choice Petri nets.

This paper differs from the above approaches because the focus is on interorganizational workflows. Only a few papers explicitly focus on the problem of verifying the correctness of interorganizational workflows [3,17]. In [3] the interaction between domains is specified in terms of message sequence charts and the actual overall workflow is checked with respect to these message sequence charts. A similar, but more formal and complete, approach is presented by Kindler, Martens, and Reisig in [17]. The authors give local criteria, using the concept of scenarios (similar to runs or basic message sequence charts), to guarantee the absence of certain anomalies at the global level. Both approaches [3,17] are not constructive, i.e., they only specify criteria for various notions of correctness but do not provide concrete design rules such as the transformation rules.

In the last decade several researchers explored notions of behavioral inheritance (also named subtyping or substitutability), see [10] for an overview. Researchers in the domain of formal process models (e.g., Petri-nets and process algebras) have tackled similar questions based on the explicit representation of a process by using various notions of (bi)simulation . The inheritance notion used in this paper is characterized by the fact that it is equipped with both *inheritance-preserving transformation rules* to *construct* subclasses [10] and *transfer rules* to *migrate* instances from a superclass to a subclass and vice versa [6]. These features are very relevant for a both constructive and robust approach towards interorganizational workflows.

We have developed a tool named *Woflan* (WOrkFLow ANalyzer [2,28]). Woflan is an analysis tool which can be used to verify the correctness of a workflow process definition. The analysis tool uses state-of-the-art techniques to find potential errors in the definition of a workflow process. Woflan is designed as a workflow management system independent analysis tool. In principle it can interface with many workflow management systems. At the moment, Woflan can interface with the workflow management systems COSA (Software Ley [25]), METEOR (LSDIS [24]), Staffware (Staffware [26]), and with the business process re-engineering tool Protos (Pallas Athena [22]). Woflan has not been designed to analyze interorganizational workflows. However, it can be used to verify the soundness property used throughout this paper, and it can also check whether a given workflow is a subclass of another workflow.

In the future we hope to extend the P2P approach in several directions. First of all, we want to address local dynamic changes. The transfer rules presented in [6] can be used to migrate workflow instances from a superclass to a subclass and vice versa. Therefore, it is possible to change the workflows in each of the domains on the fly, i.e., it is possible to automatically transfer each case to the latest version of the process. Other aspects of future work include the reconfiguration of interorganizational workflows (tasks move from one domain to another), the usage of alternative inheritance notions and the implementation of the concepts in prototypical workflow management systems, e.g., by using METEOR [5,24] or InterProcs [18].

References

1. W.M.P. van der Aalst. Verification of Workflow Nets. In P. Azéma and G. Balbo, editors, *Application and Theory of Petri Nets 1997*, volume 1248 of *Lecture Notes in Computer Science*, pages 407–426. Springer-Verlag, Berlin, 1997.
2. W.M.P. van der Aalst. The Application of Petri Nets to Workflow Management. *The Journal of Circuits, Systems and Computers*, 8(1):21–66, 1998.
3. W.M.P. van der Aalst. Interorganizational Workflows: An Approach based on Message Sequence Charts and Petri Nets. *Systems Analysis - Modelling - Simulation*, 34(3):335–367, 1999.
4. W.M.P. van der Aalst. Inheritance of Interorganizational Workflows: How to Agree to Disagree Without Loosing Control? BETA Working Paper Series, WP 46, Eindhoven University of Technology, Eindhoven, 2000.
5. W.M.P. van der Aalst and K. Anyanwu. Inheritance of Interorganizational Workflows to Enable Business-to-Business E-commerce. In *Proceedings of the Second International Conference on Telecommunications and Electronic Commerce (ICTEC'99)*, pages 141–157, Nashville, Tennessee, October 1999.
6. W.M.P. van der Aalst and T. Basten. Inheritance of Workflows: An approach to tackling problems related to change. *Theoretical Computer Science*, 2001 (to appear).
7. N.R. Adam, V. Atluri, and W. Huang. Modeling and Analysis of Workflows using Petri Nets. *Journal of Intelligent Information Systems*, 10(2):131–158, 1998.
8. Amazon.com, Inc. Amazon.com. http://www.amazon.com, 1999.
9. Barnes and Noble. bn.com. http://www.bn.com, 1999.
10. T. Basten. *In Terms of Nets: System Design with Petri Nets and Process Algebra*. PhD thesis, Eindhoven University of Technology, Eindhoven, The Netherlands, December 1998.
11. R. Benjamin and R. Wigand. Electronic markets and virtual value chains on the information superhighway. *Sloan Management Review*, pages 62–72, 1995.
12. R.W.H. Bons, R.M. Lee, and R.W. Wagenaar. Designing trustworthy interorganizational trade procedures for open electronic commerce. *International Journal of Electronic Commerce*, 2(3):61–83, 1998.
13. C.A. Ellis. Information Control Nets: A Mathematical Model of Office Information Flow. In *Proceedings of the Conference on Simulation, Measurement and Modeling of Computer Systems*, pages 225–240, Boulder, Colorado, 1979. ACM Press.
14. D. Georgakopoulos, M. Hornick, and A. Sheth. An Overview of Workflow Management: From Process Modeling to Workflow Automation Infrastructure. *Distributed and Parallel Databases*, 3:119–153, 1995.
15. R.J. van Glabbeek and W.P. Weijland. Branching Time and Abstraction in Bisimulation Semantics. *Journal of the ACM*, 43(3):555–600, 1996.
16. A.H.M. ter Hofstede, M.E. Orlowska, and J. Rajapakse. Verification Problems in Conceptual Workflow Specifications. *Data and Knowledge Engineering*, 24(3):239–256, 1998.

17. E. Kindler, A. Martens, and W. Reisig. Inter-Operability of Workflow Applications: Local Criteria for Global Soundness. In W.M.P. van der Aalst, J. Desel, and A. Oberweis, editors, *Business Process Management: Models, Techniques, and Empirical Studies*, volume 1806 of *Lecture Notes in Computer Science*, pages 235–253. Springer-Verlag, Berlin, 2000.
18. R.M. Lee. Distributed Electronic Trade Scenarios: Representation, Design, Prototyping. *International Journal of Electronic Commerce*, 3(2):105–120, 1999.
19. F. Leymann and D. Roller. *Production Workflow: Concepts and Techniques.* Prentice-Hall PTR, Upper Saddle River, New Jersey, USA, 1999.
20. T.W. Malone, R.I. Benjamin, and J. Yates. Electronic Markets and Electronic Hierarchies: Effects of Information Technology on Market Structure and Corporate Strategies . *Communications of the ACM*, 30(6):484–497, 1987.
21. M. Merz, B. Liberman, K. Muller-Jones, and W. Lamersdorf. Interorganisational Workflow Management with Mobile Agents in COSM. In *Proceedings of PAAM96 Conference on the Practical Application of Agents and Multiagent Systems*, 1996.
22. Pallas Athena. *Protos User Manual.* Pallas Athena BV, Plasmolen, The Netherlands, 1999.
23. W. Sadiq and M.E. Orlowska. Applying Graph Reduction Techniques for Identifying Structural Conflicts in Process Models. In *Proceedings of the 11th International Conference on Advanced Information Systems Engineering (CAiSE '99)*, volume 1626 of *Lecture Notes in Computer Science*, pages 195–209. Springer-Verlag, Berlin, 1999.
24. A. Sheth, K. Kochut, and J. Miller. Large Scale Distributed Information Systems (LSDIS) laboratory, METEOR project page. http://lsdis.cs.uga.edu/proj/meteor/meteor.html.
25. Software-Ley. *COSA User Manual.* Software-Ley GmbH, Pullheim, Germany, 1998.
26. Staffware. *Staffware 2000 / GWD User Manual.* Staffware plc, Berkshire, United Kingdom, 1999.
27. The White House. A Framework for Global Electronic Commerce. http://www.ecommerce.gov/framewrk.htm, 1997.
28. H.M.W. Verbeek, T. Basten, and W.M.P. van der Aalst. Diagnosing Workflow Processes using Woflan. Computing Science Report 99/02, Eindhoven University of Technology, Eindhoven, 1999.

Relaxed Soundness of Business Processes

Juliane Dehnert[1,*] and Peter Rittgen[2]

[1] Institute of Computer Information Systems, Technical University Berlin, Germany
dehnert@cs.tu-berlin.de

[2] Institute of Business Informatics, University Koblenz-Landau, Germany
rittgen@uni-koblenz.de

Abstract. Business processes play a central role in the reorganization of a company and the (re)design of the respective information system(s). Typically the processes are described with the help of a semiformal, graphical language such as the Event-driven Process Chains (EPCs) by Scheer. This approach provides a suitable medium for the communication between the participants: the domain experts and the IT specialists. But these models leave room for interpretation and hence ambiguities which makes them less suitable as a basis for the *design* of information systems. To remedy this we suggest to transform the EPCs into a formal representation (Petri nets) *preserving the ambiguities*, i.e. all possibly intended behaviour. Now formal techniques can be used to find out whether the *possible* behaviours comprise *sensible* behaviour. If so, we call the net *relaxed sound*. By not limiting the modeler compared to previous ways (e.g. [8], [3]) we take a pragmatic approach to correctness which only requires that the net represents *some* valid behaviour. This allows us to draw conclusions on mistakes in the original EPC and to make suggestions for its improvement thereby enhancing both the model's quality and its suitability for software engineering.

1 Motivation

Business processes play a central role in the reorganization of a company and the (re)design of the respective information system(s). Typically the processes are described with the help of a semiformal, graphical language such as the Event-driven Process Chains (EPCs) by Scheer [15]. Approaches of this type are suitable for the analysis phase of an IT project where the focus is on communication: reaching an agreement on how the process should look like between participants with totally different backgrounds and "knowledge cultures": CEOs, heads of department, department staff, IT experts and so on. In this phase it is imperative that the language used represents the greatest common denominator of the people involved. And more than that it should leave room for interpretation: the more ways there are to interpret a certain construct the more likely it is that an agreement is reached. The participants might not (yet) be ready to specify the "final" behaviour in detail and decide for the "correct" interpretation. But although this feature is desirable in the analysis phase of IS development it constitutes a major problem in the design phase where we need an

* This work is supported by Deutsche Forschungsgemeinschaft (reference WE 1214-3-3a, research group Petri Net Technology)

K.R. Dittrich, A. Geppert, M.C. Norrie (Eds.): CAiSE 2001, LNCS 2068, pp. 157–170, 2001.
© Springer-Verlag Berlin Heidelberg 2001

unambiguous description of the process. To remedy this problem we suggest to transform the EPCs into a formal representation (Petri nets) while *preserving the ambiguities*, i.e. all possibly intended behaviour. For this we use the workflow nets by van der Aalst based on which we can now employ established formal techniques to determine the correctness of the process in a pragmatic fashion: we analyse the net and try to find, among all the possible ways of behaviour, only the ones that are suitable, i.e. where the final state can be reached from the start state so that nothing blocks or remains undone. If enough of such execution paths exists we are satisfied and call the respective net *relaxed sound*. The execution paths leading to undesired behaviour can then be used to infer potential mistakes in the original EPC and to suggest possible improvements. An alternative approach might be to reduce the net to the well-behaved paths automatically and to use the resulting net directly as a basis for the design assuming that the EPC is of sufficient quality or a revision to costly. If the workflow net is not relaxed sound we can still do something: we can point out the parts of the EPC that lead to the net not being relaxed sound so that the modeler(s) know(s) where revision should take place. All in all the sketched approach allows us to enhance both the model's quality and its suitability for software engineering.

2 EPCs for Modeling Business Processes

Business processes have been at the heart of IS research for many years if the evidence of many publications concerned with this topic is anything to go . As a result, the amount of different approaches is equally high: IDEF, RAD, ARIS/EPC and Oracle Designer to name but a few. Despite this fact, one of these approaches plays a more predominant role, especially in practice, namely the Event-driven Process Chains (EPCs) of the Architecture of integrated Information Systems (ARIS) described in [15]. The reasons for this prevalence are manifold: on the pragmatic side, a commercial tool for EPCs (ARIS toolset) has been available for quite some time already. In addition, the great success of the company SAP suite of business applications tremendously promoted the use of this method. On the other side, EPCs have also been investigated quite thoroughly in research.

Nevertheless, there is still some argument concerning the suitability of EPCs for modeling business processes that are to be supported by an information system. Advantages such as being highly flexible and easy to learn and understand are compensated by significant disadvantages: first of all ambiguity and vagueness. It cannot be in the interest of the user if the processes described in the specification are interpreted differently by the designer. When this misunderstanding is discovered by the user it is often too late to correct the design accordingly. Where is the path that leads out of this dilemma and towards a better understanding between the participants in the software development process? We think the answer lies in a pragmatic interpretation of the correctness of an EPC as we will show in the following sections.

The language of EPCs provides the user with a set of graphical notation elements for the representation of (business) functions, events and routing constructs to describe the control flow. Functions are used to model the dynamic part of the process. Typical functions are procurement, quality assurance or processing an invoice. They are decomposed hierarchically in a separate diagram, the so-called function diagram. The behaviour of each such function is modeled as a complete EPC. Another constructive element is the event. An event either triggers a function or marks the termination of it. For example, the event *not_ok* triggers the function *complaint* whereas the event *data revised* marks the termination of *complaint*. Furthermore, to describe more complex behaviour such as sequential, conditional, parallel, and iterative routing, connectors are introduced. These fall into two categories: splits and joins. In both we have AND, XOR and OR connectors.

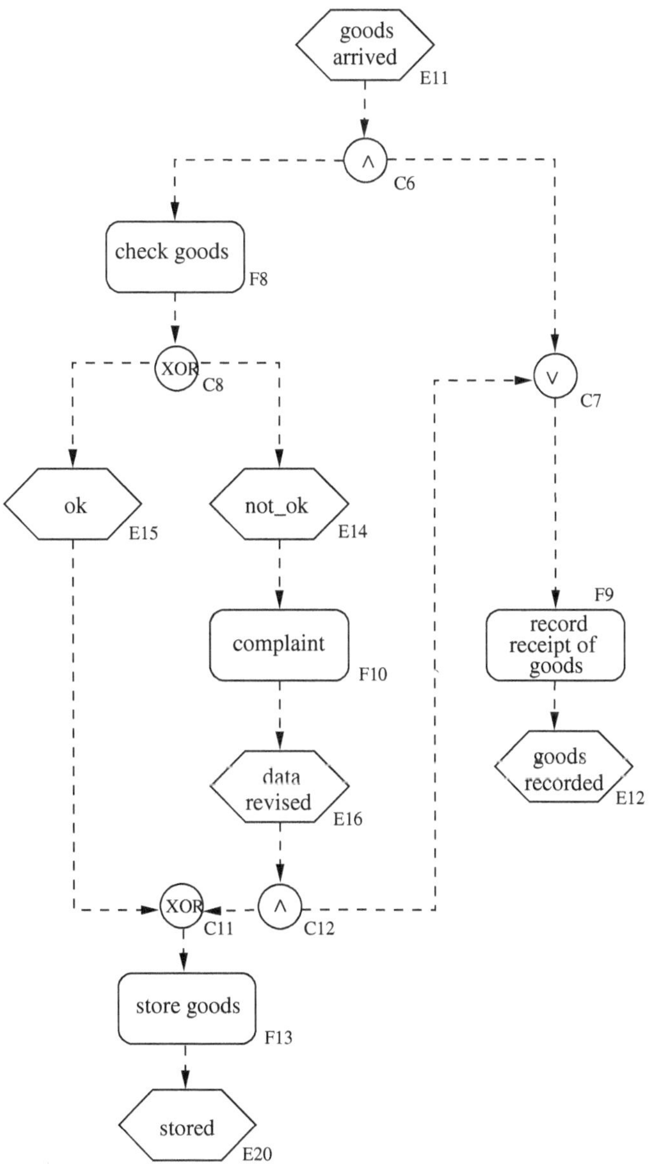

Fig. 1. Handling of incoming goods

Fig. 1 shows an EPC modeling the process "Handling of incoming goods" introduced by [8]. The process starts with the event *goods arrived.* After that the execution is split into two parallel paths (AND split *C6*), the left one checking the goods and performing the ensuing functions, the right one doing the accounting. The result of *check goods* is either *ok* or *not_ok.* In the latter case a complaint is compiled, in the former nothing happens. In either case (XOR join *C11*) the goods are stored afterwards (*store goods*). Connectors *C7* and *C12* make sure that in case of a complaint the corrected data is waited for before the receipt of goods is recorded. Otherwise the receipt can be recorded straight away.

EPCs are a semiformal method for business process modeling. Although they have been utilized quite successfully there authors did neither define a comprehensive and consistent syntax nor a corresponding semantics. A first formal description of EPCs was given in [6] (syntax) and [4] (semantics) but only for a restricted subset of EPCs. Other approaches to formalization have been developed by [14], [3], [13] and [9]. In this paper we refer mainly to the approach of [3]. Here syntax and semantics of an EPC have been described formally. The definition of an EPC includes:

- Events and functions have exactly one incoming and one outgoing arc (except start and end events).
- There are no isolated nodes.
- Connectors are either splits or joins.
- An event is always followed by a function and vice versa (modulo connectors).

As in most of the approaches mentioned the semantics of an EPC is defined by a mapping onto Petri nets. Petri nets are used because they have a clear and precise definition [10] and a similar graphical notation to that of EPCs. In addition they give us access to many existing analysis techniques and tools. The approach presented in [3] is based on the classical Petri net whereas [8], [13] and [9] use high-level variants. Alternatively we could define a formal semantics for EPCs and then define the criteria of section 4 directly on EPCs. Work to that effect is in progress.

In this paper EPCs are transformed into workflow nets but contrary to [3] we allow for ambiguity as e.g. introduced by the incorporation of the OR connector. We do this to increase the flexibility of the modeling method in the early phase of software engineering, i.e. to provide room for interpretation. This is necessary to foster the integration of the incompatible views on the common domain held by the heterogeneous parties involved in this process. Moreover it requires less modeling expertise and a less precise knowledge of the domain. Hence contrary to all existing approaches we do not resolve ambiguity. Instead, we ensure that the model is reasonable in spite of ambiguity, i.e. that it covers some reasonable behaviour that can be used as a basis for either the revision of the EPC or the design of the information system. In the following section we show how EPCs are transformed into workflow nets which are then used to assert certain properties of the EPC, most notably soundness and relaxed soundness.

3 Transformation into Workflow Nets

Workflow nets (WF nets) have been introduced by van der Aalst [2] applying Petri-Net theory to the specification of workflow processes. A WF net is a Petri net which has a unique source place (i) and a unique sink place (o). In addition a WF net requires all nodes (i.e. transitions and places) to be on a path from i to o. This ensures that every task (transition) and every condition (place) contribute to the business process.

In this paper we use the definitions of a Petri net and WF net from [2], namely:

Definition (Petri net). A Petri net is a triple (P, T, F) where:

- P is a finite set of places,
- T is a finite set of transitions,
- $F \subset (P \times T) \cup (T \times P)$ is a set of arcs (flow relation).

The function M: $P \rightarrow N$ is called marking of PN. M(p) is the number of tokens contained by the place p for the marking M. A transition t is said to be enabled by a marking M iff each input place p of t contains at least one token. An enabled transition may fire. If transition t fires then t consumes one token from each input place and produces one token in each output place. Note that the term $M_1 \xrightarrow{t} M_2$ means that the firing of transition t takes the process from state M_1 to state M_2. $M_1 \xrightarrow{*} M_2$ indicates that there is some firing sequence of transitions that leads from state M_1 to state M_2. We use (PN, M_0) to denote a Petri net with an initial state M_0. A state M' is reachable in (PN, M_0) iff $M_0 \xrightarrow{*} M'$.

Definition (strongly connected). A Petri net is strongly connected if and only if for every pair of nodes (i.e. places and transitions) x and y there is a path leading from x to y.

A WF net is a special Petri net defined as follows.

Definition (WF net): A Petri net PN = (P, T, F) is a WF net if and only if:

i. PN has two special places: i and o. Place i is a source place and place o is a sink place.
ii. If we add a transition t^* to PN which connects place o with i then the resulting Petri net is strongly connected.

The transformation of EPCs into Petri nets takes place in two steps. First we map elements of the EPC onto Petri net-modules. In the second step we provide rules to combine the different modules to form a complex process model. Our set of transformation rules is shown in Fig. 2.

Events and functions are transformed into places and transitions respectively including in- and outgoing arcs. Routing constructs such as AND split, AND join, XOR split, XOR join, OR split and OR join are mapped to small Petri net-modules. The Petri net-modules describe the behaviour of the routing constructs explicitly. This is primarily relevant for the OR, because its semantic has not been described consistently.

In Fig. 3 we see an EPC with an OR join on the left and its Petri net translation on the right side. The EPC as well as the Petri net have the semantics that E can be reached if either F_1 or F_2 or both occur. In the EPC all these different cases are described through one connector whereas in the Petri net-module all possibilities are modeled explicitly via the transitions t_a, t_b and t_c. The behaviour of the EPC and the

Petri net are equivalent, because both accept the same executions. Note that the case that E is reached twice if F_1 and F_2 occur sequentially has not been excluded.

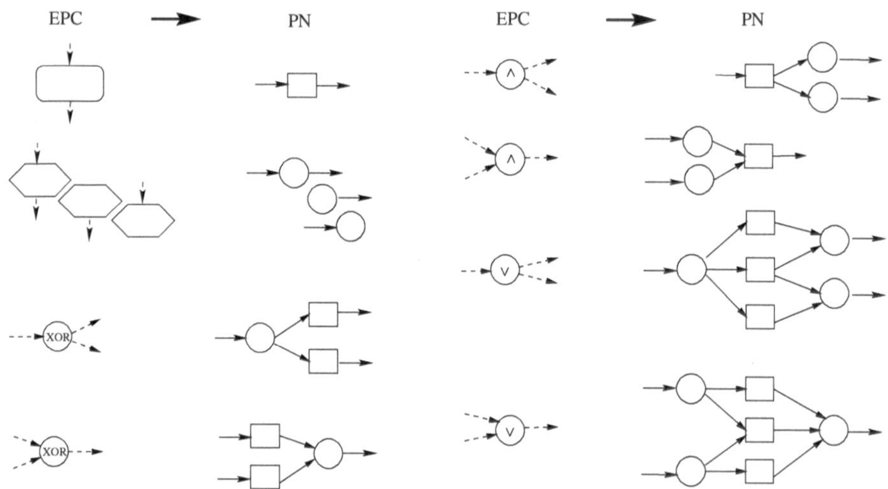

Fig. 2. Transformation rules for an EPC into a place/transition net (rule 1)

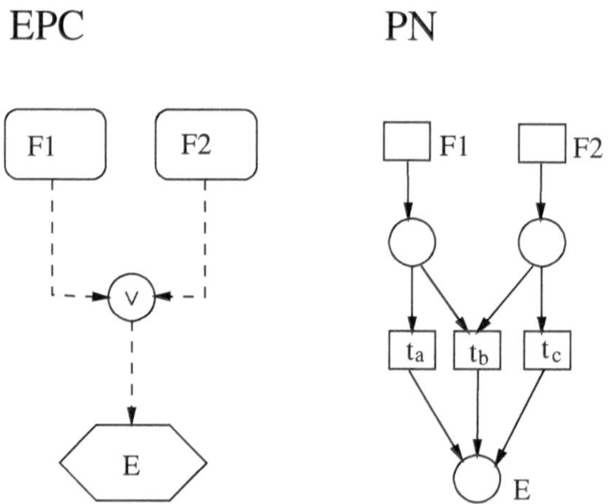

Fig. 3. Transformation of the OR-Connector

To form a coherent Petri net the single modules are (automatically) connected as follows (rule 2):

a) if input and output elements are different (place and transition) then the arcs are fused

b) if input and output elements are of the same kind (e.g. both places) then the different nodes are unified.

Fig. 4 illustrates the transformation of EPCs into place/transition nets.

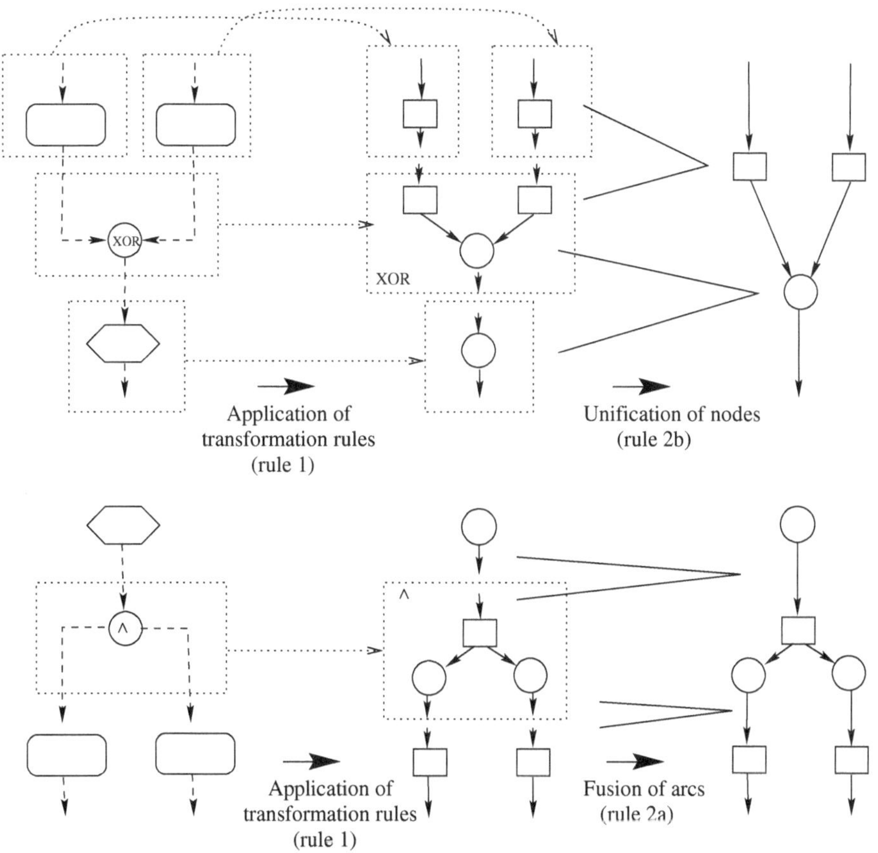

Fig. 4. Illustrating the transformation

The proposed transformation approach is slightly more general than the transformations described in [13] and [3]. The rules presented here can also be applied to transform EPCs where connectors follow each other immediately as e.g. *C6/C7* in Fig. 1.

Another advantage of this approach is that the resulting Petri net is minimal in the sense that it does not contain places or transitions not corresponding to elements of the EPC. The transformation rules by [13], [8] and [3] all contain rules which explicitly introduce new pseudo places and transitions to meet the Petri net syntax. The resulting Petri net may contain many elements which have no counterpart in the application domain.

To transform an EPC into a WF net something more has to be done. A WF net has exactly one start and one sink place. One problem when modeling with EPC is caused by an unclear concept of start and end events. There is no rule that restricts the amount of start and end events. A start (end) event is defined as an event without an incoming (outgoing) edge. Furthermore it is not clear whether the start (end) events are mutually exclusive. So translating the EPC into a Petri net does not necessarily lead to a Petri net with exactly one start and one sink place. In this case one further transformation step is required to yield a WF net. We add a new start place and a new sink place and connect them to Petri net-modules which initialize (clean up) the places representing the start and end events of the EPC in the right manner. The module introduced complements the first (last) connector on the paths from the start (end) events. For further particulars we refer the reader to [13] where this rule (rule 3) has been introduced and to the example below.

Applying the proposed rules 1 to 3, an EPC is transformed into a WF net. This transformation is unique, in the sense that to each EPC belongs exactly one WF net. An example for such a transformation is shown in Fig. 5. Here the EPC from Fig. 1 has been transformed into a WF net. For convenience we surrounded the Petri net-modules which correspond to the routing constructs of the EPC with dotted rectangles.

Transition $t10_{\_AND\text{-}Join}$ and the sink place *o* have been added due to rule 3. Transition $t10_{\_AND\text{-}Join}$ corresponds to an AND connector which complements the last connector on the paths from the end events *E12* and *E20*, namely connector *C12*. Transition $t10_{\_AND\text{-}Join}$ bundles the different path and leads to the sink place *o*.

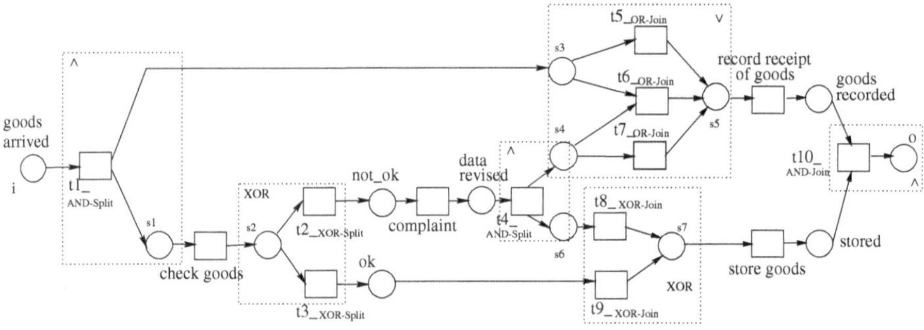

Fig. 5. WF net "handling of incoming goods"

Let us have a closer look at the Petri net-module which replaces the OR join *C7*. The Petri net-module makes the behaviour of this routing construct explicit. Transition $t5_{\_OR\text{-}Join}$ models the "straight away recording" and transition $t6_{\_OR\text{-}Join}$ models the waiting for the revision to be completed. The alternative $t7_{\_OR\text{-}Join}$ has been introduced as part of the corresponding Petri net-module, but has no expression in the original EPC. This alternative can not be chosen in the EPC, because of the AND-connector *C6* before.

By transforming the OR connector we carry the ambiguity of the OR to the WF net. The decision whether to execute transition $t5_{\_OR\text{-}Join}$, $t6_{\_OR\text{-}Join}$ or transition $t7_{\_OR\text{-}Join}$ can not be resolved locally anymore.

WF nets are a class of Petri nets for which theoretical results and efficient analysis techniques exist (cf. [1]). In the following section we will apply different criteria to check the workflow net and therefore the corresponding EPC for correctness.

4 Soundness and Relaxed Soundness

In this section we will introduce the properties of soundness and relaxed soundness as a means both to check the quality of the underlying EPC and to help with the revision of the business process if necessary. We subsequently discuss their applicability.

4.1 Soundness

Van der Aalst [1] introduced soundness as a correctness criterion for workflow nets. He argues that this criterion covers a minimal set of requirements a process definition should satisfy. Soundness ensures that the process can **always** terminate with a single token in place o and all the other places are empty. In addition, it requires that there is no dead task, i.e. any task can be executed. The following definition of soundness is taken from [1].

Definition (Soundness). A process modeled by a WF-net PN = (P,T,F) is sound if and only if:

i. For every state M reachable from the initial state i (one token in place i) there exists a firing sequence leading from state M to state o. Formally: $\forall M : \left(i \xrightarrow{*} M\right) \Rightarrow \left(M \xrightarrow{*} o\right)$
ii. State o is the only state reachable from state i with at least one token in place o. Formally: $\forall M : \left(i \xrightarrow{*} M \wedge M \geq o\right) \Rightarrow (M = o)$ (proper termination)
iii. There are no dead transitions in PN with initial marking i. Formally: $\forall t \in T : \exists M, M' : \left(i \xrightarrow{*} M \xrightarrow{t} M'\right)$

A WF net is sound if the process terminates properly in any case, i.e. termination is guaranteed and there are no spare tokens and neither deadlock nor livelock. Spare token signalize that some information was not used during execution, whereas dead- and livelock indicate situations where the execution got stuck respectively no real progress could be reached anymore.

We consider an EPC to be sound if its corresponding WF net is sound.

The WF net in Fig. 5 is not sound. There are firing sequences that do not terminate properly, e.g. the sequence: $t1_{\text{\_AND-Split}}$, check goods, $t2_{\text{\_XOR-Split}}$, complaint, $t4_{\text{\_AND-Split}}$, $t5_{\text{\_OR-Join}}$, record receipt of goods, $t8_{\text{\_XOR-Join}}$, store goods, $t10_{\text{\_AND-Join}}$.

In order to make the process sound the model has to be changed in such a way that the execution paths are restricted to sound firing sequences only where a sound firing sequence is one that terminates properly. To achieve this we have to avoid spare tokens as well as livelock and deadlock situations. These problems are, among others, the result of the incorporation of the OR connector.

Resolving the Ambiguity of the OR Connector

The ambiguous meaning of the OR connector has been discussed extensively in most formalization approaches. There are almost as many solutions as approaches.

In [11] the ambiguity of the OR connector is handled through a syntax extension on the side of the EPC. The connectors are extended by comment flags which describe the desired behaviour explicitly (wait-for-all, first-come, every-time). Wait-for-all means that the OR join waits for all paths that have been activated by the complementing opening split which the modeller has to identify as such. In the first-come (every-time) case the OR join triggers on the first path (on every path) that is completed. The first-come ignores the termination of the remaining paths. Hence the latter two do not require a complementing split. Note that this approach forces the modeler to resolve the ambiguity which he might not want to do as we already pointed out earlier.

[13] and [9] resolve the ambiguity adding places (communication channels) to the Petri net. Their task is to keep the information about the choice made by the OR split. This information is used to synchronize the corresponding OR join accordingly. [4] and [8] introduce different tokens for the same reason. All approaches mentioned impose the requirement to model in a well-structured way, i.e. every split has to be complemented by a corresponding join. Modeling with well-structured EPCs restricts the modeler considerably in his/her expressiveness and it also poses substantial requirements on the modeling expertise. Modeling with well-structured EPCs is based on a strict top-down design process which can hardly be enforced in practice.

Apart from this elements are introduced to synchronize parallel threads. Synchronization always serializes the execution. Suppose the probability that the system ends in an inconsistent state is very small. It may then be more efficient to recover (seldom) than to wait (every time). The introduction of synchronization forces the designer to think about efficiency aspects of the execution already during the modeling. Moreover it is generally problematic to introduce additional elements to the net because it thereby potentially diverges from the semantics of the original EPC. Often this leads to models which are quite different from the primary specification which in turn requires revision. This revision requires communication with the users which is complicated by the fact that the changes are often motivated by technical requirements only and have no counterpart in the application domain. Let us consider again the example "Handling of incoming goods".

We will change the WF net of Fig. 5 in such a way that a sound WF net is obtained (see Fig. 6). Firstly, transition $t7_{\_OR\text{-}Join}$ is removed. As discussed earlier it has no meaning in the original EPC and can therefore be removed without any consequences for the EPC. Then we introduce a place S_1 which takes care that transition $t5_{\_OR\text{-}Join}$ only executes if the *credit check* was *ok*. Through this construct the different threads (recording and check) have been synchronized and therefore serialized. This behaviour is not required in the original EPC. To change the EPC accordingly it has to be decomposed and composed again in a well-structured way. This change is not trivial and results in a completely different looking and behaving EPC.

Hence (strict) soundness demands that the modeler either restricts himself/herself to well-structured EPCs right from the start or he/she has to hazard the consequences of a substantial and costly revision. EPCs are a graphical modeling method which is used in the early analysis phase of an IT project where the focus is on communication. The goal of its use is reaching an agreement on how the process should look like between participants with totally different background and "knowledge cultures". The resulting models are only the base for further investigation. The applied correctness criteria should reflect the modeling knowledge and should therefore not be too strict. It should allow the modeler to postpone the more precise specification as long as possible i.e. to shift to later phases such as design and implementation.

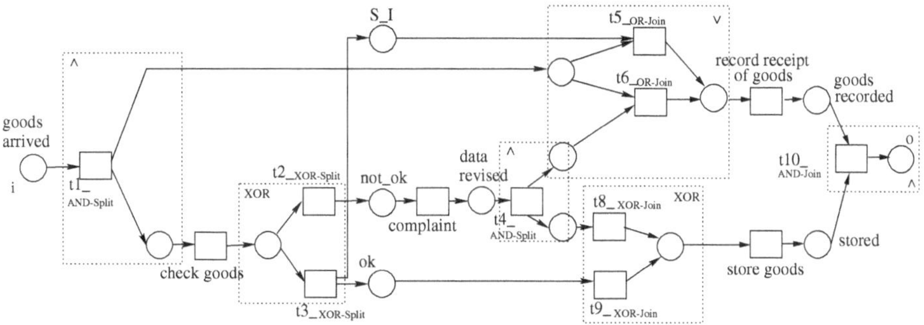

Fig. 6. Extended WF net

Hence we suggest to introduce a new relaxed soundness criterion replacing the strict version.

4.2 Relaxed Soundness

We propose to relax the soundness criterion to the new criterion *relaxed soundness* which has been introduced in [5]. Relaxed soundness is intended to represent a more pragmatic view on correctness which is weaker (in a formal sense) but more easily applicable to application-oriented modeling. It does not require to avoid situations with spare tokens or livelocks/deadlocks. It is therefore suitable to check WF nets which have been derived through the transformation of (not necessarily well-structured) EPCs containing OR connectors.

The idea behind relaxed soundness is that for each transition there exists a sound firing sequence, i.e. a sequence that takes the initial state i to the final state o. No spare tokens should be left in the Petri net in this case.

Definition (Relaxed sound).

A process specified by a WF net PN = (P, T, F) is relaxed sound if and only if every transition is in a firing sequence that starts in state i and ends in state o. Formally: $\forall t \in T : \exists M, M' : \left(i \xrightarrow{*} M \xrightarrow{t} M' \xrightarrow{*} o\right)$

Intuitively relaxed soundness means that there exist enough executions which terminate properly (i.e. without spare tokens). Enough means at least so many that every transition is covered. We argue that this criterion is closer to the intuition of the modeler. It does not force the modeler to think about all possible executions and then to care for proper termination in all cases. In spite of that relaxed soundness is still reasonable because it requires that at least all intended behaviour has been described correctly. Note that (strict) soundness implies relaxed soundness.

In terms of the EPC this means that every function can be executed reaching a desired set of end events. If the WF net is not relaxed sound then we have transitions that are not contained in any sound firing sequence. Hence the corresponding part in the EPC needs improvement. Put in other words: as a general rule we have to consider transitions that are not contained in some sound firing sequence when we are looking for parts of the process that need revision.

Let us now check the net in Fig. 5 for relaxed soundness. For this purpose we have to find a sound firing sequence for every transition. The check for relaxed soundness has been automated. The criterion can be checked with the help of the Petri net tool LoLA (**Lo**w **L**evel Petri Net **A**nalyzer) that has been implemented at the Humboldt University of Berlin [7]. It includes features such as: analysis of reachability of a given state and finding dead transitions. Recently LoLA has been extended to prove for extended computation tree logic formulae (eCTL) [12]. Within eCTL it is possible to quantify not only over states but also over state transitions. The combination of Petri nets and eCTL allows to check for relaxed soundness: for each transition t the reachability of the end state o is verified while it is required to include transition t in the path from i to o.

So far relaxed soundness can be proven only by enumeration of enough sound firing sequences. As far as the authors can judge there are no structural properties such as lifeness and boundness for soundness from which the relaxed soundness property can be derived.

The net in Fig. 5 is not relaxed sound because there is no sound firing sequence containing transition $t7_{\_OR\text{-}Join}$. As we pointed out earlier transition $t7_{\_OR\text{-}Join}$ can be left out of the WF net without loss of semantics regarding the EPC. Then the net is relaxed sound. The following two sound firing sequences contain all remaining transitions:

- $t1_{\_AND\text{-}Split}$, check goods, $t2_{\_XOR\text{-}Split}$, complaint, $t4_{\_AND\text{-}Splt}$, $t6_{\_OR\text{-}Join}$, record receipt of goods, $t8_{\_XOR\text{-}Join}$, store goods, $t10_{\_AND\text{-}Join}$
- $t1_{\_AND\text{-}Split}$, check goods, $t5_{\_OR\text{-}Join}$, $t3_{\_XOR\text{-}Split}$, $t9_{\_XOR\text{-}Join}$, record receipt of goods, store goods, $t10_{\_AND\text{-}Join}$

From this we can conclude that the EPC represents reasonable behaviour and can hence be used as a basis for software design.

5 Conclusion

We started with the assumption that business processes play a central role in reorganizing a company and (re)designing its information system(s). In doing so we typically describe the processes with the help of some semiformal, graphical language such as the Event-driven Process Chains. Modeling with EPCs usually involves ambiguities which, seen as "room for interpretation", are necessary in the early stage of analyzing a business. But for the later stages of software development we must identify the useful interpretations. This notion is formalized in our paper in terms of the relaxed soundness criterion. To make use of this criterion we first have to transform the EPC into a workflow net. This is done applying a fixed set of rules. The transformation does not resolve ambiguities but makes them explicit. The resulting WF net is used as a basis to check properties the process should satisfy. Relaxed soundness ensures that the modeled process meets some reasonable requirements. It enable us to check EPCs which contain OR connectors which typically presents a problem in other approaches.

Main aspects of future work include:

- finding a way to transform a relaxed sound net into a sound net (automatically) by reducing the net to the well-behaved paths; if this can be done the resulting net can be used directly as a basis for the design assuming that the EPC is of sufficient quality or a revision to costly,
- defining relaxed soundness directly for EPCs without requiring the intermediate step to Petri nets,
- integrating the approach into an analysis and design tool
- testing the approach in practical situations with large-scale models

In this paper we propose a new correctness criterion which is suitable to check the process model in an early phase of software engineering. Our approach allows us to draw conclusions on mistakes in the original EPC and to make suggestions for its improvement thereby enhancing both the model's quality and its suitability for software engineering.

Bibliography

1. Aalst, W.M.P. van der: Verification of Workflow Nets. In: P. Azema and G. Balbo: Application and Theory of Petri Nets 1997, Lecture Notes in Computer Science, vol. 1248, Springer, Berlin, 1997, pp. 407-426
2. Aalst, W.M.P. van der: The Application of Petri Nets to Workflow Management. The Journal of Circuits, Systems and Computers, **8** (1) 1998, pp. 21-66
3. Aalst, W.M.P. van der: Formalization and Verification of Event-driven Process Chains. Computing Science Reports 98/01, Eindhoven University of Technology, Eindhoven, 1998.
4. Chen, R., Scheer, A.-W.: Modellierung von Prozessketten mittels Petri-Netz-Theorie. Veröffentlichungen des Instituts für Wirtschaftsinformatik, Heft 107 (in German), University of Saarland, Saarbrücken, 1994
5. Derks, W., Dehnert, J., Grefen, P. and Jonker, W.: Customized atomicity specification for transactional workflow. In: Cooperative Database Systems for Advanced Applications (CODAS'01), 2001, To appear
6. Keller, G. and Teufel, T.: *SAP R/3 prozeßorientiert anwenden: iteratives Prozeß-Prototyping zur Bildung von Wertschöpfungsketten*. Addison-Wesley, Bonn, 1997.
7. Schmidt, K.: LoLA, a Low Level Petri Net Analyzer. Humboldt-Universität, Berlin. http://www.informatik.hu-berlin.de/~kschmidt/lola.htm
8. Langner, P., Schneider, C, Wehler, J.: Ereignisgesteuerte Prozessketten und Petrinetze. Report No. 196, Computer Science Department, University of Hamburg, FBI-HH-B-196/97, March 1997.
9. Moldt, D., Rodenhagen, J.: Ereignisgesteuerte Prozessketten und Petrinetze zur Modellierung von Workflows. In: Visuelle Verhaltensmodellierung verteilter und nebenläufiger Software-Systeme, vol. 24/00-I, Münster, 2000, pp. 57-63.
10. Murata, T.: Petri Nets: Properties, Analysis, and Applications. Proc. of the IEEE, **77** (4) 1989, pp. 541-580
11. Rittgen, P.: EMC - A Modeling Method for Developing Web-based Applications. International Conference of the International Resources Management Association (IRMA) 2000, Anchorage, Alaska, USA, May 21 - 24, 2000

12. Roch, S.: Extended Computation Tree Logic. In: H.D. Burkhard, L. Czaja, A. Skowron and P. Starke: Workshop Concurrency, Specification & Programming, Informatik-Bericht 140, Humboldt-Universität, Berlin, Oct. 2000, pp. 225-234.
13. Rodenhagen, J.: Darstellung ereignisgesteuerter Prozessketten (EPK) mit Hilfe von Petrinetzen. Diplomarbeit, Universität Hamburg, Fachbereich Informatik, 1996.
14. Rump, F.: Geschäftsprozeßmanagement auf der Basis ereignisgesteuerter Prozeßketten. Formalisierung, Analyse und Ausführung von EPKs. Teubner, Stuttgart, 1999
15. Scheer, A.-W.: Business Process Engineering, Reference Models for Industrial Enterprises. Springer, Berlin, 1994

Developing E-Services for Composing E-Services

Fabio Casati, Mehmet Sayal, and Ming-Chien Shan

Hewlett-Packard Laboratories
1501 Page Mill Road, 1U-4
Palo Alto, CA, 94304 USA
{casati,sayal,shan}@hpl.hp.com

Abstract. The Internet is rapidly becoming the preferred mean through which companies provide services to businesses and customers. A large number of e-services, including for instance stock trading, customized newspapers, real-time traffic report, or itinerary planning, is already available on the Web, and the type and number of e-services grows on a daily basis. In order to support the development and deployment of e-services, software vendors are developing e-services frameworks and platforms, that provide a language for *describing* an e-service, and then allow service providers to *register*, *advertise* and securely *deliver* e-services to (authorized) users. A *composite* e-service is an e-service defined by composing other basic or composite e-services. As the e-service paradigm becomes popular and more and more applications are developed or deployed as e-services, the need and opportunity for defining composite service become manifest. This paper presents a specific type of e-service (or, rather, a meta e-service) called *Composition* E-Service (CES), that allows the definition, execution, management, and monitoring of composite e-services. We first describe the advantages and the functionality of such a service. Next, we present the language used for specifying the composition, also discussing why existing workflow languages are not suitable for this purpose. Finally, we present the architecture and implementation of the CES we developed to deliver the service on top of the e-services platform *e-speak*. An analogous architecture and implementation strategy can be followed with any other e-services platform.

1 Introduction

Today, the Internet is not only being used to provide information and perform e-commerce transactions, but also as the platform through which services are delivered to businesses and customers. The explosion of the number and type of services as well as service providers requires mechanisms and frameworks that support providers in developing and delivering e-services and support consumers in finding and accessing them. Several software vendors and consortia are providing models, languages, and interfaces for describing e-services and making them available to users. Such frameworks usually allow the specification of business functions or applications in terms of their properties, which can be generic (such as the service name and location) or service-specific (such as the *car size* for a car rental service). Depending on the framework, the properties are represented by Java vectors or XML documents.

K.R. Dittrich, A. Geppert, M.C. Norrie (Eds.): CAiSE 2001, LNCS 2068, pp. 171–186, 2001.
© Springer-Verlag Berlin Heidelberg 2001

In addition, vendors also provide software platforms (called E-Services Platform, or simply ESP in the following) that allow service providers to register and advertise their services and allow authorized users to lookup and access registered services (see Fig. 1). Examples of such platforms are BEA eCollaborate, WebMethods Enterprise, Sun Jini, Microsoft .net, and HP e-speak.

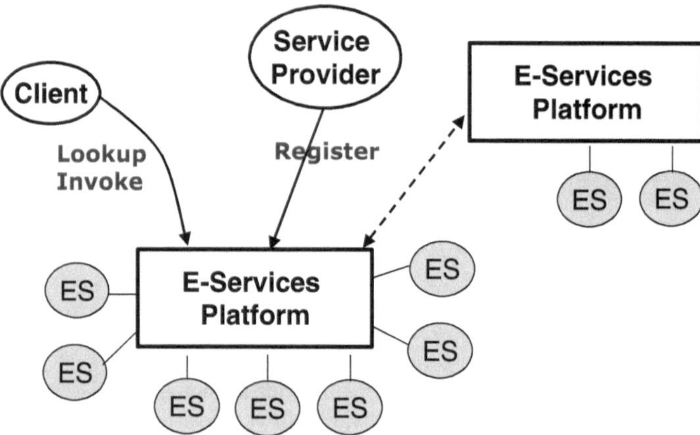

Fig. 1. E-Services platforms allow providers to register e-services and users to lookup and invoke them. Ovals labeled *ES* represent registered e-services.

These approaches enable the uniform representation, search, and access of business applications, both those used for internal operations (such as ERP operations, DBMSs, CRM, SCM, etc) and the ones that are made available to customers, typically via the Web.

The uniform representation and implementation of applications according to a homogeneous e-service framework creates the opportunity for *composing* individual, web-accessible e-services (possibly offered by different companies) into pre-packaged, value added, *composite* e-services. For instance, a provider could offer a travel reservation service by composing hotel and flight reservation services, or it could offer an itinerary planning service by composing road map services, weather services, traffic prediction services, and "utility" services to collect data from the user via the Web or send e-mail notifications.

Although composite services could be developed by hard-coding the business logic using some programming language, service providers would greatly benefit from a service composition tool that could ease the task of composing e-services, managing and monitoring them, and making them available to authorized users. This issue is similar to that of workflow applications, where the alternative is hard-coding the flow logic or using a Workflow Management System (WfMS). The advantages of service composition and workflow management tools versus hard-coding (for many practical applications) have been discussed elsewhere in the literature and will not be presented here (the interested reader is referred to [2, 4, 8]).

The traditional approach to providing a composition facility, advocated by workflow and Enterprise Application Integration (EAI) vendors, consists in offering a development environment targeted to the enterprise IT personnel. We decided to follow a different approach, that consists in providing composition functionality as an e-service itself (or, rather, a meta-service, since it is a service for developing services). By making it an e-service, the service composition facility can be advertised, discovered, delivered, managed, and protected by end-to-end security analogously to any other e-service, thereby exploiting all the advantages and features provided by the ESP. In addition, the ability of defining and deploying composite services is not limited to the ESP's owner, but can be offered to other businesses and customers, thereby relieving them from the need of maintaining a composition system that may be onerous to buy, install, and operate. In the following we will refer to this meta-service e-service as *composition* e-service, or simply CES.

In this paper we present the design, architecture, and implementation of the *CES*. We first introduce the notion of composition as an e-service and provide an overview of the functionality and behavior of such a service. Then, we discuss the characteristics of composite services, and analyze similarities and differences with respect to workflow processes. This discussion will introduce and motivate our choices in the definition of the service composition model. Next, we present the architecture and implementation of the composition e-service we have developed on top of the e-services platform *e-speak*. An analogous architecture and implementation strategy can be followed with any other e-services platform, and therefore provides a viable solution for software vendors and solution providers that need to develop a composition facility.

2 Service Composition as an E-Service

This section first briefly describes ESPs (in order to make this paper self-contained), and then introduces the basic functionality of a CES.

2.1 Basic ESP Functionality

ESPs typically allow service providers to *register* services, and allow authorized users to *lookup* and *invoke* registered services. In order to make services searchable and accessible to customers, service providers must register the service definition with the ESP, and possibly with advertising services. As part of the registration process, the service provider gives information about the service, such as the service name, the methods (operations) that can be performed on the service along with their input/output parameters, or the list of authorized users. Note that, in most service models, a service may provide several methods (operations) to be invoked as part of its *interface*. For instance, an e-music service may allow users to *browse* or *search* the catalog, to *listen* to songs, or to *buy* discs or mp3 files.

In addition, the service provider specifies who is the *handler* of the service, i.e., the application that must be contacted in order to request service executions. Depending on the service model and the ESP, the service handler can be identified by providing a URI (such as in e-speak) or by giving a proxy java object that will take care of contacting the handler (such as in Jini). Customers may look for available services by issuing *service selection queries*, that may simply search services by name, or can include complex constraints on the service properties as well as ranking criteria in case multiple services satisfy the search criteria. Service selection queries return a reference to one or more services that can be used to invoke them.

2.2 CES Functionality

This section describes the behavior of the composition e-service. A CES sits on top of an e-services platform and allows users to:

- Register and advertise definitions of composite services with the ESP and make them available to authorized users just like any other e-service. Composite services are defined in a Composite Service Description Language (CSDL), whose features will be presented later in the paper.
- Invoke (start executions of) composite services. The CES will execute the service on behalf of the user by appropriately invoking the component services as defined by the CSDL specifications.
- Monitor and manage composite services. The CES allows the modification or deletion of composite service definitions as well as running instances. Customers and service providers can monitor/track the execution of on-going instances as well as completed composite service executions.

In order to register a composite service, the service provider must give the same information needed to register a basic service (except for the handler - see below) to the CES, so that the composite service can be registered and made available to authorized users. In addition, the service provider gives the CSDL specifications to define how services should be composed[1].

Fig. 2 shows the composite service registration process for a composite service called *FoodOnWheels* (described in the following section): a provider that wants to define a new composite service invokes the *register* method of the CES by sending the service description (service information plus CSDL) as parameter. The CES then registers the composite service with the ESP in order to make it available as an e-service to the other (authorized) customers. The registration with the e-service platform is analogous to any other service registrations, and therefore the CES must provide all the required information describing the e-service and restricting access to it. In particular, it should also specify who is the handler for the service.

When a client needs a food delivery service, it queries the service repository to find out which services are available, asking the ESP to rank the services according to the specified criteria and return the best one. If the best service happens to be FoodOnWheels, then a reference to this service is returned. As for any other service, the client can then query the service description stored in the repository and perform method invocations on this service (see Fig. 3). The client has no knowledge that the service is in fact composite.

Figures 2 and 3 represent what happens "conceptually" from the CES users' viewpoint. When discussing the implementation, we will show that what happens behind the scenes is actually slightly different, but users are unaware of these differences, and the behavior of the system is as described above.

---

[1] A CES may also provide built-in, *utility* services, that provide frequently needed functionality, such as e-mail notifications or generation of web forms for collecting input data.

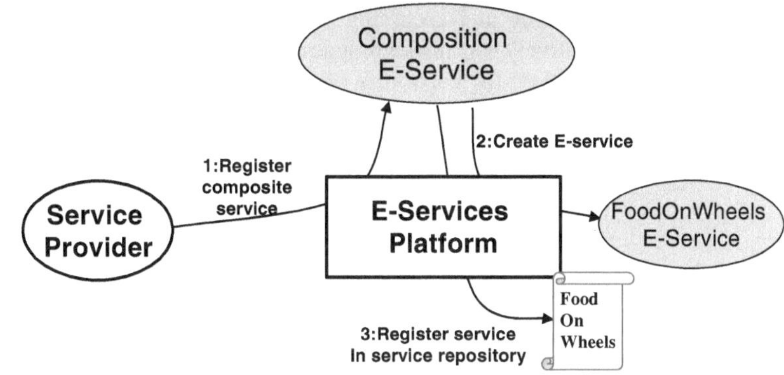

Fig. 2. Registration of a composite service, made available as an e-service

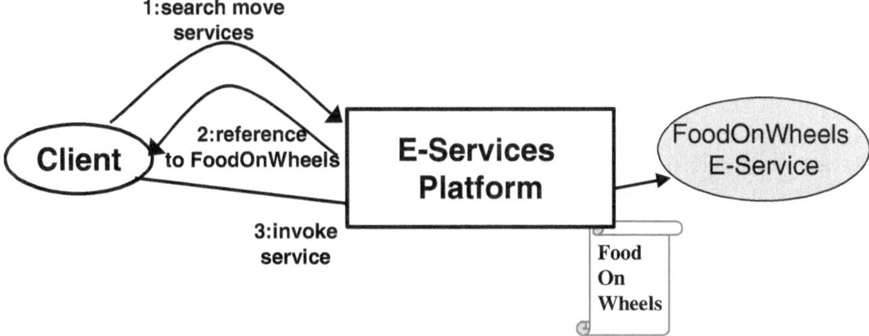

Fig. 3. Service selection and invocation

Service providers can update or delete a service definition, resulting in a corresponding update or deletion of the service registration on the ESP. Note that, technically, the definitions on the ESP are "owned" by the CES. This prevents service providers from directly updating or deleting composite service descriptions on the ESP, bypassing the CES and causing inconsistencies between information stored at the CES level and at the ESP level. The CES also allows service provider to monitor the status of service executions (note that since any composite service is itself an e-service, monitoring features provided by CES are in addition to whatever mechanism is provided by the e-services platform for service monitoring). The CES allows service providers to check how many services are in execution, at what stage they are in the execution (i.e., which path in the execution flow they have followed, which service is currently being invoked, what is the value of composite service data, etc.). CES monitoring capabilities are similar to those provided by WfMSs.

Services created by the CES also include method calls that allow users to control service executions. More specifically, users can *pause*, *resume*, and *cancel* a service execution (see Fig. 4). Note that while service providers interact with the CES, clients of composite services only interact with the services through the service reference they got as a result of the lookup, as with basic e-services.

Finally, we observe that the CES should be able to compose any service that is reachable through the ESP on top of which it is developed. Advanced ESPs such as e-speak are capable of searching and accessing e-services delivered through ESPs of different kinds, either natively or through gateways provided with the platform. Hence, we conveniently rely on the capability of the ESP to access e-services running on top of heterogeneous ESP platforms rather than re-developing the same interoperability features.

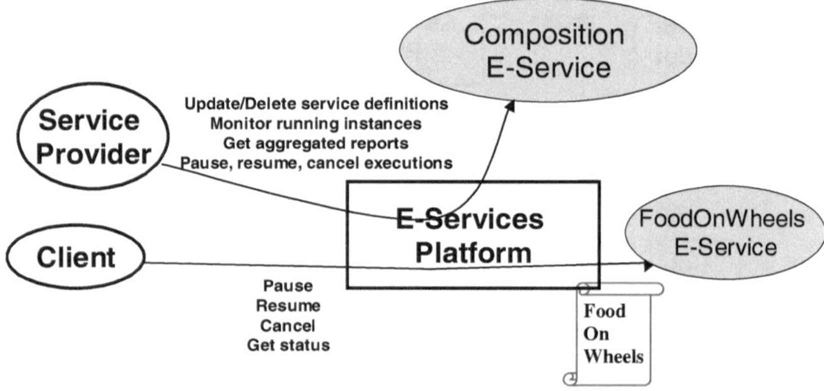

Fig. 4. Service providers can manage definitions and monitor and control executions, while service users can control executions (to the extent allowed by the provider)

3 Composite Service Definition Language

This section presents the service composition model and language. We first discuss the characteristics of a composite service and we underline the differences between workflow and e-service composition. Then, we present the composite service description language.

3.1 Workflows and E-Service Composition

This section introduces the main characteristics of composite services. In particular, we introduce them in terms of differences with respect to workflow applications. In fact, in many ways, a composite service is similar to a workflow: in order to define a composite service, the provider mainly needs to specify the *flow* of service invocations (i.e., the services to be invoked, their input and output data, and their execution dependencies). Similarly, in a workflow, the designer must specify the flow of work (i.e., the work items to be executed, their input and output data, and their execution dependencies).

Hence, an option that we had considered for CSDL was to simply use an existing workflow modeling language. However, a language and system for service composition has many different requirements with respect to workflows. We list the main differences below:

- *Service selection:* Nodes in traditional workflow graphs represent administrative or production work items, assigned to human or automated resources. Often, workflow models also impose a resource model, based on roles and/or organizational levels. Selecting a resource typically involves selecting an employee or an enterprise application by means of a (possibly rich and expressive) resource language that identifies authorized resources depending on the roles they play and on the level they belong to.
- Similarly, nodes in an e-service environment represent service invocations. As part of the service node definition, the provider specifies the service to be invoked. Although conceptually this is similar to selecting a resource for a work item, the e-service environment has very different concepts and requirements: there is typically no fixed "organizational model" or resource taxonomy. The service is selected depending on its properties, and the selection criteria are specified in the query language supported by the e-services platform, which is usually quite powerful and flexible. A service composition language should support and facilitate the definition of service selection criteria for each node in the flow, allowing also criteria that depend on the specific instance in execution (i.e., are sensible to the instance-specific data, such as the customer name or geographical location). Note that, while in principle it is possible to follow the "workflow approach" (i.e., identify and classify services in advance and then specify work assignments through some role expression), this is not required due to the presence of a (homogeneous) service repository in the ESP and of a service query and selection language. Besides not being required, the workflow approach is also not advised. In fact, the e-service environment is very dynamic and services are introduced, modified, or deleted very often. Hence, the content and structure of the repository would have to be updated all the time.
- *Input and output data*: in workflows, input and output data are typically specified by a set of variable names. The semantics is that the value of the input variables at the time the node is started is passed to the selected resource, and node execution results are inserted into the output variables. Communication between the WfMS and the resources is done through adapters, that understand the syntax and semantics of the data and perform the required data mappings.

  E-services, depending on the platform on top of which they run, typically communicate in java or XML, and these two languages dictate the rules and the syntax for data exchanges. Therefore, facility for processing Java and XML objects and transferring them to and from the invoked e-services must be provided. Also in this case, while in principle it would be possible to follow the "workflow approach" and develop adapters that bridge the composition environment and the e-services to get rid of data mapping issues (at the cost of transferring the problem onto the adapters), this is luckily not needed. In fact, e-services running on top of ESPs share the same service model and parameter passing semantics, so that it is possible to take this into account in the service composition model and provide facility for communicating with e-services as prescribed by the ESP, thereby avoiding the need for adapters. Indeed, this is a considerable advantage, given that developing adapters is difficult and tedious job, as demonstrated by the cost of commercial system integration platforms. In addition, it simplifies the use of the CES, since developers may define and deploy a composite service by simply sending a single file that includes all the business logic. There is no need of changing the configuration of several different systems, as it happens with current WfMSs[2].

- *Dynamic environment*: Unlike "traditional" business processes, composite e-services have to cope with a highly dynamic environment, where new services become available on a daily basis. In order to stay competitive, service providers should offer the best available service in every given moment to every specific customer. Clearly, it is unfeasible to continuously change the flow to reflect changes in the business environment, since these occur too frequently and modifying a composite service definition can be a delicate and time-consuming activity. Ideally, composite services should be able to transparently adapt to changes in the environment and to the needs of different customers with minimal or no user intervention. Workflow systems do not typically offer these capabilities.
- *Black boxes vs multi-methods interfaces*: typically a work item in a workflow represents the invocation of a business function. The work item is a black box from the workflow viewpoint. Instead, an e-service may have several states and state transitions, caused by method invocations. Interacting with an e-service requires operations to be performed at the service level (e.g., search and authentication) and operations to be performed at the method level (e.g., method invocations).
- *Security:* current workflow technology has very little support for security. Often, there is no encryption, and access is controlled by means of usernames and passwords. This is due to the genesis of WfMSs as systems for managing the work in a restricted and controlled environment, within a corporation. In the Internet and e-service environment the security requirements are different, and in particular e-services may require the use of certificates, which therefore should be also supported by the service composition model and language.
- *Business-to-business interactions*: a number of standards (e.g., RosettaNet, cXML, CBL) are being defined in order to support business-to-business interactions, possibly limited to specific, vertical markets (such as RosettaNet for the IT industry). Many applications that support such standards are being or have been developed, and it is likely that many service composition applications will interact with services that follow one of these standards. A CSDL should facilitate the composition of such services as well as their invocation, checking that the appropriate protocol is followed and that exceptions are thrown when deviations from the protocol are recognized. Many workflow models and systems do not provide such kind of support, although many vendors are moving in this direction.

3.2 CSDL Definition

This section presents the Composite Service Description Language. Although CSDL reuses some of the concepts developed by the workflow community, it has several innovative features that make it suitable for service composition:

---

[2] The adapter approach can still be followed, if the users so desire, by embedding the mapping semantics into suitable e-services.

1. It has a two-level service composition model, that distinguishes between invocation of services and of operations within a service. This is important since some aspects of the business logic are specific to a service and need to be specified at the service level, while others are instead specific to each method invocation, as detailed in the following.
2. The language allows the definition of how to send XML documents as input to service invocations, and of how to map XML results into composite service data items. This is important since we expect most of the interactions among e-services to occur in the form of XML documents.
3. A flexible mechanism to handle certificates is provided, to enable the definition of which certificates should be sent to a service.
4. A number of adaptive and dynamic features are provided, to cope with the rapidly evolving business and IT environment in which e-services are executed.
5. Facilities for B2B interactions are provided, in the form of service templates that can be reused by composite service designers, so that they do not need to be concerned with technical details about the standard.
6. The *entire* business logic can be defined within a single XML document, thereby making easy and practical to provide and use composition as an e-service.

CSDL originates from concepts developed in a previous HP project, called *eFlow* [3], that we have extended to take into account the characteristics of ESPs and of the e-services they support. Here we will only present the innovative aspects of CSDL.

Overview. A composite service is described as a process schema that composes other basic or composite services. A schema is modeled by a graph, which defines the order of execution among the nodes in the process. At the top level, the graph may include *service, decision,* and *event* nodes. Service nodes represent invocations of basic or composite services; decision nodes specify the alternatives and rules controlling the execution flow, while event nodes enable composite services to send and receive several types of notifications. A *composite service instance* is an enactment of a composite service schema. The same composite service may be instantiated several times, and several instances may be concurrently running.

As an example, consider a *FoodOnWheels* service, that delivers any kind of food to customers' doors. The graphical description of the composite service is shown in Fig. 5. In the figure, boxes represent service nodes while diamonds represent decision nodes. The entry points are represented by right-pointing triangles, while end points are denoted by left-pointing triangles. *FoodOnWheels* receives order from customers and, if the customer has a valid credit card, it selects one or more restaurants that provide the requested food (unless the customer specifies a preference) by accessing a restaurant selection service. Then, it picks up the food at the restaurants and delivers it to the customers at the requested time, through a food delivery service. Next, the customer's credit card is charged, by invoking a credit card payment service. A composite service is textually specified by an XML document.

A composite service may include the definition of *input*, *output*, and *local* data items (sometimes also called flow variables in the following). Input data items are parameters passed to the composite service at activation time. Output data items represent data returned to the caller at service completion. Input and output data items can also be used for routing purposes within composite service execution and for transferring data among service nodes. Local data items are neither input nor output, but are only used within the composite service to perform routing decision or to transfer data among nodes. The types of variables can be any basic Java type (e.g. String or Integer), a Java Vector, a generic Object, or an XML document. Each composite service instance has a local copy of the flow variables.

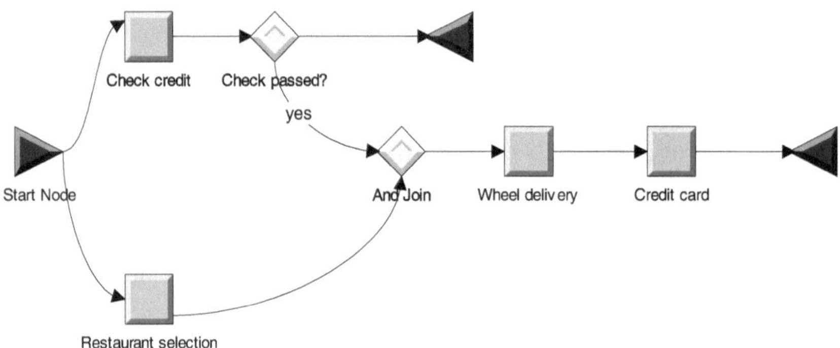

Fig. 5. Graphical representation of the FoodOnWheels composite service

Besides the graph that defines the flow of service invocations, the definition of the composite service also includes security-related specifications. In particular, the definition of a composite service includes information about the certificates to be used throughout the flow within service invocations, in case the ESP and the invoked e-service support or even require the use of digital certificates. By default, the composite service invokes component services with the privileges (i.e., the certificate) of the composite service definer. However, the designer may specify that services should be invoked with the privileges of the composite service users, or with the privileges specified by the content of a flow variable (for instance, the certificate to be used may be passed to the composite service as one of its input parameters).

Service Nodes. Service nodes represent invocations of a given service. The service to be invoked is specified by a *search recipe,* defined in the query language supported by the ESP. As the service node is started, the search recipe is executed, returning a reference to a specific service. Recipes can be configured according to the specific service instance in execution: every word in the search recipe that is preceded by a percentage sign "%" is expected to be a reference to a flow variable, and will be replaced by the value of that variable at the time the service node is started. This allows the customization of the search recipe according to the value of flow variables. Note that different activations of a service node may result in the selection of different services. However, sometimes the designer needs to specify a service node that should reuse the same service invoked by another service node. The composition service model allows this by enabling the definition of a *Service Reuse* attribute that includes the name of the service node whose service reference is to be reused.

The definition of the service node may include the certificate to be used when invoking the service's methods. The definition at the service level overrides the one done at the top (i.e., composite service) level. Since it is assumed that all invocations on the same service will use the same certificate, there is no provision for the definition of a certificate at the method invocation level.

Flow of Method Invocations. E-services, in most ESP models, will have an interface that allows several operations to be invoked on them. In order to achieve their goals, clients of these services will typically have to invoke several operations (i.e., call several methods) on the same service. Correspondingly, CSDL allows the designer to specify, within a service node, the *flow of method invocations* to be performed on a service. For instance, if we are accessing the e-music service, we may want to specify that we search for a given song (invoking the *search* method) and, if the price for the disc that includes the song is lower than a limit, then we buy the whole disc (*buyDisc* method), otherwise we simply download the mp3 file of that song only, paying the requested fee (*BuySong* method). To simplify both the language and the implementation, the method flow is specified with the same syntax (and semantics) of the top-level flow of services, with the only difference that here we are concerned with the flow of *method nodes* instead of service nodes. If only one method needs to be invoked, then the designer needs not specify the flow structure, but only a single method node. In addition, we also allow the definition of service nodes that have no method nodes inside. In fact, in a few cases, the designer might only want to execute a search recipe and get the results, possibly without invoking any method on the selected service. For instance, a node may simply need to get a service name or handle in order to pass it to another service.

Method Nodes. A method node defines the method to be invoked on a service and its input data, how to handle the reply (and specifically how to suitably map the reply message into flow variables), and how to handle exceptions that may occur during the method invocation. The name of the operation to be invoked can be statically specified, or it can be taken from the value of a flow variable, as usual specified by a string preceded by the percentage sign. The input data to be sent to the method are specified by a list of variable names or values. In case of variable names, the value of the variable at the time the node is started is sent as input to the method.

If a method invocation on a service returns a result (e.g., an integer or an XML document), then the designer needs to specify how information in the document can be extracted and inserted into flow variables. In case the method output is a (basic or complex) Java object, then the mapping is simply specified by describing the name of the flow variable to which this value should be copied. For example, method *CheckCredit* returns a Boolean value defining whether the credit check on the customer is positive or negative. In CSDL, this is defined as follows:

```
<Method-Output> <Var-Mapping Flow-Var="Confirmation" />
</Method-Output>
```

Since it is likely that most of the output data will be a string containing an XML document, CSDL provides additional support for XML, and in particular it allows the designer to specify how fragments of the XML output document can be mapped into flow variables. A flow variable name assumes the value identified by an XSL transformation or an XQL query on the output document. In the case of XQL queries, if the flow variable is of type XML, then the XQL query may actually return a set of elements, or a document. Otherwise, CSDL requires the query to identify a single element or attribute, or an exception is raised. For instance, the following mapping specifies that the XQL query `customerList/customer[0]` should be applied to the method output, and the result of the query should be put into variable "customer":

```
<Method Output>
<Var-Mapping Flow-Var="customer" Conversion-Rule=
"customerList/customer[0]" Rule-Type="XQL" />
</Method Output>
```

The definition of the query may be static or may include references to flow variables, as usual preceded by the percentage sign.

4 The *Composition* E-Service Prototype

This section presents the CES prototype, developed at HP for composing *e-speak* e-services. The same design can however be adopted for any other ESP. The prototype is built on top of a commercial workflow engine (and specifically of HP Process Manager) that handles the execution of the flow. The need of using a commercial workflow engine came from the requirement we had of building a *robust* prototype in a very short time, that ruled out the possibility of developing one by ourselves. Note that only the engine was needed for our purposes, so we removed all other HP Process Manager components to get a lighter and faster system[3].

Another key component of the architecture is the *gateway*, that enables the interaction between the workflow engine and the ESP, performing the appropriate mappings and implementing CSDL semantics that could not be supported by the workflow engine, as discussed below.

Fig. 6 shows the components of the prototype and in particular how they handle composite service registrations. The CES front-end responds to calls from service providers and clients (even if the latter are unaware of the fact that they are communicating with the CES). When a service provider registers a service, the CES front-end first translates CSDL into the language of the selected workflow engine. The translation generates a process where nodes correspond to method invocations on the ESP or on the selected e-services. However, since CSDL is in fact much richer than traditional workflow languages, the translation is a fairly complex procedure and requires the insertion of several "helper" nodes and data items that, in conjunction with the operations performed by the gateway (that has knowledge of the semantics of such helper nodes), enable the correct implementation of the CSDL semantics. Examples of issues we have to deal with in the translation include mapping the two-level (service and method) CSDL model into a single-level one and rewriting the input and output data items of nodes so that they can have all the information required to build XML documents and to map back XML replies into process data.

For instance, consider the single problem of mapping the CSDL two-level service model into a traditional workflow model. In order to map a service node, we need to insert a node that implements the search recipe (i.e., sends the service selection query to the ESP), and to define the data items needed for storing and sending certificate information. In addition, different method invocations occur in the context of the

[3] Note that using a WfMS for *implementing* the CES is not in contrast with our previous discussion on the unsuitability of a workflow language for *modeling* composite services.

same "session" with the service. Hence, we need to define and properly initialize process data items that can carry session IDs from node to node. Note that this problem could not have been solved by simply defining a subprocess, both because the need for defining service selection nodes and certificate nodes still remain, and because nodes in a subprocess do not have access to the variables of the main process (unless they are passed as input parameters, but even in that case the parameters are passed by value and not by reference). Where it was not possible to map appropriately, we encoded part of the semantics in the gateway. For instance, XQL queries are performed by the gateway. The gateway is also in charge of replacing references to flow variables in XML documents (i.e., those items preceded by the "%" symbol) with the actual value.

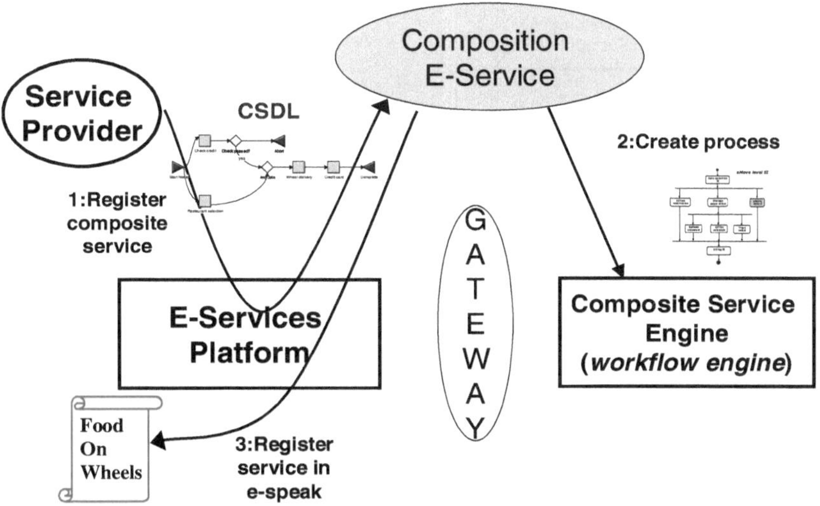

Fig. 6. Components of the CES prototype and how they handle registrations

We also mention that in this prototype we did not map adaptive features of CSDL, event nodes, and exceptions (we have only mapped deadline-related exceptions). These features will be introduced in the next version of the CES.

After the mapping has been completed and the process is installed on the workflow engine, the CES registers the new service with the e-speak ESP. As the Figure shows, the CES itself is the handler for the newly registered service. However, this does not change the validity of the scenario depicted in Figures 2 and 3. Indeed, clients simply communicate with the (composite) e-service through the reference they get, and they are not concerned with how the service is implemented on the server side.

When a client invokes a composite service (see Fig. 7), the CES starts the corresponding process in the workflow system (the mapping between the composite e-service name and the process name is defined at registration time and stored within the CES). Activities in the workflow correspond to method invocations on a given service involved in the composition. From a workflow perspective, all activities are assigned to the gateway. The gateway receives indication of what to do by the workflow engine as part of the activity definition, along with data items that provide (a) context information about the service on which method calls are being or have to be placed (e.g., service references, search recipes, certificates, and mapping information to process the XML document returned by the method and update the value of flow variables) and (b) the value of the parameters to be passed as part of the method invocation.

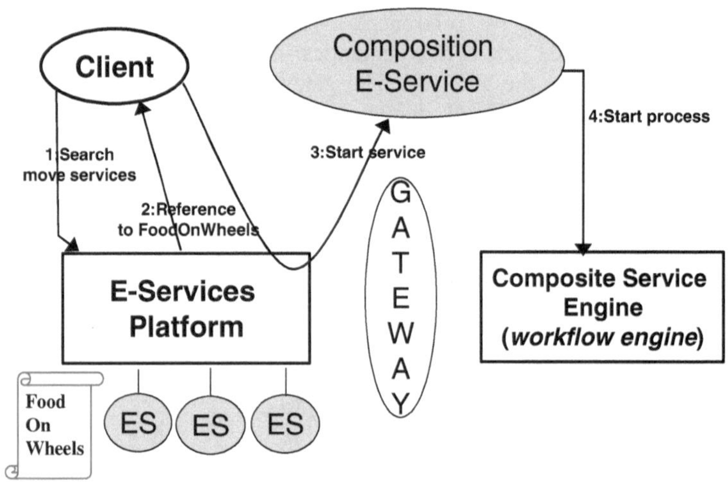

Fig. 7. Invocation of composite services

When the gateway receives work by the engine, it activates a new thread in order to process the work. The thread waits for the reply from the service, executes the mapping rules, and sends the results back to the engine. All the state information is maintained by engine, and the gateway does not persist anything. This choice is motivated by the fact that the engine logs all state changes, so there is no need for a persistent gateway.

Observe that, as experienced by WfMS vendors, building a commercial-strength workflow engine is not an easy job, especially if it includes tracking, monitoring, and business transaction functionalities and has demanding requirements in terms of availability and performance. Hence, we believe that the architecture characterized by the reuse (or possibly the adaptation) of a commercial workflow engine is the alternative that most ESP vendors will follow. Indeed, this is the path followed by HP, whose middleware offering includes e-speak and Process Manager.

We expect that in the future, as ESPs add more functionality in terms of high availability, load balancing, monitoring, and support for business transactions, the need for integrating commercial workflow engines will progressively reduce, and the development from scratch of an interpreter designed and optimized for CSDL will become realistic.

5 Related Work

To the best of our knowledge, there is no commercial composition/process management system that can perform e-service composition and satisfy the requirements stated in Section 3, neither among traditional workflow management systems (such as *MQ Wor*kflow [10], *InConcert* [7], or *Staffware2000* [11]), nor among newly developed, open, XML- and web-based systems such as *Forte' Fusion* [9] and *KeyFlow* [6].

E-services platform themselves do not provide service composition capabilities, although all vendors declared interest in moving into this space. The only exception is WebMethods, who provide a very simple composition language for composing WebMethods' services [12]. The language allows the definition of flows that are a subset of what is allowed by traditional workflow management system (basically it can only model sequential or conditional flows where services are statically bound to service nodes), and therefore does not have many of the features presented in this paper. On the other hand, it is well suited for compositions that have simple requirements and it is quite easy to use.

Within the research community, approaches that are more closely related to the work presented here have been proposed by Georgakopoulos at al. [5] and by Benatallah at el. [1]. The first paper proposes a service-oriented process model targeted at enabling cross-organizational processes. The paper also presents a service model, where services are described by a state machine that specifies the valid "logical" states of a service and the valid state transition, caused by either method invocations or by transitions performed internally by the service. The paper differs from ours in that it focuses on the service model and only briefly sketches the service composition model. Instead, we assume that the service model is provided by the ESP, and we focus on the composition. In addition, the paper does not deal with certificates and data mappings/extraction while communicating with the e-services.

Benatallah et al. propose a framework for creating and maintaining virtual enterprises, where component enterprises share e-services. The main focus of the paper is on a model for managing service communities. However the paper also deals with service composition, and proposes an ECA-rule based approach for defining the composition. Our work differs in that CSDL has a graph-based approach to specify the composition. In addition, the paper also does not deal with search recipes, certificates, and data mappings and extraction, which are critical in our approach.

6 Concluding Remarks

This paper has presented the functionality and implementation of an e-service for composing e-services. The main contributions of this paper are:

- The idea and notion of providing composition functionality as an e-service, to be used not only by the owner of the ESP, but also by any (authorized, and possibly paying) user.
- A discussion of the characteristics of composite e-services and of their differences with respect to "traditional", workflow-like composition.
- The definition of a composition model suitable for e-services.
- The description of our prototype implementation, that shows an approach that can be reused for implementing composition on top of any ESP.

This effort is the initial part of a long-term work that has the purpose of developing a lightweight engine that can execute CSDL services, based on the assumption that future ESPs will take care of providing load-balancing, monitoring, tracking, and of other functionality. In addition, we plan to integrate more concepts taken from *eFlow*, including generic nodes, multiservice nodes, and dynamic conversation selection.

References

1. B. Benatallah, B. Medjahed, A. Bouguettaya, A. Elmagarmid, and J. Beard. Composing and Maintaining Web-based Virtual Enterprises. Procs. of the VLDB-TES Workshop, Cairo, Egypt (2000)
2. F. Casati and M.C. Shan. Process Automation as the Foundation for E-Business. Procs. of VLDB2000, Cairo, Egypt (2000)
3. F. Casati, S. Ilnicki, L.J. Jin, and M.C. Shan. *eFlow*: an Open, Flexible, and Configurable System for Service Composition. Procs. of WECWIS, Milpitas, CA, USA (2000)
4. D. Georgakopoulos, M. F. Hornick, and A. P. Sheth. An Overview of Workflow Management: From Process Modeling to Workflow Automation Infrastructure. Distributed and Parallel Databases 3(2) (1995)
5. D. Georgakopoulos, H. Schuster, D. Baker, and A. Cichocki. Process-based e-service Integration. Procs. of the VLDB-TES Workshop, Cairo, Egypt (2000)
6. Keyflow Corp. Workflow Server and Workflow Designer (1999)
7. Ronni T. Marshak. InConcert Workflow. Workgroup Computing report, Vol 20, No. 3, Patricia Seybold Group (1997)
8. F. Leymann, D Roller. Production Workflows. Addison Wesley (2000)
9. J. Mann. Forte' Fusion. Patricia Seybold Group report (1999)
10. IBM. MQ Series Workflow - Concepts and Architectures (1998)
11. Staffware Corporation, Staffware2000 White Paper (1999)
12. WebMethods Inc. WebMethods Enterprise (2000)

Data Models with Multiple Temporal Dimensions: Completing the Picture

Carlo Combi and Angelo Montanari

Department of Mathematics and Computer Science, University of Udine
Via delle Scienze 206, 33100 Udine, Italy
{combi, montana}@dimi.uniud.it

Abstract. There is a widespread recognition that valid and transaction times are the fundamental temporal dimensions of any fact relative to a database. There are, however, temporal aspects of facts that cannot be naturally modeled by means of them. A remarkable limitation of valid and transaction times is that they do not allow one to distinguish between retroactive and delayed updates. A third temporal dimension, called event time, has been proposed in the literature, which makes it possible to model retroactive, on-time, and proactive updates.
In this paper, we first refine the notion of event time by showing that one event time does not suffice to model relevant phenomena, and then we introduce a further temporal dimension, that we called availability time, which can be viewed as the information system counterpart of the real-world event time. We conclude the paper by outlining current and future work directions.

1 Introduction

As claimed by Jensen and Snodgrass [12], the study of temporal databases is a significant research topic in the database community. The ambitious goal of temporal databases is to manage time-varying information in a systematic and efficient way. Hence, it is not surprising that temporal database research covers a variety of areas, ranging from the conceptual level, where a number of different temporal extensions of the basic atemporal Entity-Relationship model has been proposed (cf. [8]), to the physical one, where indexing techniques suitable for dealing with data encompassing multiple temporal dimensions have been developed (cf. [22]). The focus of this paper is on the association of times with facts, which is at the basis of temporal data management.

There is a widespread recognition that valid and transaction times are the fundamental temporal dimensions of any fact relative to a database [12]. The *valid time* of a fact is the time when the fact is true in the domain, while the *transaction time* of a fact is the time when the fact is current in the database. Hence, valid and transaction times capture the time-varying states of the domain and of the database, respectively. Pairing valid and transaction times, one can recover past states of both the domain and the database. For instance, one can retrieve the real-world evolution of a given data item as recorded in the database

K.R. Dittrich, A. Geppert, M.C. Norrie (Eds.): CAiSE 2001, LNCS 2068, pp. 187–202, 2001.
© Springer-Verlag Berlin Heidelberg 2001

at a given time point as well as the database evolution of a data item at a given time point. Jensen and Snodgrass [10] systematically analyze the semantics of all admissible combinations of valid and transaction times.

There are, however, meaningful temporal aspects of facts that cannot be naturally modeled by means of valid and transaction times. A remarkable limitation of these two temporal dimensions is that they do not allow one to distinguish between retroactive and delayed updates [15]. Kim and Chakravarthy [14] propose a third temporal dimension, called *event time*, that makes it possible to model retroactive, on-time, and proactive updates, and show how the combined use of valid, transaction, and event times allows one to maintain different past states generated by retroactive and proactive updates as well as by error corrections and delayed updates.

In this paper, we first refine the notion of event time, by showing that one event time does not suffice to model relevant phenomena. To overcome its limitations, we associate with each fact both the occurrence time of the event that initiates its validity interval and the occurrence time of the event that terminates it. Then, we introduce a further temporal dimension, that we called *availability time*, which can be viewed as the information system counterpart of the real-world event time. The availability time of a fact is the time interval during which the fact is available to and believed correct by the information system. As it will be shown, such an interval does not necessarily coincide with the transaction time interval of the fact.

The rest of the paper is organized as follows. In Section 2, we recall the basic notions of valid, transaction, and event times, and the related temporal database taxonomies. Section 3 illustrates the proposed refinement of the event time. Section 4 introduces the new concept of availability time and discusses the benefits of its usage. Section 5 outlines current and future work directions. Throughout the paper, we use examples taken from a medical scenario, which will be adopted as the target application domain. However, the proposed temporal dimensions are completely general and can be used in other application contexts as well.

2 Background

2.1 The Basic Temporal Dimensions: Valid and Transaction Times

Valid and transaction times are widely recognized as the two basic temporal dimensions of temporal databases. Furthermore, there exists a consolidated terminology about temporal databases provided with valid and/or transaction times, as witnessed by the Consensus Glossary of Temporal Database Concepts [9]. Valid and transaction times are defined as follows.

Definition 1. *Valid time: the* valid time *(VT) of a fact is the time when the fact is true in the modeled reality.*

Table 1. Database instance of the patient therapies.

Drug	VT	TT
bipuvac	[98Aug10;10:00, 98Aug10;14:00)	[98Aug10;9:00, 98Aug10;12:00)
bipuvac	[98Aug10;10:00, 98Aug10;11:15)	[98Aug10;12:00, ∞)
diazepam	[98Aug10;11:25, 98Aug10;14:00)	[98Aug10;12:00, ∞)

Definition 2. *Transaction time: the* transaction time *(TT) of a fact is the time when the fact is current in the database and may be retrieved.*

Valid time is usually provided by database users, while transaction time is system-generated and supplied. Valid and transaction times are orthogonal dimensions: each of them can be independently recorded or not and has specific properties [10].

To explain the meaning of these two basic temporal dimensions, we consider a simple example, taken from a general medical scenario.

Example 1. On August 10, 1998, the physician prescribes a bipuvac-based therapy from 10:00 to 14:00. Data about the therapy is entered into the database at 9:00. Due to the unexpected evolution of the patient state, the bipuvac infusion is stopped at 11:15 and replaced by a diazepam-based therapy from 11:25 to 14:00. The new facts are entered at 12:00. (Bipuvac and diazepam are drugs commonly used in anesthesia.)

This example can be modeled as in Table 1. Each tuple is timestamped by valid and transaction times (hereafter, without loss of generality, we will use a temporal relational data model based on the tuple timestamping approach [12]). Both valid and transaction times are represented by intervals, where the occurrence of the special symbol ∞ as the ending point of an interval means that the interval includes the current time (the ending point of an interval that includes the current time is also denoted by `NOW` or `u.c.`, for `until changed`, in the literature). There exists a variety of possible relationships between valid and transaction times of a given tuple: the starting point of the transaction time can precede the valid time (as in the first tuple of Table 1), the starting point of the transaction time can follow the valid time (as in the second tuple), the starting point of the transaction time can follow the starting point of the valid time and precede its ending point (as in the third tuple), and so on.

Depending on the modeled application domain, valid and transaction times can be related to each other in different ways. Jensen and Snodgrass [10] propose several classifications (*specializations*, in the authors' terminology) of (bi)temporal relations, based on the relationships between valid and transaction times of timestamped facts. They take into account both the relationships that exist between valid and transaction times of any given tuple of a relation as well as the relationships that exist between the valid and/or transaction times of different tuples of the relation. As an example, on the basis of the relationships that exist between the starting and/or ending points of the valid and transaction times of any given tuple of a relation, Jensen and Snodgrass characterize the following classes of relations: retroactive, delayed retroactive, predictive, early predictive, retroactively bounded, strongly retroactively bounded, delayed strongly retroactively bounded, strongly predictively bounded, early strongly predictively bounded, strongly bounded, predictively bounded, general, degenerate, retroactively determined, predictively determined [10]. For instance, given the starting point VT_s of the valid time and the starting point TT_s of the transaction time, a relation is *retroactive* if, for each tuple of the relation, the relationship $VT_s \leq TT_s$ holds, while a relation is *delayed retroactive with bound* $\Delta t \geq 0$ if, for each tuple of the relation, the relationship $VT_s \leq TT_s - \Delta t$ holds. According to Jensen and Snodgrass' classification, the relation of Table 1, taken as a whole, is *general*, because the first tuple has $VT_s > TT_s$, while the second and the third ones have $VT_s < TT_s$; if we apply the same criterion to each single tuple of the relation, we can classify the first tuple as *predictive* and the remaining two tuples as *retroactive*[1].

A problem which has been debated to some extent in the literature is whether a single valid time can always be associated with a fact or there exist situations in which multiple valid times are needed [11,16]. In the extended version of this paper [5], we argue that one valid time is enough and the problem is that of properly identifying, at the conceptual level, what constitutes a fact.

Finally, it is worth noting that we assume the valid time domain to be linearly ordered, thus excluding branching valid times. Branching valid time is needed to model situations where different perceptions of the reality exist (one for each distinct timeline). An example of this kind of situations comes from historians who hardly agree on how the world evolved. The linear vs. branching valid time alternative is, however, orthogonal to our characterization of the relevant temporal dimensions of data models, that is, the dimensions we propose in the following can be directly applied also to data models with branching valid times.

2.2 A Third Temporal Dimension: The Event Time

In [15], Kim and Chakravarthy point out the inability of valid and transaction times to distinguish between retroactive and delayed updates and propose the addition of a third temporal dimension to solve this problem.

Example 2. Consider the following two scenarios: in the first one, an on-time promotion event, that increases the rank of a physician, occurs on October 1, 1997, but its effects are recorded in the database on November 1 (delayed update); in the second one, a retroactive promotion event, whose effects are immediately recorded in the database, occurs on November 1, but the increase in rank of the physician becomes valid since October 1 (retroactive update).

In the first scenario, the increase in rank is known (and thus it produces its effects) as soon as it becomes valid (October 1), while, in the second scenario, even though the time at which the increase in rank becomes valid precedes the

[1] Notice that different classifications of the same tuple can be obtained by comparing different endpoints of its valid and/or transaction times.

time of the promotion, the increase in rank is not known until the promotion event occurs (November 1). Since two-dimensional (valid and transaction) temporal databases treat retroactive and delayed updates in the same way, they are not able to discriminate between the two scenarios. To overcome these problems, Kim and Chakravarthy [13] introduced the notion of *event time*, which was refined in [14,15].

Definition 3. *Event time: the* event time *(ET) of a fact is the occurrence time of the real-world event that generates the fact.*

Whenever the event time coincides with the starting point of the valid time, it can actually be represented by the valid time. However, such a reduction is not possible when the two times are distinct. Let us consider, for example, the hospitalization of a patient. The valid time is related to the time interval during which the patient stays in the hospital, while the event time is related to the time at which the family doctor decides the hospitalization of the patient or the patient himself decides to go to the hospital. The event time can either coincide with the starting point of the hospitalization, whenever it happens immediately after the decision, or precede the starting point of the hospitalization, in case of scheduled hospitalizations.

In [15], Kim and Chakravarthy provide two different classifications of events, based on their relationships with the corresponding valid and transaction times, respectively. On the basis of the relationship between the event time and the starting point of the corresponding valid interval, one can partition events in three classes:

- On-time events: the validity interval starts at the occurrence time of the event (e.g., hospitalization starts immediately after the family doctor decision).
- Retroactive events: the validity interval starts before the occurrence time of the event (e.g., on December 21, 1997, the manager of the hospital decided an increase of 10% of the salary of the physicians of the Pathology Department, starting from December 1, 1997).
- Proactive events: the validity interval starts after the occurrence time of the event (e.g., on January 29, 1998, the family doctor decides the patient hospitalization on February 15, 1998).

When a database is updated with a fact generated by a retroactive event, Kim and Chakravarthy call the update a *retroactive update*, while, when the generating event is a proactive event, they call the update a *proactive update* [15]. This terminology is somehow misleading, because it may happen that the database update of a fact generated by a proactive event actually occurs after the beginning of the validity interval of the corresponding fact.

A different classification can be obtained by considering the relationships between event time and transaction time:

- On-time update: the transaction time coincides with the event time. This situation happens when data values are inserted in the database as soon as they are generated.

- Delayed update: the transaction time is greater than the event time. This is the case when data values are inserted some time after their generation.
- Anticipated update (which has not been taken into account by Kim and Chakravarthy): the transaction time is less than the event time. This is the case when data values are entered into the database before the occurrence time of the event that generates them. Such a notion of anticipated update is useful to model hypothetical courses of events.

It is not difficult to realize that the two classifications are orthogonal, and thus all possible combinations of them are admissible.

3 One Event Time Is not Enough

The original notion of event time suffers from an intrinsic limitation: it implicitly assumes that it is sufficient to associate a single event (the so-called generating event) with each valid fact. The generating event is the event that initiates the validity interval of the fact. In many situations, such an assumption is acceptable. Consider the case of a relation that keeps track of the temporal evolution of the rank of a physician. Each event that either increases (promotion) or decreases (downgrading) the rank not only initiates a period of time during which the physician has the new rank, but also implicitly terminates the validity interval of the current rank. In such a case, the event time of each tuple in the rank relation is the occurrence time of the promotion or downgrading event that initiates its validity interval, while the occurrence time of the event that terminates the validity interval is the event time of the tuple describing the next rank of the physician. On the contrary, whenever there are gaps in temporal validity (consider the case of a relation modeling admissions to hospital, which only records information about inpatients), or only incomplete information about the effects of an event is available (this is the case when we know that an event that changes the rank of the physician occurred, but we do not know if it is a promotion or a downgrading event), or, finally, the expected termination of a validity interval is (must be) revised, while its initiation remains unchanged (suppose that a prescribed therapy needs to be stopped due to an unexpected evolution of the patient state), we need to distinguish between the occurrence times of the two events that respectively initiate and terminate the validity interval of the fact.

The following example, which slightly revises Example 1, should clarify the point.

Example 3. On August 10, 1998, at 8:00, the physician prescribes a bipuvac-based therapy from 10:00 to 14:00. Data about the therapy is entered into the database at 9:00. Due to the unexpected evolution of the patient state, at 11:00 the physician decides a change in the patient therapy. Accordingly, the bipuvac infusion is stopped at 11:15 and replaced by a diazepam-based therapy from 11:25 to 14:00. The new facts are entered at 12:00.

This example can be modeled as in Table 2. The combined use of valid, transaction, and event times allows us to characterize the times at which the

Table 2. Database instance of the patient therapies with one event time (case 1).

Drug	VT	ET	TT
bipuvac	[98Aug10;10:00, 98Aug10;14:00)	98Aug10;8:00	[98Aug10;9:00, 98Aug10;12:00)
bipuvac	[98Aug10;10:00, 98Aug10;11:15)	98Aug10;8:00	[98Aug10;12:00, ∞)
diazepam	[98Aug10;11:25, 98Aug10;14:00)	98Aug10;11:00	[98Aug10;12:00, ∞)

Table 3. Database instance of the patient therapies with one event time (case 2).

Drug	VT	ET	TT
bipuvac	[98Aug10;10:00, 98Aug10;14:00)	98Aug10;8:00	[98Aug10;9:00, 98Aug10;12:00)
bipuvac	[98Aug10;10:00, 98Aug10;11:15)	98Aug10;11:00	[98Aug10;12:00, ∞)
diazepam	[98Aug10;11:25, 98Aug10;14:00)	98Aug10;11:00	[98Aug10;12:00, ∞)

physician takes his/her decisions about the therapy, the time intervals during which therapies are administered, and the times at which facts are entered into the database.

However, the proposed representation suffers from a major weakness: it seems that, at 8:00, the physician prescribes a bipuvac-based therapy from 10:00 to 11:15 and that, at 11:00, he/she prescribes a diazepam-based one. This is not correct because at 8:00 the physician actually prescribes a bipuvac-based therapy from 10:00 to 14:00, which is revised at 11:00[2]. The alternative representation given in Table 3 also does not work. According to such a representation, it seems that, at 11:00, the physician prescribes a bipuvac-based therapy from 10:00 to 11:15, thus making the prescription a retroactive event (which is clearly meaningless is the considered domain).

The problem illustrated by the above example motivated us to suitably refine the concept of event time. We distinguish between the occurrence time of the event that initiates the validity interval of a fact (*initiating event time*) and the occurrence time of the event that terminates the validity interval of a fact (*terminating event time*).

Definition 4. *Event time (revisited): the* event time *of a fact is the occurrence time of a real-world event that either initiates or terminates the validity interval of the fact.*

For any fact, we denote its initiating and terminating event times by ET_i and ET_t, respectively. Obviously, it holds that $ET_i \leq ET_t$. The possible relationships between event and valid times of a fact are depicted in Figure 1. Providing each fact with these two event times makes it possible to correctly model the considered example, as shown in Table 4. Notice that, unlike what has been pointed out in [12], there are more than one event time per fact (the initiating

[2] Notice that the first tuple cannot be used to explain the situation, because it has been logically deleted, and thus it does not model the current state of the world.

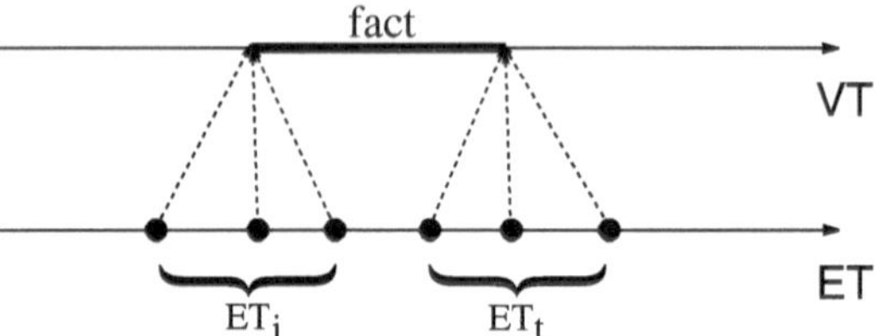

Fig. 1. On the relationships between event (ET_i, ET_t) and valid (VT) times of a fact.

Table 4. Database instance of the patient therapies with two event times.

Drug	VT	ET_i	ET_t	TT
bipuvac	[98Aug10;10:00, 98Aug10;14:00)	98Aug10;8:00	98Aug10;8:00	[98Aug10;9:00, 98Aug10;12:00)
bipuvac	[98Aug10;10:00, 98Aug10;11:15)	98Aug10;8:00	98Aug10;11:00	[98Aug10;12:00, ∞)
diazepam	[98Aug10;11:25, 98Aug10;14:00)	98Aug10;11:00	98Aug10;11:00	[98Aug10;12:00, ∞)

and terminating event times), even when there is only one decision maker (the physician in the above example).

This distinction between initiating and terminating event times comes from ideas underlying classical formalisms in the area of reasoning about actions and change [4]. In particular, a similar classification of events is given in the model of change of Kowalski and Sergot's Event Calculus (EC) [17]. The notions of event, property, time point, and time interval are the primitives of the formalism: events happen at time points and initiate and/or terminate time intervals over which properties hold. From a description of the events which occur in the real world and the properties they initiate or terminate, EC derives the validity intervals over which properties hold [3]. In temporal deductive databases, this approach has been adopted, for example, by Sripada in [24], where EC is extended to manage the transaction time. The main differences between our database-oriented approach and the EC ontology are: (i) EC explicitly records event occurrences and derives validity intervals of properties at query time, while we directly record properties (i.e., facts) and their validity intervals, together with the occurrence times of the events that initiate and terminate them; (ii) EC only deals with on-time events, while we also consider proactive and retroactive events[3].

The choice of adding event time(s) as a separate temporal dimension has been extensively debated in the literature [21]. The basic issue is whether or not events and facts must be dealt with in a different way in temporal databases. If we consider events just as a special class of facts (instantaneous facts), we do not need event time(s) at all: events are explicitly recorded in the database,

[3] For the sake of simplicity, we only considered atomic events. Whenever the initiation (resp. termination) of the validity of a fact is the combined effect of a set of events, we need to replace atomic events by composite events. A preliminary formalization of a calculus of macro-events can be found in [2].

as all the other relevant facts, and their occurrence time is (captured by) the valid time. A major weakness of this approach is that the data model does not provide the relationships between events and initiated/terminated facts with any built-in semantics, but delegates such a task to the application. On the contrary, conventional relations model the reality relevant to a given domain as a set of temporally-extended facts, which are delimited by the occurrence times of their initiating and terminating events [11]. Events are not considered first-class citizens and thus are not recorded into the database. In such a case, the addition of event time(s) is needed to deal with retroactive and/or proactive events that occur at times which do not coincide with the starting (resp. ending) point of the valid time of the fact they initiate (resp. terminate). In our opinion, this second approach provides a model of the domain which is both more adequate and more concise. On the one hand, it explicitly represents the cause-effect relationships that capture the behaviour of the domain (adequacy). On the other hand, it completes the temporal description of a fact by representing the occurrence times of its initiating/terminating events, without imposing the explicit storage of other information about them, which is often neither needed nor possible (conciseness)[4].

4 A New Temporal Dimension: The Availability Time

Event times are strictly related to the modeled world, as they allow one to represent the occurrence times of events (decisions, happenings, actions, etc.) that initiate or terminate meaningful facts of the considered domain. In this section, we focus on the temporal dimensions relevant to the information system. By *information system* we mean the set of information flows of an organization and the human and computer-based resources that manage them. From such a point of view, we may need to model the time at which (someone/something within) the information system becomes aware of a fact as well as the time at which the fact is stored into the database. While the latter temporal aspect is captured by the transaction time, the former has never been explicitly modeled.

We first show that such a time neither coincides with any of the two event times nor with the transaction time. Let us consider the following scenario.

Example 4. Due to a trauma that occurred on September 15, 1997, Mary suffered from a severe headache starting from October 1. On October 7, Mary was visited by a physician. On October 9, the physician administered her a suitable drug. The day after, the physician entered acquired information about Mary's medical history into the database. On October 15, the patient told the physician that her headache stopped on October 14; the physician entered this data into the database on the same day.

[4] In those few cases where further information about initiating/terminating events is needed (for instance, with regards to Example 3, we may be interested in storing the name and the specialty of the physician(s), who prescribed the therapies), event times should be paired with other event-related attributes.

Table 5. Database instance after the first update.

symptom	VT	ET_i	ET_t	TT
headache	[97Oct1, ∞)	97Sept15	null	[97Oct10, ∞)

Table 6. Database instance after the second update.

symptom	VT	ET_i	ET_t	TT
headache	[97Oct1, ∞)	97Sept15	null	[97Oct10, 97Oct15)
headache	[97Oct1, 97Oct14)	97Sept15	97OCt9	[97Oct15, ∞)

This scenario involves several temporal dimensions: the time interval during which Mary suffered from headache, the occurrence time of the trauma, the occurrence time of the drug administration, the times at which the physician became aware of Mary's headache onset and cessation, and finally the times at which the database is updated.

Valid and transaction times are respectively used to record the time interval during which Mary suffered from headache and the database update times, while the occurrence times of trauma and drug administration are modeled by means of the initiating and terminating event times (cf. Tables 5 and 6). There is no way of representing the times at which the physician became aware of Mary's headache onset and cessation. Indeed, while information about Mary's headache cessation is entered in the database as soon as it became available, information about the headache onset became available three days before its registration into the database.

In general, the time at which information becomes available precedes, but not necessarily coincides with, the time at which it is recorded in the database, and thus we cannot use transaction time to model it.

Indeed, in many applications, where data insertions are grouped and executed in batches, possibly on the basis of previously filled report forms, transaction time cannot be safely taken as the time when a fact has been acquired by the information system. In some cases, it may happen that the order according to which facts are known by the information system differs from the order in which they are stored into the database. Since there are many application domains, including the medical one, where decisions are taken on the basis of the available information, no matter whether or not it is stored in the database, we need to introduce a new temporal dimension, that we call *availability time*, to deal with it. We informally define the *availability time* (AT) as the time at which a fact becomes available to the information system. Later on (cf. Definition 5), we will generalize and make more precise such a definition. Figure 2 shows how transaction and availability times are related to the database and information systems, respectively. Notice that the information system includes, but does not necessarily coincide with, the database system. In our medical example, for instance, the information system includes both the database and the physician.

By exploiting the availability time, we are able to completely model the previous example. As shown in Table 7, the availability time of the first tuple,

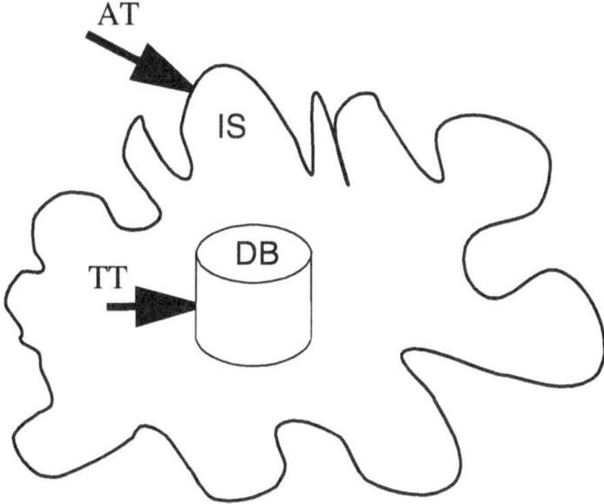

Fig. 2. Availability (AT) and transaction (TT) times and their relationships with the database (DB) and information systems (IS).

Table 7. Database instance after the second update (revisited).

symptom	VT	ET_i	ET_t	AT	TT
headache	[97Oct1, ∞)	97Sept15	null	97Oct7	[97Oct10, 97Oct15)
headache	[97Oct1, 97Oct14)	97Sept15	97Oct9	97Oct15	[97Oct15, ∞)

which is the time at which the physician learns about Mary's headache onset (October 7), strictly precedes the beginning of the transaction time interval (October 10), while the database time and the beginning of the transaction time interval of the second tuple coincide (October 15).

Let us suppose that we are interested in evaluating the quality of the care provided by the physician. In case we lack information about the availability time (cf. Table 6), we only know that Mary's headache started on October 1, due to some event occurred on September 15 (as of October 10), and stopped on October 14, thanks to (presumably) some therapeutic action occurred on October 9 (as of October 15). As of October 10, we can conclude only that the physician became aware of Mary's headache some time in between October 1 (the starting point of the validity interval) and October 10 (the starting point of the transaction time). This uncertainty makes it difficult to assess the quality of the care, because it is not possible to unambiguously relate clinical data (headache onset and cessation) to therapeutic strategies (to decide the proper therapeutic action, on the basis of the available knowledge about the patient history).

In particular, if we assumed that the physician became aware of the headache on October 1, we would observe a delay of more than one week between the awareness of Mary's problem and the execution of the action that solved the problem. On the contrary, if we assumed that the physician became aware of Mary's headache on October 10, we should conclude that Mary recovered from her headache without any intervention from the physician (the event that terminates the validity time interval of the headache happened on October 9).

Table 8. Database instance after error correction.

symptom	VT	ET_i	ET_t	AT	TT
headache	[97Oct1, ∞)	97Sept5	null	97Oct7	[97Oct10, 97Oct15)
headache	[97Oct1, 97Oct14)	97Sept5	97Oct9	97Oct15	[97Oct15, 97Oct21)
headache	[97Oct1, 97Oct14)	97Sept15	97Oct9	97Oct20	[97Oct21, ∞)

Remark. *The concept of availability time allows us to clarify the relationships between Kim and Chakravarthy's (initiating) event time and the related notion of decision time, which has been originally proposed by Etzion and his colleagues in [6] and later refined in subsequent work, e.g., in [7,19,20]. The* decision time *of a fact is the time at which the fact is decided in the application domain of discourse. More precisely, it can be defined as the occurrence time of a real-world event, whose happening induces the decision of inserting a fact into the database. In the conceptual framework we propose, the decision time models the temporal aspects of an event, initiating the validity interval of a fact, that occurs within the information system, and thus is immediately known by it. In these circumstances, the initiating event time and the availability time coincide (and are before than or equal to the starting point of the transaction time interval).*

Up to now, we have used the availability time to model the time at which relevant knowledge becomes available to the system, without considering the possibility that it acquires erroneous data. In the following, we will show how the notion of availability time can be generalized to take into account such a possibility. Let us now consider the following variant of the previous example.

Example 5. Due to an insertion mistake (or to an imprecision in Mary's talk), the trauma has been registered as happened on September 5, 1997. Only on October 20, the mistake was discovered. The day after, the physician entered the correct data into the database.

The database instance resulting from the error correction is shown in Table 8. A limitation of this representation is that information about the time interval during which any fact is known and *believed correct* cannot be obtained from the corresponding tuple. As an example, to conclude that the second tuple was considered correct until October 20, we need to access the availability time of the third tuple. Unfortunately (cf. [1]), it is not possible, in general, to determine how a given tuple of a relation has been updated, whenever several modifications and/or insertions are performed by means of a single transaction (in such a case, all the updated tuples share the same transaction time). Even worse, when the discovery of an error forces us to (logically) delete some tuples, which are not replaced by new ones (suppose that, on October 20, we discover that the trauma did not happen at September 5, but we do not know yet when it actually happened), there is no way of keeping trace of the time at which the error has been discovered.

That leads us to consider the availability time as an interval for a fact. The starting point of the availability time is the time at which the fact becomes available to the information system, while its ending point is the time at which

Table 9. Database instance after error correction (revisited).

symptom	VT	ET_i	ET_t	AT	TT
headache	[97Oct1, ∞)	97Sept5	null	[97Oct7, 97Oct15)	[97Oct10, 97Oct15)
headache	[97Oct1, 97Oct14)	97Sept5	97Oct9	[97Oct15, 97Oct20)	[97Oct15, 97Oct21)
headache	[97Oct1, 97Oct14)	97Sept15	97Oct9	[97Oct20, ∞)	[97Oct21, ∞)

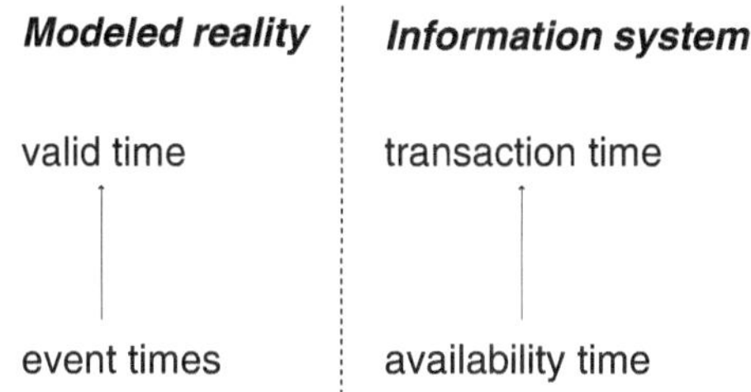

Fig. 3. Temporal dimensions, modeled reality, and information system.

the information system realizes that the fact is not correct (as for transaction time, being the ending point equal to ∞ means that the fact is currently believed correct). By modeling availability time as an interval, we are able to correctly represent the previous example, as shown in Table 9.

Definition 5. *Availability Time: the* availability time *(AT) of a fact is the time interval during which the fact is known and believed correct by the information system.*

As for the relationships between database and transaction times of a fact, it holds that $AT_s \leq TT_s$ and $AT_e \leq TT_e$: a fact can be inserted into the database only when it is known, or later, and it can be logically removed from the database only when it is recognized incorrect, or later.

Remark. *Notice that the availability time can be viewed as the transaction time of the information system. If the information system outside the database is considered a database (in general, this is not case; in our medical examples, for instance, the information system includes both databases and physicians), then the availability time of a fact is when the fact is inserted into (deleted from) the information system. This parallel between availability time and transaction time makes it immediately clear why the availability time has to be an interval. This point of view on the availability time resembles the notion of temporal generalization (cf. [10]) which allows several levels, and thus several transaction times, in a database.*

Figure 3 summarizes the relationships between valid, transaction, event, and availability times, the modeled reality, and the information system.

Transaction time is append-only: facts previously stored into the database cannot be changed. The acquisition of new knowledge about the domain results into the addition of new facts, whose transaction time interval includes the current time, to the database; the (logical) deletion of incorrect knowledge is obtained by "closing" the transaction times of the corresponding facts; modifications can be performed by combining the two previous actions. In contrast, valid and event times, being related to times of the represented real world, can be located either in the past or in the future, and they can be modified freely. As regards the availability time, it is append-only in its nature, because facts previously known and believed correct by the information system cannot be changed. However, from the point of view of the database system, availability time would be really append-only only if there were no errors in data entry. Since we cannot exclude such a possibility, previous states of the information system, according to the availability time, can be revised by entering new data. Furthermore, even assuming data entry without errors, database and transaction times may "append" facts according to two different orders. A systematic analysis of the relationships between the different temporal dimensions is provided in [5].

5 Conclusions and Further Work

In this paper, we proposed a new, fully symmetric conceptual data model with multiple temporal dimensions, that revises and extends existing temporal models. We first described a refinement of the concept of event time, that replaces the single event time, originally proposed by Kim and Chakravarthy, by a pair of *initiating* and *terminating event times*. In such a way, we are able to represent real-world situations where two different events, with (possibly) different occurrence times, are related to the starting and ending points of the validity interval of a fact. Then, we introduced a new temporal dimension, named *availability time*, which captures the time interval during which a fact is known and believed correct by the information system. The availability time allows us to distinguish the time at which a fact becomes available to the information system from the time at which it is entered into the database. This capability can be exploited, for example, to analyze the quality of decision making in information systems where data insertion is performed in batches and thus some delay is possible between data availability and data insertion.

Whether a special support for event and availability times must be provided or not is debatable. This issue can be considered as an instance of the more general problem of deciding whether or not built-in supports for temporal dimensions in databases must be provided (since the proposal of valid and transaction times, there has been a dispute about that). In this paper, we have shown that, from a conceptual point of view, event and availability times are temporal dimensions that capture some general aspects of applications which cannot be naturally managed by using valid and transaction times, and thus event and availability times cannot be reduced to simple user-defined times. We believe that capturing the application-independent meaning of temporal information is a main step towards the conceptual modeling of temporal information systems [23]. Highlighting these aspects at the conceptual level is, indeed, a basic task both when designing information systems by standard (atemporal) methodologies and when designing and adopting fully-fledged temporally-oriented methodologies [8]. The design issues, we are faced with when dealing with the problem of defining a query language for temporal databases (cf. [5]), confirm that all the four temporal dimensions must somehow be taken into account, also in the case we decide to manage them by a classical atemporal database system.

As for current and future research, we are working at a generalization of the temporal logic reconstruction of valid and transaction temporal databases outlined in [18], that copes with both event and availability times. Another problem we are dealing with is the identification and characterization of suitable temporal normal forms. Many temporal dependencies among data may indeed arise when relations are provided with multiple temporal dimensions. Finally, it goes without saying that suitable indexing techniques are needed to effectively manage temporal databases with four temporal dimensions. To this end, we are currently analyzing existing indexing techniques for multidimensional data. The existence of general constraints among the different temporal dimensions hints at the possibility of tailoring existing indexing techniques and search algorithms to our specific context.

Acknowledgments

We would like to thank Alberto Policriti, Richard Snodgrass, and the anonymous reviewers for their useful suggestions and comments on the work, Alberto Pasetto, chair of the Anesthesiology and Intensive Care Unit of the University of Udine, and his staff, who helped us to select meaningful clinical examples and gave us the opportunity to test a prototype of a database system with multiple temporal dimensions in a real-world, data-intensive, clinical unit, and our former students Ivan Andrian, Silvia Riva, and Fabio Valeri that contributed to the development of the temporal data model, to the study of temporal functional dependencies, and to the definition and implementation of the temporal query language T4SQL, respectively.

References

1. G. Bhagrava and S.K. Gadia. The Concept of Error in a Database: an Application of Temporal Databases. Proc. of the International Conference on Management of Data (COMAD), McGraw-Hill, New York, 106–121, 1990.
2. I. Cervesato and A. Montanari. A Calculus of Macro-Events: Progress Report. In A. Trudel, S. Goodwin (eds.), Proc. of the 7th International Workshop on Temporal Representation and Reasoning (TIME). IEEE Computer Society Press, Los Alamitos, 47–58, 2000.
3. L. Chittaro and A. Montanari. Efficient Temporal Reasoning in the Cached Event Calculus. *Computational Intelligence*, 12: 359–382, 1996.
4. L. Chittaro and A. Montanari. Temporal representation and reasoning in artificial intelligence: Issues and approaches. *Annals of Mathematics and Artificial Intelligence*, 28(1-4): 47–106, 2000.
5. C. Combi and A. Montanari. Data Models with Multiple Temporal Dimensions: Completing the Picture (revised version). *Research Report 40/00, Dipartimento di Matematica ed Informatica, Universita' di Udine*, December 2000.

6. O. Etzion, A. Gal, and A. Segev. Temporal Support in Active Databases. In Proc. of the Workshop on Information Technologies & Systems (WITS), 245–254, 1992.
7. O. Etzion, A. Gal, and A. Segev. Extended Update Functionality in Temporal Databases. In O. Etzion, S. Jajodia, and S. Sripada (eds.), *Temporal Databases - Research and Practice*, LNCS 1399, Springer, Berlin Heidelberg, 56–95, 1998.
8. H. Gregersen and C.S. Jensen. Temporal Entity-Relationship Models: a Survey. *IEEE Transactions on Knowledge and Data Engineering*, 11(3): 464–497, 1999.
9. C. Jensen, C. Dyreson (Eds.) et al. The Consensus Glossary of Temporal Database Concepts - February 1998 Version. In O. Etzion, S. Jajodia, and S. Sripada (eds.), *Temporal Databases - Research and Practice*, LNCS 1399, Springer, Berlin Heidelberg, 367–405, 1998.
10. C. Jensen and R. Snodgrass. Temporal Specialization and Generalization. *IEEE Transactions on Knowledge and Data Engineering*, 6: 954–974, 1994.
11. C. Jensen and R. Snodgrass. Semantics of Time-Varying Information. *Information Systems*, 21(4): 311–352, 1996.
12. C. Jensen and R.T. Snodgrass. Temporal Data Management. *IEEE Transactions on Knowledge and Data Engineering*, 11: 36–44, 1999.
13. S.K. Kim and S. Chakravarthy. Semantics of Time-Varying Information and Resolution of Time Concepts in Temporal Databases. In R.T. Snodgrass (ed.), Proc. of the International Workshop on an Infrastructure for Temporal Databases, Arlington, TX, G1–G13, 1993.
14. S.K. Kim and S. Chakravarthy. Modeling Time: Adequacy of Three Distinct Time Concepts for Temporal Databases. In Proc. of the 12th International Conference of the Entity-Relationship Approach, LNCS 823, Springer, Berlin Heidelberg, 475–491, 1993.
15. S.K. Kim and S. Chakravarthy. Resolution of Time Concepts in Temporal Databases. *Information Sciences*, 80: 91–125, 1994.
16. S. Kokkotos, E.V. Ioannidis, T. Panayiotopoulos, and C.D. Spyropoulos. On the Issue of Valid Time(s) in Temporal Databases. *SIGMOD Record*, 24(3): 40–43, 1995.
17. R. Kowalski and M. Sergot. A Logic-Based Calculus of Events, *New Generation Computing*, 4: 67–95, 1986.
18. A. Montanari and B. Pernici. Towards a Temporal Logic Reconstruction of Temporal Databases. In R.T. Snodgrass (ed.), Proc. of the International Workshop on an Infrastructure for Temporal Databases, Arlington, TX, BB1–BB12, 1993.
19. M. Nascimento and M. Dunham. Indexing a Transaction-Decision Time Database. In Proc. of ACM Symposium of Applied Computing (SAC), ACM Press, New York, 166–172, 1996.
20. M. Nascimento and M. H. Eich. On Decision Time for Temporal Databases. In S. Goodwin and H. Hamilton (eds.), Proc. of the 2nd International Workshop on Temporal Representation and Reasoning (TIME), 157–162, 1995.
21. G. Özsoyoglu and R. T. Snodgrass. Temporal and Real-Time Databases: A Survey. *IEEE Transactions on Knowledge and Data Engineering*, 7(4): 513–532, 1995.
22. B. Salzberg and V.J. Tsotras. Comparison of Access Methods for Time Evolving Data. *ACM Computing Surveys*, 31(2): 158–221, 1999.
23. Y. Shahar and C. Combi. Editors' Foreword: Intelligent Temporal Information Systems in Medicine. *Journal of Intelligent Information Systems (JIIS)*, 13(1-2): 5–8, 1999.
24. S.M. Sripada. A logical framework for temporal deductive databases. In Proc. 14th Very Large Data Bases Conference, Morgan Kaufmann, San Francisco, 171–182, 1988.

Querying Data-Intensive Programs for Data Design

Jianhua Shao, Xingkun Liu, G. Fu, Suzanne M. Embury, and W.A. Gray

Department of Computer Science
Cardiff University
Cardiff, CF24 3XF,Wales, UK
{J.Shao, S.M.Embury, W.A.Gray@cs.cf.ac.uk}

Abstract. A data-intensive program is one in which much of the complexity and design effort is centred around data definition and manipulation. Many organisations have substantial investment in data design (data structures and constraints) coded in data intensive programs. While there is a rich collection of techniques that can extract data design from database schemas, the extraction of data design from data intensive programs is still largely an unsolved problem. In this paper, we propose a query-based approach to this problem. Our approach allows users (maintainers or reverse engineers) to express a complex extraction task as a sequence of queries over the source program. Unlike conventional techniques, which are designed for extracting a specific aspect of a data design, our approach gives the user the control over what to extract and how it may be extracted in an exploratory manner. Given the variety of coding styles used in data intensive programs, we believe that the exploratory feature of our approach represents a plausible way forward for extracting data design from data intensive programs. We demonstrate the usefulness of our approach with a number of examples.

1 Introduction

To many organisations, knowledge of the data designs used within their information systems is vitally important. Without knowing precisely how data is organised, organisations will not be able to update their systems correctly and hence cannot support business change effectively. Yet, when working with legacy systems, this important design knowledge cannot always be assumed to be available. Typically, after many years of development and upgrades to a system, some of this design knowledge is lost, largely as a result of poor documentation and staff turnover. Consequently, legacy system maintenance is a difficult,costly and error-prone process. It is desirable, therefore, to consider how data design may be recovered from the legacy systems themselves. This has been an important area of research in information system engineering [10,2].

It is useful to begin by clarifying what we mean by data design. For many members of the reverse engineering community, "data design" means a data model of some sort. However, data models by themselves can only give a partial picture of how data is organised within an information system. To see the

K.R. Dittrich, A. Geppert, M.C. Norrie (Eds.): CAiSE 2001, LNCS 2068, pp. 203–218, 2001.
© Springer-Verlag Berlin Heidelberg 2001

full picture, details of data constraints must be present too. While some data models (e.g. relational schemas, ER models) can encode some constraints (e.g. uniqueness or cardinality constraints), in general there will always be some data constraints which cannot be represented in this way. These more general constraints can only be represented by application code. This would suggest that approaches to the reverse engineering of data designs should analyse both schema information *and* source code. However, with one or two exceptions [1,20,11], the potential value of source code in datareverse engineering has so far largely been ignored.

In this paper, we focus on the problems of extracting data design (i.e. data structure and constraints) from *data intensive programs*. A data intensive program (DIP) is an application program (usually associated with a database or data bank of flat files) that is rich in elements for either defining or manipulating data. A typical example of a DIP is an order handling transaction, written in COBOL and executing against a CODASYL database system. DIPs make up a significant proportion of the legacy code in use in industry today, and they represent a rich and largely untapped vein of information for reverse engineering of both data structure and constraints.

However, the use of DIPs as the raw material for reverse engineering of data design is hindered by the following difficulties:

- The kind of source code that is found in real legacy systems is very different from the programs that are found in textbooks. Programmers make use of clever implementation tricks that can obscure semantics, and which are very difficult to anticipate. In addition, many program languages offer a variety of ways of coding any given data structure, all equally valid. This makes it difficult to be able to predict in advance the set of patterns that can be expected to represent a given data structure or constraint. For example, in COBOL, REDEFINE statements are often used to denote sub structures within a larger one. However, they can also be used to save on memory usage. Any reverse engineering tool which assumes that a REDEFINE indicates only the former semantics will produce inaccurate results. This variety of coding techniques and standards present in DIPs demands great flexibility of any reverse engineering technique. It must be possible to apply such techniques selectively, and under the full control of the user.
- By their very nature, DIPs typically operate on data structures in their lowest level form (typically as COBOL record structures). Therefore, we not only have to contend with a great diversity in the encoding of functionality, but also in the representation of data structures. Because of this, we cannot expect to find definitive patterns from which we can definitely infer the presence of some data structure or constraint. Instead, we must try to collect evidence for and against some hypothesised data design. In this situation, it is unlikely that any single, fixed algorithm will be sufficient for our needs. We must be able to apply a range of different algorithms, each capable of detecting a specific form of "evidence", and we must be able to tailor them, to suit the characteristics of the situation in hand.

- Database schemas can represent some limited forms of constraint: referential integrity, uniqueness constraints, etc. However, the constraints present in DIPs can be arbitrarily complex. Again, this variety of semantic forms means that we cannot expect one or two algorithms to be capable of detecting the presence of all of them. Instead, we require an environment in which we can experiment with a range of different techniques, operating over the same input source code, in order to locate and build up a collection of candidate constraints, and to collect evidence for or against them.

How well do current approaches to reverse engineering from source code fit with these requirements? Broadly, there are three approaches to the construction of a reverse engineering toolkit. In the first, reverse engineering tools operate directly on the source code itself. In the second, the source code is translated into a machine-friendly intermediate format, and the reverse engineering tools operate on this special format rather than on the code itself. In the third approach, the transformed source code is stored in a database (called a *program base*), and the reverse engineering tools use its query interface in order to carry out their analyses.

We believe that this last approach provides the most suitable environment for reverse engineering data design from DIPs. Despite the fact that it overcomes many of the limitations of the other approaches in this context, it has not yet been employed for the reverse engineering of data designs. We propose an architecture based around the notion of a program base. Our approach has the following advantages. Firstly, it gives the user control over the extraction process. Unlike the techniques which are designed for extracting a specific aspect of a data design, our approach gives the user control over what to extract and how it may be extracted in an exploratory manner. Secondly, our approach allows extraction reuse. Once a successful extraction strategy (a sequence of query and manipulation) is developed, it can be reused or tuned for some future extraction tasks. Finally, the approach we propose is well suited for extracting data constraints that are fragmented and distributed in DIPs. Given the variety of coding styles used in DIPs, we believe that the exploratory feature of our approach represents a plausible way forward for successfully extracting data design from legacy data intensive programs.

The paper is organised as follows. Section 2 gives a discussion of the related work. We explain our approach in detail in Section 3. In Section 4,we demonstrate the usefulness of our approach by presenting examples from the analysisof source code taken from a legacy system currently in use at British Telecommunications (BT). Finally, conclusions are drawn in Section 5.

2 Related Work

After many years of research into data and software reverse engineering, a rich collection of techniques have been developed for the recovery of software and data designs from legacy systems [10,2,5]. In terms of data design extraction, the majority of this research has focussed on the recovery from structural schema-based information [16,4,6,9,12,18]. The techniques proposed take a database schema (relational, hierarchical or networked) as input and produce a conceptual data model as output. Typically, they assume the availability of some semantic knowledge about the schema: keys, candidate keys and some data dependencies [4]. For example, the technique proposed by Chiang, Barron and Storey [4] can be used to extract accurate EER models from a set of relations, but only when the input relations are in Third Normal Form. For DIPs, these assumptions are often unrealistic, since much of this required semantic knowledge is not explicitly specified in the program.

In terms of source code analysis, much existing work is in the area of software reverse engineering, which focuses on the extraction of software design from programs. This work can be categorised into the following three architectural styles.

Direct Analysis. In this first architectural style, reverse engineering techniques are implementedto operate directly on the source code itself. The result of the analysis isthen output in a form that is suitable for human consumption.

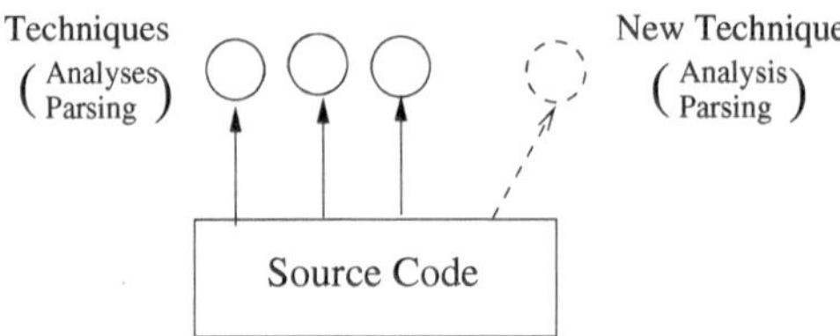

Since each technique is designed to work on the source code directly, it is necessary that the technique understands the syntax of the source code. This means that each direct analysis tool has to contain a built-in parser for the source language in which the source code is written. For example, a COBOL analysis technique proposed by Andersson [1] is quite powerful in resolving complex data structures and deriving limited data constraints from COBOL programs, but it must have a built-in COBOL parser if it is to work. Most program slicing techniques are of this type too [19,11] – they integrate the syntactical understanding of a program and the process of extracting a slice into one tool. This results in one major limitation of this type of technique. When a new tool is to be implemented, a new parser must be implemented within the tool, even if it is functionally equivalent to the parsers that already exist within the previously developed tools for the same language. Similarly, if an existing tool is to be used to analyse source code in a different language, it must be reimplemented with a new built-in parser. This results in a considerable duplication of effort whenever several tools are to be developed for use on the same source code.

Indirect Analysis. The second architectural style overcomes these limitations by using a single parserto translate the input program into a common machine-friendly format, on which all the reverse engineering tools operate.

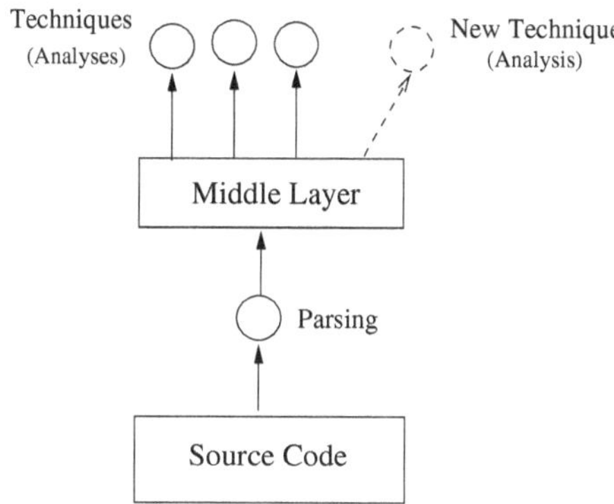

Now, when a new tool is developed, there is no need to re-implement the part of the tool that parses the source code. Similarly, existing techniques can be made to operate on source programs in a different language simply by implementing a new translator from that language to the common format.For example, the wide spectrum language (WSL) used in Maintenance Assistant [20] is sufficiently general to allow the semantics of several different source languages to be represented.

While it is a significant improvement, this approach still has some limitations. The techniques associated with this architecture are usually designed to do one "complete" extraction, i.e., to produce some pre-determined design pattern as output. While this might be thought to be entirely appropriate, it has the effect that the individual tools tend to be independent of one another and are not *composable*. In other words, it is usually not possible to use the output of one tool as the input of another, in order to perform some more complex analysis step. And yet this ability to combine tools is crucial if the flexible experimentation required for extraction of data designs from DIPs is to be supported in an economic manner.

For example, Yang and Chu have developed a system which translates a COBOL program into the WSL and then transform the WSL representation of the program into an ER model [20]. However, all the transformations implemented in their system are designed to work on some intended COBOL constructs in a prescribed way, and there is no room for the user to experiment with a variety of different transformations easily. That is, if one is to experiment with a new transformation that is not already available in the system, then the new transformation must be programmed in a general purpose programming language and brought into the system first. This can be time consuming and requires a good knowledge of the system. This seriously limits the usefulness of their system. For example, a REDEFINE statement is treated as defining an independent entity in their system, and no other interpretation of this statement may be attempted.

Query-Based Analysis. The third architectural style brings yet greater flexibility by using a database management system to store the source code in its intermediate, parsed form.

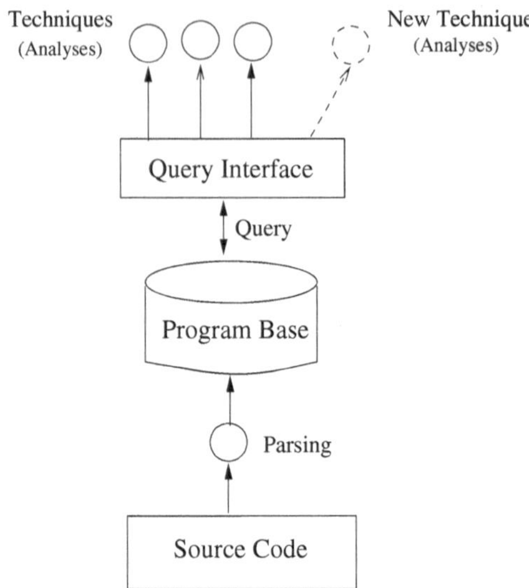

The main benefit of this is that analyses of programs now take the form of queries over this database and manipulation of query results[1]. Since the results of these queries and manipulations are database objects themselves (either retrieved from the original program's parse tree or generated from the manipulation of query results), they can be reused by other queries and manipulations. For example, one query might identify the parts of the program which are suitable for analysis by a particular group of reverse engineering techniques, while other queries/manipulations implement that family of analysis techniques. This allows more complex analysis tools to be created by composing existing ones.

In addition to this benefit, since the tools are written in a high level (and often declarative) language, they are much simpler to write, and easier to modify. This allows the user to formulate hypotheses, test them and then modify them in the light of their experience and query result. Moreover, since the results of analyses can also be stored in the database, time is saved that would otherwise be spent repeatedly calculating the same results. Given the size of many DIPs, and the complexity of many reverse engineering techniques, this can be a significant improvement in the overall usability of the reverse engineering environment. A further advantage is that the results of several different analyses can be compared with one another, to help the user make the correct decision when different analyses produce conflicting results.

The idea of querying source code for information is of course not new. For example, the commonly-used string searching tools, such as `grep` and `awk`, can be regarded as the simplest forms of source code querying. They are easy to use, but are only suitable for tracking some lexical structures in the source code. Paul and Prakash, on the other hand, developed an interesting algebra for querying

[1] Note that we have used the term "query" here and in the following rather loosely to include both retrieval and update operations.

source code stored as an abstract syntax tree [15]. With the algebra, it is possible and easy to compose primitive operations to form a non-trivial operation. For example, the following query retrieves all the function names used in file `dfa.c`:

$$retrieve_{func-name}(func - list(pick_{file-name(x1)='dfa.c'}(FILE)))$$

and is composed of 3 primitive operations.

There are other query-based systems which use different source code representations [3] or different query interfaces [13]. They all provide environments similar to the kind that we believe are most suitable forreverse engineering of data design from DIPs. Despite the fact that it overcomesmany of the limitations of the other approaches in this context, the query-based approachhas not yet been employed for the reverse engineering of data designs.

3 Our Approach

In this section we describe our approach to query-based architecture for reverse engineering of data design and demonstrate its usefulness with a number of examples. The work reported here is part of the BRULEE project[2] which aims to develop a software architecture to support the extraction and presentation of business rules buried within legacy systems. For more detailed discussion on business rules and the BRULEE project, the reader is referred to [17,7].

3.1 The Architecture

The system architecture is shown in Figure 1, which consists of three distinct parts: a preparation part, the Program Base and an extraction part.

In the preparation part, our main concern is to parse a DIP to create an abstract syntax tree, which is then stored in the Program Base. A parser generator is also included in our architecture to facilitate the production of parsers for different programming languages.

The Program Base is an object-oriented database management system. Given the variety and complexity of program constructs that may be found in DIPs and the fact that much of our source code analysis is highly navigational in nature, we consider that this form of DBMS is an appropriate one – it allows us to handle complex program constructs with ease and efficiency. The parsed program constructs are stored in the Program Base in an object-oriented format. For example, the statement `MOVE A TO B` in COBOL is stored in the database as

```
MoveStatement(2-2-2-391, 'MOVE', LeftArg, 'TO', RightArg)
LeftArgument(2-2-2-546, IdentifierList)
:
```

It is easy to see how all such objects link up to form an abstract syntax tree.

[2] Business RULe Extraction and Exploitation

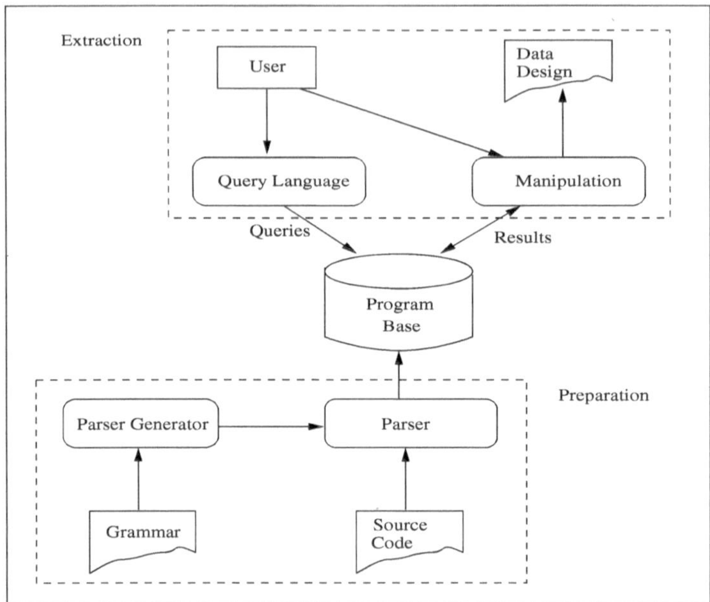

Fig. 1. The system architecture

In the extraction part, the user analyses the source code stored in the Program Base in order to identify any patterns or evidence that will contribute to the reconstruction of the data design buried in the source code. The way in which the source code is analysed is very different from a non-query based approach. Here, the user is in control and can explore the source code by posing any number of queries against the Program Base. By examining the query results, the user can then determine whether any of the results contribute to the data design to be recovered and what analysis should be performed next. A complex strategy for extracting data design is developed in this exploratory fashion.

Consider the following example. Intuitively, we could hypothesize that a `MOVE A TO B` statement, where `A` and `B` are attributes belonging to two different record types, might indicate that two record types are somehow related. Knowledge of such relationships is useful in order to understand a data design in a system. For example, with some additional semantics (extracted from the program or obtained from a domain expert), we may be able to establish that the two record types are in fact "referentially related" or forming a sub-type/super-type relationship. To verify this hypothesis (i.e. whether two record types are related via a `MOVE` statement), we can formulate the following sequence of queries/manipulations (we have used a slightly edited version of the query to make it easy to understand):

```
S1 = select(P, MOVE, left-arg = identifier);
For each M of S1
```

```
3        S2 = select(P, RECORD, field = M's left-arg);
4        S3 = select(P, RECORD, field = M's right-arg);
5        S3 = extend(S3, RECORD, related = S2);
6        S2 = extend(S2, RECORD, related = S3);
7        select(S2,RECORD);
8        select(S3,RECORD);
9    EndFor
```

These statements are largely self-explanatory. `select(src, cls, pred)` means a selection of program constructs of class `cls` from source `src` with an optional predicate `pred`, and `extends(src, cls, att=vlu)` creates new objects which are the extension of objects of class `cls` in source `src` with a new attribute value pair `att = vlu`. Finally, we use `S =` to represent an insertion of objects into `S`.

What the above code achieves is this. First, we select, from the program (`P`), all the `MOVE` statements whose left argument is a variable (`1`). For each such statement, we then select all the records that contain the left and right variable, respectively (`3,4`). Following that, we extend copies of objects in `S2` and `S3` with a new attribute `related` (`5,6`), where the association of two records through the `MOVE` statement is recorded. Note that the original records will remain intact. Finally, a set of related records is selected from `S2` and `S3` (`7,8`), which may be analysed further.

This approach has a number of advantages.

- Firstly, it is very simple and easy to develop a compact expression to verify a hypothesis that the user may have. This is important because when analysing a DIP, we often rely on some heuristics to search for plausible evidence, rather than definitive patterns from which we can definitely infer the presence of some data structures and constraints. For example, in addition to the `MOVE` statements, a nested record structure in COBOL could also (but not always) indicate an implicit relationship held between the record and sub-record. We believe, therefore, that this exploratory form of manipulation is essential for handling DIPs.
- Secondly, our approach allows the user to build up a data design gradually and incrementally. This is important because, given the complexity of a DIP, trying to recover a data design at a click of a single button would be too ambitious. In the above example, for instance, it is possible that after the result is retrieved and examined, we may decide to refine the extraction process by carrying out any or all of the following analyses, depending on what is actually contained in result set and whether there are are other forms of input available for us to analyse:
  - analysis of data to establish an inclusion dependency between the two related record types.
  - querying the program base for further evidence of relationships (e.g the implicit relationships defined by the nested records) to complete or confirm the set of relationships that have already been established.
  - filtering out the records that are related at the record level (i.e. COBOL `01` level) if we are only interested in attribute level relationships.

  - consulting human experts for confirmation of each of the identified relationships.

  These are just examples. Many more possible analyses and manipulation can be envisaged. However, it would not be easy to build all such possible steps into one algorithm. Nor would it be economic to do so because not all these steps are required in all data design extraction tasks. We therefore believe that incremental extraction represents a realistic and manageable way of extracting and building up a data design from a complex legacy DIP.
- Thirdly, our approach allows extraction reuse. Once the user has constructed a successful application over the program base to extract some aspects of a data design from a DIP, the expertise invested in and the experience gained from that process may be reused in future data design extraction tasks. That is, we may reuse the extraction code developed for one task (tuned as necessary) in another. This is a useful property because it can reduce the amount of effort required to develop new applications to extract data design from legacy systems.
- Finally, our approach is particularly useful for extracting various data constraints or business rules from DIPs. The extraction of such constraints are largely ignored by most existing data design extraction techniques. This is probably because until now general constraints have not been included explicitly as part of a data design, hence extracting them from legacy systems has not been considered. However, the need to understand and to model constraints in an explicit way is beginning to be appreciated, and the ability to extract them from legacy systems is becoming increasingly important. The exploratory nature of our approach makes it particularly suitable for extracting constraints that are deeply coded in various kinds of program statement and dispersed across the whole program.

4 Experiments of Our Approach

Having explained our approach in general, we now present a number of experiments that we have conducted to demonstrate the usefulness of our approach. The reverse engineering environment provided by our system is shown in Figure 2.

The system is designed based on visual programming principles. There are a set of tools available (shown at the bottom of the frame) and these tools may be composed to form a reverse engineering process (shown as connections in the frame). Each tool represents a specific function, and can be set parameters and executed. For example, the `File` tool allows the user to specify the input file and its execution will open the file. Users may use any type of tool as many times as they like in a single reverse engineering process, provided that the compositions make sense. Once the tools are connected or composed, the execution of one tool will trigger the execution of all its predecessors. For example, the execution of the `Querying` tool will trigger the execution of the `Loading` tool in Figure 2, asking it to supply a program base for querying. The `Loading` tool in turn triggers the

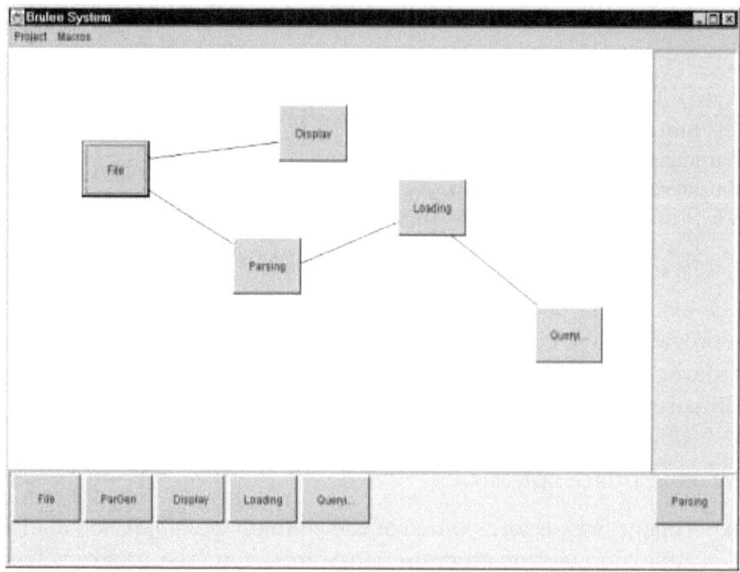

Fig. 2. Reverse engineering environment

execution of the `Parsing` tool, asking it to supply a parsed program tree, and so on. Such an environment makes the reverse engineering of data design both interactive and flexible. It should be noted, however, that we are still at an early stage of software development. The tools shown here are not adequate yet for complex reverse engineering tasks, but our system is generic enough to allow new tools to be added in easily.

4.1 Experiment Setup

For our experiments, we have set up the preparation part of our architecture (see Figure 1) as follows.The source code is a data intensive program from an existing legacy system: a COBOL application from BT. The program, which is part of a much larger COBOL application that we are studying, is just under 22,000 lines of code. For confidentiality, we cannot discuss the application itself here. We have used SableCC [8] as our parser generator. The grammar used to generate the required COBOL parser consists of over 2000 lines of BNF definitions.

The COBOL source program is parsed into a strongly typed abstract syntax tree, represented as a set of Java classes. There are over 1000 classes generated.The abstract syntax tree generated by the parser is stored in Objectivity [14], an object-oriented database management system. In total, we have about 115,000 objects (program constructs) stored in the database. For this experiment, we have not performed any transformation on the abstract syntax tree generated. That is, the abstract syntax tree is stored in Objectivity as is generated by the parser.

Table 1. Querying single class of program constructs

Program Construct	Predicate	Number of Found
RECORD definition		77
MOVE Statement		2088
IF Statement		792
IF Statement	condition *contains* 'C000-JOB-NO'	8
MOVE Statement	left-arg = 'C000-JOB-CMPLN-DATE	2

For the extraction part, currently, we use the query language and the user interface provided by Objectivity.Users issue queries to the underlying program base via Java, the binding language for Objectivity, and the result of query evaluation is returned to the user as an `Iterator` in Java.

4.2 Source Code Exploration

In this experiment, we performed one of the simplest possible code analyses: the retrieval of a single class of program constructs from the program base. This type of operation is often the starting point in source code analysis and is useful for gaining an initial understanding of the source code in an exploratory manner. Table 1 shows the result of some of the queries we performed.

The results of these queries can be useful in a number of ways. First, they allow the user to examine the result set to determine what analysis could or should be performed next. For example, we noticed in our examination of the `MOVE` set that some `MOVE` statements involve the attributes of IDMS records, and these `MOVE` statements have a quite recognisable pattern. For example, the following is one of them

```
MOVE C000-JOB-CMPLN-DATE TO I1503-INST-START-DATE
```

where an IDMS record is labelled with an `Ixxx` prefix. Since these `MOVE` statements cannot contribute to the investigation of how two COBOL record types may be linked together via a `MOVE` statement, they can be removed from further processing in this case. Thus, examining the `MOVE` set helps to develop a strategy to analyse the `MOVE` statements further.

Second, the user can quickly test out some hypothesis that he or she may have. Consider the last entry in Table 1 for example. If the user is expecting the value of `C000-JOB-CMPLN-DATE` to be assigned to only one variable (only to an IDMS record, for example), then our query will reveal that this is not the case:

```
MOVE C000-JOB-CMPLN-DATE TO WS-CHKPNT-CMPLN-DT
MOVE C000-JOB-CMPLN-DATE TO I1503-INST-START-DATE
```

This could indicate that the data design may have evolved beyond the user's knowledge about the system.

Finally, some simple "impact analysis" can be supported. For example, the last but one entry in Table 1 can be used to help the analysis of the effect on the `IF` blocks if changes are made to `C000-JOB-NO`.

It is worth noting that it is in this experiment of retrieving various program constructs from the program base that the use of an object oriented database system is fully justified. It is true that simple tools such as `grep` could also be used to locate various program fragments. However, where the retrieval of the located fragments from a program is concerned, an abstract syntax tree stored as a set of strongly typed objects in an object oriented database makes it a much easier task.

4.3 Incremental Extraction

In this second experiment, we demonstrate the use of our approach to extract a data design in an incremental fashion. Consider the discovery of possible relationships between two record types again. Assume that our initial hypothesis is that all `MOVE` statements in the source code could contribute to such relationships. We therefore set up the following to verify our hypothesis (we describe our queries and manipulations below in diagrams to highlight the *incremental* characteristics of our approach):

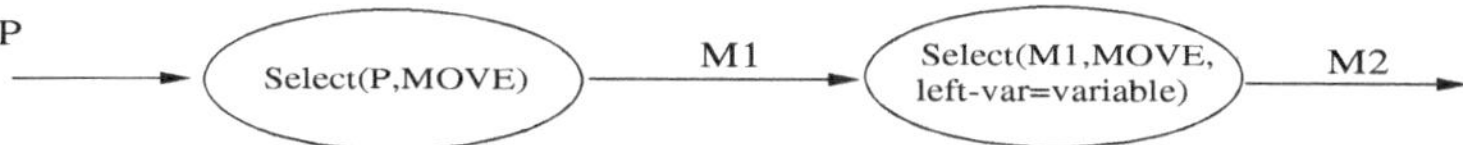

We select `MOVE` statements from the program first to produce `M1` (`select(P,MOVE)`) and then select from `M1` the set of `MOVE` statements that have variables as the left argument (`select(M1,MOVE,left-arg=variable)`). This produces the final result `M2`. However, having examined the `MOVE` statements contained in `M2`, we find that not all the variables used in those statements are the attributes of records; some are IDMS bindings. We therefore refine our process accordingly to include further manipulations as follows:

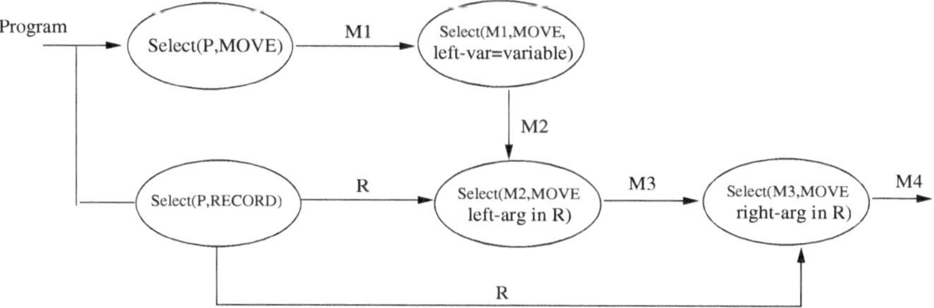

That is, we select `RECORD` objects from the program to produce `R` and use `R` in conjunction with `M2` to select a set of `MOVE` statements that contain only the attributes of records as the left argument. This produces `M3`. Finally, we use `M3` and `R` together to produce `M4`.

This is a relatively simple example, but it is interesting to observe how we allow the extraction process to be constructed "incrementally" and how the operations are composed one with the other. It is also interesting to note how the sizes of the subsets of relevant program constructs gradually reduce to provide a focus on the constructs that are truly relevant: from M1 containing 2208 MOVE statements to M4 containing just 79.

4.4 Find Constraints

Most existing data design extraction techniques can extract data structures [4], relatively few can extract integrity constraints [12] and almost none can extract more general constraints. In this experiment, we show how our approach can help to recover constraints associated with data objects. Let's now consider a concrete example. Suppose that we are interested in finding constraints associated with attribute WS-AC-INST-PROD. In a DIP, constraints are typically fragmented across the whole program and coded in various program constructs. Thus, there is need to locate the relevant fragments and then to assemble them back to a form of constraint. In the following, we will illustrate the process of extracting conditions embedded in IF statements that govern the update of WS-AC-INST-PROD. The following steps were taken:

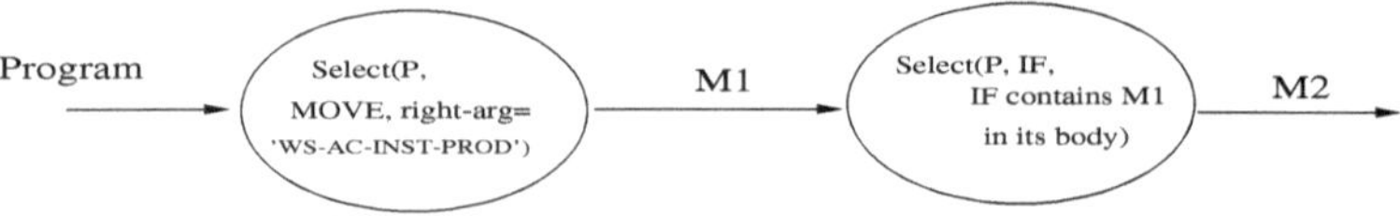

We first select all the MOVE statements containing WS-AC-INST-PROD as the right argument (i.e. the target of an assignment). We then select all those conditional statements (IF statements) that involve the update of this attribute in their bodies. For the given example, we retrieved 7 such blocks and one of them is shown below:

```
IF I1506-CP-TOTAL-QTY > 0
   MOVE I1506-PRODUCT-ID TO I3076-PRODUCT-ID
   OBTAIN CALC PRODUCT
   IF DB-REC-NOT-FOUND
      CONTINUE
   ELSE
      PERFORM Y998-IDMS-STATUS
      IF I3076-MNEMONIC = 'CCOD'
         MOVE I1506-PRODUCT-ID TO WS-AC-INST-PROD
      ELSE
         CONTINUE
      END-IF
   END-IF
END-IF
```

If we rewrite the above into a more understandable form, then we have the following:

```
IF    (I1506-CP-TOTAL-QTY > 0) AND
      (NOT DB-REC-NOT-FOUND)   AND
      (I3076-MNEMONIC = 'CCOD')
THEN
      UPDATE WS-AC-INST-PROD WITH I1506-PRODUCT-ID
```

These extracted fragments (plus other fragments that may be extracted from the program) can then be analysed further for the derivation of a constraint.

5 Conclusions

In this paper we have presented a query-based approach to extracting data design from data intensive programs and demonstrated its usefulness with experiments using legacy code in use at BT. The key strength of this approach is its flexibility: the user can explore the source code arbitrarily, test any hypothesis easily and develop his or her own extraction strategy incrementally, all in the light of his or her experience and intermediate query results. These features, we argue, are essential for reverse engineering a DIP successfully. The exploratory nature of our approach makes it particularly suitable for extracting constraints or business rules that are deeply coded in various kinds of program statement and dispersed across the whole program.

While the paper has suggested a promising approach to extracting data design from source code, the work is still at an early stage and a number of interesting research issues still remain. Firstly, while we can always develop a data design extraction application over the program base by using any query language provided by the underlying DBMS, it is desirable to have a set of primitive operations with which the development of an extraction could be made easier. Designing a meaningful set of such primitive is not a trivial task. Secondly, the extraction of general constraints or business rules from DIPs is still a challenge. We need to design heuristics that can be used to locate such constraints in the source code and to develop techniques that can relate and interpret the recovered fragment constraints meaningfully. Finally, the design of efficient internal structure for storing the source code is worth further investigation.

Acknowledgement

This work is supported by Grant GR/M66219 from the U.K. Engineering and Physical Sciences Research Council. We are grateful to Nigel Turner, Kay Smele and Marc Thorn of BT for their help in working with BT legacy systems. We like to thank Malcolm Munro of Durham University for sharing his experience of developing a COBOL parser with us. Finally, we acknowledge the energy and creativity of Dr. Chris Pound, who initiated many of the ideas underpinning the BRULEE project.

References

1. M Andersson. Searching for semantics in COBOL legacy applications. In S. Spaccapietra and F. Maryanski, editors, *Data Mining and Reverse Engineering: Searching for Semantics*, pages 162–183. Chapman & Hall, 1998.
2. M.R. Blaha. Dimentions of Database Reverse Engineering. In *Proc of Working Conference on Reverse Engineering*, pages 176–183, 1997.
3. Y. Chen, M.Y Nishmoto, and V. Ramamooethy C. The C Information Abstraction System. *ITTT Trans on Software Engineering*, 16(3):325–334, 1990.
4. R.H.L. Chiang, T.M Barron, and V.C Storey. Reverse Engineering of relational databases: Extraction of an EER model from a relational database. *Data & Knowledge Engineering*, 12(2):107–141, 1994.
5. E.J. Chikofsky and J.H. Cross. Reverse Engineering and Design Recovery: A Taxonomy. *IEEE Software*, 7(1):13–17, 1990.
6. M.M. Fonkam and W.A. Gray. An Approach to Eliciting the Semantics of Relational Databases. In *Proceedings of Entity-Relationship Approach - ER'92*, pages 463–480, 1992.
7. G Fu, X.K Liu, J Shao, S.M Embury, and W.A Gray. Business Rule Extraction for System Maintenance. In *to appear in the Proceedings of Re-engineering Week*, 2000.
8. E. Gagnon. SableCC. An Object-Oriented Compiler Framework. In *Master's Thesis, School of Computer Science, McGill University, Canada*, 1998.
9. J.-L Hainaut. Specification preservation in schema transformations - application to semantics and statistics. *Data & Knowledge Engineering*, 19(2):99–134, 1996.
10. J-L Hainaut et al. Contribution to a Theory of Database Reverse Engineering. In *Proc of IEEE Working Conference on Reverse Engineering*, 1993.
11. J. Henard et al. Program Understanding in Database Reverse Engineering. In *Proceedings of DEXA98*, pages 70–79, 1998.
12. H Mannila and K-J Raiha. Algorithms for inferring functional dependencies from relations. *Data & Knowledge Engineering*, 12(2):83–99, 1994.
13. H.A. Muller, M.A. Orgun, S.R. Tilley, and J.S Uhl. A Reverse engeering Approach to Subsystem Structure Identification. *Software Maintenance: Research and Practice*, 5(4):181–201, 1993.
14. Inc. Objectivity. Objectivity for Java Guide, Release 5.2. In *Part Number: 52-JAVAGD-0*, 1999.
15. S. Paul and A. Prakash. A Query Algebra for Program Databases. *IEEE Trans. on Software Engineering*, 22(3):202–216, 1996.
16. W.J. Premerlani and M.R. Blaha. An Approach for Reverse Engineering of Relational Databases. *Communication of ACM*, 37(5):42–49, 1994.
17. J. Shao and C Pound. Extracting Business Rules from Information System. *BT Technical Journal*, 17(4):179–186, 1999.
18. Z. Tari, O Bukhres, J Stokes, and S Hammoudi. The Reengineering of Relational Databases based on Key and Data Correlations. In S. Spaccapietra and F. Maryanski, editors, *Data Mining and Reverse Engineering: Searching for Semantics*, pages 184–216. Chapman & Hall, 1998.
19. M. Weiser. Programmers use slices when debugging. *Communications of the ACM*, 25(7):446–452, 1982.
20. H. Yang and W.C. Chu. Acquisition of Entity Relationship Models for Maintenance - Dealing with Data intensive Programs In A Transformation System. *Journal of Information Science and Engineering*, 15(2):173–198, 1999.

Consistency Management of Financial XML Documents

Andrea Zisman[1] and Adamantia Athanasopoulou[2]

[1] City University, Department of Computing, Northampton Square, London EC1V 0HB, UK
a.zisman@soi.city.ac.uk

[2] Singular International SA, R&D Department, 31 Polytechneiou St., Thessaloniki 54626, Greece
aath@si.gr

Abstract. In the financial domain a large number of inconsistent documents are produced every day. Up to now, many of the consistency management activities are executed manually, generating significant expense and operational risks. In this paper we present an approach for consistency management of financial XML documents. The approach includes the activities of consistency checking and consistency handling. It is based on consistency rules, used to express relationships among elements and documents, and resolution actions, used to restore the documents to a consistent manner. A prototype tool has been developed to demonstrate and evaluate the approach.

1 Introduction

A large number of financial documents are produced every day, either as a result of financial transactions involving multiple actors with different views and opinions, or as a result of accessing and manipulating data in trading systems. These systems are generally created and administered independently, differing physically and logically. The heterogeneity of these systems are exhibited in the use of different programming languages, the availability of different platforms and operating systems, and the various ways of storing, manipulating and exchanging data. Inevitably, the produced documents are often inconsistent.

It is important to manage these inconsistencies to allow interoperability and integration of front-, middle- and back-offices, and to support tasks like trade confirmation, trade settlement, and trade collateral matching. Up to now, many of the consistency management activities are executed manually, generating significant expense and operational risks.

For example, consider a trade confirmation process, where each party produces a document with its own view of a trade that has been agreed to over the phone. In a normal scenario, the parties exchange these documents via fax or electronically in order to check inconsistencies of the documents' content, such as settlement date, rate, amount, name of the parties. The parties produce consistent versions by checking

K.R. Dittrich, A. Geppert, M.C. Norrie (Eds.): CAiSE 2001, LNCS 2068, pp. 219–233, 2001.
© Springer-Verlag Berlin Heidelberg 2001

documents manually and discussing changes via phone and fax before executing the settlement.

With the development of the Internet and eXtensible Markup Language (XML) [4], new standard data interchange mechanisms for the financial domain have been proposed. Examples of these standards are the Financial product Markup Language (FpML) [15], Financial Information Exchange Protocol Markup Language (FIXML) [16], Network Trade Model (NTM) [19], and Open Trading Protocol (OTP) [22]. These standards have been extensively used to support many financial activities, ranging from Internet-based electronic dealing and confirmation of trades, to interchange data between front-, middle-, and back-office services. The result is the creation of various documents representing instances of the standards, generated by different applications and person everyday. However, consistency management of these documents is still an open question.

In this paper we propose an approach for consistency management of financial XML documents. Our approach is simple, lightweight, and concentrates on inconsistencies in the documents' instance level. It is based on *consistency rules*, used to express relationships among elements, and *resolution actions*, used to restore the documents to a consistent state. When dealing with financial documents the resolution of inconsistencies is necessary and important to avoid mismanagement of data.

The proposed approach tackles the activities of *consistency checking* and *consistency handling*, in the consistency management process [31]. Consistency checking is concerned with the tasks of specifying *consistency rules* and identifying inconsistent elements, by checking for violations of the consistency rules. Consistency handling is related to the tasks of specifying *resolution actions* for dealing with inconsistencies, selecting the resolution actions to be executed, and restoring the documents to a consistent format by applying the resolution actions[1].

The work presented in this paper extends previous work for consistency management proposed in [10][34]. In the previous work the authors proposed an approach to allow identification and detection of inconsistencies in distributed XML documents, based on consistency rules, where the related elements are associated through hyperlinks named *consistency links*. The approach presented in this paper complements this former work by describing a way of handling inconsistencies, and applying the whole approach in a specific domain, i.e. finance.

The rest of this paper is organized as follows. Section 2 describes the approach being proposed. Section 3 presents a formalism to express the consistency rules, a classification for the different types of rules, and examples of these various types. Section 4 describes a formalism to express the resolution actions and one example of these actions. Section 5 addresses the implementation of XRule tool to support the approach. Section 6 discusses related work. Finally, section 7 summarizes the approach and suggests directions for future work.

---

[1] As outlined in [31], in a consistency management process the consistency handling actions depend on the type of inconsistencies and can be of various types. Examples are actions that modify documents, restore or ameliorate inconsistencies, notify users about inconsistencies, and perform analysis. In this paper we concentrate our work on actions that restore the documents, called hereafter as *resolution actions*.

2 The Approach

Our approach is based on XML and related technologies such as XPath [7] and XSLT [6]. Fig. 1 presents an overview of the approach. The main component of our approach is called *XRule* and is composed of a *consistency checker* and a *consistency handler*.

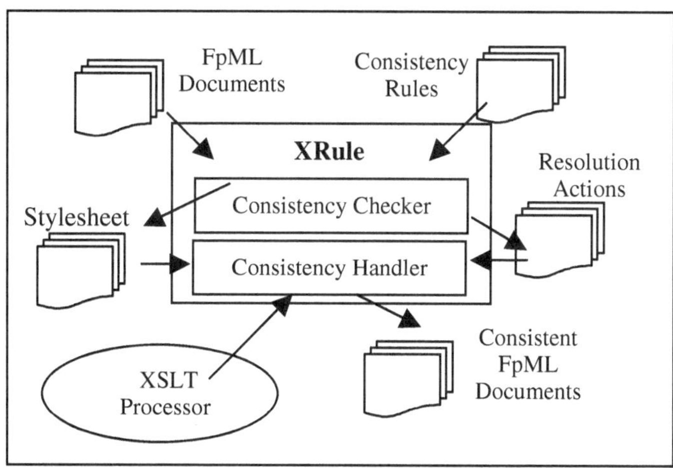

Fig. 1. An overview of the approach

The consistency checker is responsible for identifying inconsistencies in the participating XML documents. It receives as input financial XML documents (e.g. FpML instance documents) and consistency rules previously defined. The consistency rules describe the relationship that should hold between the participating documents. In the next step, the consistency checker verifies for violation of the consistency rules in the XML documents, i.e. the relationships that do not hold. In cases where inconsistencies are detected, the consistency checker generates resolution actions related to these inconsistencies and a XSLT stylesheet document.

The resolution actions specify the parts inside the XML documents that are inconsistent and to which value they should be changed in order to eliminate inconsistencies. The XSLT stylesheet describes how to transform one XML document into another XML document. The stylesheet and resolution actions are used by the consistency handler to restore the XML documents into a consistent state. The restoration of the documents is supported by XSLT processor, which transforms inconsistent XML documents into 'new' consistent XML documents[2].

The resolution actions and stylesheets are dynamically created during the consistency management process. This is due to the fact that it is not possible to know which parts of the participating documents are inconsistent before executing the consistency checks. In addition, depending on the type of inconsistency and on the inconsistent element, the user needs to interact with the system to specify a 'new value' to which the inconsistent element should be modified.

---

[2] A detailed description of XSLT [6] is beyond the scope of this paper.

3 Consistency Rules

In this section we present the syntax used to express the consistency rules and the different types of consistency rules that we can express using the syntax. The consistency rule syntax is similar to the formalism proposed in [34]. However, it involves logical quantifiers (forall, exists) and is described in terms of XML [4] and XPath [7] syntax. The reasons for using XML and XPath are (a) to provide an open and standard way of expressing the rules; (b) to facilitate and standardise the construction and execution of a consistency rule interpreter; (c) to aid generation of resolution action; and (d) to facilitate access to and modification of XML documents.

A consistency rule is composed of two parts. Part 1 is related to relevant sets of *elements* in various documents to which the rule has to be verified. Part 2 is concerned with *conditions* expressing the relationships between the elements in part 1 that have to be tested.

Fig. 2 illustrates the Document Type Definition [4] for the consistency rule syntax. It contains a root element called ConsistencyRule composed of six element contents and one attribute *id.* Attribute *id* uniquely identifies the consistency rule. The element contents are described below.

- Description – it contains a natural language description of the consistency rule;
- Source & Destination – they contain XPath expressions for identifying sets of elements to be checked against the consistency rule. It is possible to have more than one type of Destination elements to be checked against the same type of Source elements. This occurs when a consistency rule refers to more than two types of element sets in the participating documents. Thus, for each type of Destination element set in a rule there is a unique identification represented by attribute *dest_id*, which is referenced as an attribute in element Condition;
- Condition – it is composed of six attributes:

- *expsourcequant* - a quantifier that can have value "forall" or "exist", which is used to specify if the condition has to be satisfied for all elements, or at least one element, respectively, in the Source element set;
- *expsource* - an expression related to the Source element set;
- *op* - an operator associating expsource with expdest, which can have the following values: *equal, not_equal, greater_than, less_than, less_equal, greater_equal,* and *sum*;
- *dest_ref* - a reference to the unique identification of a Destination set;
- *expdestquant* - a quantifier that can have value "forall" or "exist", which is used to specify if the condition has to be satisfied for all elements, or at least one element, respectively, in the Destination element set;
- *expdest* - an expression related to the Destination element set.

- Operator – this element is related to the situation in which the rule is composed of more than one condition. It contains an attribute value, which can have the Boolean content "AND" or "OR".

The proposed consistency rule syntax allows the representation of different types of consistency rules. We classify these types based on the facts that (a) the consistency management process is executed by comparing the participating documents and the elements composing the Source and Destination sets pair wise;

and (b) in XML documents data can be represented either as *elements* or *attributes*[3]. Therefore, the different types for consistency rules that can be represented by using the syntax shown in Fig. 2 are related to the comparison of documents, elements, and mixture of documents and elements.

```
<!ELEMENT ConsistencyRule (Description, Source,
    Destination+, Condition, (Operator, Condition)*)>
<!ATTLIST ConsistencyRule  id   ID #REQUIRED>
<!ELEMENT Description (#PCDATA)>
<!ELEMENT Source (XPath)>
<!ELEMENT Destination (XPath)>
<!ATTLIST Destination dest_id  ID #REQUIRED>
<!ELEMENT Xpath (#PCDATA)>
<!ELEMENT Condition EMPTY>
<!ATTLIST Condition
        expsourcequant  CDATA  #REQUIRED
        expsource       CDATA  #REQUIRED
        op              CDATA  #REQUIRED
        dest_ref        CDATA  #REQUIRED
        expdestquant    CDATA  #REQUIRED
        expdest         CDATA  #REQUIRED>
<!ELEMENT Operator  EMPTY>
<!ATTLIST Operator  value (AND|OR)  "AND" >
```

Fig. 2. Consistency rule syntax

Table 1 summarizes a general classification for the consistency rules. In the table the names in the rows and columns are related to the different types of components being compared: the Source element set and Destination element set, respectively. The Source and Destination sets reference participating documents and XML elements. The content of each position in the table refers to different consistency rules described below.

This general classification can be refined to a more specific classification where we consider the cardinality of the Source and Destination sets. Table 2 presents a specialized classification for the consistency rules, based on the fact that either all elements or at least one element, and either all documents or at least one document in the Source and Destination sets are compared.

In order to illustrate, we present examples of some of the different types of consistency rules related to finance. The rules are specified in XML, based on the DTD syntax of Fig. 2. For the examples we assume documents related to FpML standard [15].

---

[3] XML has no rules related to when data should be represented as element or attribute. For instance, in the XML Metadata Interchange (XMI) standard [21] all components of a UML model are represented as elements. In this text we use the term element meaning both XML elements and attributes.

Table 1. General classification of consistency rules

Destination Set / Source Set	Element	Document
Element	Type 1	Type 3
Document	Type 2	Type 4

Table 2. Specialised classification of consistency rules

Destination Set / Source Set	∀ Element	∃ Element	∀ Document	∃ Document
∀ Element	Type 1.1	Type 1.2	Type 3.1	Type 3.2
∃ Element	Type 1.3	Type 1.4	Type 3.3	Type 3.4
∀ Document	Type 2.1	Type 2.2	Type 4.1	Type 4.2
∃ Document	Type 2.3	Type 2.4	Type 4.3	Type 4.4

Type 1: Existence of related elements

This type of rule is related to the existence of related elements in different documents or in the same document.

Example: (Type 1.2) - For every two Foreign Exchange (FX) swap trade documents F_1 and F_2, the party references in F_1 has to be the same as the party references in F_2.

```
<ConsistencyRule id="R1.2">
<Description> For every two FX swap trade documents representing a trade, the party
references in each of the documents have to be the same </Description>
<Source> <XPath> /fpml:FpML/fpml:Trade/fpml:tradeIDs/fpml:TradeIDs/tid:TradeID/
    tid:partyReference  </XPath> </Source>
<Destination dest_id="pR"> <XPath>  /fpml:FpML/fpml:Trade/fpml:tradeIDs/
    fpml:TradeIDs/tid:TradeID/tid:partyReference </XPath> </Destination>
<Condition  expsourcequant="forall"
            expsource="."
            op="equal"
            dest_ref="pR"
            expdestquant="exists"
            expdest="."  />    </ConsistencyRule>
```

Type 2: Existence of elements due to the existence of documents

This type of rule is related to the situation in which the existence of one or more documents requires the existence of elements in another document.

Example: (Type 2.4) - For every Foreign Exchange (FX) swap trade documents F_1 and F_2, related to a trade involving two parties, the names of the parties must exist in the documents.

```
<ConsistencyRule id="R2.4">
<Description> For every FX swap trade documents, related to a trade involving two parties, the names of the parties must exist in the documents.
</Description>
<Source> <XPath> /fpml:FpML/fpml:Trade </XPath> </Source>
<Destination dest_id="pR"> <XPath> /fpml:FpML/fpml:Trade </XPath> </Destination>
<Condition  expsourcequant="exists"
        expsource="./fpml:tradeIDs/fpml:TradeIDs/tid:TradeID[1]/tid:partyReference"
        op="equal"
        dest_ref="pR"
        expdestquant="exists"
        expdest="./fpml:tradeIDs/fpml:TradeIDs/tid:TradeID[1]/tid:partyReference"/>
<Operator value="OR"/>
<Condition  expsourcequant="exists"
        expsource="./fpml:tradeIDs/fpml:TradeIDs/tid:TradeID[1]/tid:partyReference"
        op="equal"
        dest_ref="pR"
        expdestquant="exists"
        expdest="./fpml:tradeIDs/fpml:TradeIDs/tid:TradeID[2]/tid:partyReference"/>
<Operator value="AND"/>
<Condition  expsourcequant="exists"
        expsource="./fpml:tradeIDs/fpml:TradeIDs/tid:TradeID[2]/tid:partyReference"
        op="equal"
        dest_ref="pR"
        expdestquant="exists"
        expdest="./fpml:tradeIDs/fpml:TradeIDs/tid:TradeID[1]/tid:partyReference"/>
<Operator value="OR"/>
<Condition  expsourcequant="exists"
        expsource="./fpml:tradeIDs/fpml:TradeIDs/tid:TradeID[2]/tid:partyReference"
        op="equal"
        dest_ref="pR"
        expdestquant="exists"
        expdest="./fpml:tradeIDs/fpml:TradeIDs/tid:TradeID[2]/tid:partyReference"/>
</ConsistencyRule>
```

Type 3: Existence of documents due to the existence of elements

This type of rule is related to the situation in which the existence of one or more elements requires the existence of a document.

Example: (Type 3.2) - For every exchange rate in a FX swap trade document F_1, there must exist a document F_2 with all the existing exchange rates.

```
<ConsistencyRule id="R3.2">
<Description> For every exchange rate in a FX swap trade document, there must exist
a document with a list of all the exchange rates.</Description>
<Source> <XPath> /fpml:FpML/fpml:Trade/fpml:product/decendant::fxs:exchangeRate
</XPath> </Source>
<Destination dest_id="eR"><XPath>/DocumentType/Rate/r:FixedRate/
</XPath></Destination>
<Condition  expsourcequant="forall"
            expsource="."
            op="equal"
            dest_ref="eR"
            expdestquant="exists"
            expdest="."/>
</ConsistencyRule>
```

Type 4: Existence of related documents

This type of rule is related to the situation in which the existence of a document requires the existence of another document.
Example: (Type 4.2) - For every FX swap trade document F_1 referencing two parties pR_1 and pR_2, and produced by party pR_1, there must exist a FX swap trade document F_2 produced by party pR_2, with the same party reference name.

```
<ConsistencyRule id="R4.2">
<Description> For every FX swap trade document referencing two parties and produced by one
party, there must exist a FX swap trade document produced by the other party, with the same
party reference names </Description>
<Source> <XPath> /fpml:FpML/fpml:Trade </XPath> </Source>
<Destination dest_id="ST"> <XPath> /fpml:FpML/fpml:Trade </XPath> </Destination>
<Condition  expsourcequant="forall"
      expsource="./fpml:tradeIDs/fpml:TradeIDs/tid:TradeID[1]/tid:partyReference"
      op="equal"
      dest_ref="ST"
      expdestquant="exists"
      expdest="./fpml:tradeIDs/fpml:TradeIDs/ tid:TradeID/tid:partyReference"/>
<Operatorvalue="AND"/>
<Condition  expsourcequant="forall"
      expsource="./fpml:tradeIDs/fpml:TradeIDs/tid:TradeID[2]/tid:partyReference"
      op="equal"
      dest_ref="ST"
      expdestquant="exists"
      expdest="./fpml:tradeIDs/fpml:TradeIDs/tid:TradeID/tid:partyReference"/>
</ConsistencyRule>
```

With the proposed syntax it is also possible to represent consistency rules that check for the existence of unrelated documents or elements. Due to limitation of space we do not present here examples of these types of consistency rules.

4 Resolution Actions

In this section we present the syntax used to express the resolution actions. Similar to the consistency rules, and for the same reasons, the syntax to express the resolution action is also described in terms of XML [4] and XPath [7] syntax.

A resolution is composed of two parts. Part 1 is related to *fragment parts* of the participating documents that are inconsistent. Part 2 is concerned with *values* that should be replaced in the fragment parts of part 1, to convert the document to a consistent state.

Fig. 3 illustrates the Document Type Definition [4] for the resolution action syntax. It contains a root element called Action composed of two element contents. The two element contents are described below. Fig. 4 presents an example of a resolution action for consistency rule R4.2 (Type 4) shown in section 3

- DocumentFragment - it contains XPath expressions identifying the part of the document where inconsistency was detected;
- NewValue – it contains the 'new' value to which the fragment part has to be modified to restore the document to a consistent mode.

```
<!ELEMENT Action (DocumentFragment, NewValue)>
<!ELEMENT DocumentFragment (XPath)>
<!ELEMENT NewValue (#PCDATA)>
<!ELEMENT XPath (#PCDATA)>
```

Fig. 3. Resolution action syntax

```
<Action>
<DocumentFragment> <XPath>
   fpml:FpML/fpml:Trade/fpml:tradeIDs/fpml:TradeIDs/tid:TradeID[1]/
   tid:partyReference </Xpath> </DocumentFragment>
<NewValue>ABC Trust
</NewValue>
</Action>
```

Fig. 4. Example of a resolution action

Based on the action a stylesheet is created to support the restoration of an inconsistent document. Fig. 5 presents an example of a XSLT stylesheet related to the resolution action shown in Fig. 4. The stylesheet is composed of two template rules. The first template rule matches all the attributes and nodes in the original XML document (source tree), and creates a new XML document (result tree) by copying these attributes and nodes. The second template rule is related to the content of the resolution action. It matches the nodes related to the inconsistent part of the document, and replaces them with the 'new value' specified in the resolution action.

```
<?xml version="1.0" encoding="UTF-8"?>
<xsl:stylesheet version="1.0"  xmlns:xsl=http://WWW.w3.org/1999/XSL/Transform>
<xsl:output method="xml"/>

<xsl:template match="@*|node( )"/>
  <xsl:copy>
    <xsl:apply-templates select="@*|node( )"/>
  </xsl:copy>
</xsl:template>

<xsl:template  match="fpml:FpML/fpml:Trade/fpml:tradeIDs/tid:TradeIDs/tid:TradeID[1]/
        tid:partyReference"/>
  ABC Trust
</xsl:template>
</xsl:stylesheet>
```

Fig. 5. An example of the XSLT stylesheet

5 The XRule Tool

In order to evaluate our approach, we developed a prototype tool called *XRule,* which supports the consistency management process. The tool was implemented as proof of concept and developed in JDK 1.1.6. It uses Apache [1] Xerces Java Parser 1.1.2, for parsing XML documents, and Apache Xalan Java XSLT Processor 1.1, as the XSLT processor.

The prototype contains the implementation of essential features that enable us to evaluate and prove the feasibility of our approach. The main goals of our tool are to perform consistency checks and restore the documents to a consistent state. The document restoration is based on interaction with the user. This interaction is necessary to identify the correct instance value of the inconsistent document part.

Fig. 6 presents the initial screen of the XRule tool. It contains a list of all available consistency rules. For the prototype we have implemented ten different types of rules. We grouped these rules into two categories*:* *single document rules*, for the rules related to only one document, and *pair of document rules*, for the rules related to two or more documents.

After selecting a consistency rule, the documents to be checked for consistency are specified by the user. These documents can be located in the same machine where XRule tool is being used or accessed over the Web. The result of the consistency checking process is presented to the user, as shown in Fig. 7. For the example we consider the execution of only one trade on 26/05/1999. Therefore, there are only two documents checked for consistency rule 7 in Fig. 6. The first document (Source) is related to the trade summary; the second document (Destination) is related to the trade itself.

When an inconsistency is found, the application presents the fragment parts that are inconsistent in both documents, with their respective values, as shown on the top of Fig. 7. The two participating documents are also displayed on the screen to allow the user to browse the documents, if necessary.

In the case where the user wants to execute the consistency handling process, s/he selects to "proceed". The application presents to the user the screen shown in Fig. 8 with the condition in the related consistency rule that does not hold for the participating documents. After selecting the condition the application automatically displays the related inconsistent values in the fragment of the participating documents. The user specifies which document is inconsistent, by selecting either the *Source* or *Destination* value, and specifies the 'new value' that should replace the inconsistent one. For our example the user assumes that the Destination document is inconsistent, i.e. the document related to the trade executed on 26/05/1999.

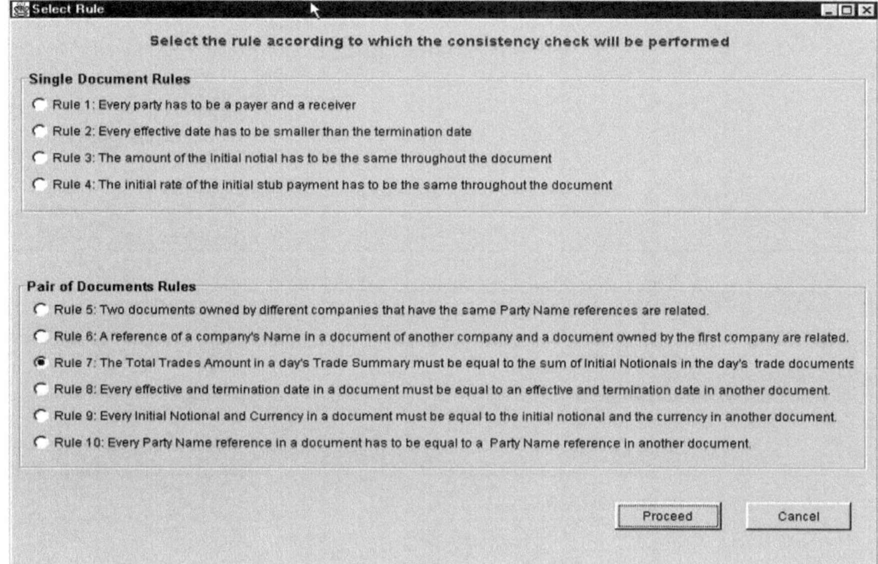

Fig. 6. Consistency rules

Based on the information specified by the user, XRule generates the resolution action document and the XSLT stylesheet document. The tool executes the restoration process, and a consistent document is generated and presented to the user.

6 Related Work

Many approaches have been proposed for consistency management. In particular, approaches for software engineering documents and specifications. A complete and up to date survey can be found in [31].

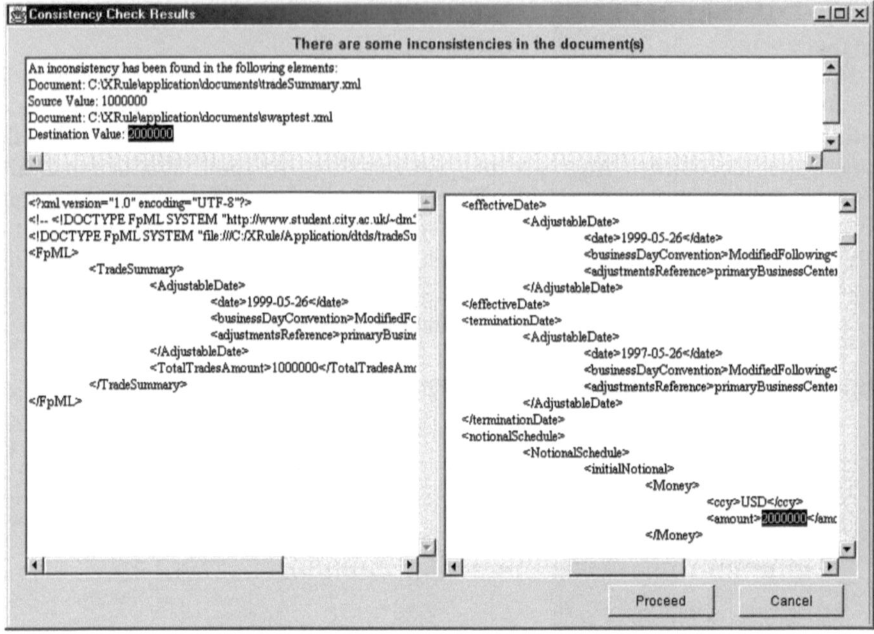

Fig. 7. Result of the consistency checking process

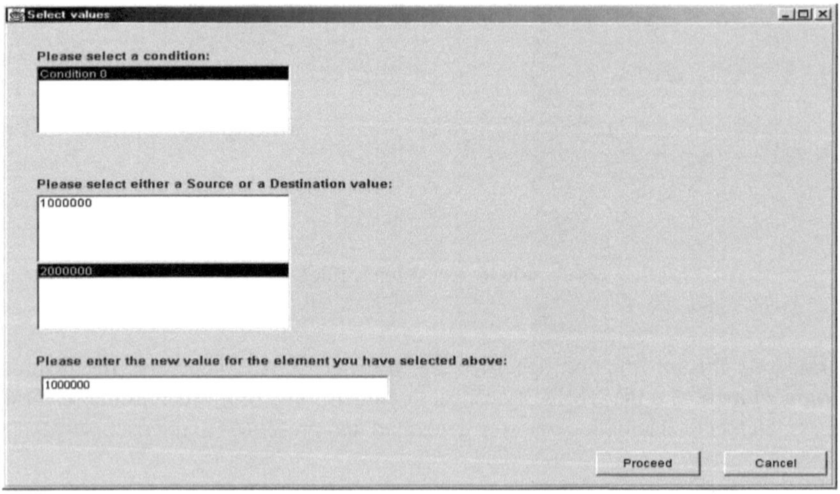

Fig. 8. Consistency handling process

In [8][9][13] inconsistency is seen as a logical concept and the authors proposed a first-order logic-based approach to consistency management in the ViewPoints framework. However, this approach has not been implemented in a distributed setting.

Spanoudakis and Finkelstein [27][28] suggested a method called *reconciliation* to allow detection and verification of overlaps and certain types of inconsistencies between specifications expressed in an object-oriented framework. When managing inconsistency, overlap detection is an activity that precedes consistency rule construction [14]. We are investigating the use of reconciliation method to check consistency of meta-level XML documents, i.e. XML Schema documents [12].

In [8][13][18][29][33] the authors proposed logic-based approaches for consistency checking, where some formal inference technique is used to detect inconsistencies in software models expressed in formal modeling languages. Our work complements the work in [10], where the authors developed a technique for detecting inconsistencies in distributed documents with overlapping content, based on XML and related technologies.

The strategies proposed for consistency handling can be divided into two groups. One group is related to the approaches that use actions, which modify the documents by repairing or ameliorating the inconsistencies [8][9][20][23][33]. The other group is concerned to approaches that notify the stakeholders about inconsistencies and perform analysis that would safe further reasoning from documents [3][13][18].

Van Lamsweerde et al. [32][33] proposed a formal framework for various types of inconsistencies that can occur during requirements engineering process. The idea is to manage conflicts at the goal level in the context of the KAOS requirements engineering methodology. This is achieved by introducing new goals or by transforming specifications of goals into new specifications free of conflicts.

In addition, methods for consistency tracking have been proposed in [11][13]. On the other hand, specification and application of consistency management policies are presented in [11][13][24][28].

Identification and resolution of semantic and syntactic conflicts are also issues in the multidatabase system domain. Many approaches have been proposed in the literature [5][17][25][26]. A survey of different approaches to detect and resolve conflicts can be found in [2].

Although the existing approaches have contributed to a better understanding of the consistency management problem, an approach which deals with consistency management of distributed financial documents have not yet been proposed.

7 Conclusion and Future Work

In this paper we presented an approach to consistency management of XML financial documents. The approach supports the activities of consistency checking and consistency handling. It uses XML and related technologies to allow Internet-scale distribution, standardisation of the consistency management process, and access to XML documents.

We proposed the use of *consistency rules* and *resolution actions* to support the management process. We developed a prototype tool as proof of concept to evaluate the ideas of the work and demonstrate the feasibility and applicability of the approach.

Although the approach has been proposed for dealing with inconsistencies in the financial domain, it can be deployed in other settings where consistency management is necessary and important. Examples are found in the health care domain, scientific domain, and business domains, among others.

Before large-scale experimentation and use, we are expanding the prototype to allow consistency management of heterogeneous financial documents such as FIXML [16], FpML [15], and OTP [22], for meta-level (XML Schema) and instance documents. We are also extending our work to allow different types of consistency rules and resolution actions. In particular, rules and actions involving complex financial calculations like derivative models, and semantic aspects of the data. In [31] we proposed an approach for monitoring financial information where we present a syntax to describe complex financial calculations. We also plan to expand the approach to support other important activities of the consistency management process, such as consistency tracking, consistency diagnosis, and consistency policy [31].

References

[1] Apache. http://xml.apache.org/index.html.

[2] C. Batini, M. Lenzerini, and S.B. Navathe. A Comparative Analysis of Methodologies for Database Schema Integration. *ACM Computer Surveys*, 18(4), pages 323-364, December 1986.

[3] B. Boehm and H. In. Identifying Quality Requirements Conflicts. *IEEE Software*, pp. 25-35, March 1996.

[4] T. Bray, J. paoli, C.M. Sperberg-McQueen, E. Maler. Extensible Markup Language (XML) 1.0. W3C Recommendation, http://www.w3.org/TR/2000/REC-xml-20001006, World Wide Web Consortium.

[5] M.W. Bright, A.R. Hurson, and S. Pakzard. Automated Resolution of Semantic Heterogeneity in Multidatabases. *ACM Transaction on Database Systems*, 19(12), pages 212-253, June 1994.

[6] J. Clark. XSL Transformations (XSLT) Version 1.0. Recommendation http://www.w3.org/TR/1999/REC-xslt-19991116, World Wide Web Consortium.

[7] J. Clark and S. DeRose. XML Path Language (XPath). Recommendation http://www.w3.org/TR/1999/REC-xpath-19991116, World Wide Web Consortium.

[8] S. Easterbrook, A. Finkelstein, J. Kramer, and B. Nuseibeh. Co- ordinating Distributed ViewPoints: the anatonomy of a consistency check. In *Concurrent Engineering Research & Applications,* CERA Institute, USA 1994.

[9] S. Easterbrook and B. Nuseibeh. Using ViewPoints for Inconsistency Management. *IEE Software Engineering Journal*, November 1995.

[10] E.Ellmer, W. Emmerich, A. Finkelstein, D. Smolko, and A. Zisman. Consistency Management of Distributed Documents using XML and Related Technologies. UCL-CS Research Note 99/94, 1999. Submitted for publication.

[11] W. Emmerich, A. Finkelstein, C. Montangero, S. Antonelli, and S. Armitage. Managing Standards Compliance. *IEEE Transactions on Software Engineering*, 25(6), 1999.

[12] D.C. Fallside. XML Schema Part 0: Primer. Working Draft http://www.w3.org/TR/2000/WD-xmlschema-0-20000407, World Wide Web Consortium.

[13] A. Finkelstein, D. Gabbay, A. Hunter, J. Kramer, and B. Nuseibeh. Inconsistency Handling in Multi-Perspective Specifications. *IEEE Transactions on Software Engineering,* 20(8), pages 569-578, August 1994.

[14] A. Finkelstein, G. Spanoudakis, and D. Till. Managing Interference. Joint Proceedings of the SIGSOFT'96 Workshops – Viewpoints'96: An International Workshop on Multiple Perspectives on Software Development, San Francisco, ACM Press, pages 172-174, October 1996.

[15] FpML. Financial product Markup Language. http://www.fpml.org.

[16] FIXML. Financial International Exchange Markup Language. http://www. fix.org.

[17] J. Hammer and D. McLeod. An Approach to Resolving Semantic Heterogeneity in a Federation of Autonomous, Heterogeneous Database Systems. *International Journal of Intelligent and Cooperative Information Systems*, 2(1), pages 51-83, 1993.

[18] A. Hunter and B. Nuseibeh. Managing Inconsistent Specifications: Reasoning, Analysis and Action. *ACM Transactions on Software Engineering and Methodology*, 7(4), pp. 335-367, 1998.

[19] Infinity. Infinity Network Trade Model. http://www.infinity.com/ntm.

[20] B. Nuseibeh and A. Russo. Using Abduction to Evolve Inconsistent Requirements Specifications. *Australian Journal of Information Systems*, 7(1), Special Issue on Requirements Engineering, ISSN: 1039-7841, 1999.

[21] OMG (1998). XML Metadata Interchange (XMI) - Proposal to the OMG OA&DTF RFP 3: Stream-based Model Interchange Format (SMIF). Technical Report AD Document AD/98-10-05, Object Management Group,m 492 Old Connecticut Path, Framingham, MA 01701, USA.

[22] OTP. Open Trading protocol. http://www.otp.org.

[23] W. Robinson and S. Fickas. Supporting Multiple Perspective Requirements Engineering. *In Proceedings of the 1st International Conference on Requirements Engineering (ICRE 94)*, IEEE CS Press, pp. 206-215, 1994.

[24] W. Robinson and S. Pawlowski. Managing Requirements Inconsistency with Development Goal Monitors. *IEEE Transactions on Software Engineer*, 25(6), 1999.

[25] E.Sciore, M. Siegel, and A. Rosenthal. Using Semantic Values to Facilitate Interoperability Among Heterogeneous Information Systems. *ACM Transactions on Database Systems*, 19(2), pages 254-290, June 1994.

[26] M. Siegel and S.E. Madnick. A Metadata Approach to Resolving Semantic Conflicts. *In proceedings of the 17th International Conference on Very Large DataBases,* pages 133-145, Barcelona, Spain, 1991.

[27] G. Spanoudakis and A. Finkelstein. Reconciliation: Managing Interference in Software Development. *In Proceedings of the ECAI '96 Workshop on Modelling Conlicts in AI*, Budapest, Hungary, 1996.

[28] G. Spanoudakis and A. Finkelstein. A Semi-automatic Process of Identifying Overlaps and Inconsistencies between Requirements Specifications. *In Proceedings of the 5th International Conference on Object-Oriented Information Systems* (OOIS 98), pages 405-425, 1998.

[29] G. Spanoudakis, A. Finkelstein, and D. Till. Overlaps in Requirements Engineering. *Automated Software Engineering Journal*, vol. 6, pp. 171-198, 1999.

[30] G. Spanoudakis and A. Zisman. Information Monitors: An Architecture Based on XML. *In Proceedings of 6th International Conference on Object Oriented Information Systems - OOIS 2000*, London, December 2000.

[31] G. Spanoudakis and A. Zisman. Inconsistency Management in Software Engineering: Survey and Open Research Issues. *Handbook of Software Engineering and Knowledge Engineering*, 2000. (To appear).

[32] A. van Lamsweerde. Divergent Views in Goal-Driven Requirements Engineering. *In Proceedings of the ACM SIGSOFT Workshop on Viewpoints in Software Development,* San Francisco, pages 252-256. October 1996.

[33] A. van Lamsweerde, R. Darimont, and E. Letier. Managing Conflicts in Goal-Driven Requirements Engineering. *IEEE Transaction on Software Engineering.* November 1998.

[34] A. Zisman, W. Emmerich, and A. Finkelstein. Using XML to Specify Consistency Rules for Distributed Documents, *In Proceedings of the 10th International Workshop on Software Specification and Design (IWWSD-10),* Shelter Island, San Diego, California, November, 2000.

The Relationship Between Organisational Culture and the Deployment of Systems Development Methodologies

Juhani Iivari[1] and Magda Huisman[2]

[1] Department of Information Processing Science, University of Oulu, P.O.Box 3000, 90014 Oulun yliopisto, Finland
iivari@rieska.oulu.fi

[2] Department of Computer Science and Information Systems, Potchefstroom University for CHE, Private Bag X6001, Potchefstroom, 2531, South Africa
rkwhmh@puknet.puk.ac.za

Abstract. This paper analyses the relationship between organisational culture and the perceptions of use, support and impact of systems development methodologies (SDMs) interpreting organisational culture in terms of the competing values model. The results show that organisations with different culture differ in their perceptions concerning the support provided by SDMs and in their perceptions concerning the impact of SDMs on the quality of developed systems and the quality and productivity of the systems development process. The results depend, however, on the respondent groups (developers vs. managers). The findings also suggest that the deployment of SDMs is primarily associated with the hierarchical culture which is oriented toward security, order and routinisation. Also managers' criticality towards the deployment of SDMs in organisation with high rational culture (focusing on productivity, efficiency and goal achievement) is noteworthy.

1 Introduction

This paper investigates the relationship between organisational culture and the deployment of systems development methodologies (SDMs). There are a number of reasons for the selection of this topic. Firstly, SDMs have formed one of the central topics in Information Systems and Software Engineering. In spite of the huge effort devoted to their development and the pressure to adopt them [9] their practical usefulness is still a controversial issue ([9],[11]). Secondly, very little is known about the actual usage of SDMs. A recent survey of research on SDMs [33] identifies only 19 papers addressing SDM usage, of which 12 have been published since 1990. The recent surveys indicate, however, quite consistently that many organisations claim that they do not use any SDMs (e.g. [5],[13],[27]). Thirdly, most literature on the use of SDMs is descriptive. It does not attempt to explain the use of SDMs.

This study goes beyond the existing literature in the sense that it analyses the relationship between organisational culture and the deployment of SDMs. There are sev-

K.R. Dittrich, A. Geppert, M.C. Norrie (Eds.): CAiSE 2001, LNCS 2068, pp. 234–250, 2001.
© Springer-Verlag Berlin Heidelberg 2001

eral reasons for the selection of organisational culture as the focus of this study. Firstly, organisations tend to develop specific cultures. These form a context in which systems development (SD) and the deployment of SDMs takes place. Secondly, organisational culture is a rich concept, comprising of symbols, heroes, rituals and values [15]. Therefore SDMs can be conceived to be part of an organisational culture. Wastell [30], for example, argues that a SDM may provide an organisational ritual with the primary function to serve as a social defence against the anxieties and uncertainties of SD rather than as an efficient and effective means of developing systems. Thirdly, the role of organisational culture as a significant source of organisational inertia is well known ([4],[28]). There is also an emerging interest in the influence of culture on the acceptance of IT [25]. Our assumption is that organisational culture may also be influential in the acceptance of SDMs, too.

The organisational culture of a large organisation cannot be expected to be homogenous but it consists of a number of subcultures [29]. Recognising this plurality, we decided to focus on cultures of IS departments, because they can be expected to be most closely associated with the behaviour of IS developers. This study applies a specific model of organisational culture, a competing values model [31], to analyse the relationship between the organisational culture and the usage of SDMs. The next section explains the competing values model in greater detail. Section 3 introduces the research design and section 4 the results. Section 5 discusses the findings and makes some concluding comments.

2 Organisational Culture, the Competing Values Framework and Deployment of Systems Development Methodologies

2.1 Organisational Culture

Organisational culture is a versatile concept that has been used in several meanings ([1],[29]). Despite the differences, there seems to be an agreement that an organisational culture includes several levels with a varying degree of awareness by the culture-bearers. Schein [28] suggests that the deepest level consists of patterns of basic assumptions that the organisational members take as granted without awareness. At the surface level lie artefacts as the visible and audible patterns of the culture. The intermediate level covers values and beliefs, concerning what 'ought' to be done. Similarly, [15] proposes a model for manifestations of organisational culture, including symbols, heroes, rituals and values. In this framework symbols are at the most superficial level and values at the deepest level.

Research into organisational culture has mostly been qualitative. This is related with the distinction whether a culture is seen to be unique to each organisation or whether it is seen to include significant universal aspects [7]. Studies emphasising uniqueness have often been qualitative, whereas those assuming universality have more often been quantitative. Despite the dominance of the qualitative tradition, there has been some efforts to develop quantitative "measures" for organisational culture (e.g. [15]). This paper applies a specific quantitative model of organisational culture, the competing values framework ([7],[23],[24]). As its name suggests, it focuses on values as core constituents of organisational culture.

2.2 The Competing Values Framework

The competing values framework is based on two distinctions: change vs. stability and internal focus vs. external focus. Change emphasises flexibility and spontaneity, whereas stability focuses on control, continuity and order. The internal focus underlines integration and maintenance of the sociotechnical system, whereas the external focus emphasises competition and interaction with the organisational environment [7]. The opposite ends of each of these dimensions pose competing and conflicting demands on the organisation.

Based on the two dimensions, one can distinguish four organisational culture types. The *group culture* (GC) with change and internal focus has a primary concern with human relations and flexibility. Belonging, trust, and participation are core values. Effectiveness criteria include the development of human potential and member commitment. The *developmental culture* (DC) with change and external focus is future-oriented considering what might be. The effectiveness criteria emphasise growth, resource acquisition, creativity, and adaptation to the external environment. The *rational culture* (RC) with stability and external focus is very achievement-oriented, focusing on productivity, efficiency and goal achievement. The *hierarchical culture* (HC) with stability and internal focus is oriented toward security, order and routinisation. It emphasises control, stability and efficiency through following regulations ([7],[23]). Each of the culture types has its polar opposites [7]. The GC, which emphasises flexibility and internal focus, is contrasted with the RC, stressing control and external focus. The DC, which is characterised by flexibility and external focus, is opposed by the HC, which emphasises control and internal focus.

The four cultural types are ideal types in the sense that organisations are unlikely to reflect only one culture type [7]. In fact, the competing values model stresses a reasonable balance between the opposite orientations, even though some culture types may be more dominant than others [7]. This imposes paradoxical requirements for effective organisations ([3],[20],[22]).

Cooper [6] applied the competing values model to understand IT implementation. He proposed that different ISs may support alternative values, and that when an IS conflicts the organisational culture (values), the implementation of the system will be resisted. These implementation problems may lead to underutilisation of the system, if implemented, and to the adaptation of the system to the existing culture. The latter may lead to more conservative development as initially planned. This paper applies the model to analyse the deployment of SDMs to be discussed next.

2.3 Deployment of Systems Development Methodologies

This paper uses the term "methodology" to cover the totality of SD approaches (such as the structured approach, information modelling approach, object-oriented approach), process models (such as the linear life-cycle, prototyping, evolutionary development, spiral models), specific methods (e.g. MSA, IE, NIAM, OMT, UML, ETHICS) and specific techniques.

Deployment encapsulates the post-implementation stages of the innovation diffusion process, where the innovation is actually being used and incorporated into the organisation [26]. This focus on the deployment of SDMs is necessary because adopted SDMs might not be used effectively, might not be used at all or might not have the intended consequences. The deployment of SDMs can be analysed from several perspectives. In this paper we focused on the use of SDMs, the perceived support it provides, and its impact on the developed system and the development process. Accordingly, the following seven perspectives were selected:

1. maximum intensity of methodology use (vertical use)
2. methodology use across the organisation (horizontal use)
3. perceived methodology support as production technology
4. perceived methodology support as control technology
5. perceived methodology support as cognitive & co-operation technology
6. perceived impact on the quality of developed systems
7. perceived impact on the quality and productivity of the development process

The first and second perspectives were suggested in [21]. The third, fourth and fifth ones are adapted from [14]. [14] developed and empirically tested a functional model for IS planning and design aids that distinguishes two major functional categories: production technology and co-ordination technology. The functionality of *production technology* "directly impacts the capacity of individual(s) to generate planning or design decisions and subsequent artifacts or products". The *co-ordination technology* defined as "functionality that enables or supports the interactions of multiple agents in the execution of a planning or design task" comprises control functionality and co-operative functionality. The *control functionality* "enables the user to plan for and enforce rules, policies or priorities that will govern or restrict the activities of team members during the planning and design process". The *co-operative functionality* enables the user "to exchange information with another individual(s) for the purpose of influencing (affecting) the concept, process and product of the planning/design team"[1].

[1] [14] also identifies organisational technology consisting of two additional functionalities: support functionality "to help an individual user understand and use a planning and design aid effectively" and infrastructure defined as "standards that enable portability of skills, knowledge, procedures, or methods across planning or design processes". The support functionality can be interpreted as a functionality' in the sense that it supports the utilisation of all the basic functionalities. One of the findings of [14] was that the support functionality was difficult for respondents to clearly differentiate. The infrastructure component resulted from the feedback during the study and its differentiation was not tested empirically in [14]. We see infrastructure functionalities such as standards to support cooperation.

2.4 Organisational Culture and Systems Development Methodologies

The competing values model can also be applied to IS departments, emphasising that the effectiveness of an IS department imposes paradoxical requirements of balancing opposite cultural orientations. In the following organisational culture is confined to this specific context. The paradoxical nature implies that the relationship between the organisational culture and SDM deployment may be quite complicated. The direction of causality may be in either of the two directions, i.e. culture influences the deployment of SDMs or vice versa. This paper takes a view that it is interactive. The relationship may also be either reinforcing or complementary. The former implies that a SDM reinforces the existing culture and the latter that it complements it in some way. To exemplify the former case, organisations with a HC may use SDMs as means of imposing security, order and routinisation. On the other hand, one can conceive that organisations with a DC, for example, may also perceive SDMs as means of imposing necessary security, order and routinisation.

Because of the above complexities we are not prepared to put forward any specific hypotheses about the relationship between organisational culture and SDMs. Instead we will focus on the research question. *Does organisational culture, when applied to IS departments, have any relationship with the deployment of SDMs*?

3 Research Design and Method

To analyse organisational cultures of IS departments we decided to focus on the culture perceptions of IS developers rather than of IT managers in order not to associate culture with IT managers' view of the desirable culture to be imposed on the IS department. In the case of the deployment of SDMs we decided to study both IS developer's and IT managers' perceptions. One reason for this is the possible bias brought by the research design where the same respondents (i.e. IS developers) assess both the organisational culture and deployment of SDMs. Our research design allows intergroup analysis where the culture is assessed by IS developers and the deployment by IT managers. One should note, however, that the purpose of this study is not a systematic comparison of IT managers' and IS developers' perceptions (some of these are reported in [16]).

3.1 The Survey

This study is part of a larger survey on SDM use in South Africa, which was conducted between July and October 1999. The 1999 IT Users Handbook (the most comprehensive reference guide to the IT industry in South Africa) was used and the 443 listed organisations were contacted via telephone to determine if they were willing to participate in the study. 213 organisations agreed to take part. A package of questionnaires was sent to a contact person in each organisation who distributed it. This package consisted of one questionnaire to be answered by the IT manager, and a number of questionnaires to be answered by individual IS developers in the organisation. The number of developer questionnaires was determined for each organisation during the telephone contacts. The response rate is given in Table 1. The responses came from organisations representing a variety of business areas, manufacturing (33%) and finance/banking/insurance (15%) as the major ones. At the individual level the respondents reported considerable experience in SD, 22% between 3 and 5 years, 23% between 5-19 years and 38% more than 10 years.

Table1. Response rate of survey

	Number Distributed	Number Returned	Response Rate (%)
Organisations	213	83	39.0
Developers	893	234	26.2
Managers	213	73	34.3

3.2 Measurement

All the questions, except organisational culture and horizontal methodology use, were addressed to both developers and managers. Only developers evaluated the organisational culture, and the horizontal methodology use was assessed only by managers.

Multiple items were used to measure the perceived support provided by SDMs and its impact on the developed system and the development process. This resulted in a large data set. In order to reduce the data set, factor analysis using the principal components method with varimax normalised rotation on the data was performed. The Kaiser criterion was used to determine the number of factors to retain. This was followed by reliability analysis (Cronbach's alpha) on the items of each of the factors identified. We used a cutoff value 0.6 for acceptable reliability.

Organisational culture was measured using the instrument suggested in [34]. It includes 12 items, three items measuring each culture orientation. The reliability of the 3-item measure for the GC was 0.68 and the reliability of the measure for the DC was 0.69. Reliability analysis indicated that one item of the 3-item measure of the HC and one item of the 3-item measure for the RC decreased substantially the reliability. After deleting the two items, the reliability of the 2-item measures for the HC and for the DC were 0.71 and 0.71, respectively. The indexes for each of the culture types was computed as averages of the two or three items included in the measure.

Vertical methodology use was measured as the maximum intensity of organisational usage of 29 listed methods, possible other standard (commercial) methods and possible in-house developed methods. *Horizontal methodology use* was measured using two items, namely the proportion of projects that are developed in the IS department by applying SDM knowledge, and the proportion of people in the IS department that apply SDM knowledge regularly. The reliability of these two items was 0.89.

As explained above the distinction between perceived methodology support as production technology, perceived methodology support as control technology and perceived methodology support as cognitive and co-operation technology was adapted from [14]. However, the nature of the survey did not allow the use of their detailed questions to measure these functionalities, but this study adopted a shorter version.

Perceived methodology support as production technology was measured using eleven items. Factor analysis using the developer data gave only one factor and using the manager data three factors: "Support for organisational alignment" with five items, "Support for technical design" with three items and "Support for verification and validation" with two items. The following analysis uses the more refined factor structures. The reliability of the first factor was 0.90/0.91, of the second factor 0.85/0.82, of the third factor 0.83/0.86.[2]

Perceived methodology support as control technology was measured using nine items. Separate factor analyses based on the developer data and the manager data gave only one factor. Its reliability was 0.94/0.92. *Perceived methodology support as cognitive and co-operation technology* was measured using eleven items. Separate factor analyses based on the developer data and the manager data gave very similar factor structures, comprising two factors: "Support for the common conception of systems development practice" with nine items and "Support for the evaluation of systems development practice" with two items. The reliability of the former factor was 0.92/0.92 and of the second factor 0.79/0.92.

Perceived methodology impact on the quality of the developed systems was measured using eight items adopted from ISO 9126 standard [19]. Separate factor analyses based on both developer data and manager data gave only factor. Its reliability was 0.95/0.93. *Perceived methodology impact on the quality and productivity of the development process* was measured using ten items. Factor analysis using the developer data gave only on factor. Factor analysis based on the manager data gave two factors: "Productivity effects and morale" with five items, and "Quality effects, goal achievement and reputation" with five items. The reliability of the first factor was 0.89/0.90 and of the second factor 0.88/0.86.

3.3 Data Analysis

Data analysis was performed using Statistica (version 5) software. Indexes of the four organisational culture types for each organisation were calculated as averages of the developer responses from that organisation. In the case of all other variables, individual developer and manager responses were aggregated separately to the organisational level calculating the aggregated responses as means of individual responses. The developer and manager responses were analysed separately.

To derive empirically the organisational culture, cluster analysis is used. After identifying clusters ANOVA/MANOVA is used to analyse the differences in the perceptions of use, support, and impact of SDMs. Finally, multiple regression analysis is used to investigate in more detail the relationship between the four cultural orientations and the perceptions of use, support and impact of SDMs.

---

[2] The figure before the slash refers to the developer data and the figure after the slash to the manager data.

Table 2. Results of cluster analysis

	Cluster 1 (n = 26)	Cluster 2 (n = 18)	Cluster 3 (n = 13)	Cluster 4 (n = 12)
Group culture (GC)	3.1	3.6	2.7	3.9
Development culture (DC)	2.9	3.4	2.2	3.8
Hierarchical culture (HC)	2.9	1.9	2.3	3.4
Rational culture (RC)	3.5	3.7	2.4	4.0
Interpretation of clusters	Moderate rationally oriented culture	Moderate non-hierarchical culture	Weak group-oriented culture	Strong comprehensive culture

4 Results

4.1 Organisational Culture of IS Departments

In order to derive the organisational culture of an IS department, cluster analysis was conducted as K-means clustering using the four indicators of culture as clustering variables. Experimenting with alternative number of clusters (3-5), a four cluster solution turned out as the easiest to interpret (Table 2). It shows that IS departments can have one of the following cultures: a moderate rationally oriented culture, a moderate non-hierarchical culture, a weak group-oriented culture, or a strong comprehensive culture.

4.2 Differences in the Deployment of SDMs among IS Departments with Different Organizational Culture

ANOVA/MANOVA was used to test whether any differences exist in the SDM deployment among IS departments within the different culture clusters. Neither vertical methodology use nor horizontal methodology use differed between the four clusters.

As explained in section 3.2, the perceived support provided by SDMs is measured using the following three perspectives: Perceived support as production technology, perceived support as control technology, and perceived support as cognitive & co-operation technology. Table 3 shows the perceived methodology support as production technology in the four cultural clusters. The results of MANOVA indicated that the vector consisting of support for organisational alignment, support for technical design and support for verification and validation differ between the culture clusters for both the developers' perceptions (Wilks' Lambda = 0.75 at the level of $p \leq 0.10$) and the managers' perceptions (Wilks' Lambda = 0.70 at the level of $p \leq 0.10$). When we consider the individual factors used to measure perceived support as production technology, F values shows that managers see support for organisational alignment to differ significantly between the clusters, whereas developers see support for the technical design and support for verification and validation to differ between them.

Table 3. Differences in perceived SDM support as production technology among the organisational culture clusters

	Support for organisational alignment	Support for technical design	Support for verification and validation
Moderate rationally oriented culture	De: 3.4 Ma: 3.8	De: 3.4 Ma: 3.5	De: 3.3 Ma: 3.0
Moderate non-hierarchical culture	De 3.2 Ma: 2.8	De: 3.3 Ma: 3.1	De: 3.0 Ma: 2.3
Weak group-oriented culture	De: 3.1 Ma: 4.0	De: 3.0 Ma: 3.4	De: 2.7 Ma: 3.0
Strong comprehensive culture	De: 3.8 Ma: 3.5	De: 3.7 Ma: 3.5	De: 3.7 Ma: 3.1
F	De: 2.22' Ma: 4.25*	De: 2.88* Ma: 0.71	De: 4.11* Ma: 1.49

'$p \leq 0.10$ *$p \leq 0.05$ **$p \leq 0.01$ ***$p \leq 0.001$

Table 4. Differences in perceived SDM support as cognitive & cooperation technology among the organisational culture clusters

	Support for the common conception of SD practice	Support for the evaluation of SD practice
Moderate rationally oriented culture	De: 3.2 Ma: 3.6	De: 3.3 Ma: 3.7
Moderate non-hierarchical culture	De: 3.0 Ma: 3.0	De: 3.1 Ma: 2.3
Weak group-oriented culture	De: 3.0 Ma: 3.2	De: 2.9 Ma: 3.6
Strong comprehensive culture	De: 3.5 Ma: 3.4	De: 3.3 Ma: 3.5
F	De: 1.59 Ma: 1.61	De: 0.78 Ma: 4.85**

'$p \leq 0.10$ *$p \leq 0.05$ **$p \leq 0.01$ ***$p \leq 0.001$

Perceived methodology support as control technology did not differ significantly between the four clusters in any respondent groups. However, organisation with a strong comprehensive culture generally reported the highest values and organisations with a moderate non-hierarchical culture the lowest values.

Table 4 depicts the perceived methodology support as cognitive & co-operation technology in the four cultural clusters. The results of MANOVA indicate that the vector consisting of support for the common conception of SD practice and support for the evaluation of SD practice differ between the culture clusters for the managers' perceptions (Wilks' Lambda = 0.72 at the level of $p \leq 0.05$), but not for the developers' perceptions. When we consider the individual factors used to measure perceived support as cognitive & co-operation technology, F values show that managers' perceptions differ significantly in the four clusters, especially the perceptions of the support for the evaluation of SD practice.

Table 5. Differences in perceived SDM impact on the quality of the developed system and the quality and productivity of the development process among the organisational culture clusters

	Impact on the quality of developed systems	Productivity effects and morale	Quality effects, goal achievement and reputation
Moderate rationally oriented culture	De: 3.4 Ma: 3.7	De: 3.2 Ma: 3.6	De: 3.3 Ma: 3.7
Moderate non-hierarchical culture	De 3.4 Ma: 3.1	De: 3.3 Ma: 3.2	De: 3.2 Ma: 3.4
Weak group-oriented culture	De: 3.0 Ma: 3.6	De 2.9 Ma: 3.4	De: 3.1 Ma: 3.8
Strong comprehensive culture	De: 3.8 Ma: 3.3	De: 3.6 Ma: 3.0	De: 3.7 Ma: 3.1
F	De: 3.02* Ma: 1.33	De: 2.72' Ma: 0.80	De: 2.32' Ma: 1.21

'$p \leq 0.10$ *$p \leq 0.05$ **$p \leq 0.01$ ***$p \leq 0.001$

Perceived impact of SDMs was measured using two perspectives, namely the perceived impact on the quality of the developed system, and the perceived impact on the quality and productivity of the development process. When we consider the perceived impact on the quality of the developed system in the second column of Table 5, we find that only developer perceptions differed between the four clusters.

The last two columns of Table 5 report the perceived methodology impact on the quality and productivity of the development process. The results of MANOVA indicate that the vector consisting of productivity effects and morale, and quality effects, goal achievement and reputation, differ between the culture clusters for the developers' perceptions (Wilks' Lambda = 0.82 at the level of $p \leq 0.10$), but not for the managers' perceptions. F values show significant ($p \leq 0.10$) differences between the clusters in the case of developer perceptions, when productivity effects and morale, and quality effects, goal achievement and reputation are considered individually.

4.3 The Relationhips between Culture Orientations and the Deployment of SDMs

The above analysis indicates that organisations with different organisational cultures perceive the methodology support for SD differently. Developers also see significant differences in the SDMs' impact on the quality of the developed systems and the productivity and quality of the SD process. Because the empirically derived clusters of culture were synthetic, it is difficult to conclude which of the four culture orientations may explain differences between the clusters. To test this regression analysis was used considering each of the use, support areas and impact dimensions as the dependent variable and the four indicators of organisational culture as the independent variables.

Multiple regression analysis assumes (1) interval or ratio scale measurement, (2) linearity, (3) homoscedasticity, i.e. the constancy of the residuals across the values of the predictor variables, (4) independence of residuals, (5) normality of residuals and (6) no multicollinearity [12]. Billings and Wroten [2] assess that the assumption of equal interval is not critical concluding that carefully constructed measures employing reasonable number of values and containing multiple items will yield data with sufficient interval properties. Linearity of the relationships was tested visually using the standardised residual and partial regression plots. None of the variables violated this assumption. Homoscedasticity was tested visually, using the standardised residual and observed values plots. None of the variables violated this assumption. Independence of residuals was assessed using the Durbin-Watson statistics which ranges from 0-4, with the value 2 indicating that there is no autocorrelation. In the case of the manager data the values varied between 1.65 and 2.06, and in the case of the developers between 1.67 and 2.13, with the exception of vertical use which had a value of 1.42. Normality of residuals was assessed using the modified Kolmogorov-Smirnov test (Lilliefors). Violations were detected ($p < 0.05$) in the regressions with vertical use as the dependent variable for both the manager and the developer data. Multicollinearity was tested using the tolerance values. The lowest tolerance value in the case of the developer data was 0.43 and in the case of the manager data it was 0.40. These values far exceeded the cutoff value of 0.01 as suggested by [12]. Taken together, the specific assumptions of multiple regression analysis were reasonable satisfied.

Table 6. The relationship between culture orientations and methodology use

	Vertical methodology use	Horizontal methodology use
	ß	ß
Group culture (GC)	De: -0.24 Ma: 0.14	De: - Ma: -0.06
Developmental culture (DC)	De: -0.03 Ma: 0.20	De: - Ma: 0.09
Hierarchical culture (HC)	De: 0.25' Ma: 0.19	De: - Ma: 0.10
Rational culture (RC)	De: -0.02 Ma: -0.34'	De: - Ma: -0.21
R^2 Adjusted R^2	De: 0.12' Ma: 0.07 De: 0.06 Ma: 0.00	De: - Ma: 0.04 De: - Ma: -0.04

'$p \leq 0.10$ *$p \leq 0.05$ **$p \leq 0.01$ ***$p \leq 0.001$

Table 6 reports the relationship between the strength of each culture dimension and methodology use. It shows that the four dimensions of culture are weak predictors of use. Only when developers assessed the vertical use the four dimensions explained methodology use to a significant degree ($p \leq 0.10$). In that case the dimension of HC was significantly associated with the vertical use.

Table 7 shows the relationship between the cultural dimensions and factors of perceived methodology support as production technology. It indicates that the four culture dimensions explain significantly the methodology support as production technology as perceived by developers. Among the cultural dimensions especially the strength of the HC and partly also the DC are positively associated with the perceived methodology support as production technology. The RC on the other hand is predominantly negatively associated with manager perceptions of methodology support as production technology, even though most of the regression coefficients are not statistically significant.

Table 7. The relationship between culture orientation and perceived support as production technology

	Support for organisational alignment	Support for technical design	Support for verification and validation
	ß	ß	ß
Group culture (GC)	De: -0.02 Ma: -0.21	De: 0.03 Ma: -0.09	De: 0.00 Ma: -0.19
Developmental culture (DC)	De: 0.18 Ma: 0.02	De: 0.33' Ma: 0.33	De: 0.39* Ma: 0.11
Hierarchical culture (HC)	De: 0.17 Ma: 0.26'	De: 0.07 Ma: 0.20	De: 0.41** Ma: 0.32*
Rational culture (RC)	De: 0.10 Ma: -0.27	De: 0.05 Ma: -0.35'	De: -0.15 Ma: -0.13
R^2	De: 0.12 Ma: 0.18'	De: 0.17* Ma: 0.09	De: 0.29*** Ma: 0.11
Adjusted R^2	De: 0.06 Ma: 0.10	De: 0.11 Ma: 0.01	De: 0.24 Ma: 0.03

'$p \leq 0.10$ *$p \leq 0.05$ **$p \leq 0.01$ ***$p \leq 0.001$

Table 8 depicts the relationship between the culture-orientations and both perceived methodology support as control and perceived methodology support as cognitive & co-operation technologies. It shows that the four cultural dimensions, and especially the strength of the HC, explain a significant part of variance of the methodology support as control technology as perceived by developers. It also shows that the four cultural dimensions do not explain the methodology support as cognitive & co-operation technology. Despite that, the HC exhibits the most significant regression coefficient also here.

Finally, Table 9 depicts the relationship between culture-orientations and perceived methodology impact on the quality of developed systems and the quality and productivity of the SD process. It shows that the four cultural dimensions explain very weakly the quality of developed systems and the quality and productivity of SD process. It shows again that the RC is negatively associated with methodology impact on the quality of developed systems and the quality and productivity of SD process as perceived by managers.

Table 8. The relationship between culture orientations and perceived SDM support as control and cognitive & co-operation technologies

	Support as control technology	Support for the common conception of SD practice	Support for the evaluation of SD practice
	ß	ß	ß
Group culture (GC)	De: -0.13 Ma: -0.19	De: 0.06 Ma: -0.09	De: 0.16 Ma: -0.14
Developmental culture (DC)	De: 0.20 Ma: -0.03	De: -0.01 Ma: 0.33	De: -0.03 Ma: 0.18
Hierarchical culture (HC)	De: 0.36* Ma: 0.15	De: 0.28' Ma: 0.20	De: 0.13 Ma: 0.22
Rational culture (RC)	De: 0.04 Ma: -0.11	De: 0.02 Ma: -0.18'	De: 0.02 Ma: -0.20
R^2	De: 0.19* Ma: 0.09	De: 0.09 Ma: 0.07	De: 0.05 Ma: 0.15
Adjusted R^2	De: 0.13 Ma: 0.01	De: 0.03 Ma: -0.02	De: -0.02 Ma: 0.07

'$p \leq 0.10$ *$p \leq 0.05$ **$p \leq 0.01$ ***$p \leq 0.001$

Table 9. The relationship between culture-orientations and perceived methodology impact on the quality of developed systems and the quality and productivity of systems development process

	Impact on the quality of developed systems	Productivity effects and morale	Quality effects, goal achievement and reputation
	ß	ß	ß
Group culture (GC)	De: 0.07 Ma: -0.07	De: 0.12 Ma: -0.16	De: 0.13 Ma: -0.10
Developmental culture (DC)	De: 0.11 Ma: 0.17	De: 0.25 Ma: 0.41'	De: 0.04 Ma: 0.19
Hierarchical culture (HC)	De: 0.07 Ma: 0.16	De: -0.04 Ma: 0.10	De: 0.03 Ma: 0.16
Rational culture (RC)	De: 0.20 Ma: -0.29	De: 0.09 Ma: -0.34'	De: 0.20 Ma: -0.39'
R^2	De: 0.13' Ma: 0.06	De: 0.15' Ma: 0.10	De: 0.10 Ma: 0.11
Adjusted R^2	De: 0.07 Ma: -0.02	De: 0.09 Ma: 0.02	De: 0.04 Ma: 0.03

'$p \leq 0.10$ *$p \leq 0.05$ **$p \leq 0.01$ ***$p \leq 0.001$

5 Discussion and Final Comments

Despite the differences between the respondent groups, the above results provide some support for the conjecture that the deployment of SDMs differ in the four empirically derived cultural clusters. They indicate that methodology use did not differ in the four cultural clusters. Also in the case of methodology support as control technology no differences were found. In other cases at least one of the two respondent groups reported cultural differences. It is quite difficult to explain these differences. This leads to a further research question whether different functionalities of SDMs, for example, may differ in their cultural sensitivity.

Table 10 summarises the significant ($p \leq 0.10$) coefficients identified in regression analyses (+ for positive and– for negative)[3]. One can clearly see that the HC orientation is most consistently associated with SDM deployment, when assessed by developers: the more hierarchical the culture is perceived to be, the more SDMs are used and the more support they are perceived to provide. The DC is also found to have some positive associations with the methodology deployment, although not systematically. Quite interestingly, the more RC orientation, the more critical management seems to be with regard to the methodology support and impact.

Table 10. Summary of the results of regression analyses

	SDM use	Support as production technology	Support as control technology	Support as cognitive & co-ordination technology	Impact on quality of developed systems	Impact on quality and productivity SD process
GC						
DC		De: ++				Ma: +
HC	De: +	De: + Ma: ++	De: +	De: +		
RC	Ma: -	Ma: -		Ma: -		Ma: - -

The relationship between the HC orientation and methodology deployment may be interpreted as reinforcing, implying either that a HC promotes methodology deployment or that SDMs as such are perceived as part and parcel of the HC. As Table 10 indicates the association between the HC and methodology support is confined to developer perceptions. Hypothesising that the HC promotes methodology deployment, the question is why the association between the strength of HC and perceived methodology deployment differs between developers and managers. One answer in the case of this study may be that developers, who assessed the organisational culture, interpreted SDMs as manifestations of the HC.

Referring to the DC, its positive association with the developers' perceptions of the methodology support as production technology can be explained either as SDMs' direct support for creativity and adaptation to the external environment or alternatively as indirect support for creativity and adaptation through increased order and routinization. The former direct support can be interpreted as reinforcing whereas the

[3] The number of + and– signs shows many times the significant beta coefficient was found.

latter is more complementary. This study does not allow testing these alternative explanations in more detail.

The negative association between the strength of the RC and methodology deployment in the case of manager perceptions shows management's critical attitude towards SDMs in organisations which are highly achievement-oriented, focusing on productivity, efficiency and goal achievement. When contrasted with developers, the likely explanation for this is that managers emphasise these goals more than developers. The question is, of course, whether these results really reflect the weakness of SDMs when evaluated on the criteria of productivity and efficiency. Even though one would not agree with the claim, it is also obvious that it is extremely difficult to demonstrate the contribution of SDMs to productivity and efficiency. In view of this uncertainty managers in organisations with a rationally oriented culture may take a more critical attitude towards SDMs. It may also be that the strong emphasis on productivity and efficiency leads to focus on short-run impacts, whereas SDMs' benefits accrue more slowly [8]. In an extreme case, the question may be about IT managers' disappointment with SDMs when projects start to fall behind schedules. It is well-known that SDMs are not very helpful solving these crisis situations and that projects then easily fall into a chaotic *ad hoc* style of SD without any SDM [17]. Obviously, there is a clear need for additional research on the reasons underlying managers' critical perceptions in rationally oriented organisations.

What are the practical implications of the results? Assuming that the HC supports methodology deployment, the results imply that in organisations with a strong HC orientation the chances of getting SDMs accepted are higher than in organisations with a weak HC orientation. In the latter case, one should pay special attention to measures of introducing SDMs. In organisations with a DC orientation, the results suggest as one possibility to emphasise SDMs' support for creativity and adaptation to the external environment. If the SDM to be introduced does not support them directly, it may be deliberately engineered to comprise these features. A second option is to introduce a SDM as an effective means to make less creative aspects of SD work more ordered and routine, thus freeing developers' time for more creative work.

An alternative interpretation of the association between the HC orientation and methodology deployment is that SDMs as such are manifestations of the HC. If an organisation does not wish to move into that direction, one should pay special attention to means of avoiding the hierarchical flavour of SDMs when introducing them. One means to make SDMs less bureaucratic is to introduce them as general approaches [18] rather than as complicated conglomerates of numerous techniques with massive documentation. This higher level granularity may also make SDMs more useful as [10] concludes.

References

1. Allaire, Y. and and Firsirotu, M.E., Theories of organizational culture, *Organization Studies*, Vol. 5, No. 3, 1984, pp. 193-226

2. Billings, R.S. and Wroten, S.P., Use of path analysis in industrial/organizational psychology: Criticism and suggestions, *Journal of Applied Psychology*, Vol. 63, No. 6, 1978, pp. 677-688
3. Cameron, K.S., Effectiveness as paradox: Consensus and conflict in conceptions of organizational effectiveness, *Management Science*, Vol. 32, 1986, pp. 539-553
4. Cameron, K.S. and Freeman, S.J., Cultural congruence, strength, and type: Relationships to effectiveness, in Woodman, R.W. and Pasmore, W.A. (eds.), *Research In Organizational Change and Development*, Volume 5, JAI Press Inc, Greenwich, CT, 1991, pp. 23-58
5. Chatzoglou, P.D. and Macaullay, L.A., Requirements capture and IS methodologies, *Information Systems Journal,* Vol. 6, 1996, pp. 209-225
6. Cooper, R.B., The inertial impact of culture on IT implementation, *Information & Management*, Vol. 27, No. 1, 1994, pp. 17-31
7. Denison, D.R. and Spreitzer, G.M., Organizational culture and organizatioanl development: A competing values approach, in Woodman, R.W. and Pasmore, W.A (eds.), *Research in Organizational Change and Development*, Volume 5, JAI Press Inc, Greenwich, CT, 1991, pp. 1-21
8. Fichman, R.G., Kemerer, C.F., Adoption of software engineering process innovations: the case of object-orientation, *Sloan Management Review*, Winter 1993, pp. 7-22
9. Fitzgerald, B., Formalized systems development methodologies: a critical perspective, *Information Systems Journal,*Vol. 6, pp. 3-23, 1996
10. Fitzgerald, B., The use of systems development methodologies in practice: a field study, *Journal of Information Systems*, Vol. 7, No.3, 1997, pp. 201-212
11. Glass, R.L., *Software Creativity*, Prentice Hall, Englewood Cliffs, NJ, 1995
12. Hair, J.F.Jr., Anderson, R.E., Tatham, R.L. and Black, W.C., *Multivariate Data Analysis with Readings*, Macmillan, New York, NY, 1992
13. Hardy, C.J., Thompson J.B. and Edwards H.M., The use, limitations and customization of structured systems development methods in the United Kingdom, *Information and Software Technology,* Vol. 37, No. 9, 1995, pp.467-477
14. Henderson, J.C. and Cooprider, J.G., Dimensions of I/S planning and design aids: A functional model of CASE technology, *Information Systems Research,* Vol. 1, No. 3, 1990, pp. 227-254
15. Hofstede, G., Neuijen, B., Ohayv, D.D. and Sanders, G., Measuring organizational cultures: A qualitative and quantitative study across twenty cases, *Administrative Science Quarterly*, Vol. 35, 1990, pp. 286-316
16. Huisman, M. and Iivari, J., Perceptual congruence in the deployment of systems development methodologies, *Proceedings of the 11th Australian Conference on Information Systems,* Brisbane, Australia, 2000
17. Humphrey, W.S., *Managing the Software Process*, Addison-Wesley, Reading, Massachusetts, 1989
18. Iivari, J., Hirschheim, R. and Klein, H., A Paradigmatic Analysis Contrasting Information Systems Development Approaches and Methodologies, *Information Systems Research*, Vol. 9, No. 2, 1998, pp. 164-193
19. ISO, Information technology– Software product evaluation– Quality characteristics and guidelines for their use, ISO/IEC DIS 9126, ISO, 1990
20. Lewin, A.Y. and Minton, J.W., Determining organizational effectiveness: Another look, and an agenda for research, *Management Science*, Vol. 32, 1986, pp. 514-538
21. McChesney, I.R. and Glass, D., Post-implementation management of CASE methodology, *European Journal of Information Systems,* Vol. 2, No. 3, 1993, pp. 201-209

22. Quinn, R.E. and Cameron, K.S., Organizational life cycles and shifting criteria of effectiveness, Some preliminary evidence, *Management Science*, Vol. 29, 1983, pp. 33-51
23. Quinn, R.E. and Kimberly, J.R., Paradox, planning, and perseverance: Guidelines for managerial practice, in Kimberly, J.R. and Quinn, R.E. (eds.), *New Futures: The Challenge of Managing Organizational Transitions*, Dow Jones-Irwin, Homewood, ILL, 1984, pp. 295-313
24. Quinn, R.E. and Rohrbaugh, J., A spatial model of effectiveness criteria: Towards a competing values approach to organizational analysis, *Management Science*, Vol. 29, No. 3, 1983, pp. 363-377
25. Robey, D. and Boudreau, M.-C., Accounting for the contradictory organizational consequences of information technology: Theoretical directions and methodological implications, *Information Systems Research*, Vol. 10, No. 2, 1999, pp. 167-185
26. Rogers, E.M., *Diffusion of Innovations,* Fourth edition, The Free Press, New York, NY, 1995
27. Russo, N.L., Hightower, R. and Pearson, J.M., The failure of methodologies to meet the needs of current development environments, in Jayaratna, N. and Fitzgerald, B. (eds.), Learned from the Use of Methodologies: Fourth Conference on Information Systems Methodologies, 1996, pp. 387-394
28. Schein, E.H., *Organizational Culture and Leadership*, Jossey-Bass, San Fransisco, CA, 1985
29. Smircich, L., Concepts of culture and organizational analysis, *Administrative Science Quarterly*, Vol. 28, No. 3, 1983, pp. 339-358
30. Wastell, D.G., The fetish of technique: methodology as a social defence, *Information Systems Journal*, Vol. 6, No. 1, 1996, pp. 25-40
31. Woodman, R.W. and Pasmore, W.A. (eds.), *Research in Organizational Change and Development*, JAI Press, Greenwich, CT, 1991
32. Wynekoop, J.L. and Russo, N.L., System development methodologies: unanswered questions and the research-practice gap, in DeGross, J.I *et al.* (eds.), *Proceedings of the Fourteenth International Conference on Information Systems*, Orlando, FL, 1993, pp.181-190
33. Wynekoop, J.L. and Russo, N.L., Studying system development methodologies: an examination of research methods, *Information Systems Journal,* Vol. 7, 1997, pp. 47-65
34. Yeung, A.K.O., Brockbank, J.W. and Ulrich, D.O., Organizational culture and human resource practices; An empirical assessment, in Woodman, R.W. and Pasmore, W.A. (eds.), *Research In Organizational Change and Development*, Volume 5, JAI Press Inc, Greenwich, CT, 1991, pp. 59-81

Increasing Reusability in Information Systems Development by Applying Generic Methods

Silke Eckstein, Peter Ahlbrecht, and Karl Neumann

Information Systems Group,
TU Braunschweig
P.O.box 3329, D-38023 Braunschweig
{s.eckstein, p.ahlbrecht,k.neumann}@tu-bs.de

Abstract. Increasing the reuse of parts of the specification and implementation of complex software systems, as for example information systems, may lead to substantial progress in the development process. This paper focuses on reusing parts of specifications with the help of generic methods and explores two aspects: the parameterization concepts of the languages UML and TROLL, and how formal parameters in such concepts can be restricted if needed.

1 Introduction

The development of large information systems is by far no trivial task, since the probability of errors grows significantly with increasing complexity. However, approaches are being made to realize large systems on the basis of generic methods (cf. e.g. [5]) and thereby to reduce the complexity of the development process. The term "generic methods" also includes parameterized programming and related approaches at the specification level [9].

Information systems are mostly rather complex programs, which are frequently custom tailored for special applications. Their central part is usually a database system, and additional functionality, especially the user interface, is realized as an extensive application program. In recent years database technology moved from centralized systems to distributed databases and client/server systems [3,40], and the swift acceptance of the World Wide Web led to the situation that nowadays users are expecting to be able to use their familiar web browser as the interface to various information systems [21,2].

In this paper we focus on the specification of such information systems and particularly on parameterization aspects. In general, we propose to apply UML [29,43] together with the formal specification language TROLL [27,25] developed in our group, in order to utilize the advantages of both a semi–formal graphical language and a formal one, as discussed for instance in [48] for an older version of TROLL and a predecessor of the UML, the Object Modeling Technique (OMT) [42]. Regarding parameterization, both languages do provide such concepts, and in this paper we study how they correspond to each other.

To this end we first give an overview of current research activities related to generic methods, putting special emphasis on parameterization concepts for

K.R. Dittrich, A. Geppert, M.C. Norrie (Eds.): CAiSE 2001, LNCS 2068, pp. 251–266, 2001.
© Springer-Verlag Berlin Heidelberg 2001

describing variants of a system. We then introduce the formal object oriented specification language TROLL before sketching our area of application. In Sect. 5 we introduce and investigate the parameterization concepts of TROLL und UML and focus on one hand on the parameter granularity and on the other hand on parameter constraints. Finally we summarize our paper.

2 Generating Software

As a long-term objective the possibility to describe families of information systems with the support of specification libraries is as desirable as generating concrete runnable systems from such a description. However, while the concept of generators has been in use in some fields of software engineering, for instance as scanner and parser generators in the area of compiler construction or the generation of user interfaces [44], to name a very recent approach, there is so far no uniform theory and methodology on generating information systems. Nevertheless the automatic implementation of such reactive systems is considered possible, not only based on specifications given at a high level of abstraction, but even starting from the results of the requirements analysis [26].

[5] discusses various generators for highly specialized software systems, for example Genesis as a generator for database management systems, Ficus for generating distributed file systems, ADAGE as a generator for avionic software, and SPS for software on signal processing. According to [4], all these generators belong to the so-called GenVoca approach, in which the generator constructs a software system with respect to the following points: a system consists of modules, which can be assigned to different fields of application, and which form distinctively larger units than given by classes or functions. Modules are implemented by components, and different components for the same task provide the same interface. The combination of modules and their adaptation to meet special requirements is achieved by parameterization. In addition to these points the possibility to check configurations for validity should also be given [30].

The different types of generators can be classified according to a variety of criteria. We may, for example, distinguish between whether they are compositional or transformational. While the former compose the desired software out of prefabricated components, the latter actually generate the code themselves. Another classification criterion is the question of whether the software is generated statically or dynamically, i.e. of whether a generated complete system may be reconfigured at runtime or not. Regarding parametric languages it is also possible to differentiate between generators that switch from the source language to a different target language, and those which maintain the same language for input and output. Generators of the latter type will replace formal parameters by provided elements during instantiation, thus producing non-parametric code of the same language. Domain-specific generators exist which are capable of creating systems of a special type — examples are scanner and parser generators — and there are also generators for "arbitrary" systems like, for instance, compilers for programming languages.

Under the term *generic methods* we should subsume not only generators and their application for creating software, but also languages which permit a description of families of specification and implementation modules by means of generic elements — so–called formal parameters — in such a way that these descriptions may then be adjusted according to given special requirements by binding the formal parameters to actual values.

In the theory of abstract datatypes the idea of parameterization is a well known concept which has already been investigated some time ago. Here it suffices to mention the frequently used example of a stack for storing elements of an unspecified type. C++, for instance, supports the implementation of such a parameterized abstract datatype with its template construct. In Java, however, a corresponding language element is currently still not available, but attempts to add such a parameterization concept are being made (cf. e.g. [38,7]). Compared to C++ these approaches also provide the possibility to define (syntactical) properties of the parameters. In the literature this possibility is also termed *bounded parametric polymorphism* [1] or *constrained genericity* [36].

Parameterization concepts have also been investigated and formalized in the field of algebraic specification of abstract datatypes (cf. e.g. [19,35]). Here, it is also possible to define semantic properties by providing axioms which the actual parameters have to meet, in addition to being able to describe the signature of the parameters. [20] transferred these results on to larger units, namely modules, which may themselves have other modules as parameters. Quite a few specification languages have been developed in this area, which we will not discuss in further detail, but refer this to [49], which provides a useful overview.

We only mention OBJ [23], as this specification language had a major impact on the development of LILEANNA [22], even though the latter also incorporates implementation aspects in addition to specification concerns and the formal semantics of the languages differ. LILEANNA, too, facilitates having entire modules as parameters and has been used during the already mentioned ADAGE project for the generation of avionic software. Furthermore, the language distinguishes between a vertical and horizontal composition and, accordingly, also between horizontal and vertical parameters. Vertical composition permits the description of a system in separate layers, while horizontal composition supports structuring of the single layers [24].

In all the approaches which we have discussed so far the structuring elements like classes or abstract datatypes may be generic. The structure which they describe, however, is and remains unchangeable. In contrast to this the *collaboration–based design* [46] assumes that the collaborations among different classes, i.e. class–spanning algorithms, represent the reusable units, which should be adaptable to various concrete class structures. In [37] these collaborations are described relatively to an abstract class graph, which again represents the interface to a concrete class graph. This setup is termed *structure generic components*.

Irrespective of the type of components, classes or modules, they have to be stored in libraries to be available for reuse. Here, powerful tool support is needed (cf. e.g. [10,31,11]) to make effective reuse possible.

3 The TROLL–Approach

The object oriented language TROLL [27,25] was developed for specifying information systems at a high level of abstraction. It has been successfully utilized in an industrial environment to develop an information system in the area of computer aided testing and certifying of electrical devices [32,28].

In this approach informations systems are regarded as being communities of concurrent, interacting (complex) objects. Objects are units of structure and behavior. They have their own state spaces and life cycles, which are sequences of events. A TROLL specification is called *object system* and consists of a set of datatype and a set of object class specifications, a number of object declarations and of global interaction relationships.

Object classes describe the potential objects of the system by means of structure, interface, and behavior specifications. Attributes are used to model the state spaces of the objects. Together with the actions they form the objects' local signatures. Attributes have datatypes and may be declared as hidden, i.e. only locally visible, optional or constant. They can be derived, meaning their values are calculated from the values of other attributes, or initialized.

The second part of the signature determines the actions of the objects. Each action has a unique name and an optional parameter list. Input and output values are distinguished and each parameter has a certain datatype. The visibility of actions can be restricted in such a way that they do not belong to the object class' interface but can be used internally only. Actions that create objects of a class, so called birth actions, and actions that destroy objects, so called death actions, are explicitly marked.

While the signature part of an object class determines the interface of the objects, the behavior part constitutes the reactions of the objects to calls of the actions declared in the interface. The admissible behavior of the objects can be restricted by means of initialization constraints and invariants.

Complex objects may be built using aggregation and with the consequence that the component objects can exists only in the context of the complex one and that the superior object can restrict the behavior of its components. By means of specialization hierarchies base classes may be extended with further aspects.

The set of potential instances is determined by object declarations at the object system level. At runtime concrete instances can be created through the calling of birth actions of the respective object classes. All instances of a system are concurrent to each other and synchronize when they interact. Interactions are global behavior rules that together with the local ones describe the behavior of the system.

Semantics is given to TROLL specifications using different techniques: the static structure of an object system is semantically described with algebraic methods, and to describe properties of distributed objects a special kind of temporal logic has been developed. This logic is called *Distributed Temporal Logic* (DTL) [17,16] and is based on n-agent logics. Each object has a local logic allowing it to make assertions about system properties through communication with other objects. Objects are represented by a set of DTL formulae interpreted over labelled prime event structures [39]. Interaction between concurrent objects is done by synchronous message passing. The system model is obtained from its object models, whereby object interaction is represented by shared events. An exhaustive description of the model theory is given in [18,15].

Presently, an extension of TROLL with module concepts is under investigation, aiming on one hand at providing more sophisticated structuring concepts and on the other hand at supporting reuse of specifications by means of concepts for parameterization [12,13]. Regarding theoretical foundations, module theory is being addressed e.g. in [33,34], where in particular DTL has been extended to MDTL *(Module Distributed Temporal Logic)*.

4 Area of Application

The starting point for our investigations on generic information systems are web-based information systems which can be generated in order to facilitate the administration of tutorials. In a tutorial, students are supervised in small groups which are guided by a tutor. The students have to complete exercises handed out by one of the lecturer's assistants in teams of two. If they achieve a certain percentage of the total of points assigned to the exercises, they obtain a certificate for the course at the end of term.

Looking from an organizational point of view this means that the students have to form teams of two, and register for the tutorial providing certain personal information (name, registration number, etc.). During term time the tutors keep account of the points which the teams and students of their group receive, so that by the end of term the certificates can be printed and signed by the lecturer.

As a first basic structuring of the administration system three layers can be identified (cf. also [41]): one for presenting information, one for storing it, and the third to facilitate the exchange of data between the former two. These tasks are taken over by the packages *Presentation* and *Storage* and the class *Controller*, respectively, which are components of the package *TutorialAdministrationSystem*, which again represents the entire system. In the following a more detailed discussion will be given only for the package modeling the user interface. Figure 1 shows the corresponding class diagram for the package *Presentation*.

In this figure, the possibilities to access the single web pages are modeled by compositions. The parts cannot be created before and die at the latest with the death of the composite object. A *StandardPage*, for example, can be viewed only after a successful log-on on a *StartPage*. The specialization of the *StandardPage* into the *Assistant*–, *Tutor*– and *StudentPage* states that access to the *Tutorial*–

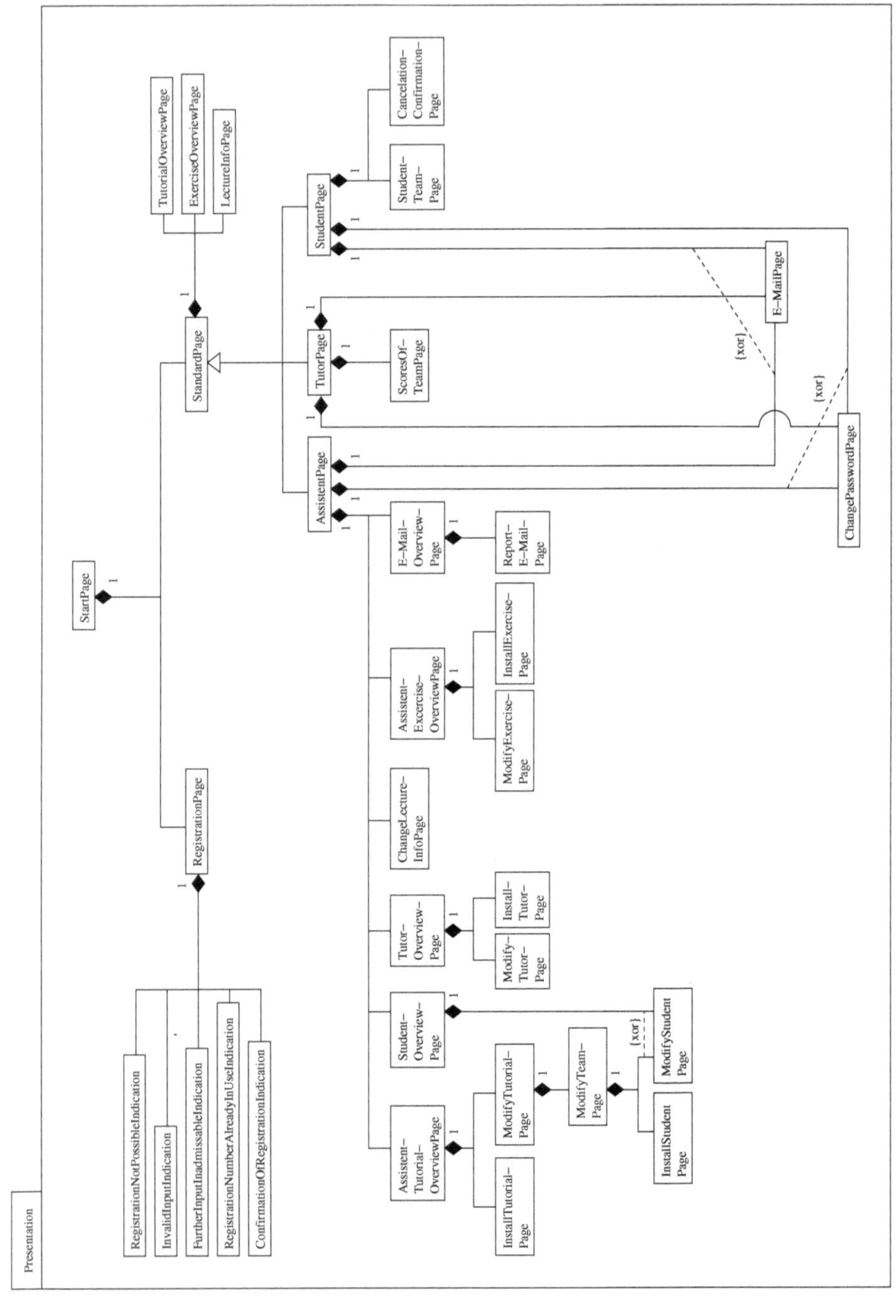

Fig. 1. Detailed class diagram of package *Presentation*

and *ExerciseOverviewPage* as well as to the *LectureInfoPage* is also possible from the specialized pages. In contrast to this, changing the password or sending e-mail can only be done from the specialized pages.

Access to the system is possible via a *StartPage* only. There, a user can choose to register for the tutorials or retrieve general information on them. Someone already registered with the system may, in accordance with his user status, retrieve further information or perform additional operations. Students for example may change their address and retrieve the score of their team. Identification is accomplished by prompting for a login and password at the *StartPage*. An unregistered student may enroll himself and others using the *RegistrationPage* as many times as admissible to enter the data of the team members. A detailed description of the registration procedure using activity diagrams can be found in [14]; due to space restrictions we refrain from elaborating on it here.

On the one hand, the static part of the *Presentation* package thus describes the information which the system has to provide for the different groups of users. On the other hand it also models the possibilities for navigating between the user interfaces, as e.g. [45] recommends for the design of web sites.

The requirements which have been outlined here so far apply to many tutorials with different deviations. For example, the size of the teams and the required minimum score may vary. The tutorials may be held several times per week or less, and even completely irregularly. Furthermore, they do not have to take place at the same time, which should be reflected in the application procedure: With different contact hours of the tutors it would be nice to offer the students the possibility to choose a time which suits their timetable, requiring a completely different algorithm from the one which merely distributes them equally over the available tutorials in the case of simultaneous contact hours. Finally, the number of teams per group may be restricted, for instance for practicals which use equipment that is only available in a limited number.

5 Specifying Generic Information Systems

The application area introduced in Sect. 4 has been a starting point for a case study addressing certain aspects from the subject "generating information systems". We studied parameterization concepts at the specification level or, more precisely, parameterization concepts of the UML, which can be used to describe variants of a system, and implemented a generator program, which produces runnable systems from prefabricated components.

In this paper we concentrate on the specification aspects. In particular, we present the parameterization concepts of TROLL, compare them to those of the UML and discuss some results of our case study.

The UML not only allows to parameterize classes, as e.g. C++ does, but also arbitrary model elements. Such parameterized classes, collaborations, packages etc. are called templates or template classes, template packages and so forth. TROLL provides exactly one parameterizable model element, the so–called module, which can, however, consist of one or more classes together with their structural and communication relationships, or even entire subsystems. The latter are comparable to UML–packages.

There are different types the formal parameters of a UML–template can belong to: If actual arguments of a parameter are supposed to be values of a certain datatype, the parameter is specified in the form *name: type*. If, on the other hand, the actual argument is supposed to be a class or a datatype itself, it suffices to state the formal parameter's name. In this case we also talk about datatype parameters. Furthermore parameters may even represent operations. To represent a template, a small dashed rectangle containing the formal parameter is superimposed on the upper–right corner of the respective model element. In TROLL formal parameters always have to be declared together with a type. Valid types for formal parameters are datatypes or classes themselves, the statement "type", if the actual value is supposed to be a datatype or a class, or the statement "module", if the actual value is supposed to be a bunch of classes.

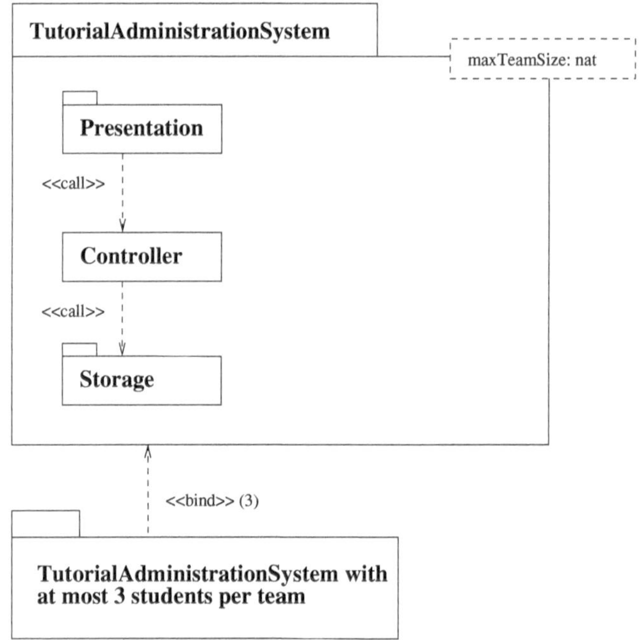

Fig. 2. Parameterized *TutorialAdministrationSystem*

Let us start examining these concepts by means of a small example from our application area. We stated in Sect. 4 that the students should complete exercises in teams of two. It may be desirable to leave the exact team size open at first and generalize the *tutorial administration system* by specifying the maximal number of students per team (*maxTeamSize*) as a parameter. In order to obtain a concrete or actual model element the parameters have to be instantiated.

Here, the formal parameter is bound to the value "3". The corresponding UML specification is shown in Fig. 2 and in TROLL this looks like the following:

```
module TutorialAdministration
  parameterized by maxTeamSize: nat;
  subsystem Presentation ... end_subsystem:
  object class Controller ... end;
  subsystem Storage ... end_subsystem:
end-module;
```

Instantiation is expressed as follows:

```
instantiate module TutorialAdministration
  as TutorialAdministrationSystem_with_at_most_3_students_per_team;
  bind maxTeamSize to 3;
```

In the course of the case study it turned out to be beneficial to use lager units as parameters than those provided by the UML. The reason for this is that the number of parameters may increase rapidly and consequently the specification becomes unintelligible. Using UML as the specification language, we chose to employ the concept of packages and allow them to be used as parameters. As mentioned earlier, in TROLL modules are allowed to be used as parameters for other modules and hence parameters can be as large and as complex as needed.

In [14] these results are illustrated by means of a somewhat more complex part of our application area. There, two different procedures to registrate for the tutorials are discussed. For the first one it is assumed that all tutorials take place at the same time. Consequently, the students can be distributed on the tutorials automatically, whereby a level partition is guaranteed. In the second procedure the students are allowed to choose their groups on their own. In [14] three variants are discussed of how to parameterize the specification in such a way that the different registration procedures are taken into account. In the following we give an overview of these three variants, which differ in the granularity of the employed parameter types. Due to space limitations we refrain from presenting the respective translation to TROLL.

Allowing packages as types for parameters we can model the sketched scenario as follows (Fig. 3): The *registrationPage* and all its depending classes are comprised in a package called *registration*, which represents a formal parameter. This parameter can be instantiated with packages that contain the classes needed for the respective variants of registration. The controller is modelled as a formal parameter of type class. The expected actual arguments are classes, which offer operations needed for the interactions between the presentation and the storage component and which differ in the registration procedure. By instantiating the parameters *registration* and *controller* with suitable packages and classes unparameterized specifications result, which describe systems realizing the respective variant of registration.

A variant to model the requested scenario without using packages would look like Fig. 3 with the *Registration: package* being removed. Essentially, actual

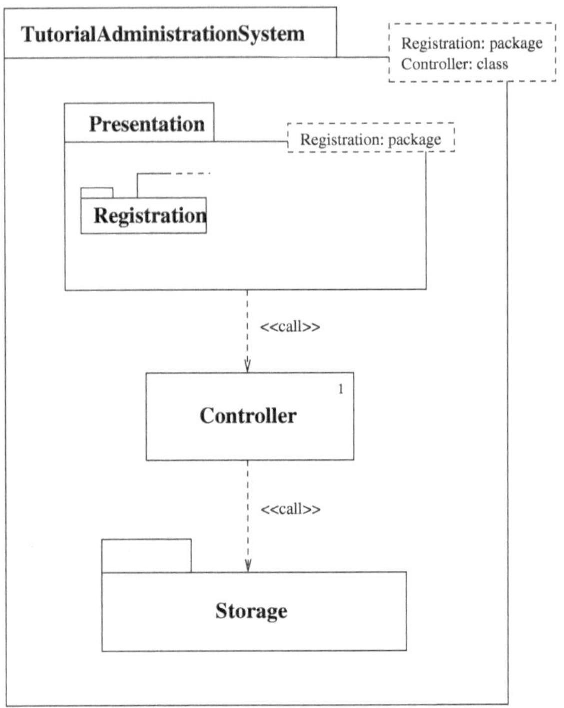

Fig. 3. Variant 1: Packages and classes as parameters

arguments for this parameter are equal to the ones sketched above, but the specification of the presentation package becomes somewhat more complex. As in this variant no packages are allowed to be used as parameters, it is not possible to group all required classes into one unit and exchange them together. Accordingly, this part of the specification has to be so general that it can deal with all actual controller classes which the formal parameter may be instantiated with and thus becomes more complex and a number of additional integrity constraints arises.

Figure 4 shows the last variant we are going to present here. Only data values and operations are used as actual arguments and thus the parameterization exhibits a much finer granularity than before. For instantiating the template two numerical values and an operation have to be provided. The numerical values determine the multiplicity values *maxTeamSize* and *maxNumberOfTeamsPerTutorial*, while the operation determines the core algorithm of the registration procedure.

For this variant, too, the above remarks with respect to the complexity of the presentation package hold, as only the specification of the controller has changed. Comparing the three parameterization variants one can see that in the third one the list of formal parameters may quickly become long and unintelligible while at the same time one can easier determine which parts of the specification are actually variable.

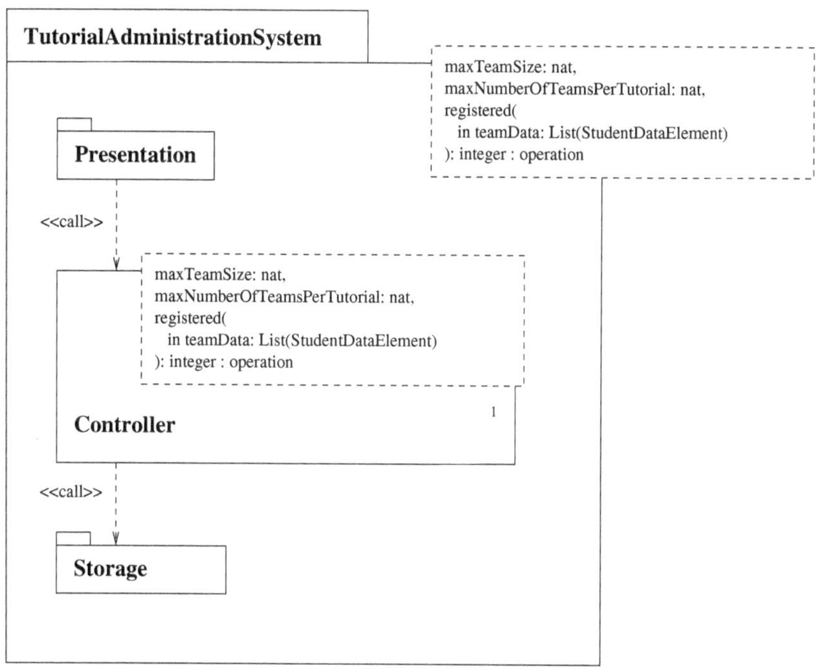

Fig. 4. Variant 3: Values and operations as parameters

It is true that variants one and two are easier to understand, but they require more redundancy between the different actual arguments a certain formal parameter may be bound to. For example considerable parts of the controller specification would be identical for each actual argument, resulting in problems with the maintenance of the specification. Presumably best results will be reached with a gradually applied parameterization of mixed granularity.

Independent from the (permissible) parameter granularity it is often necessary to restrict the set of potential actual arguments by means of additional rules. Such rules can for instance determine which combinations of actual arguments for different formal parameters can be used together, or which requirements the actual arguments have to fulfill in general in order to produce a correct non-parameterized specification. Depending on the type of parameter, different kinds of rules may be applied.

Five types of parameters can be distinguished. Table 1 gives an overview of the possibilities to restrict the respective types of parameters. The simplest case are value parameters as, for example, the maximum team size mentioned above. These parameters are roughly described by their datatype and can be made more concrete by a further restriction of their range.

All other types of parameters are settled at a higher level of abstraction, where datatypes make no sense. Instead, the expected signature can be specified. For classes, operations and packages it may be useful to fix the complete

Table 1. Possible parameter restrictions

	value parameter	datatype parameter		operation	package/ module
		basis datatype	class		
type specification	×				
signature			×	×	×
signature part		×	×		×
range restriction	×			×	
pre– and postcond.				×	
choice			×	×	×

signature. In the case of operations this would be the operation name as well as the names and datatypes of the input and output parameters. In the case of classes this would be the operations the respective class is expected to provide, and in the case of packages it would be the classes with their signatures.

Besides specifying the complete signature, one may also want to fix only that part of it that is relevant in the respective template. For example, it could be sufficient to state that the required datatype has to provide a compare operation for its elements. For classes it can be useful to specify a base class where the class given as an argument has to be derived from. While it does not make much sense to specify an incomplete signature for an operation, it sometimes makes sense to restrict the range of its return values. One may for instance think about an operation to produce random numbers, where the interval of the output values shall be restricted to a certain range. Furthermore, operations can be characterized by the specification of pre– and postconditions. All restrictions that can be constituted for operations that are parameters in their own right can also be constituted for operations as parts of class or package parameters.

Other types of restrictions are needed in the case that the actual arguments a parameter can be bound to should not be any arguments which fulfill the restrictions, but only such arguments which are provided by a library. Here, rules regarding permisible combinations of actual arguments are of interest. Referring to our case study one may think of the situation where all possible registration algorithms and all variants of the controller class belonging to them are provided by a library. In the case that for instance algorithm x works together with controller classes a and b only, while algorithm y needs controller class c to cooperate with, this should be stated as a parameter rule.

The UML does not explicitly provide language constructs for the specification of parameter rules in templates. As also mentioned in [6], which develops a classification for stereotypes similar to the one given here for types of formal parameters, with the OCL (Object Constraint Language) being part of the UML [47], we have a formalism at hand with which such rules can in principle be stated. However, if we employ UML together with Troll in order to utilize the advantages of both, we could shift the specification of parameter constraints to Troll and use the power of a formal specification language for this purpose.

In our example from above such a parameter rule would be specified as follows:

```
module TutorialAdministration
  parameterized by maxTeamSize: nat;
  parameter constraint 1 < maxTeamSize < 6;
  subsystem Presentation ... end_subsystem:
  object class Controller ... end;
  subsystem Storage ... end_subsystem:
end-module;
```

Here, the range of `maxTeamSize` is restricted. In general all kinds of rules that have been sketched above can be specified.

6 Conclusions

Further demands on the already complex task of developing information systems motivate using generic methods, which facilitate and support reusing parts of a specification and implementation. Having outlined this in the first chapter of this paper, we gave an overview of current research activities related to generic methods putting special emphasis on parameterization concepts for describing variants of a system. Chapter 3 provided a brief introduction to the general concepts of the object oriented language TROLL, to its building blocks for specifications as well as to its formal semantics. Following this, we sketched an application area for information systems at universities, namely the administration of tutorials, and showed that also in this area use of generic methods is desirable and useful.

With this as the foundation, in chapter 5 we outlined a specification for a tutorial administration systems, introducing and investigating on the parameterization concepts of TROLL und UML. Our first main focus here was on the granularity which the formal parameters of parameterized elements of specifications should have, and it turned out that presumable best results with regard to intelligibility and redundancy will be achieved by a gradually applied parameterization of mixed granularity.

The other main focus of chapter 5 was to discuss how restrictions on formal parameters, which are frequently needed with generic methods, can be expressed using rules. We gave an overview of different types of formal parameters together with the possibilities to restrict them, and provided examples of where such restrictions might be desired. Even though the UML provides the sublanguage OCL, which could — after some extensions — be used to express such restrictions, we propose to use TROLL instead and in general to use the UML only for the description of the overall architecture, shifting the specification of details to the textual language. To profit significantly from this joint application of a semi formal graphical language with a formal textual one, it is essential to give a precise definition of their combination [8], which is one of the further steps we plan to do.

References

1. M. Abadi and L. Cardelli. *A Theory of Objects.* Springer–Verlag, New York, 1996.
2. S. Abiteboul, P. Buneman, and D. Suciu. *Data on the Web: From Relations to Semistructured Data and XML.* Morgan Kaufmann Publishers, San Francisco, 2000.
3. P. Atzeni, S. Ceri, S. Paraboschi, and R. Torlone. *Database Systems: Concepts, Languages and Architectures.* McGraw–Hill, London, 1999.
4. D. Batory. Software Generators, Architectures, and Reuse. Tutorial, Department of Computer Science, University of Texas, 1996.
5. D. Batory. Intelligent Components and Software Generators. Invited presentation to the "Software Quality Institute Symposion on Software Reliability", Austin, 1997.
6. S. Berner, M. Glinz, and S. Joos. A Classification of Stereotypes for Object–Oriented Modeling Languages. In *Proc. 2nd Int. Conf. on the UML*, LNCS 1723, pages 249–264. Springer, Berlin, 1999.
7. G. Bracha, M. Odersky, D. Stoutamire, and P. Wadler. Making the future safe for the past: Adding genericity to the java programming language. In *Proc. Int. Conf. on Object Oriented Programming Systems, Languages and Applications (OOPSLA'98)*, pages 183–200, 1998.
8. S. Brinkkemper, M. Saeki, and F. Harmsen. Assembly Techniques for Method Engineering. In B. Pernici and C. Thanos, editors, *Proc. 10th Int. Conf. on Advanced Information Systems Engineering (CAiSE'98), Pisa*, LNCS 1413, pages 381–400. Springer, Berlin, 1998.
9. K. Czarnecki and U.W. Eisenecker. *Generative Programming — Methods, Tools, and Applications.* Addison–Wesley, Boston, 2000.
10. E. Damiani, M.G. Fugini, and C. Belletini. A hierarchy–aware approach to faceted classification of object–oriented components. *ACM Transactions on Software Engineering and Methodology*, 8(3):215–262, 1999.
11. K.R. Dittrich, D. Tombros, and A. Geppert. Databases in Software Engineering: A RoadMap. In A. Finkelstein, editor, *The Future of Software Engineering (in conjunction with ICSE 2000)*, pages 291–302. ACM Press, 2000.
12. S. Eckstein. Modules for Object Oriented Specification Languages: A Bipartite Approach. In V. Thurner and A. Erni, editors, *Proc. 5th Doctoral Consortium on Advanced Information Systems Engineering (CAiSE'98), Pisa.* ETH Zürich, 1998.
13. S. Eckstein. Towards a Module Concept for Object Oriented Specification Languages. In J. Bārzdiņš, editor, *Proc. of the 3rd Int. Baltic Workshop on Data Bases and Information Systems, Riga*, volume 2, pages 180–188. Institute of Mathematics and Informatics, University of Latvia, Latvian Academic Library, Riga, 1998.
14. S. Eckstein, P. Ahlbrecht, and K. Neumann. From Parameterized Specifications to Generated Information Systems: an Application. (In German). Technical Report 00–05, Technical University Braunschweig, 2000.
15. H.-D. Ehrich. Object Specification. In E. Astesiano, H.-J. Kreowski, and B. Krieg-Brückner, editors, *Algebraic Foundations of Systems Specification*, chapter 12, pages 435–465. Springer, Berlin, 1999.
16. H.-D. Ehrich and C. Caleiro. Specifying Communication in Distributed Information Systems. *Acta Informatica*, 36:591–616, 2000.
17. H.-D. Ehrich, C. Caleiro, A. Sernadas, and G. Denker. Logics for Specifying Concurrent Information Systems. In J. Chomicki and G. Saake, editors, *Logics for Databases and Information Systems*, chapter 6, pages 167–198. Kluwer Academic Publishers, Dordrecht, 1998.

18. H.-D. Ehrich and A. Sernadas. Local Specification of Distributed Families of Sequential Objects. In E. Astesiano, G. Reggio, and A. Tarlecki, editors, *Recent Trends in Data Types Specification, Proc. 10th Workshop on Specification of Abstract Data Types joint with the 5th COMPASS Workshop, S.Margherita, Italy, May/June 1994, Selected papers*, LNCS 906, pages 219–235. Springer, Berlin, 1995.
19. H. Ehrig and B. Mahr. *Fundamentals of Algebraic Specification 1.* Springer, Berlin, 1985.
20. H. Ehrig and B. Mahr. *Fundamentals of Algebraic Specification 2: Module Specifications and Constraints.* Springer, Berlin, 1990.
21. A. Gal, S. Kerr, and J. Mylopoulos. Information Services for the Web: Building and Maintaining Domain Models. *Int. Journal of Cooperative Information Systems (IJCIS)*, 8(4):227–254, 1999.
22. J. Goguen. Parameterized Programming and Software Architecture. In IEEE Computer Society, editor, *Proceedings Fourth International Conference on Software Reuse*, pages 2–11, 1996.
23. J. Goguen and G. Malcolm, editors. *Software Engineering with OBJ: Algebraic Specification in Action.* Kluwer, Boston, 2000.
24. J. Goguen and W. Tracz. An Implementation Oriented Semantics for Module Composition. In G.T. Leavens and M. Sitaraman, editors, *Foundations of Component–Based Systems*, pages 231–263. Cambridge University Press, 2000.
25. A. Grau, J. Küster Filipe, M. Kowsari, S. Eckstein, R. Pinger, and H.-D. Ehrich. The TROLL Approach to Conceptual Modelling: Syntax, Semantics and Tools. In T.W. Ling, S. Ram, and M.L. Lee, editors, *Proc. 17th Int. Conf. on Conceptual Modeling (ER'98)*, pages 277–290. Springer, LNCS 1507, 1998.
26. D. Harel. From Play–In Scenarios to Code: An Achievable Dream. In T. Maibaum, editor, *Proc. 3rd Int. Conf. on Fundamental Approaches to Software Engineering (FASE 2000)*, pages 22–34. Springer, LNCS 1783, 2000.
27. P. Hartel. *Conceptual Modelling of Information Systems as Distributed Object Systems. (In German).* Series DISDBIS. Infix–Verlag, Sankt Augustin, 1997.
28. P. Hartel, G. Denker, M. Kowsari, M. Krone, and H.-D. Ehrich. Information systems modelling with TROLL — formal methods at work. *Information Systems*, 22(2–3):79–99, 1997.
29. M. Hitz and G. Kappel. *UML@Work.* dpunkt, Heidelberg, 1999.
30. S. Jarzabek and P. Knauber. Synergy between Component-Based and Generative Approaches. In O. Nierstrasz and M. Lemoine, editors, *Software Engineering - ESEC/FSE'99*, pages 429–445. Springer, LNCS 1687, 1999.
31. M. Jeusfeld, M. Jarke, M. Staudt, C. Quix, and T. List. Application Experience with a Repository System for Information Systems Development. In R. Kaschke, editor, *Proc. EMISA (Methods for Developing Information Systems and their Applications)*, pages 147–174. Teubner, 1999.
32. M. Krone, M. Kowsari, P. Hartel, G. Denker, and H.-D. Ehrich. Developing an Information System Using TROLL: an Application Field Study. In P. Constantopoulos, J. Mylopoulos, and Y. Vassiliou, editors, *Proc. 8th Int. Conf. on Advanced Information Systems Engineering (CAiSE'96)*, LNCS 1080, pages 136–159, Berlin, 1996. Springer.
33. J. Küster Filipe. *Foundations of a Module Concept for Distributed Object Systems.* PhD thesis, Technical University Braunschweig, 2000.
34. J. Küster Filipe. Fundamentals of a Module Logic for Distributed Object Systems. *Journal of Functional and Logic Programming*, 2000(3), March 2000.
35. J. Loeckx, H.-D. Ehrich, and M. Wolf. *Specification of Abstract Data Types.* John Wiley & B. G. Teubner, New York, 1996.

36. B. Meyer. *Object–oriented Software Construction.* Prentice Hall, New York, 1988.
37. M. Mezini and K. Lieberherr. Adaptive Plug–and–Play Components for Evolutionary Software Development. In *Proc. of the 1998 ACM SIGPLAN Conference on Object–Oriented Programming Systems, Languages and Applications (OOPSLA '98)*, volume 33 (10) of *SIGPLAN Notices*, pages 97–116, Vancouver, 1998.
38. A.C. Myers, J.A. Bank, and B. Liskov. Parameterized Types for Java. In *Proc. of the 24th ACM Symposium on Principles of Programming Languages*, pages 132–145, Paris, 1997.
39. M. Nielsen, G. Plotkin, and G. Winskel. Petri Nets, Event Structures and Domains, Part 1. *Theoretical Computer Science*, 13:85–108, 1981.
40. M.T. Özsu and P. Valduriez. *Principles of Distributed Database Systems.* Prentice Hall, Upper Saddle Rive, 2. edition, 1999.
41. G. Preuner and M. Schrefl. A Three-Level Schema Architecture for the Conceptual Design of Web-Based Information Systems: From Web-Data Management to Integrated Web-Data and Web-Process Management. *World Wide Web Journal, Special Issue on World Wide Web Data Management*, 3(2), 2000.
42. J. Rumbaugh, M. Blaha, W. Premerlani, F. Eddy, and W. Lorensen. *Object–Oriented Modeling and Design.* Prentice Hall, New York, 1991.
43. J. Rumbaugh, I. Jacobson, and G. Booch. *The Unified Modeling Language Reference Guide.* Addison–Wesley, 1999.
44. P. di Silva, T. Griffiths, and N. Paton. Generating User Interface Code in a Model Based User Interface Development Environment. In V. di Gesu, S. Levialdi, and L. Tarantino, editors, *Proc. Advanced Visual Interfaces (AVI 2000)*, pages 155–160. ACM Press, New York, 2000.
45. O. De Troyer. Designing Well–Structured Websites: Lessons to Be Learned from Database Schema Methodology. In T.W. Ling, S. Ram, and M.L. Lee, editors, *Proc. 17th Int. Conf. on Conceptual Modeling (ER'98), Singapore*, pages 51–64. Springer, LNCS 1507, 1998.
46. M. VanHilst and D. Notkin. Using Role Components to Implement Collaboration Based Designs. In *Proc. of the 1996 ACM SIGPLAN Conference on Object–Oriented Programming Systems, Languages and Applications (OOPSLA '96)*, volume 28 (10) of *SIGPLAN Notices*, pages 359–369, San Jose, 1996.
47. J. Warmer and A. Kleppe. *The Object Constraint Language: Precise Modeling with UML.* Addison–Wesley, Reading, 1999.
48. R. Wieringa, R. Jungclaus, P. Hartel, T. Hartmann, and G. Saake. OMTROLL – Object Modeling in TROLL. In U.W. Lipeck and G. Koschorreck, editors, *Proc. Intern. Workshop on Information Systems — Correctness and Reusability IS-CORE '93, Technical Report, University of Hannover No. 01/93*, pages 267–283, 1993.
49. M. Wirsing. Algebraic Specification Languages: An Overview. In E. Astesiano, G. Reggio, and A. Tarlecki, editors, *Recent Trends in Data Type Specification*, pages 81–115. Springer, LNCS 906, 1995.

An Assembly Process Model for Method Engineering

Jolita Ralyté[1,2] and Colette Rolland[1]

[1] CRI, Université Paris 1 Sorbonne, 90, rue de Tolbiac, 75013 Paris, France
[2] MATIS, Université de Genève, rue du Gén. Dufour 24, 1211 Genève, Switzerland
{ralyte, rolland}@univ-paris1.fr

Abstract. The need for a better productivity of system engineering teams, as well as a better quality of products motivates the development of solutions to adapt methods to the project situation at hand. This is known as situational method engineering. In this paper we propose a generic process model to support the construction of a new method by assembling method chunks generated from different methods that are stored in a method base. The emphasis is on the guidance provided by the process model, as well as on the means underlying guidelines such as similarity measures and assembly operators. The process model is exemplified with a case study.

1 Introduction

We are concerned with *Situational Method Engineering* (SME). SME aims at defining information systems development methods by reusing and assembling different existing method fragments. The term *method fragment* was coined by Harmsen in [5] by analogy with the notion of a software component. Similarly to the component driven construction of software systems, SME promotes the construction of a method by assembling reusable method fragments stored in some method base [20], [6], [16], [13]. As a consequence SME, favours the construction of modular methods that can be modified and augmented to meet the requirements of a given situation [5], [21]. Therefore, a method is viewed as a collection of method fragments that we prefer to call *method chunks* [15], [13] to emphasise the coherency and autonomy of such method modules. New methods can be constructed by selecting fragments/chunks from different methods, which are the most appropriate to a given situation [3], [10]. Thus, method fragments/chunks are the basic building blocks, which allow to construct methods in a modular way.

The objective of our work is to propose a complete approach for method engineering based on a method chunk assembly technique. In previous papers [16], [13] we presented a *modular method meta-model* allowing to represent any method as an assembly of the reusable method chunks. In this paper we are dealing with the method chunk assembly process. We present a generic process model, the *Assembly Process Model* (APM), to guide the assembly of method chunks using different strategies depending on the type of situation in which the assembly activity has to be carried out. Chunk assembly is the support of situational method engineering and therefore we propose a *Method Engineering Process Model* (MEPM) providing several different ways to assemble chunks with the objective of constructing new

K.R. Dittrich, A. Geppert, M.C. Norrie (Eds.): CAiSE 2001, LNCS 2068, pp. 267–283, 2001.
© Springer-Verlag Berlin Heidelberg 2001

methods or enhancing the existing methods by new models and/or new ways of working. Whereas the APM views the assembly of method chunks 'in the small', the MCPM takes a broader view where assembling method chunks is part of a larger method engineering process. As a consequence, the APM is embedded in the MEPM.

Both process models, namely the APM and the MEPM, are expressed using the same notations provided by a process meta-model. A *process meta-model* is an abstraction of different process models, i.e. a process model is an instance of a process meta-model. In this paper, we use the strategic process meta-model presented in [19] and [1]. Following this meta-model, a process model is presented as a *map* and a set of associated *guidelines.* Such representation of the process model allows us to provide a strategic view of different processes. Indeed, this view tells what can be achieved - the *intention*, and which *strategy* can be employed to achieve it. We separate the strategic aspect from the tactical aspect by representing the former in the method map and embodying the latter in the guidelines. By associating the guidelines with the map, a smooth integration of the strategic and the tactical aspects is achieved.

This paper is organised as follows: section 2 highlights the need for different strategies for assembling method chunks to form a new method and motivates different ways of method engineering based on method chunk assembly. The former is encapsulated in the APM whereas the latter is captured in the MEPM. In section 3, we take the view of method engineering 'in the large' and present the MEPM. The MEPM includes the APM, which is presented in section 4. Section 5 illustrates the approach with an example demonstrating the process step by step. Section 6 draws some conclusions around our work.

2 Chunk Assemblies and Method Engineering

The attempts to define assembly processes [3], [11], [22] highlight the assembly of method fragments as rather independent and supplementary to one another. A typical example would be to adding a given way of working some new activity borrowed from another method and/or adding to the product model of one method a new concept borrowed from another method. In such a case, the assembly mainly consists in establishing links between the 'old' elements and the 'new', added ones. We found cases quite different where elements to assemble are overlapping. This led us to the identification of two assembly strategies: the *association* and the *integration.*

As shown in Figure 1, the first strategy is relevant when the method chunks to assemble do not have elements in common. This might occur when the end product of one chunk is used as a source product by the second chunk. For example, the chunk producing use cases and the chunk constructing the system object structure can be assembled to get a method with a larger coverage than any of the two initial ones. The assembly process is therefore mainly dealing in making the bridge. The second strategy is relevant to assemble chunks that have similar engineering objectives but provide different ways to satisfying it. In such a case, the process and product models are overlapping and the assembly process consists in merging overlapping elements. The integration strategy will be necessary, for example, to assemble two different chunks dealing both with a use case model construction. These two strategies are embedded in the APM presented in section 4.

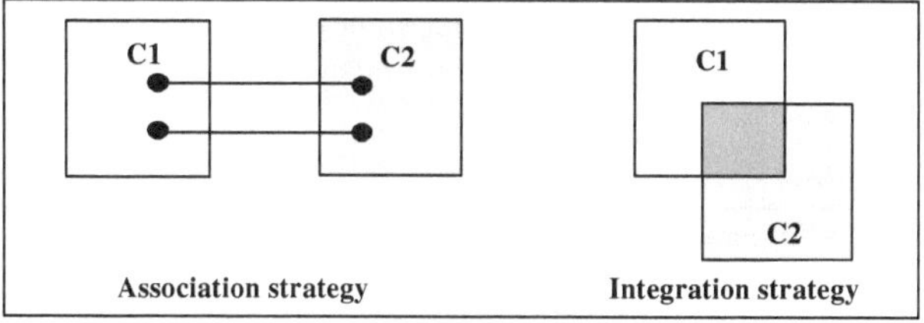

Fig. 1. Strategies to assemble method chunks

The assembly of method chunks is a means, a technique to construct a method in a reuse driven fashion. However, method construction is not restricted to chunk assembly. It includes, for example, the elicitation of requirements for the method to construct, amend or enhance. Besides, the ways the assembly technique will be used differ depending of the objective assigned to the situational method engineering project. There are many different reasons for constructing a new method. We identified three of them:

1. To define a brand new method to satisfy a set of situational requirements;
2. To add alternative ways-of-working in a method to its original one;
3. To extend a method by a new functionality.

Each of these delineates a specific strategy for method engineering that we have embedded in the MEPM. The first strategy is relevant in situations where either there is no method in use or the one in use is irrelevant for the project (or class of projects) at hand. The second strategy is relevant when the method in use is strong from the product point of view but weak from the process viewpoint. Enhancing the process model of the existing method by one or several new ways of working is thus the key motivation for method engineering. The third strategy is required in situations where the project at hand implies to add a new functionality to the existing method which is relevant in its other aspects. We present the MEPM in the next section.

3 The Method Engineering Process Model (MEPM)

Figure 2 shows our proposal to engineer a method through an assembly of method chunks. We use the strategic process meta-model [19], [1] to represent our process model as a *map* with associated *guidelines*. A *map* is a directed labelled graph with *intentions* as nodes and *strategies* as edges. The core notion of a map is the *section* defined as a triplet <*source intention, target intention, strategy*>. A map includes two specific strategies, *Start* and *Stop* to start and stop the process, respectively. As illustrated in Figure 2, there are several paths from *Start* to *Stop*. A map therefore includes several process models that are selected dynamically when the process proceeds, depending on the current situation. Each section is associated to a *guideline* that provides advice to fulfil the target intention following the section strategy. Furthermore, a section can be *refined* as an entire map at a lower level of granularity.

The MEPM represented by a map in Figure 2 includes two intentions *Specify method requirements* and *Construct method*. The latter corresponds to the method engineering's essential goal, whereas the former is the prerequisite for the latter. The formulation of this intention, *Specify method requirements* means that our approach is requirements-driven. In order to construct a new method, we propose to start by eliciting the requirements for the method engineering activity.

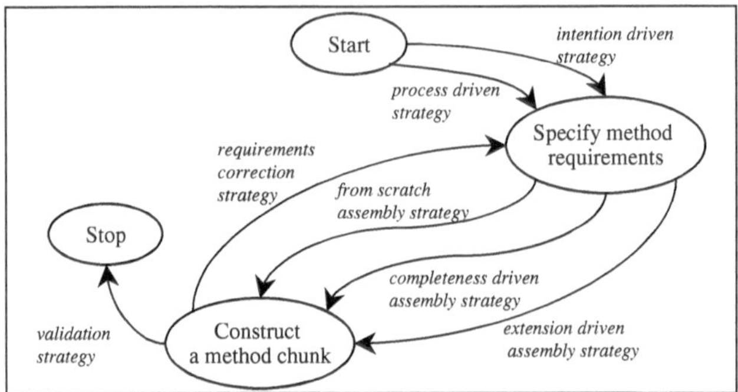

Fig. 2. The method engineering map (MEPM)

The map identifies two different strategies to *Specify method requirements*, namely the *intention driven strategy* and the *process driven strategy*. Both lead to a set of requirements expressed as a map that we call the *requirements map*. However, each strategy corresponds to a different way of eliciting the requirements. The former is based on the inventory of engineering goals whereas the latter infers these goals from an analysis of the engineering activities that must be supported by the method.

Once the requirements have been elicited, the intention *Construct method* can be achieved. The MEPM proposes three different assembly strategies: *from scratch*, *enhancement driven* and *extension driven*, to help the method engineer to achieve this intention. The three strategies correspond to the three method engineering situations that were identified and motivated in the previous section. The *from scratch assembly strategy* corresponds to situations where a brand new method has to be developed, whereas the two others, *enhancement driven assembly strategy* and *extension driven assembly strategy*, are relevant when a method already exists. As indicated by the names of the strategies, the three proposed ways are based on a method chunk assembly technique developed in the next section. Backtracking to the requirements definition is possible thanks to *the requirements correction strategy*. Finally, the *validation strategy* helps verifying that the assembly of the selected method chunks satisfies all requirements and ends the method engineering process.

According to the map meta-model used above to model the MEPM, each section of the map is associated to a *guideline* providing advice on how to proceed to achieve the target intention and can be *refined* as a map of a finer level of abstraction. For the sake of space we do not present the guidelines associated to every section of the MEPM (see [14] for details) but concentrate on the refinement of the section <*Specify method requirements, Construct method, from scratch assembly strategy*> dealing with the assembly of method chunks. The refined map of this section models the method chunk assembly process. The APM is presented in detail in the next section.

4 The Process Model for Method Chunk Assembly (APM)

This section presents the *assembly process model* guiding the selection of method chunks matching a set of situational requirements and their assembly to form a new method. It is a generic process model in the sense that it includes a number of strategies to retrieve and assemble chunks providing solutions for the different engineering situations the method engineer may be faced with. In particular the map includes two strategies *(integration strategy and association strategy*) to assemble chunks that we identified from the literature and case studies and introduced in section 2. The APM is presented as a map, the *assembly map*, in Figure 3.

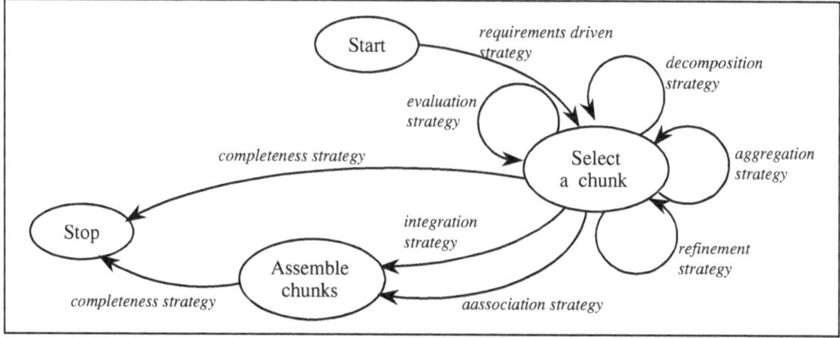

Fig. 3. The assembly map (APM)

As shown in Figure 3, the APM proposes several different ways to select chunks matching requirements as well as different strategies to assemble them. It is based on the achievement of two key intentions: *Select a chunk* and *Assemble chunks*. The achievement of the first intention leads to the selection of chunks that match the requirements from the method base. The second intention is satisfied when the selected chunks have been assembled in a coherent manner.

The process starts by the selection of the *requirements driven strategy*. The associated guideline helps the method engineer to select candidate chunks that are expected to match the requirements expressed in the requirements map (the result of the achievement of the intention *Specify method requirements* in MEPM). It suggests to formulate queries to the method base giving values to the attributes of the descriptors and interfaces of chunks (see [13], [14]) to identify the chunks that are likely to match part or the totality of the requirements map.

Any time a chunk has been retrieved, the assembly map suggests to validate this candidate chunk by applying the *evaluation strategy*. The *evaluation strategy* helps in evaluating the degree of matching of the candidate chunk to the requirements. This is based on similarity measures between the requirements map and the map of the selected chunk. We present these similarity measures in section 4.1.

The *decomposition, aggregation and refinement* strategies help to refine the candidate chunk selection by analysing more in depth if the chunk matches the requirements. The *decomposition* strategy is relevant when the selected method chunk is an aggregate one having some component parts that may not be required. This strategy helps to select those, which are the adequate ones. The *aggregation* strategy is relevant when the candidate chunk partly covers the requirements. This strategy suggests to search for an aggregate chunk containing the candidate chunk based on the assumption that the aggregate method chunk might provide a solution for the missing requirements. The *refinement* strategy proposes to search for another chunk satisfying the same intention but providing a set of guidelines richer than those of the candidate chunk.

When at least two chunks have been selected, the method engineer can progress to the assembly of these chunks following one of two strategies, namely the *integration strategy* and the *association strategy*. As discussed in section 2.1, the choice of the strategy depends of the presence /absence of overlap between the chunks to assemble. If the two chunks help to achieve the same intention in the system engineering process and to construct the same or a similar product, the *integration strategy* must be chosen. If the selected chunks do not overlap in terms of intention to achieve and product to construct, then the *association strategy* must be selected. In the first case where chunks partially overlap, the integration of these chunks produces a new method whose product and process models are 'richer' than those of the initial chunks. With the second strategy for assembling chunks dealing with different aspects of system design, and supplementing one another, the result is a new method providing a larger coverage of design activities. We present the two strategies in more detail in sections 4.2 and 4.3. The two assembly strategies use *assembly operators*, *similarity measures* and *quality validation rules* [12], [14]. Their use will be exemplified in section 5. Just as a reminder, let us mention that there are two types of operators to assemble process models parts and product models parts, respectively. The similarity measures are used to compare chunks before their assembly and to identify whether they are overlapping. This will help to choose the right strategy between the *integration strategy* and the *association strategy*.

In order to check if the current chunk assembly matches the requirements, the method engineer shall use the *completeness strategy*. If the response is a positive one, the assembly process ends. In the reverse case, other chunks have to be selected and assembled to gain the required method completeness.

4.1 Similarity Measures

Measures to estimate the similarity of conceptual schemas have been proposed by several authors, for different purposes: in order to identify reusable components [4] and to select these components [8]. Generally, these approaches measure the closeness between entities of different conceptual schemas by evaluating the common properties and links with other entities [4]. The global complexity of the schemas is also taken into account. Bianco et al. [2] proposes similarity metrics to analyse heterogeneous data base schemas. These metrics are based on the semantic and structural similarity of elements of these schemas. In our approach we use measures inspired from those proposed by Castano [4] and Bianco et al.[2]. We distinguish two types of measures: those, which allow to measure the similarity of the elements of *product models* and those which allow to measure the closeness of *process models* elements. We present them in turn.

Product Models Similarity Measures. We use semantic and structural measures to compare elements of product models.

The *Name Affinity (NA)* metric allows us to measure the semantic similarity of concepts belonging to different product models. To apply this measure, the concepts of the chunks must be defined in a thesaurus of terms. Moreover, the synonymy (SYN) and the hyperonymy (HYPER) relationships between these concepts must be defined. The SYN relation connects the terms t_i and t_j ($t_i \neq t_j$) which are considered as synonyms. This is a symmetrical relation. For example, <*Goal SYN Objective*>. The HYPER relation connects two terms t_i and t_j ($t_i \neq t_j$) where t_i is more general than t_j. This is not a symmetrical relation. The inverse relation is the hyponymy (HYPO). For example, <*Scenario HYPER Exceptional scenario*>.The thesaurus of terms is a network where the nodes are the terms and the edges between the nodes are the terminological relations. Every terminological relation R $\in$ {SYN , HYPER /HYPO} has a weight σ_R. For example, $\sigma_{SYN} = 1$ and $\sigma_{HYPER/HYPO} = 0.8$. Therefore, the name affinity metric is defined as follows:

$$NA(\mathrm{n(e_{ij}),n(e_{kl})}) = \begin{cases} 1 & \text{if } < \mathrm{n(e_{ij})}\ SYN\ \mathrm{n(e_{kl})} > \\ \sigma_{1R} * \ldots * \sigma_{(m-1)R} & \text{if } \mathrm{n(e_{ij})} \rightarrow {}^{m}\mathrm{n(e_{kl})} \\ 0 & \text{else} \end{cases}$$

where $\mathrm{n(e_{ij})} \rightarrow {}^{m}\mathrm{n(e_{kl})}$ is a length of the path between e_{ij} and e_{kl} in the thesaurus and $m \geq 1$, σ_{nR} shows the weight of the nth relation in $\mathrm{n(e_{ij})} \rightarrow {}^{m}\mathrm{n(e_{kl})}$.

The semantic similarity is not sufficient to determine if two concepts are similar. We also need to compare their structures. The measure of the structural similarity of concepts is based on the calculation of their common properties and their common links with other concepts. Thus, to obtain the *Global Structural Similarity of Concepts (GSSC)* we need to measure the structural similarity of their properties and the structural similarity of their links with other concepts. These measures are respectively called *Structural Similarity of Concepts (SSC)* and *Adjacent Similarity of Concepts (ASC)*. The formulas are as follows:

$$\mathrm{GSSC}(c_1,c_2) = \frac{\mathrm{SSC}(c_1,c_2) + \mathrm{ASC}(c_1,c_2)}{2} \qquad \mathrm{SSC}(c_1,c_2) = \frac{2 * (\text{Number of common properties in } c_1 \text{ and } c_2)}{\sum_{i=1}^{2} \text{Number of properties in } c_i}$$

$$\mathrm{ASC}(c_1,c_2) = \frac{2 * (\text{Number of commun adjacent concepts to } c_1 \text{ and } c_2)}{\sum_{i=1}^{2} \text{Number of adjacent concepts ato } c_i}$$

Process Models Similarity Measures. In this section we propose metrics to compare elements of process models i.e. of maps. Elements to compare are intentions, sections and maps themselves to evaluate the global similarity of maps.

We use two kinds of semantic similarity : the *Semantic Affinity of Intentions (SAI)* and the *Semantic Affinity of Sections (SAS)*. The *SAI* is used to measure the closeness of two intentions. This metric is based on the comparison of the two parameters composing the intention: verb and target by using the SYN relation. The *SAS* measures the closeness of two map sections. It is based on the measure of the *SAI* of its source intentions, the *SAI* of its target intentions and the application of the SYN relation between their strategies. The two formulas are defined as follows:

$$SAI(i_i, i_j) = \begin{cases} 1 & \text{if } (i_i.\text{verb SYN } i_j.\text{verb}) \wedge (i_i.\text{target SYN } i_j.\text{target})) \\ 0 & \text{else} \end{cases}$$

$$SAS(< i_i, i_j, s_{ij} >, < i_k, i_l, s_{kl} >) = \begin{cases} 1 & \text{if } SAI(i_i, i_k) = 1 \wedge SAI(i_j, i_l) = 1 \wedge s_{ij} \text{ SYN } s_{kl} \\ 0 & \text{else} \end{cases}$$

The structural similarity measures are needed to compare the structures of two maps and to identify their overlapping parts. We use two kinds of structural measures: the *Structural Similarity by Intentions (SSI)* and the *Structural Similarity by Sections (SSS)*. The *SSI* is used to measure the proportion of similar intentions in two maps. This is based on the calculation of the *SAI* of their intentions. The *SSS* allows us to measure the proportion of similar sections in two maps. Sometimes, we also need to compare the proportion of similar sections for a couple of intentions which exist in the two maps. For this we introduce the *Partial Structural Similarity (PSS)* metric. The three measures are defined as follows:

$$SIS(m_1, m_2) = \frac{2 * \text{Number of similar intentions in } m_1 \text{ and } m_2}{\sum_{i=1}^{2} \text{Number of intentions in } m_i}$$

$$SSS(m_1, m_2) = \frac{2 * \text{Number of similar sections in } m_1 \text{ and } m_2}{\sum_{i=1}^{2} \text{Number of sections in } m_i}$$

$$PSSS(m_1 :< i_{1i}, i_{1j} >, m_2 :< i_{2k}, i_{2l} >) = \frac{2 * \text{Nb. of similar sctions between } < i_{1i}, i_{1j} > \text{ and } < i_{2k}, i_{2l} >}{\text{Nb. of sect. between } < i_{1i}, i_{1j} > + \text{ Nb. of sect. between } < i_{2k}, i_{2l} >}$$

m_1, m_2 : the maps; $m_1 :< i_{1i}, i_{1j} >$: a couple of intentions in the map m_1

4.2 Chunk Assembly by Integration

Figure 4 shows the process model corresponding to the assembly by integration strategy, the *integration map*. This map is a refinement of section <*Select a chunk, Assemble chunks, integration strategy*> of the APM (Figure 3).

The assembly process by integration, or the *integration process* for short, consists in identifying the common elements in the chunks product and process models and merging them. The maps of the these chunks must have some similar intentions and their product models must conceptualise the same objects of the real world by using similar concepts. As shown in Figure 4, the method engineer can start the assembly process by the integration of the process models followed by the integration of the product models or vice versa. At every moment he can navigate from the process models integration to the product models integration and vice versa.

Let us first consider the assembly of chunks process models, i.e. maps. It might be necessary to make some terminology adjustments of maps before their integration The mechanism of integration merges similar intentions that must have the same name. This is not necessarily the case in the initial chunks selected for assembly: intentions having the same semantics may have different names whereas semantically different intentions my be named exactly the same. The guideline of the *name unification strategy* helps to identify a couple of similar intentions requiring some name unification. The *SAI* (Sect. 4.1) measure is used to detect that the intentions are similar and then, the RENAME_INTENTION operator [12], [14] is recommended to unify their naming. Either directly or after having proceeded to the unification of names, the method engineer can move to the intention *Construct the integrated process model* following the *merge strategy* which recommends the use of the MERGE_INTENTION operator for each couple of similar intentions.

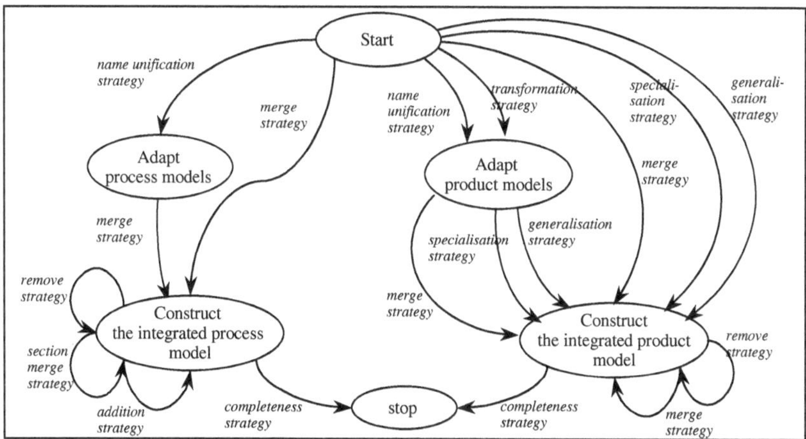

Fig. 4. Integration process map

The integration of product models is based on the identification of couples of similar concepts to be merged. Again, this might require naming revision or can be done directly. Two concepts to be merged must have the same semantics. In addition, if their structures are identical, they must have the same name. Vice-versa, if their structures are different, they must be named differently. For this reason, the product models integration may also be preceded by an adaptation step following the *name unification strategy* or the *transformation strategy*. The *name unification* strategy must be selected to solve the problem of naming ambiguity of concepts belonging to the different product models. The associated guideline uses the *NA* and *GSSC* measures (Sect. 4.1) to identify a couple of such concepts and proposes to rename one of them by applying the RENAME_CONCEPT operator. The *transformation* strategy must be selected when the same real world object is modelled differently in the two product models. For example, the object may be presented by a concept in one model and by a link between two concepts or by a structural property of another concept in the other model. The associated guideline helps to identify the couples of elements that need to be unified (concept and link, concept and property, or link and property) and to apply one of the product assembly operators OBJECTIFY_LINK or OBJECTIFY_PROPERTY according to the situation.

The same product integration strategies (*merge, generalisation* and *specialisation*) are possible to fulfil the intention *Construct the integrated product model* independently of the starting intention, the *Start* intention or the *Adapt product models* intention. The guidelines associated to the respective sections are identical. The *merge strategy* is applicable to merge concepts with similar semantics and similar structure. The corresponding guideline helps to identify a couple of similar concepts by applying the *NA* and *GSA* measures and to apply the product assembly operator MERGE_CONCEPT. The *generalisation strategy* shall be used when the two concepts have the same semantics but different structures: the GSA measure helps evaluating if the difference of their structures forbid their merging. The guideline associated to the corresponding sections proposes to generalise the two concepts into a new one by using the GENERALISE operator. The two initial concepts must have different names before their generalisation; therefore the name unification strategy shall be required first. Finally, the *specialisation strategy* is required when one concept represents a specialisation of the other concept. The associated guideline introduces a specialisation link between the two concepts by applying the SPECIALISE operator.

At any step of the integration process, it could be necessary to improve the current solution. The integration process map (Figure 4) proposes three strategies: *remove*, *addition* and *merge section*, to refine the integrated process model. The *remove strategy* deals with the need to remove elements in the integrated model. Many different reasons can justify such removals; for example to remove a useless or redundant guideline. The guideline associated to this section suggests the use of the REMOVE_SECTION operator to perform this operation. Some new guidelines can also be required to complete the integrated process model, particularly if the integrated product model integrates generalisation and/or specialisation of concepts. The integrated process model needs to be extended in these cases. The *addition strategy* helps doing so by applying the ADD_SECTION operator. Finally, the *merge section strategy* suggests to merge sections which are duplicates by applying the MERGE_SECTION operator.

Similarly, it can be necessary to improve the current version of the integrated product model. For example, the *remove strategy* allows to eliminate concepts, links or properties of the integrated product model by applying one of the operators REMOVE_CONCEPT or REMOVE_LINK or REMOVE_PROPERTY according to the situation at hand. To end the integration process the method engineer is invited to apply the quality rules and to verify the coherence and the completeness of the obtained product and process models following the *completeness strategy*.

4.3 Chunk Assembly by Association

In this section we consider the assembly of method chunks carried out following the *association strategy*. Figure 5 shows the process model corresponding to this assembly strategy represented by a map which is a refinement of the section <*Select a chunk, Assemble chunks, association strategy*> of the APM presented in Figure 3.

The assembly process by association, the *association process* for short, consists in connecting chunks such that the first one produces a product which is the source of the second chunk. Thus, the association process may consist in simply ordering chunks processes and relating chunks products to one another. The *association process* is simpler than the *integration process*. The association of product models is achieved by establishing links between concepts or adding elements connected to other concepts. The association of the process models consists in ordering the process activities provided by the two different models and possibly adding some new activity.

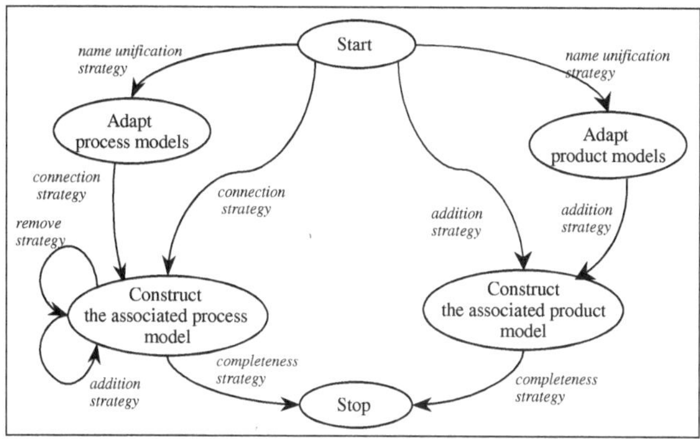

Fig. 5. The association process map

As in the case of the integration driven assembly, the association of chunks may also require the unification of their terminology. The *name unification strategy* is provided to unify names in maps and product models. Maps of chunks in this case should not have similar intentions as their product models should not contain similar concepts. The *SAI, NA* and *GSSC* (Sect. 4.1) measure must be applied to the suspected intentions and concepts and the `RENAME` operators must be applied if necessary.

If the two maps do not have any naming problems, the construction of the associated process model can start directly whereas the *Adapt process models* intention has to be fulfilled first in the reverse case. Then, the *connection strategy* is needed to carry out the association. The associated guidelines suggest a plan of action in three steps: (1) to determine the order in which the chunk processes must be executed; (2) to identify in the map of the first ordered chunk the intention that results in the product which is the source to the second chunk process, and (3) to merge this intention with the *Start* intention of the second chunk by applying the `MERGE_INTENTION` operator.

A similar set of strategies is proposed in Figure 5 to deal with the association of product models. The product models may by associated the *addition strategy* which advises to identify the concepts in the product models which can be connected by a link or by introducing an intermediary concept. These corresponding guidelines recommend the use of the product assembly operators `ADD_LINK` or `ADD_CONCEPT` depending of the situation at hand.

5 Application Example

In this section we illustrate the use of the method engineering process model with an example. We show how the method engineering map and its refined maps guide a method engineer step by step to construct a new method by retrieving and assembling method chunks.

Let us suppose that a method engineer has to construct a method supporting the elicitation of functional system requirements in a goal-driven manner, to conceptualise them using textual devices such as scenarios or use cases, to validate them in an animated fashion and finally to document them. According to the method engineering map presented in Figure 2, the first intention to achieve is to *Specify method requirements*. The *process driven strategy* looks adapted to the situation at hand as the requirements are expressed above in a process-oriented way. Assume that the application of this strategy leads to the requirements map presented in Figure 6.

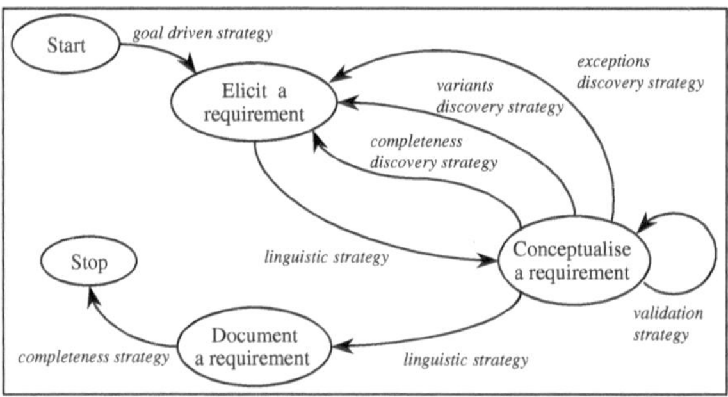

Fig. 6. The requirements map of the application example

Once the requirements for the new method have been elicited, the MEPM (Figure 2) suggests to *Assemble Chunks*. As the objective is to construct an entirely new method, the *from scratch assembly strategy* proposed in this method engineering map is chosen. The refined map of this assembly process (Figure 3) proposes to start with the selection of method chunks matching part or the totality of the requirements map. The guideline suggests to formulate queries to the method base in order to retrieve candidate method chunks. These queries give values to the different attributes of chunk interfaces and chunk descriptors [13], [14] as for example, design activity = *requirements engineering*, reuse intention = *discover functional system requirements*, situation = *problem description*.

Let's assume that method engineer selects the *L'Ecritoire* chunk as a candidate one. The process and product parts of this chunk are shown in Figure 7. This method chunk provides guidelines to discover functional system requirements expressed as goals and to conceptualise these requirements as scenarios describing how the system satisfies the achievement of these goals [17], [18]. Several different strategies are provided by the chunk to support goal elicitation, scenario writing and scenario conceptualisation. The method engineer wants to get a quantitative evaluation of the fit of *L'Ecritoire* to the requirements map. Therefore, he selects the *evaluation strategy* (Figure 3), which helps him to compare the map of the candidate chunk with the requirements map. The map similarity measures SAI, SAS, SSI and PSS (Sect. 4.1) are used. For example, owing to the SAI measure we detect that the intentions *Elicit a requirement* (requirements map) and *Elicit a goal* (*L'Ecritoire* map) are similar because they use the same verb and their targets *requirement* and *goal* are synonyms. The measure SSI, calculated as follows: *SSI (Requirements map, L'Ecritoire map) = (2*2 similar intentions) / (6 intentions in two maps) = 2/3*, shows that a large part of the requirements map is covered by the *L'Ecritoire* map. To validate this assumption, we search for similar sections by applying the SAS measure. For example, the SAS calculated as follows: *SAS (Requirements map: <Conceptualise a requirement, Elicit a requirement, variants discovery strategy>, L'Ecritoire* *map: <Conceptualise a scenario, Elicit a goal, alternative discovery strategy>) = 1*, shows that the concerned sections are similar. Next, for each couple of similar intentions we apply the PSS measure to verify if the strategies between these intentions are also similar. For instance, the *PSS (Requirements map: <Conceptualise a requirement, Elicit a requirement>, L'Ecritoire map: <Conceptualise a scenario, Elicit a goal>) = (2* 2 similar strategies) / (6 strategies) = 2/3* shows that the map of the *L'Ecritoire* matches a part of the requirements map. The degree of matching is satisfactory enough to select the *L'Ecritoire* chunk.

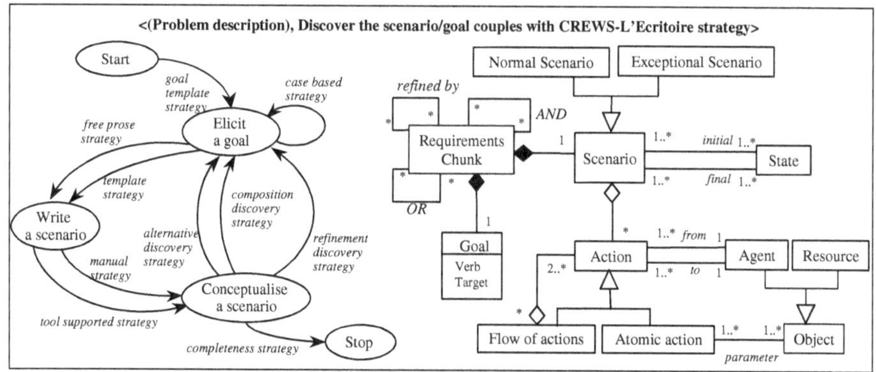

Fig. 7. The *L'Ecritoire* method chunk

The requirements coverage is not complete and the method engineer must continue the search. However, he knows the properties of the chunks he is looking for and can formulate precise queries. The required chunks must have the following values in their interfaces : situation = *goal* or *scenario*, intention = *to discover exceptional requirements (or goals).*

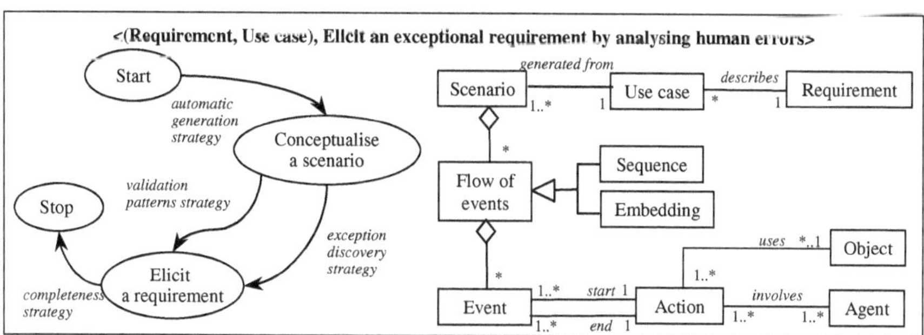

Fig. 8. The *SAVRE* method chunk

The *SAVRE* method chunk [23], [9] presented in Figure 9 is one of the chunks retrieved by the query. This chunk provides guidelines to discover exceptions in the functioning of a system under design caused by human errors. It generates scenarios corresponding to the system requirements and identifies, through an analysis of these scenarios, possible exceptions caused by human errors (*exception discovery strategy*).

The chunk also includes validation patterns to validate the requirements (*validation patterns strategy*).

The matching measures convinced the method engineer to make the decision to assemble the two selected method chunks, thus to move in the APM (Figure 3) to *Assemble chunks.* The two chunks have the same broad objective, to discover system requirements, and their process and product models overlap (they contain similar intentions and concepts). Thus, the *integration strategy* to assemble these chunks is adequate.

Following the integration map shown in Figure 4, the method engineer understands that he first needs to adapt the product and process models of the two chunks. It is only after the necessary terminological adaptations that he will be able to proceed to their integration. As an example, he selects the *name unification strategy* in the integration map (Figure 4) and changes the name of the intention *Elicit a requirement* in the *SAVRE* map into *Elicit a goal* by applying the `RENAME_INTENTION` operator. Then, he progresses to the construction of the integrated process model with the *merge strategy* to integrate the two maps. He applies the `MERGE_INTENTION` operator on the couples of identical intentions. The merged intentions are represented in grey in Figure 10. By selecting the *addition strategy* in the integration map he adds the *transformation strategy* to the integrated map. This new strategy permits the coupling of the two types of scenarios (the ones in *L'Ecritoire* and the ones in *SAVRE*) in the same integrated product and the transformation of *L'Ecritoire* scenarios into *SAVRE* scenarios and vice versa.

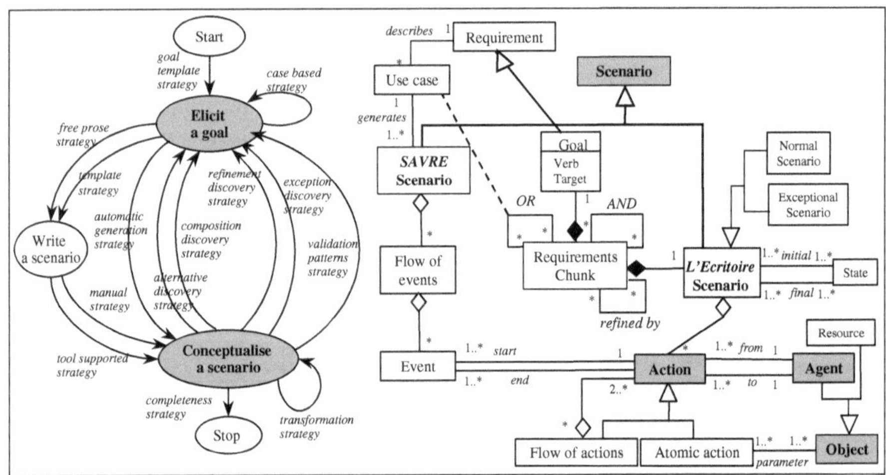

Fig. 9. The process and the product models of the integrated method chunk

The integration of the two product models also requires some adaptations. For example, the two product models contain the *scenario* concept. The two *scenarios* have the same semantics, but their structures are different. Thus, the method engineer renames the *scenario* concept in the *L'Ecritoire* into *L'Ecritoire scenario* and in the *SAVRE* into *SAVRE scenario.* Then he selects the *generalisation strategy* in the integration map to integrate the two scenario concepts by applying the `GENERALISE` operator. The notion of the *Agent* is the same in the two product models. Thus, the *merge strategy* can be selected to help applying the MERGE_CONCEPT operator on these two concepts. The result of the integration of *two* method chunks is illustrated in Figure 10.

The requirements coverage is still not completed and the method engineer continues the search for chunks that can fill in the gap between the requirements map and the integrated chunk. There is a need for validating the requirements. Thus, the method engineer formulates a new query asking for chunks with the intention to *validate the requirements* in their interface. Among the retrieved method chunks, the method engineer retrieves the *Albert* chunk [7] presented in Figure 11. This chunk proposes guidelines to validate requirements in an animated manner. It can transform scenarios describing requirements into an *Albert* specification and then, supports the animation of these scenarios by activating the tool called *Animator*.

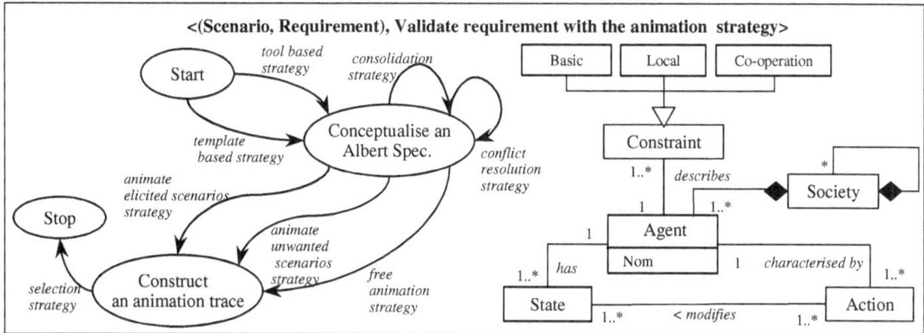

Fig. 10. The *Albert* method chunk

The maps of the two chunks to assemble do not have similar intentions. Thus, there is no need to adapt the maps before their association. The method engineer selects the *connection strategy* in the association map (Figure 5) to the construction of the associated process model. Following the associated guideline, he identifies that the achievement of the intention *Conceptualise a scenario* in the integrated map constructs a product (a scenario) which is a source product for the *Albert* chunk. The operator MERGE_INTENTION is used on the intention *Conceptualise a scenario* and the *Start* intention of the *Albert* map. Some refinements are necessary on the associated map. For example, it seems reasonable to forbid a progression from the intention *Conceptualise a scenario* to *Stop*. By selecting the *remove strategy* in the association process map (Figure 5) the method engineer applies the operator REMOVE_SECTION on this section.

The construction of the associated product model consists in the adaptation of the *Agent*, *State* and *Action* concepts and addition of the links between the corresponding concepts. The end result is shown in Figure 12 (only the final map).

In a similar manner the selection of additional chunks to cover the entire requirements map and their assembly with the current integrated chunk will continue till the completeness strategy ensures that the result is satisfactory enough to stop the assembly process.

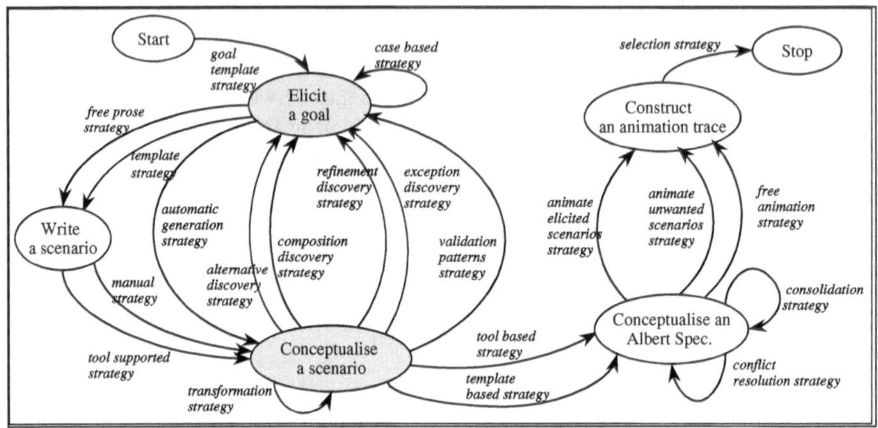

Fig. 11. The end result of the chunk assembly

6 Conclusion

In this paper we look at situational method engineering from a process perspective and propose two embedded generic models to support:

- method construction, and
- method chunk assembly.

Both are concerned with engineering methods matching a set of requirements through a method chunk assembly technique. The former deals with assembly 'in the large' whereas the latter offer solutions 'in the small'.

The process models are represented as maps with associated guidelines. This allows us to offer flexibility to the method engineer for carrying out the engineering activity. Besides, guidelines provide a strong methodological support, thanks to some formally defined techniques. Metrics to evaluate the distance between two method chunks and a set of operators to perform the assembly tasks are the two most important techniques.

The approach is currently used in a professional environment in the context of a rather large project (§10 millions). Results are encouraging, the experience is positive, even if it highlights the need for improvements among which is a software environment to support the process.

References

1. Benjamen A., *Une Approche Multi-démarches pour la modélisation des démarches méthodologiques*. Thèse de doctorat en informatique de l'Université Paris 1, 1999.
2. Bianco G., V. De Antonellis, S. Castano, M. Melchiori, *A Markov Random Field Approach for Querying and Reconciling Heterogeneous Databases*. Proc. of the 10th Int. Workshop on Database and Expert Systems Applications (DEXA'99), Florence, Italy, September 1999.

3. Brinkkemper S., M. Saeki, F. Harmsen, *Assembly Techniques for Method Engineering*. Proc. of the 10th Conf. on Advanced Information Systems Engineering, Pisa Italy, 1998.
4. Castano S., V. De Antonellis, *A Constructive Approach to Reuse of Conceptual Components*. Proc. of Advances in Software Reuse: Selected Papers from the Second International Workshop on Software Reusability, Lucca, Italy, March 24-26, 1993.
5. Harmsen A.F., S. Brinkkemper, H. Oei, *Situational Method Engineering for Information System Projects*. Proc. of the IFIP WG8.1 Working Conference CRIS'94, pp. 169-194, North-Holland, Amsterdam, 1994.
6. Harmsen A.F., *Situational Method Engineering*. Moret Ernst & Young , 1997.
7. Heymans P., E. Dubois, *Scenario-Based Techniques for Supporting the Elaboration and the Validation of Formal Requirements*. Requirements Engineering Journal, 3 (3-4), 1998.
8. Jilani L.L., R. Mili, A. Mili, *Approximate Component Retrieval : An Academic Exercise or a Practical Concern ?* Proceedings of the 8th Workshop on Istitutionalising Software Reuse (WISR8), Columbus, Ohio, March 1997.
9. Maiden N.A.M., *CREWS-SAVRE: Scenarios for Acquiring and Validating Requirements*. Journal of Automated Software Engineering, 1998.
10. Plihon V., J. Ralyté, A. Benjamen, N.A.M. Maiden, A. Sutcliffe, E. Dubois, P. Heymans, *A Reuse-Oriented Approach for the Construction of Scenario Based Methods*. Proc. 5th Int. Conf. on Software Process (ICSP'98), Chicago, Illinois, USA, 14-17 June 1998.
11. Punter H.T., K. Lemmen, *The MEMA model : Towards a new approach for Method Engineering*. Information and Software Technology, 38(4), pp.295-305, 1996.
12. Ralyté J., C. Rolland, V. Plihon, *Method Enhancement by Scenario Based Techniques*. Proc. of the 11th Conf. on Advanced Information Systems Engineering CAISE'99, Heidelberg, Germany, 1999.
13. Ralyté J., *Reusing Scenario Based Approaches in Requirement Engineering Methods: CREWS Method Base*. Proc. of the First Int. Workshop on the RE Process - Innovative Techniques, Models, Tools to support the RE Process, Florence, Italy, September 1999.
14. Ralyté J., *Ingénierie des méthodes par assemblage de composants*. Thèse de doctorat en informatique de l'Université Paris 1. Janvier, 2001.
15. Rolland C., N. Prakash, *A proposal for context-specific method engineering*, IFIP WG 8.1 Conf. on Method Engineering, pp 191-208, Atlanta, Gerorgie, USA, 1996.
16. Rolland C., V. Plihon, J. Ralyté, *Specifying the reuse context of scenario method chunks*. Proc. of the 10th Conf. on Advanced Information Systems Engineering, Pisa Italy, 1998.
17. Rolland C., C. Souveyet, C. Ben Achour, *Guiding Goal Modelling Using Scenarios*. IEEE Transactions on Software Engineering, 24 (12), 1055-1071, Dec. 1998.
18. Rolland C., C. Ben Achour, *Guiding the construction of textual use case specifications*. Data & Knowledge Engineering Journal, 25(1), pp. 125-160, March 1998.
19. Rolland C., N. Prakash, A. Benjamen, *A multi-model view of process modelling*. Requirements Engineering Journal, p. 169-187,1999.
20. Saeki M., K. Iguchi, K Wen-yin, M Shinohara, *A meta-model for representing software specification & design methods*. Proc. of the IFIP WG8.1 Conference on Information Systems Development Process, Come, pp 149-166, 1993.
21. van Slooten K., S. Brinkkemper, *A Method Engineering Approach to Information Systems Development*. In Information Systems Development process, N. Prakash, C. Rolland, B. Pernici (Eds.), Elsevier Science Publishers B.V. (North-Holand), 1993.
22. Song X., *A Framework for Understanding the Integration of Design Methodologies*. In: ACM SIGSOFT Software Engineering Notes, 20 (1), pp. 46-54, 1995.
23. Sutcliffe A.G., N.A.M. Maiden, S. Minocha, D. Manuel, *Supporting Scenario-based Requirements* Engineering. IEEE Transactions on Software Engineering, 24 (12), 1998.

Process Reuse Architecture

Soeli T. Fiorini, Julio Cesar Sampaio do Prado Leite, and Carlos José Pereira de Lucena

Pontifícia Universidade Católica do Rio de Janeiro – PUC-Rio
Rua Marquês de São Vicente 255, 22451-041 Gávea-RJ
Rio de Janeiro, Brasil
{soeli,julio,lucena}@inf.puc-rio.br

Abstract. This paper proposes a systematic way to organize and describe processes, in order to reuse them. To achieve that, a process reuse architecture has been developed. This architecture is based on processes and their types (standard, pattern, usual and solution), on process frameworks, based on the theory of application framework and on different kinds of process modeling languages, which are specified in XML, to describe each type of process. In order to facilitate the reuse and retrieval of information, we use facets, reuse guidelines, as well a process patterns taxonomy. Some processes of requirements engineering have been analyzed so that it was possible to create a process framework and a web tool has been developed to enable a case study to validate the proposed architecture.

1. Introduction

The growing interest in Process Engineering is an evolution of studies focused on products which verified that the quality of the developed software has a strong relation with the process which is used to elaborate it. In 1984, at the First International Workshop on Software Processes - ISPW a group of researchers learned about the new area of process technology, which emerged at the time. Afterwards, more than twenty workshops and conferences have been held (ICSP, IPTW, EWSPT, FEAST e PROFES)

In the last few years research on software reuse has focused on process related aspects, such as: software process programming [1] [2], software engineering environments based on processes [3], user interface guided by defined processes [4], software processes improvement [5] [6] [7] [8] and definition of process patterns [9] [10] [11] [12] [13] [14]. Although some problems, directly or indirectly related, have already been solved, others remain waiting for better proposals.

Nowadays, while attempting to improve processes, many companies try to reach levels of process maturity [5], based on their improvement and definition. However, those improvements are expensive, and they rely on the involvement of managers and teams, process data, of people who know process modeling, training, and cultural change. Several factors, imposing difficulties, make companies spend long periods of time to define some processes [15], and some of them give up in the middle of the maturity process. A frequently mentioned process to accelerate this within a company is to replicate one organizational process in other projects. At this point, the process

K.R. Dittrich, A. Geppert, M.C. Norrie (Eds.): CAiSE 2001, LNCS 2068, pp. 284–298, 2001.
© Springer-Verlag Berlin Heidelberg 2001

descriptions are very important because they allow the knowledge to be reused. However, the creation of a process, which is used in several other projects, is a hard task [19].
Models such as CMM [5] or the paradigm of the software programming process [1] aim at following one defined process and at having one standard; nevertheless, they rarely have the concept of reuse of process and artifacts, as a central objective [16]. Many organizations have to begin the definition of their processes based on some kind of existent tacit knowledge or solely on bibliographies which demand a lot of time to be understood, because they often contain inconsistencies.
Another possibility is to begin the definition based on models or norms, such as ISO and CMM; these however, are either too generic or need interpretation and are not always organized in the form of a process. Efforts made to measure the definition of processes to projects, report that it is necessary, for a key area (CMM), between 800 and 1000 hours/people or more, depending on the amplitude and depth of the process [17].

In order for the processes to be reused, the companies need to express common elements and variables within one process. Frameworks [18], [19] provide a mechanism to obtain such reutilization [17] and are very appropriate to domains where various similar applications are built several times, from scratch. Researches on patterns [20] [21] also have shown that they are effective tools for reutilization.
The use of application framework and patterns has taken place in many areas, such as design, interface, code, organizational and analysis; however, it seldom happens in processes. The utilization in processes is still isolated [22], [23] and [9], with a not very clear definition and needs to define what processes and patterns frameworks really are, and they do not have specific techniques for its development. Therefore, by integrating many concepts, we present a process reuse architecture, based on kinds of processes (standard, pattern, usual and solution), on process frameworks, based on the application frameworks theory, and on different types of process modeling languages, which are specified in XML, to describe each type of process. A tool, which implements this architecture and its elements, is presently in its late stage of development.

Section 2 presents the process reuse architecture and its elements. Section 3 presents how to use the architecture (processes and frameworks) to generate a solution process – a process instance. Section 4 presents related works and we conclude in section 5.

2. Process Reuse Architecture

Architecture in its strict meaning is the art of building. Thus, the process reuse architecture proposed here is an organization of processes and related elements. The objective of the architecture is to organize processes to enable their reutilization. By using the organization and structuring of information together, we try to facilitate the access and reutilization of the processes. Because it is reusing processes, it has all the benefits known in the technology of reuse, making it possible to use acquired knowledge and experience, which have been already established by other organizations or researchers [28]. A process is defined here as a set of interrelated activities that receive inputs and produce outputs.

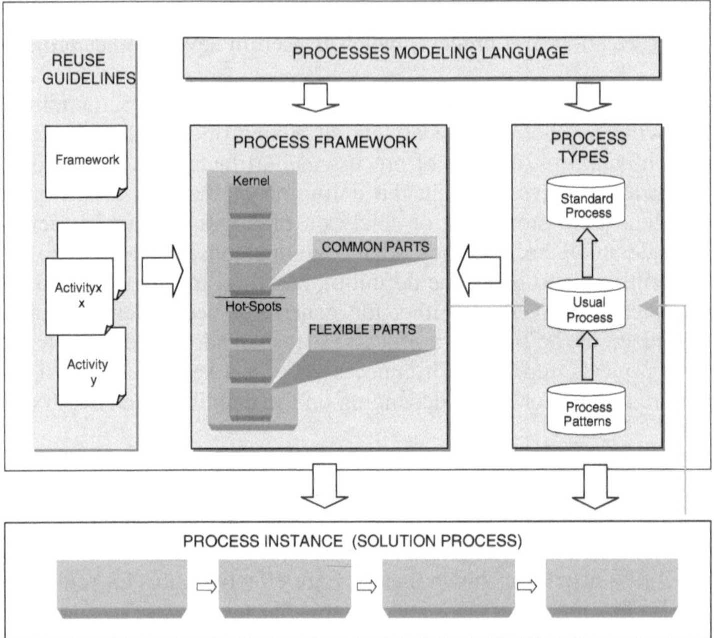

Fig. 1. Process reuse architecture.

Figure 1 illustrates the proposed architecture. The architecture elements are the process modeling language, types of processes (standard, pattern, usual e solution), process framework and reuse guidelines.

Process Modeling Language

A process modeling language is a formal notation used to express process models. A process model is the description of a process, expressed in a process modeling language [24]. To express the process model we have chosen XML [25]. XML specifies a marking meta-language that allows the structured representation of several types of information. This information can be sent/received and processed on the Web in an even way. Because the processes naturally have a hierarchical structure, we adopted the XML, taking the advantage of its benefits to work with documents structuring. The XML's DTD – Document Type Definition – contains or points to the declarative marks, which define a grammar or set of rules to a certain class of documents. This way, in a DTD we define languages to model the processes. It is important to emphasize that each one of the process types has a representation and therefore they can contain different attributes in their description, although many attributes appear more than once. The languages for the process patterns are both an evolution of design patterns templates [20], [21] e [26] and research on software reuse. For the other kinds of processes, the base was previous results of process modeling [27], [28] and [29], including the framework idea. All kinds of processes have a great deal in common as far as the modeling is concerned, and that allows the reuse to be carried out more amply. For example, all of them have activities

Process Types

Processes will form the base of the architecture. The types have the purpose of distinguishing processes, characterizing different sources of information used in the process definition. All the processes are stored in a database. When the reuse takes place, a process instance is stored in the usual process database and an XML file can be generated, enabling the process engineer to store one copy in his/her directory. The types of process defined are:

- **Standard Process:** it is a standard (for example, CMM or ISO) form of process. It is a base for process definition, according to specifics process improvement and quality assurance standards. It has a normative purpose.
- **Process Pattern:** it is a process that deals with problems related to processes. According to the definition of patterns, it can not be new or hypothetical [20] [21].
- **Usual Process:** is any existing process that is neither standard nor a pattern. It is not standard because it does not have the normative purpose and it is not a pattern because it has not been tested (applied) a considerable number of times to solve one recurrent problem.
- **Solution Process**: it is an instance of either the framework or of any other process (pattern, usual or standard), or of the combination of those.

Process Framework

A process framework models thc behavior of the processes in a given domain [41]. A process framework is defined as a reusable process. Its kernel models the common activities among the processes of the domain. Hot -spots model the specific and flexible parts of the processes of the domain and they are reified during the instantiation of the framework. The hotspots are activities or other elements of the process (techniques, tools...), that define characteristics or specific paths of a solution process (process instance). The hotspots are instantiated by the process engineer, redefining the description of activities or elements, both on the macro and detailed levels. The framework itself will have to be represented as a process, identifying the parts that are hot spots and common activities. The instantiation of one processes framework generates a solution process. In the next Section we detail the use of the framework

Reuse Guidelines

The objective of the reuse guidelines is help reuse and adapt the framework and it also shows both flexibilization and common points. The reuse guidelines are available for the framework, given an overview of whole process, and for the macro activities. On the upper side of the guide is the process or the activity name. The guide design (Figure 2), follows a structure based on hot-spot cards [18], features model [30] and components reuse concepts – 3C – [31], whose details follow:

REUSE GUIDELINE

Framework or Activity Name

CONCEPT ( ) Framework

( ) Macro Activity

( ) Kernel ( ) Hot spot

CONTENTS

Activities	Features	Components		P. Standard/ P.Pattern
Activity Name 1	E	Component	x	
Activity Name n	OP, RE, PA	Component	y	

CONTEXTO ( )General ( ) Specific

Fig. 2. Reuse guidelines

The 3Cs Model

3Cs Model of reuse design provides a structure, which has been effective in the design of reusable assets [22]. The model indicates three aspects of a reusable component – its concept, its contents and its context. The concept indicates the abstract semantics of the component; the contents specify its implementation and the context specifies the environment, which is required to use the component.

The "Concept" section of the reuse guidelines provides information as to whether the guide is one of framework as a whole (giving a general view of the process), or a specific activity, on macro level. It also provides information to identify, in the case of activities, whether they are hot spots or common point. (kernel).

The "Contents" section identifies activities/ components. Activities are elements of the framework and components are other activities (from database) that can complement process description or fill the framework hot spots - when reusing the framework. For a framework, the macro-activities are the components themselves. For the macro-activities, the detailed activities are the components. It also provides information to identify the features of each activity (see next item).

For each activity it should be possible to access its basic data and the components related. For example, "Elicit Requirements" is related to "Collect facts". In the same section, available process patterns and process standard items, such as practice in the CMM, for the macro activities are listed.

The "Context" section defines whether the framework is generic or specific. As far as the activities are concerned, the section describes the applicable techniques and when they should be applied.

Features Model
The features are present in their original proposal, the FODA (Feature Oriented Domain Analysis) [30] project, aiming at "through the model, capturing the final user's view of the applications requirements to a determined domain". In our proposal we use the characteristics as a semantic model, where they are the process activities attributes. They are useful to organize the framework, allowing it to identify the flexibilization and common points. The domain engineer captures, from the analyzed processes, certain relevant features, so that the process engineer can have a reference about the process activities in that domain when he reuses processes.

Some examples of features are:

[CO]mmon: the characteristic indicates that the activity is common, in the set of processes, which are analyzed in the domain;

[OP]tional: the characteristic indicates that the activities are not mandatory in the domain;

[S]pecific: when the activity is specific of few processes in the domain. They represent particularities of certain processes;

[IM]plicit: the characteristic indicates that the activity in the analyzed processes in the domain is implicitly present in the process descriptions, although it does not appear clearly as an activity.

Hot-Spots Card
According to Pree [18] a hot-spot card describes hot spots. Comparing what the hot-spot card specifies and what the reuse guidelines provides, we have, respectively:

- The hot spot name – in the heading of the guide it is possible to visualize the name of the process or activity. However, it is only in the "Concept" section of the guide that we find information on the flexibilization points (common point or hot spot).
- A concise term describing the functionality – In the "Contents" section of the guide the component names and the components lead to the objective/description through links.
- Degree of flexibilization – the "Context" section of the guide defines the flexibilization.

Operations

The process engineer can carry out the basic operations: inclusion, exclusion, searches and update associated with the process database. During the research he can choose a process, which might be reused as it is, or not. If the process does not meet the engineer's expectations, he or she can adapt the chosen process or create a new one, by reusing activities from several processes available in the database. The framework tailoring is not manual. We instantiated processes originally from the databases, from where they are retrieved by the use of facets and other resources. Our proposal is not to enact processes, but to define them.

Process Classification and Retrieval

There ought to be a mechanism to classify the content of the databases and to assist the user in the queries of processes, to find the process that better matches his or her needs. We define several process elements that are used in the process classification and retrieval. For example, we can retrieve using the usual process classification (fundamental, organizational or support) or using the taxonomy created for process pattern based on problems addressed (requirements elicitation, requirements analysis, ...). In this section we will describe two classification methods: facets [32] for all process types and Pattern Society [28] for process pattern.

Facets

Prieto-Díaz [32] proposes a scheme of classification through facets to catalog software components where it is believed that the components can be described according to their functions, how they perform such functions and details of implementation. Facets are considered as perspectives, points of view or dimensions of a particular domain. The elements or classes that comprise a facet are called Terms. It has been demonstrated that facets help in organizing and retrieving in component libraries.

The facets are used as describers of activities, focusing on the action that they carry out and on objects, which are manipulated by the activity. For example, the activity "elicit requirements" has the action "to elicit" and as object "Requirements". Both the classification and the retrieval specify the actions and/or objects when carried out. All the actions may have a list of synonyms so that they improve the retrieval.

Process Patterns Society

One of the problems that we face when we search for a pattern is to locate the appropriate pattern and to understand its solution. Many patterns are proposed individually, and others come organized in some languages for a specific purpose. Therefore, it is essential that there be a form to classify patterns in order to retrieve and use them. The term "Process Patterns Society" expresses taxonomy to organize the process patterns. In the textual description of the pattern solution, many times several links appear, for example, reference to other patterns. Our approach gives more structure for the pattern solution. The society is a way to organize them and make it look like a "patterns network map", which facilitates the search and understanding of the presented solution. It is something similar to a Catalog used in Design Patterns or Patterns System [33]. In our work, with the taxonomy of the proposed society, it is possible to organize the process patterns as: individual, family or community (see figure 3).

- Individual Process Pattern: is a process pattern, which presents a solution to a problem, without making reference to other patterns.
- Family Process Pattern: is a process pattern whose solution comprises one or more patterns.
- Community Process Pattern: is a set of process patterns, individuals and families of a certain domain.

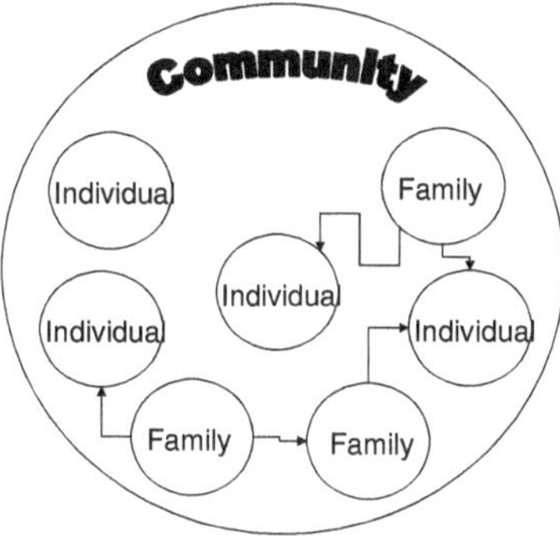

Fig. 3. Process pattern society

RAPPeL – A Requirements Analysis Process Pattern Language for Object Oriented Development [23] is a pattern language with several patterns. One of them is the Requirements Validation pattern and the other one is the Prototypes pattern. Using RAPPeL´s Requirements Validation pattern as an example, we will explain our approach. This pattern has the following problem and solution description:

Problem: "*How to verify that the specified behavioral requirements are correct and complete?*"

Solution: "*Have all interested parties thoroughly read the requirements specification. Conduct review meetings on sections of the requirements specification. Have a secretary take note of every issue raised during the reviews in the Issues List. Follow up on all issues raised.*

Build prototypes and review them with users. Again, record every issue raised in the Issues List and have follow up on each issue. Continue verification of requirements during system development through each iteration.

If needed, establish an arbitration group to reconcile disagreements on requirements.

Distribute prototypes to customers and conduct surveys and usability studies."

Using our approach we are looking for activities (components) at the solution and emphasize them. As a result the Requirements Validation pattern, described below, has a solution composed by two components (we have describe only the first component in our structured language):

1) **Component Name:** Conduct (action) reviews meeting (object)
   Synonym: Realize, execute
   Pre-condition: Have a documented requirements specification
   Input: Requirements specification
   Recommendation: Conduct review meetings on sections of the requirements specification. Have a secretary take note of every issue raised during the reviews in the Issues List. Follow up on all issues raised. Use cases are excellent in structuring the specification of the behavioral requirements for study and validation. Customers can 'role play' the various use cases using the domain objects to get a better feel on whether the system is doing what is expected.

Restriction: it is a human process with no automated support
Post-Condition: requirements specification reviewed.
Output: Issues List

2) **Component Name:** Build (action) prototypes (object) – individual pattern

The Requirements Validation pattern solution makes a reference to the Prototypes pattern (build prototypes), so it is a process pattern family. Component names are organized using *action* and *object* facets, which are used to retrieve information. Sometimes we complement the recommendation with other information described in patterns sections other than solution section (”Use cases are ....” is not present into the original solution).

3. Generating the Solution Process

In order to generate the solution process from the proposed architecture, we can reuse processes by using the framework, or we can surf throughout the database of the process (standard, usual and patterns), looking for one which is adequate to our needs. The solution process, which is generated, is saved as a usual process and can be visualized and used in the XML version (Figure 4).

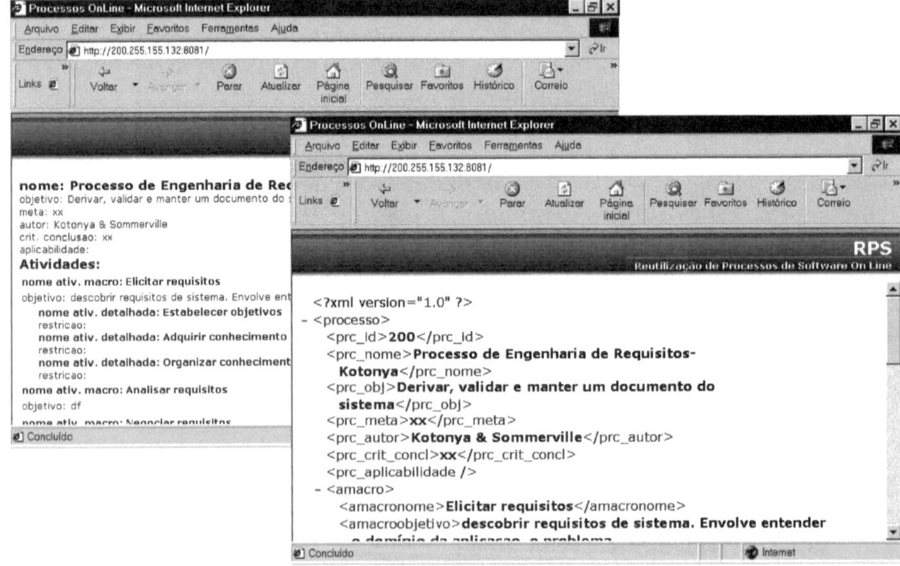

Fig. 4 XML view

In order to reuse from the database, we use the search resource, throughout which we can choose any element of the process to carry the search, such as tools, techniques, and templates. Also facets can be used and in order to locate patterns we can base our search on process patterns society. A web tool, which implements the proposed architecture, allows searching and the results can be compared. For example, we can compare a process made by an author with another one, activity by activity. The processes can be partially or totally reused (some activities). The solution process for example, may contain activities from any kind of process: standard, usual and patterns. This is possible because all process types have the same kernel – the "activities detailed" – that are called "items" in process standards and "components" in process patterns. The kernel is the activity's name, description, inputs, outputs, pre-conditions and post-conditions.

To generate the solution process, based on framework, we must utilize the reuse guidelines. The following table illustrates the initial part of a guide, for the requirements engineering process – giving the general view of the process (framework). The guide partially shows two macro activities: optional [OP] – hotspot – and a common [CO] and essential [E] – kernel. What follows is a description of how it is carried out and reused, based on the guide.

REUSE GUIDELINES						
REQUIREMENT PROCESS ENGINEERING						
CONCEPT						
(x) Framework		( ) Macro Activity	( ) Detailed Activity			
		[ ] kernel [ ] Hotspot				
CONTENTS						
Objective: derive, validate and maintain a requirement document.						
Activities	Features	Components	Process Standard/ Pattern			
			CMM N2 ( )All	*CMM N3 ( ) All*	*CMMI Cont.*	*Pattern*
Analyze the Problem	*OP,AN, FO*	Analyze the Problem – Rational Define Scope – Alister FO - Context Analysis– Bracket (outside the requirement process)			*GP1.1*	
Elicit Requirements	*CO, E*	Elicit requirements – Kotonya/ Julio/ Loucopoulos/ Sommerville Collect facts – Alister Capture Requirements – Alcazar Identify Requirements – Brackett Understand Necessities of Stakeholders (it has Elicit requests of the Stakeholders)/ Analyze the Problem – Rational			*GP2.7 and SG1*	*Define initial requirem ent*

Reusing Activities

Whenever a process engineer selects a framework, its activities are showed together with its reuse guideline (table below). The instance that will store the new process, based on the framework, needs to be selected (from usual process combo box). Using the features into the guide it is possible to know what activities are common (CO) and essential or optional (OP). The common macro activity description can be complemented with component description. The optional activities (hot spots) will be deleted or will be filled. We can fill the activities looking to the reuse guidelines for one component or process pattern that matches the process engineer needs. When a component or process pattern is reused to fill an optional activity its detailed activities are reused too. In the case of common macro activities it needs detailing. That means that it is necessary to inform which detailed activities are part of the macro activity. To do so, we must use the macro activity reuse guidelines. In this example, it would be the guide for the reuse of the "Elicit Requirements" activity. The "Context" section of the reuse guidelines (not shown in the example) provides the techniques that can be used and when they should be used.

Reusing standards
We try to use standards in our work, such as CMM, ISO/IEC 12207 or CMMI in a different way. Usually, a standard is chosen and processes are defined from it. On the framework, each activity can have one or more items from the associated standard; for example, "Elicit Requirement" has two associated CMMI standard items. In this proposal, we start from a process (framework) and we verify if the existing description complies with the standard.

Obviously, some standard items are not totally covered by the framework, which makes it necessary to act in the usual form. However, we provide support to carry it out in this form.

The standards usually have items that refer to activities themselves and others that refer to institutionalization matters, resources and organization. Those standard items that refer to activities and do not have references on the framework, are automatically created as activities in the solution process, when the option is "all" the standard. For those standard items, which have a reference activity on the framework, the process engineer should refer to two standard aids: interpretation and recommendation. The interpretation is a specialist's knowledge about the standard and it aims simply at facilitating the understanding, minimizing doubts about how to use the standard item. The recommendation is a tip on how the item can be applied in the activity, making it comply with the standard. The standard items that do not refer to activities are linked to the definition in other elements of the process, such as tools, method, technique and training, when necessary.

Reusing patterns
Patterns work as an aid for defining the process activities. Patters related to the activities are listed beside and can be consulted. For example, "Elicit Requirements" has an associated pattern, which is "Defining initial requirements". When we wish to use a pattern in the activity definition, we can chose to reuse the pattern solution as it is, or we can create a new one, based on the pattern solution.

4. Related Work

In the literature little has been found regarding the description of process patterns. Coplien [34] defines process and organization patterns as managerial practices,

Ambler [9] shows process patterns in a particular way, classifying process patterns into three levels, but recognizes that his descriptions does not follow any established form. Landes at. all [35] use quality patterns as a "primitive representation to describe experiences in a structured manner", which is our goal too. The experience is classified, in more details than in our approach, using some facets and some keywords. But they do not have a pattern organization (pattern society) similar to ours and a structured way to describe solution inside the pattern. On the other hand, they have subcategories for patterns (theory patterns, practice patterns and lessons patterns) to model the evolution of experience over time. Vasconcelos [12] describes an environment and forms for process patterns, not clarifying, however, the difference between solution process and process pattern. Vasconcelos also uses the concept of activity pattern for what we call components, and he does not use a hypertext to improve the description of the patterns. In [36] we can see patterns "related to the system testing process", but the description form is traditional. In the article "A Pattern Language for Pattern Writing" [37] a lot of patterns related to pattern structure are shown, pattern naming and referencing, patterns for making patterns understandable and language structure patterns, which coincide with our main idea.

Related to our general idea about the architecture, some works are similar in some aspects, without process types and guidelines and with focus on tailoring process to project. Henninger [38] [39] has an environment for reuse process that tries to reduce the gap between the process that is defined and the one that is in use, by combining an organizational learning meta-process with a rule-based process engine. With a similar objective Rolland et. all [40] use a contextual model approach constructing process models dynamically. Hollenbach and Frakes [22] have a framework organized in several sections for process definition. Each section has some attributes similar to and different from ours. They have a tailoring methodology.

Concerning our Patterns Society schema for patterns classification, it is somehow similar to a Pattern Catalog (collection of related patterns) or a Pattern Systems (cohesive set of related patterns which work together to support the construction and evolution of whole architectures) [33].

5. Conclusion

The result of Rollenbach and Frakes's research [22] shows that at least a tenfold improvement in time and effort to create a project or business unit process description occurs when we instantiate a reusable process instead of building the process from scratch. Our approach involves to building processes from the process data base and frameworks. We use processes as a way to store and reuse knowledge in organizations.

In this article we define an architecture based on the concepts of Standard Process, Process Pattern, Process and Solution Process. It is important to emphasize that each one of them has a representation and therefore they can contain different attributes in their description, although many attributes appear more than once. Process patterns are both an evolution of design patterns templates and research on software reuse. The other approaches are based on current research on process modeling. We achieve more expressiveness through a powerful process data base that integrates this kind of processes, because we can establish relationships between, for example, problems (described in process pattern) and usual process of the same domain. Also, we propose a taxonomy for the classification and retrieval in which patterns can be perceived individually, as a family and as a community and activities can be organized using facets. It is very important for process engineers and SEPG managers when defining process (in software process improvement programs) because they will have one central information source together with one schema of classification and retrieval to help them. This reduces process definition time.

In short, the contributions are:

- An architecture of processes reuse gathering different concepts (patterns, standards and frameworks), which enables the reuse in a systematic and practical way.
- A more structured language, specified in XML, to define processes, patterns and standards.
- By using XML (Extensible Markup Language) and XSL (Extensible Stylesheet Language), it is possible to achieve reuse, of the process content, structure and presentation.
- A processes organization that allows for the creation of a process which can be the base for the generation of a standard process and whose definition can comply with certain models and norms, which are useful in certification programs.
- A framework for Requirements Engineering and its use definition.
- A systematic scheme of process database organization and access to search and selection.

Our future work involves the development of case studies to validate the architecture. To exemplify the proposed architecture we are finishing a WEB tool. The tool will make possible that a collection of patterns be available for use. We will start with process patterns, process standards and usual process for requirements engineering and will use a questionnaire driven method to measure the level of reuse that potential users of the tool may have achieved [41].

References

[1] Osterweil, L.: Software Process are Software too, Proceedings of the 9th International Conference on Software Engineering (1987)

[2] Ambriloa, V., Conradi, R., Fuggetta, A.: Assessing Process-Centered Software Engineering Environments, ACM Transactions of Software Engineering and Methodology, 6 (3), (1997) 283-328

[3] Karrer, A S. and Scacchi, W.: Meta-Environment for Software Production, Information and Operation Management Departament, University of Southern California (1994)

[4] Guimenes, I. M. S.: Uma introdução ao Processo de Engenharia de Software, Universidade Estadual de Maringá, Paraná, Julho (1994)

[5] Paulk, C.M., Curtis, B., Chrissis, M. B., Weber, V.C.: Capability Maturity Model for Software, Ver. 1.1, Software Engineering Institute, Carnegie Mellon University, CMU/SEI-93-TR-24, ESC-TR-93-177 (1993)

[6] Konrad, M. and Paulk, M.: An Overview of SPICE's Model for Process Management, Proceedings of the International Conference on Software Quality, October, Austin, TX, (1995) 291-301
[7] The Trilhum Model, http://www2.umassd.edu/swpi/BellCanada/trillium-html/trillium.html
[8] Haase, V. et al, Bootstrap: Fine-Tuning Process Assessment, IEEE Computer, Vol. 27, Number 17, July (1994) 25-35 – http://www.bootstrap-institute.com/
[9] Ambler, S. W.: Process Patterns – Building Large-Scale Systems Using Object Technology, Cambridge University Press/ SIGS Books (1998)
[10] Wills, A. C.: Process Patterns in Catalysis, http://www.trireme.com/catalysis/procPatt
[11] Berger, K. at all : A Componentware Development Methodology based on Process Patterns, OOPSLA 98 (1998)
[12] Vasconcelos, F. and Werner, C.M.L: Organizing the Software Development Process Knowledge: An Approach Based on Patterns, International Journal of Software Engineering and Knowledge Engineering, Vol. 8, Number 4, (1998) 461-482
[13] Fontoura, M. F. M. C.: A Um Ambiente para Modelagem e Execução de Processos (An Environment to Model and Enact Processes), Master Thesis, PUC/RJ (1997)
[14] Coplien, J. O.: A Generative Development Process Pattern Language, in PloP (1995)
[15] SEI, Process Maturity Profile of the Software Community 1999 Year End Update, SEMA 3.00, Carnegie Mellon University, March (2000)
[16] Henninger, S.: An Environment for Reusing Software Processes, International Conference on Software Reuse, (1998) 103-112
[17] Hollenbach, C., Frakes, W.: Software Process Reuse in an Industrial Setting, Fourth International Conference on Software Reuse, Orlando, Florida, IEEE Computer Society Press, Los Alamitos, CA (1996)
[18] Pree, W.: Framework Patterns: Sigs Management Briefings (1996)
[19] Buschmann, F., Meunier, R.: A System of Patterns, in PLoP (1995)
[20] Coplien, J. O.: Software Patterns, SIGS Books & Multimedia, USA (1996)
[21] Gamma, E., Helm, R., Johnson, R., e Vlissides, J.: Design Patterns: Elements of Reusable Object-Oriented Design, Addison-Wesley (1995)
[22] Hollenbach, C., Frakes, W.: Software Process Reuse in an Industrial Setting, *Fourth International Conference on Software Reuse*, Orlando, Florida, IEEE Computer Society Press, Los Alamitos, CA (1996)
[23] Whitenack, B., RAPPeL: A Requirements-Analysis-Process Pattern Language for Object-Oriented Development, *in PloP* (1995)
[24] Finkelstein, A., Kamer, J., Nuseibech, B.: *Software Process Modelling and Tecnology*, Research Studies Press Ltda, 1994
[25] http://www.w3.org/TR/REC-xml
[26] Alexander, C.: The Timeless Way of Building, New York: Oxford University Press (1979)
[27] Fiorini, S.T., Leite, J.C.S.P., Lucena, C.J.P.: *Describing Process Patterns*, Monografias em Ciência da Computação no. 13/99, PUC-Rio, Agosto (1999)
[28] Fiorini, S.T., Leite, J.C.S.P, Lucena, C.J.P.: Reusing Process Patterns, *Proceedings of the Workshop on Learning Software Organizations*, Oulu 2000
[29] Fiorini, S.T., Leite, J.C.S.P, Macedo-Soares, T.D.L.: Integrating Business Processes with Requirements Elicitation, *Proceedings of the 5th Workshops on Enabling Technologies: Infrastructure for Collaborative Enterprises* (WET ICE'96) (1996)
[30] Kang, K. et al., Features-Oriented Domain Analisys – Feasibility Study, *In SEI Technical Report* CMU/SEI-90-TR-21, November (1990)
[31] Lator, L.,Wheeler, T., Frakes, W.:Descriptive e Prescriptive Aspects of the 3C's Model: SETA1 Working Group Summary. *Ada Letters*, XI(3), pp 9-17 (1991)
[32] Prieto-Díaz, R.: Implementing Faceted Classification for Software Reuse, Software Engineering Notes, Vol. 34, no. 5, May , (1991) 88-97

[33] Appleton, B.,*Patterns and Software: Essential Concepts and Terminology*, http://www.enteract.com/~bradapp/
[34] Coplien, J, O, Schmidt, D. C.: Pattern Languages of Program Design, (First PLoP Conference), Addison-Wesley (1995)
[35] Landes, D. Schneider, K., Houdek, F.: Organizational Learning and Experience Documentation in Industrial Software Projects, *Proceedings of First Workshop OM-98* (Building, Maintaining, and Using Organizational Memories), Brighton, UK, August (1998)
[36] DeLano, D. Rising, L.: System Test Pattern Language, http://www.agcs.com/patterns/papers/systestp.html
[37] Meszaros, G. and Doble, J.: A Pattern Language for Pattern Writing, http://hillside.net/patterns/Writing/pattern_index.html
[38] Henninger, S.: An Environment for Reusing Software Processes, International Conference on Software Reuse, (1998) 103-112
[39] Henninger, S.: Using Software Process to Support Learning Organizations, Proceedings of the Workshop on Learning Software Organizations, Kaiserslaurten, Germany (1999)
[40] Rolland, C., Prakash and Bejamen, A.: A Multi-Model View of Process Modeling, Requirements Engineering, 4, (1999) 169-187
[41] Fiorini, S.T., Leite, J. C. S. P.: *Arquitetura para Reutilização de Processos* (Process Reuse Architecture), Tese de Doutorado (PhD Thesis), Pontifícia Universidade Católica do Rio de Janeiro – PUC-Rio, Departamento de Informática, (2001).

Using a Metadata Software Layer in Information Systems Integration*

Mark Roantree[1], Jessie B. Kennedy[2], and Peter J. Barclay[2]

[1] School of Computer Applications, Dublin City University, Dublin, Ireland
mark.roantree@compapp.dcu.ie
[2] School of Computing, Napier University, Edinburgh, Scotland

Abstract. A Federated Information System requires that multiple (often heterogenous) information systems are integrated to an extent that they can share data. This shared data often takes the form of a federated schema, which is a global view of data taken from distributed sources. One of the issues faced in the engineering of a federated schema is the continuous need to extract metadata from cooperating systems. Where cooperating systems employ an object-oriented common model to interact with each other, this requirement can become a problem due to the type and complexity of metadata queries. In this research, we specified and implemented a metadata software layer in the form of a high-level query interface for the ODMG schema repository, in order to simplify the task of integration system engineers. Two clears benefits have emerged: the reduced complexity of metadata queries during system integration (and federated schema construction) and a reduced learning curve for programmers who need to use the ODMG schema repository.

1 Introduction

Many database applications require a mechanism by which 'generic' applications can determine a database's structure at runtime, for functions such as graphical browsers, dynamic queries and the specification of view schemata. This property, often referred to in programming languages as *reflection*, has been a feature of databases for many years, and the 2.0 specification of the ODMG metamodel [3] has provided a standard API for metadata queries in object-oriented databases. As part of our research into federated databases, we specified and implemented a global view mechanism to facilitate the creation of views for ODMG databases, and the subsequent integration of view schemata to form federated schemata. Please refer to [6,10] for a complete background on federated databases. In this paper we do not concentrate on the topic of federated databases but instead focus on the construction of a metadata interface to ODMG information systems. This paper is structured as follows: the remainder of this section provides a brief overview of the nature of our research, the importance of metadata to our view mechanism, and the motivation for this research; in Sect. 2 the main concepts in

* Supported by Forbairt Strategic Research Programme ST/98/014

K.R. Dittrich, A. Geppert, M.C. Norrie (Eds.): CAiSE 2001, LNCS 2068, pp. 299–314, 2001.
© Springer-Verlag Berlin Heidelberg 2001

this form of research are discussed, together with an informal description of the metadata layer; in Sect. 3 the pragmatics of the language are presented through a series of examples; in Sect. 4 we present details of the implementation; and finally in Sect. 5 we offer some conclusions.

In this paper we use the term *view* (or ODMG view) to refer to an ODMG subschema which may contain multiple classes, and is defined on an ODMG database, or on another view which has been defined on an ODMG database.

1.1 Background and Motivation

The main focus of our research was to extend the ODMG 2.0 model to provide views in a federated database environment. This work yielded the specification and implementation of a global view mechanism, using the ODMG model as the common model for a *federation* of databases. The concept of a federation of databases [10] is one where heterogeneous databases (or information systems) can communicate with one another through an interface provided by a common data model. In our case, the common data model is the ODMG model, the standard model for object-oriented databases since 1993 [3]. The most common architecture for these systems is as follows: data resides in many (generally heterogeneous) information systems or databases; the schema of each Information System (IS) is translated to an O-O format, and this new schema is called the component schema; view schemata are defined as shareable subsets of the component schema; the view schemata are exported to a global or federated server where they are integrated to form many global or federated schemata. Our focus was to extend the ODMG model so that it was possible to define the view schemata on top of each component schema, and define integration operators which facilitated the construction of federated schemata. This extension provided a layer of ODMG views on top of the component schema. However, it was also necessary to provide a mapping language which could bind the component schema to its local model representation. This facilitates the translation of ODMG queries (to the their local IS equivalent), and enables data transfer to the ODMG database when views are defined.

The classes which comprise the database schema are used to model the real world entities which are stored in the database. Additionally, there is a set of metaclass instances which are used to describe the database classes. Thus, we can think of an ODMG database as having two distinct sets of classes: those which reside in the *database schema*, and the abstract classes (metaclasses) which reside in the *schema repository*. Whenever we process and store a *view* definition, a set of metaclass instances are stored in the database. Where a view definition involves multiple classes, each with their own extents, this combination of meta-objects can become quite complex. Thus, the request to display a view, or extract data from the local IS often requires powerful query facilities in order to retrieve the required meta-information. In Fig. 1 the role of the schema repository within a federated database environment is illustrated. Both the *Component Database* and *Federated Database Server* are ODMG databases. The schema repository contains a description of the database classes, hence the

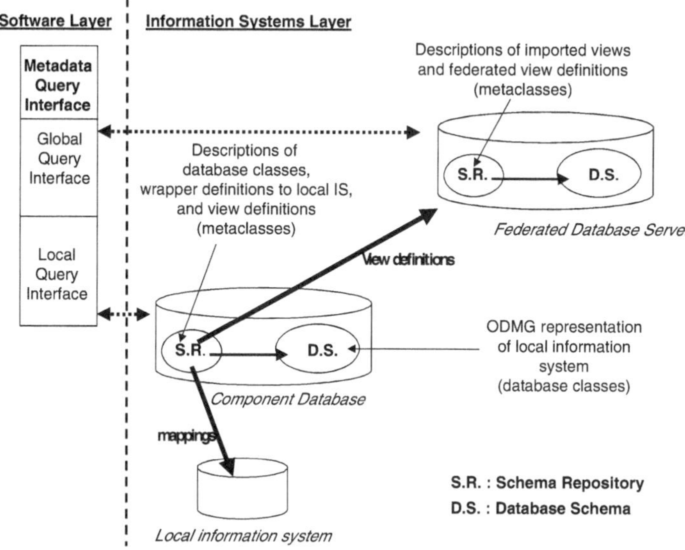

Fig. 1. Metadata architecture within ODMG information systems.

arrow towards the database schema. However, in this type of architecture, the schema repository will contain a large amount of additional data required by view definitions.

This research involved extending the ODMG metamodel in order to construct view schemata. However, due to the complex nature of both the base metamodel and extensions, many of the OQL queries which were required both to retrieve base and view class metadata would necessitate long expressions. For this reason we specified some language extensions to OQL for the specific purpose of easy retrieval of metadata from the schema repository. The contribution of this work is to provide a software metadata layer which facilitates the easier expression of ODMG metadata queries. In fact all metadata queries can be expressed in a single line. In addition, we believe it is possible to improve the performance of metadata queries as some of our experiments (discussed briefly in Sect. 4) have shown.

2 Metadata Objects and Queries

In this section we provide a description of the main concepts involved in this research: the metadata objects and the queries used to manipulate metadata. Metadata objects are used to describe both the structural elements of the participating systems (base metadata objects), and the virtual elements which are defined using a view language and are mapped to base metadata objects. Metadata queries are categorized into groups representing the type of metadata information required.

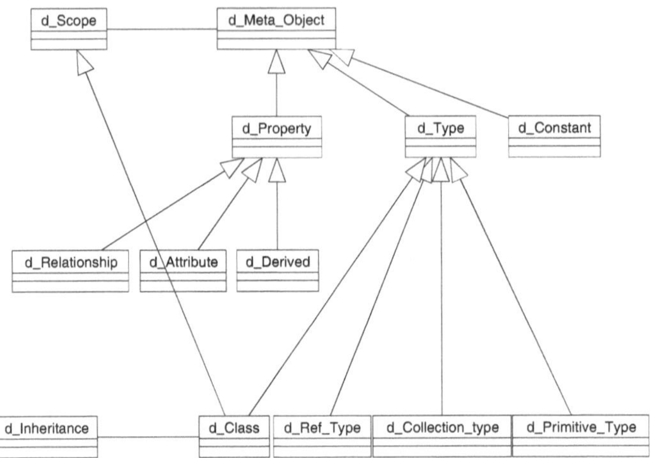

Fig. 2. The ODMG metamodel (subset).

2.1 Metadata Elements

In Fig. 2 a brief outline of the ODMG schema repository interface (or ODMG *C++* metamodel) is illustrated. For integration engineers the main points of interest are classes and their properties, and issues such as scoping and inheritance. The illustration attempts to show a hierarchy of metaclasses with the `d_Scope`, `d_Inheritance` and `d_Meta_Object` metaclasses at the top of the hierarchy. Otherwise, all metaclasses derive from `d_Meta_Object` (identified by arrows) and multiple inheritance occurs where some metaclasses derive from `d_Scope`. By deriving from a `d_Scope` object, meta-objects can place other meta-objects inside a (conceptual) container class. For example, a `d_Class` object derives from `d_Scope`, and can use its properties to bind a set of `d_Attribute` objects to it. Specifics of the roles of each of the metaclasses is best described in [5].

In an ODMG database that contains a view mechanism, it is necessary to be able to distinguish between base and virtual classes, and for virtual classes one must be capable of obtaining the same information regarding structure as can be obtained for base classes. The view mechanism, its specification language, and implementation are described in [8][9]. This paper assumes that view definitions have already been processed and stored, and a requirement exists to retrieve meta-information in order to display views or process global queries. A view or wrapper is represented by the meta-objects outlined below.

1. **Subschema or Wrapper Construction.** The definition of a virtual subschema requires the construction of a `v_Subschema` instance.
2. **Class Construction.** Where it is necessary to construct new virtual classes, a `v_Class` object is instantiated for each new virtual class.
3. **Attribute and Relationship Construction.** A `v_Attribute` instance and a `v_Primitive_Type` instance is constructed for each attribute property, and

a `v_Relationship` and `v_Ref_Type` instance is constructed for every relationship property.

4. **Inheritance Construction.** This type of meta-object `connects classes to subclasses.`
5. **Class Scope Update.** When `v_Attribute` and `v_Relationship` instances are constructed, it is necessary to associate these properties with a specific (virtual) `v_Class` instance. This is done by updating the `v_Scope` object which the `v_Class` object inherits from.
6. **Subschema Scope Update.** When `v_Class` instances are constructed, it is necessary to associate these virtual classes with a specific `v_Subschema` instance.

2.2 Metadata Query Language

The Schema Repository Query Language (SRQL) is an extension to ODMG's Object Query Language, and has been implemented as a software layer which resides between the client database application and the ODMG database. The language comprises fifteen productions detailed in an appendix in [7]. In this section we provide an informal description of the types and usage of query language expressions. The language resembles OQL in the fact that it employs a select expression. However, SRQL expressions employ a series of keywords, and are always single line expressions. As shall be demonstrated in the next section, this has practical advantages over using standard OQL to retrieve metadata information.

- **Subschema Expressions**. This type of query is used to retrieve subschema objects, which are container objects for all elements contained within a view definition. The `subschema` keyword identifies this type of expression.
- **Class Expressions**. This type of query can be used to retrieve specified base or virtual class objects, the entire set of base class objects, the entire set of virtual class objects, or the set of virtual classes contained within a specified schema. The `class` keyword identifies the type of expression, with the qualifier `virtual` specified for virtual classes, and the qualifier `in` used when retrieving virtual classes for a specific subschema (or view).
- **Attribute Expressions**. This type of query is used to retrieve single base or virtual attribute objects, the entire set of base or virtual attribute objects, or all attributes for a specific base or virtual class, by specifying the `attribute` keyword. The query can also be expressed as a shallow retrieval (only those attributes for the named class) or a deep retrieval (attributes for the named class and all derived classes). The qualifiers (`virtual` and `in`) are used in attribute query expressions, and a further qualifier `inherit` is used to determine between shallow and deep query expressions.
- **Relationship Expressions**. This type of query is semantically identical to attribute queries. Syntactically, the `attribute` keyword is replaced with the `relationship` keyword.

- **Link and Base Expressions**. These queries return the meta-objects to which virtual objects are mapped. In a view mechanism, each virtual element which has been generated as a result of a view definition must map to an equivalent base or virtual element. For example, a virtual attribute object may map to another virtual attribute object, which in turn maps to a base attribute object. The `link` query expression will return either a virtual class, attribute or relationship object if the specified object is mapped to a virtual element, or NULL, if it is mapped directly to a base element. The base query expression will always return either a base class, attribute or relationship object, nut never NULL as *all* virtual elements must eventually map to a base element.
- **MetaName and MetaCount Expressions**. Both query expressions take a single SRQL expression as an argument and return the names of the meta-objects and the count of the meta-objects respectively.
- **Type Expressions**. This query is used to return the type of (base or virtual) attribute or relationship meta-objects. Each ODMG attribute and relationship type is taken from a predefined set of types.

3 Pragmatics of SRQL Usage

Although the ODMG model provides a specification for access to the schema repository, it is quite complex and often not easy to formulate OQL metadata queries as we shall later demonstrate. Since metadata queries can be regarded as a small static group of queries, we have developed a query sub-language for the ODMG schema repository. This query language is based on OQL but extends the base language with a series of constructs which are specifically employed in metadata querying. For this reason, we called this metadata sub-language, the Schema Repository Query Language (SRQL).

3.1 Sample Metadata

In our previous work [9] we described how view schemata can be defined using our `subschema` statement. The resulting view can have any number of base or (newly derived) virtual classes, and some of these classes are connected using inheritance or relationship links. Where a view contains both base and virtual classes, it is not possible to connect classes from both sets. In this case, the view contains disjoint schema subsets. A view definition is placed inside an Object Definition Language file (ODL file), passed through the *View Processor*[1], and the result is the storage of the view definition as a set of meta-objects in the database's schema repository. It is these meta-objects which are queried by system integrators as they seek to discover similarities and differences between schemata which are due to be merged, as it is generally view schemata that are merged, rather than the entire base schemata of participating systems. Most of

[1] This is the same process as is used for defining the base schema.

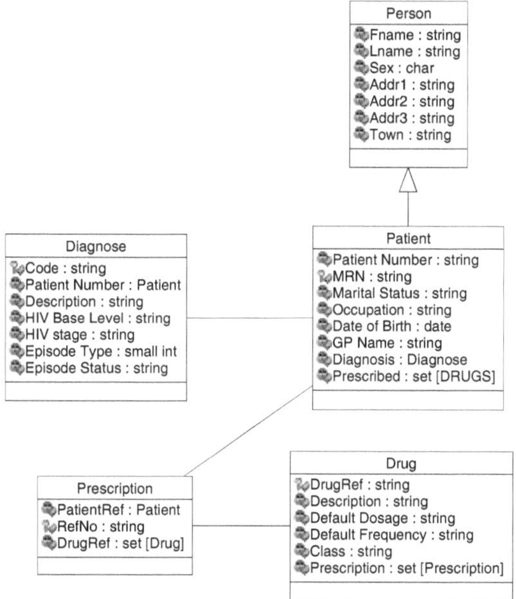

Fig. 3. The v3 view of the PAS database.

our work has involved healthcare systems, and one of these systems is the *Patient Administration System* (PAS) database at a hospital in Dublin. This database contains a wide range of information including a patient's demographic data, treatment of each illness, consultant information, and details of prescribed medication. For the purpose of the examples used later in this paper, assume that the view illustrated in Fig. 3 as patients, details of a particular illness, and the medication prescribed, is stored as *v3* in the database. The types of meta-objects constructed for this view were described in Sect. 2.1.

As it stands, the ODMG metamodel or its C++ API provides an 'open' standard for retrieving ODMG metadata, and a basis for developing a method to insert data into the schema repository. This provides a powerful mechanism both for creators of dynamic software applications (views and dynamic querying), and federated database engineers, whose role it is to extract schema information from participating systems. This work necessitates executing many metadata queries during the course of the schema integration process.

3.2 Metadata Query Samples

In this section, we argue the requirement for the proposed language by illustrating a series of metadata queries using conventional OQL, and demonstrate how these queries might be simplified using an extension to OQL. In addition to using 'pure' OQL syntax which appears to treat all derived attributes as local to each specialized class [3] (pp. 84), we have also opted to use the C++ bindings as provided by the Versant O-O database product [12][2]. Our motivation is to ensure that these OQL queries can actually be expressed in at least one vendor product, and to demonstrate the mappings between 'native' OQL and a typical O-O database vendor.

Example 3.1: *Retrieve a base class object called 'Person'.*

This query is expressed by again querying its `d_Meta_Object` superclass, but in this example, a set of references to `d_Class` objects should be generated as output.

Example 3.1(a): ODMG OQL
select C **from** d_Class
where C.name = 'Person'
Example 3.1(b): Vendor OQL
select oid **from** d_Class
where d_Meta_Object::name = 'Person'

The mapping between the ODMG specification and the C++ implementation is also clear as shown by the vendor OQL expression in 3.1(b). Its SRQL equivalent is provided in *3.1(c)* below.

Example 3.1(c): SRQL
select class Person

The result of this query will be the set of base class (`d_Class`) instances which have the `name` attribute value of *Person*. For base class instances, this will always be a single object reference, but for virtual classes it is possible for many to share the same name, providing they belong inside different subschemata. This is explained in [9] where each subschema has its own scope and class hierarchy, and thus class names can be repeated across different subschema definitions.

Example 3.2: *Retrieve virtual class object called 'Person' from subschema v3.*

Assume we now wish to retrieve a specific class *Person* from subschema *v3*.

Example 3.2(a): ODMG OQL
select C **from** v_Class
where C.name = 'Person'
and C.SchemaContainer =
(**select** S **from** v_SubSchema **where** S.name = 'v3')
Example 3.2(b): Vendor OQL
s_oid = (**select** oid **from** v_Subschema
where v_Meta_Object::name = 'v3');
select oid **from** v_Class
where v_Meta_Object::name = 'Person'
and SchemaContainer = s_oid;

[2] With one small exception: we use the term *oid* instead of the vendor term *SelfOid*, as we feel that oid is more generic. Before executing any of the queries in this section the term *oid* should be replaced with *SelfOid*.

Example 3.2(c): SRQL
select virtual class v3.Person

In this example, the SRQL makes the OQL query easier as we use a subschema qualifier to specify the correct class. Although the query can be expressed easily in OQL, it was necessary to break the vendor query into two segments as it was not possible to pass object references from an inner query. With our SRQL approach in Example 3.2(c), any implementation is hidden behind the language extensions.

Example 3.3: *Retrieve a virtual attribute 'name' from class 'Person' in subschema v3.*

In this example we again have the problem of first selecting the correct `v_Class` reference and then selecting the appropriate `v_Attribute` object reference. Assume that the virtual class is from the same subschema (*v3*) as the previous example.

Example 3.3(a): ODMG OQL
select A **from** v_Attribute
where A.in_class in
(**select** C **from** v_Class
where C.name = 'Person'
and C.SchemaContainer in
(**select** S **from** v_SubSchema **where** S.name = 'v3'))
Example 3.3(b): Vendor OQL
s_oid = (**select** oid **from** v_Subschema
where v_Meta_Object::name = 'v3');
c_oid = (**select** oid **from** v_Class
where v_Meta_Object::name = 'Person'
and SchemaContainer = s_oid);
select oid **from** v_Attribute
where v_Meta_Object::name = 'name'
and in_class = c_oid);

In Example 3.3(a) the pure OQL version of the query is expressed by simply adding another layer to the nested query. However, the vendor product in 3.3(b) requires three separate queries, and thus, it will be necessary to embed the OQL inside a programming language such as C++ or Java. Since this is the most likely scenario for an O-O database program, it does not raise any major problems, but it does demonstrate the unwieldy nature of using some of the OQL implementations when building O-O database software. In Example 3.3(c) the SRQL version of the query is a simple expression.

Example 3.3(c): SRQL
select virtual attribute v3.Person.name

These examples demonstrate how a query language based on OQL could be used to simplify querying operations against the schema repository. These types of queries are crucial to schema integrators who require metadata information in order to determine the structural makeup of a schema, before subsequently restructuring and merging different schemata. Initial queries when connecting to a database for the first time will be: *what are the names of export schemata? What are the names of classes within a specific export schema? How many attributes does a particular class contain? What is the type of attribute x?* Previous queries assumed that this type of data had already been acquired. In the following examples we will illustrate these types of queries, and will now express queries in OQL and SRQL only as the problem regarding vendor-specific versions of OQL has already been shown.

Example 3.4: *Retrieve all classes within subschema v3*

In Examples 3.4(a) and (b) the syntax for both expressions to retrieve all references to `v_Class` objects within the subschema *v3* is illustrated. As these queries are simpler than those in previous examples, the OQL expressions are fairly straightforward.

Example 3.4(a): ODMG OQL
select C **from** v_Class
where C.SchemaContainer.name = 'v3'
Example 3.4(b): SRQL
select virtual class in v3

The keyword `virtual` is used to distinguish between base and virtual classes, and similar to the `select subschema` expression, this *predicate* can be dropped in circumstances where all instances are required. In Example 3.4(c) five possible formats are illustrated. The semantics for the selection of base classes is clear: in example (i) the complete set of `d_Class` object references is returned, and in example (ii) a single `d_Class` reference is returned. For virtual classes, there are three possibilities with examples (iii) and (v) similar to their base query equivalents. However, example (iv) is different: all virtual classes called *Person* are returned.

Example 3.4(c): class selection formats
(i) **select class**
(ii) **select class** Person
(iii) **select virtual class**
(iv) **select virtual class** Person
(v) **select virtual class** v3.Person

A subschema can comprise both base and virtual classes [9]. The `in` keyword was used in the previous section to select classes within a specified subschema. If base classes are required, the keyword `virtual` is dropped. Both formats are illustrated in Example 3.4(d).

Example 3.4(d): retrieve classes within a specified subschema

select class in v3
select virtual class in v3

Example 3.5: *retrieve all relationships within Person within the v3 schema.*

In this example we require a reference to all relationship objects inside the *Person* class.

Example 3.5(a): ODMG OQL

```
select R from v_Relationship
where R.defined_in_class in
( select C from v_Class
where C.name = 'Person' and C.SchemaContainer =
(select S from v_SubSchema where S.name = 'v3')
```

Example 3.5(b): SRQL

```
select virtual relationship in v3.Person
```

In Example 3.5(b) it is clear that the SRQL format is far easier to express than the base OQL query. Additionally, queries regarding inheritance can be a little unwieldy due to the complexity of the O-O model.

Example 3.6: *Retrieve all attributes, including derived ones for the class e.*

Suppose it were necessary to retrieve all attributes for class *e*, which is derived from classes *a*,*b*,*c*, and *d* (in subschema *v3*). (Please refer to [5] for a description on inheritance in the ODMG metamodel.)

Example 3.6(a): ODMG OQL

```
select A from v_Attribute
where A.in_class in
( select C from v_Class
where C.name = 'e'
and C.SchemaContainer in
(select S from v_SubSchema where S.name = 'v3')
union
select A from v_Attribute
where A.in_class in
( select i.inherits_to from v_Inheritance
where i.inherits_to.name = 'e'
and i.inherits_to.SchemaContainer in
(select S from v_SubSchema where S.name = 'v3') )
```

In Example 3.6(a) the OQL query to return the required `v_Attribute` references for the class *e* is illustrated. In the first segment (before the union operator is applied) it is necessary to provide nested queries to obtain the correct `v_Subschema` instance, and then the correct `v_Class` instance, before the attributes for class *e* are retrieved. In the second segment, it is necessary to retrieve all `v_Class` references which are superclasses of class *e*, and perform the same operations on these classes.

Example 3.6(b): SRQL
select virtual attribute in v3.Person **inherit**

Using the SRQL, the query expression is very simple: the keyword `inherit` is applied to the end of the expression to include the additional `v_Attribute` objects in the result set.

The `select attribute` and `select relationship` expressions can take different forms as illustrated in Example 3.6(c). Only examples (iii) and (vii) will definitely return a collection containing a single object reference[3].

Example 3.6(c): attribute selection formats
(i) **select attribute**
(ii) **select attribute** age
(iii) **select attribute** Person.sex
(iv) **select virtual attribute**
(v) **select virtual attribute** age
(vi) **select virtual attribute** Person.sex
(vii) **select virtual attribute** v3.Person.sex

Finally, the area of query transformation and the resolution of mappings between virtual and (other virtual objects and) base objects requires a different form of metadata query expression. Suppose it is necessary to retrieve the base attribute to which a particular virtual attribute is mapped.

Example 3.7: *retrieve mapped attribute (without SRQL)*

Assume that *v3.Person.name* is mapped to *Person.Fullname* in the base schema. It is necessary to retrieve the mapped attribute name to assist in the query transformation process. Assuming the query expressed in Example 3.3 returns an object reference *R* (the *name* attribute in *Person* class in *v3*), then Example 3.7(a) can be used to retrieve its mapped base attribute.

Example 3.7(a): ODMG OQL (requires result set Q)
select A.VirtualConnector **from** v_Attribute
where A **in** Q
Example 3.7(b): SRQL (full query expression)
select link attribute v3.Person.name

In Example 3.7(b) the entire query expression is illustrated. Whereas the basic OQL expression requires three nested queries, the entire expression using SRQL can be expressed in a single `select link` statement (identical syntax to Example 3.7(b)). The resolution of mappings becomes even more complex when there are a series of mappings from virtual entities to the base entity, eg. where a number of subschema definitions are stacked on top of each other. To retrieve the base attribute in this type of situation requires an unwieldy

[3] or possibly *NULL*.

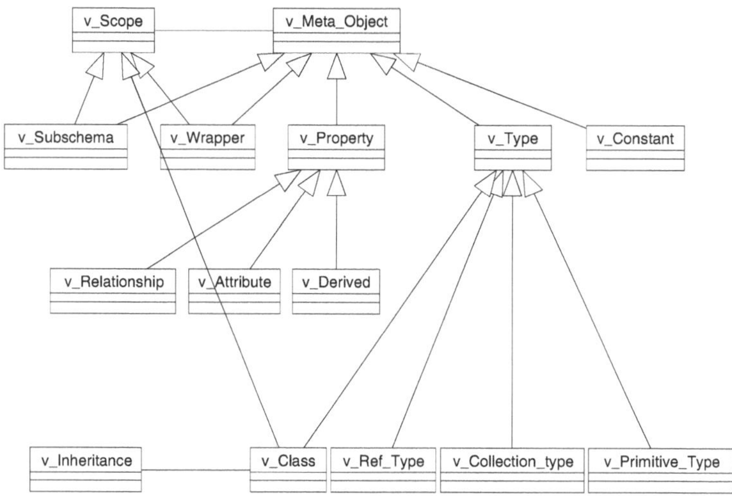

Fig. 4. ODMG metamodel extensions (subset)

OQL expression, whereas a single `select` statement will suffice if we provide the appropriate extensions to OQL.

In [7] we provide the production rules for the schema repository query language. The same publication contains the syntax for further queries such as the retrieval of subschema and wrapper objects, the retrieval of attribute and relationship type data, and the retrieval of object names and object counts.

4 Implementation of Metadata Interface Layer

The extensions to the C++ metamodel make provision for virtual schemata, any component virtual classes, and their properties and types. The ODMG 2.0 standard uses the '*d_*' prefix to denote metaclasses and to avoid confusion with standard metaclasses we employ a '*v_*' prefix to denote virtual metaclasses. The schema repository for ODMG databases was illustrated in Fig. 2. Two design goals were identified before planning our extension to the ODMG metamodel: virtual metaclasses need contain only mapping information to base (or virtual) metaclasses; yet virtual subschemata must contain enough information for high-level graphical tools to browse and display virtual types. This has been well-documented in the past: for example [11] stating that modern database systems require a richer means of querying data than that offered by simple ASCII-based query editors.

In Fig. 4 the segment of metamodel extensions which contains the most useful metaclasses is illustrated. For the purpose of clarity, the base metaclasses are not shown but the top layer in the hierarchy (containing `v_Scope`, `v_Inheritance` and `v_Meta_Object` metaclasses) is placed at the same level in the metaclass hierarchy as their base metaclass equivalents. We deliberately chose to extend the metamodel by adding virtual classes rather than extending existing metaclasses, in order to keep our base metamodel compatible with the ODMG standard. In Fig. 4 a list of the metaclass types added to the set specified by the ODMG is illustrated. Please refer to [9] for a description of virtual metaclasses and their functions.

4.1 Metadata Software Layer

An object library was built using Visual C++ 6.0 for the Versant O-O database running on NT platforms. It is assumed that client applications may be either software modules or user interfaces that have a requirement for dynamic queries. In Example 4.1 a query is required to return the names of all classes inside a subschema named *v3*. The object library parses the expression, opens the appropriate database, and generates the result set as a collection of objects of type *Any*[4], to which the program has access. The objects in each collection can then be 'repackaged' as objects of a specific metaclass depending on the type of query.

Example 4.1 Repository Query

```
database PAS
srql {
MetaName ( select virtual class in v3 ) ; };
```

Internally, the program accesses the metadata query layer by creating an instance of type `srql` and passing the query string to the constructor. The same `srql` instance provides access to the result set.

Ideally, all ODMG databases should provide a standard interface for functions such as opening the database and accessing the schema repository. However, this is not the case, and it was necessary to develop an adaptor for the Versant ODMG database for all I/O actions. A suitable adaptor will be necessary for all ODMG implementations. However, we have isolated those functions which are non-generic and placed them inside the `srql` class implementation. Thus, to use the metadata software layer with other implementations of ODMG databases, it is necessary to implement only those functions isolated within the `srql` class. Specifically, these functions are opening and closing the database, query execution, the construction of the result set, and the transfer of data from database-specific (eg Versant) objects to C++ objects. The SRQL parser, and the *semantic actions* for each production are all generic, and thus require no modification.

When developing semantic actions (*C++* program code) for each production, it was possible to take one of two routes: map the SRQL expression to the equivalent OQL expression (shown informally in Sect. 3), and then to the vendor implementation of OQL (VQL for the Versant product in our case); or

[4] Most databases and template software will use an *Any* (or similarly named) class as an abstract class for all possible return types. In this object library, the small number of possible return types keeps the type conversion simple.

alternatively, use *C++* to retrieve the objects from the database directly. Our initial prototype used the former approach but this lead to problems as not all SRQL (or OQL) queries could be expressed in the vendor implementation of the language. For this reason we adopted the latter solution and bypassed the ODMG OQL and vendor specific OQL query transformations. By doing so we had more control over the performance of queries as we could take advantage of the structure of the schema repository, and construct intermediate collections and indices depending on the type of query expression. Space restrictions prevent a further discussion on this subject here but can found in a forthcoming technical report.

The parser component was developed using ANTLR [1] and the semantic actions for each of the query expressions were written in C++. Thus many of the federated services (such as querying and data extraction from local ISs) use the SRQL for metadata retrieval. All of the SRQL examples in this paper were tested using our prototype: the Versant O-O database was used to test vendor OQL queries; and a research O-O database [4] (the closest implementation of OQL to the ODMG specification) was used to test each of the native OQL queries. All of this work is based on the ODMG 2.0 specification [3].

5 Conclusions

In this paper we described the construction of a metadata software layer for ODMG databases. Due to the complex nature of the schema repository interface, we defined simple constructs to facilitate the easy expression of metadata queries. It was felt that these extensions provide a far simpler general purpose interface to both the ODMG schema repository, and the extended repository used by our view mechanism. In particular, these extensions can be used by integration engineers who have a requirement to query metadata dynamically, the builders of view (and wrapper) software, and researchers and developers of high level query and visualization tools for ODMG databases.

Although the ODMG group has not addressed the issue of O-O views, their specification of the metamodel provides a standard interface for metadata storage and retrieval, a necessary starting point for the design of a view mechanism. Our work extended this metamodel to facilitate the storage of view definitions, and provided simple query extensions to retrieve base and virtual metadata. A cleaner approach would have been to re-engineer the schema repository interface completely rather than implement a software layer to negotiate the complexity problem, and this has been suggested by some ODMG commentators. Perhaps this could be regarded as a weakness in our approach, but we chose to retain the standard (now at version 3.0), and attempt to make its metadata interface more usable. Standards are an obvious benefit to systems interoperability, but quite often that standard can be improved.

The complex nature of the ODMG metamodel requires a substantial learning curve for programmers and users who require access to metadata: we believe that we have reduced this learning curve with our metadata language extensions. By implementing a prototype view system, we have also shown how this metadata query service can be utilized by other services requiring meta-information.

References

1. ANTLR Reference Manual. *http://www.antlr.org/doc/* 1999.
2. Booch G., Rumbaugh J., and Jacobson I. *The Unified Modelling Language User Guide.* Addison-Wesley, 1999.
3. Cattell R. and Barry D. (eds), *The Object Database Standard: ODMG 2.0.* Morgan Kaufmann, 1997.
4. Fegaras L., Srinivasan C., Rajendran A., and Maier D. $\lambda - DB$: An ODMG-Based Object-Oriented DBMS. *Proceedings of the 2000 ACM SIGMOD*, May 2000.
5. Jordan D. *C++ Object Databases: Programming with the ODMG Standard.* Addison Wesley, 1998.
6. Pitoura E., Bukhres O. and Elmagarmid A. Object Orientation in federated database Systems, *ACM Computing Surveys*, 27:2, pp 141-195, 1995.
7. Roantree M. A Schema Repository Query Language for ODMG Databases. *OASIS Technical Report OAS-09*, Dublin City University, (www.compapp.dcu.ie/~oasis), July 2000.
8. Roantree M. Constructing View Schemata Using an Extended Object Definition Language. *PhD Thesis.* Napier University, March 2001.
9. Roantree M., Kennedy J., and Barclay P. Defining Federated Views for ODMG Databases. *Submitted for publication*, July 2000.
10. Sheth A. and Larson J. Federated Database Systems for Managing Distributed, Heterogeneous, and Autonomous Databases. *ACM Computing Surveys*, 22:3, pp 183-236, ACM Press, 1990.
11. Subieta K. Object-Oriented Standards: Can ODMG OQL be extended to a Programming Language? *Proceedings of the International Symposium on Cooperative Database Systems for Advanced Applications*, pp. 546-555, Japan, 1996.
12. Versant Corporation. *Versant C++ Reference Manual 5.2*, April 1999.

Distributed Information Search with Adaptive Meta-Search Engines

Lieming Huang, Ulrich Thiel, Matthias Hemmje, and Erich J. Neuhold

GMD-IPSI, Dolivostr. 15,
D-64293, Darmstadt, Germany
{lhuang, thiel, hemmje, Neuhold}@darmstadt.gmd.de

Abstract. With the flourishing development of E-Commerce and the exponential growth of information produced by businesses, organizations and individuals on the Internet, it becomes more and more difficult for a person to find information efficiently by only using one or a few search engine(s), due to the great diversity among heterogeneous sources. This paper proposes a method for building adaptive meta-search engines with which (1) users can express their information needs sufficiently and easily, and (2) those sources that may answer user queries best can be selected, and the mapping between user queries and the query capabilities of target sources is performed more accurately. Experiments show that this adaptive method for constructing a meta-search engine can achieve more precision than traditional ways. The method can be applied to all kinds of information integration systems on the Internet, as well as on corporate Intranets.

1 Introduction

The revolution of the World Wide Web (WWW or Web for short) has set off the globalization of information access and publishing. Organizations, enterprises, and individuals produce and update data on the web everyday. With the explosive growth of information on the WWW, it becomes more and more difficult for users to accurately find and completely retrieve what they want. Although there are thousands of general-purpose and specific-purpose search engines and search tools, such as AltaVista , Yahoo! , etc., most users still find it hard to retrieve information precisely. Why? Here are some reasons:

1. Users do not know which sources can best answer their information needs. Most users only use a few well-known generic search engines, but each search engine has limited coverage and is not sufficiently capable of coping with information in specific domains.
2. Even though there are some users who collect the URLs (Uniform Resource Locator) of almost all relevant information sources in their bookmarks, still, considering that the user interfaces of these sources differ a lot, it is difficult to expect a user to be familiar enough with the user interfaces and function-

K.R. Dittrich, A. Geppert, M.C. Norrie (Eds.): CAiSE 2001, LNCS 2068, pp. 315–329, 2001.
© Springer-Verlag Berlin Heidelberg 2001

alities of all sources to use them sufficiently. For this reason, users usually only use a few their favorite sources.

3. The precision of some search engines is too low. A search engine may return thousands or even millions of hits for a user's query. Therefore, picking out the wanted hits is time-consuming and a terrible chore.

How can we overcome such difficulties? An adaptive meta-search engine will effectively and efficiently help all kinds of users find what they want. Such a tool can do several things:

1. It integrates any number of differing information sources (such as search engines, online repositories, etc.). Based on the differences with respect to domain, functionalities, performances, etc., these integrated sources are classified into different groups. When users input a query, the meta-search engine will select those sources that may answer the user query best.
2. It provides users with a uniform user interface to multiple internal and external knowledge sources, thus making the great diversity of various sources transparent to users, improving search efficiency, and reducing the cost of managing information. In order to let users input specific queries and make these queries more close to the target sources, the adaptive mechanism is employed to dynamically construct the user interface.
3. By means of this adaptive mechanism, it can translate user queries into selected sources more accurately. Therefore, users can get more complete and precise results.

In this paper we discuss how we designed such an adaptive meta-search engine. The remainder of this paper is organized as follows. We start by briefly introducing some related work. In section 3 we discuss the architecture of an adaptive meta-search engine and its major components. In section 4, we introduce some experiments carried out for testing the efficiency of a meta-search engine based on this architecture. Finally, section 5 concludes this paper.

2 Related Work

In the Internet there are a lot of meta-search engines such as SavvySearch[1], Dogpile[2], ProFusion[3], Ask Jeeves[4], askOnce[5], to name just a few. Although they integrate a lot of WWW search engines, most of their user interfaces are too simple. They only use a "Least-Common-Denominator" (LCD) user interface, discarding some of the rich functionalities of specific search engines. It is difficult for users to input complicated queries and retrieve specific information. "From the users' perspective the integration

[1] http://www.savvysearch.com

[2] http://www.dogpile.com

[3] http://www.profusion.com

[4] http://www.askjeeves.com

[5] The document company, Xerox. http://www.xerox.com

of different services is complete only if they are usable without losses of functionality compared to the single services and without handling difficulties when switching from one service to another"[6]. In order to avoid losing important functions of search engines, both generality and particularity should be considered when designing a meta-search engine. Some other meta-search engines display the searching controls of all search engines on one page or on several hierarchically organized pages, e.g. All-in-One[7] displays many original query interfaces on a single page. Many efforts have been put into research on information integration, such as [1], [2], [3], [4], [5], [6], [7], [9], etc. However, they do not consider using the adaptive mechanism to construct information integration systems. Compared with previous work, our meta-search engine model has several advantages: (1) the dynamically generated, adaptive user interface will benefit the progressively self-refining construction of users' information needs; (2) conflicts among heterogeneous sources can be coordinated efficiently; (3) user queries will match the queries supported by target sources as much as possible.

3 Constructing an Adaptive Meta-Search Engine

In this section, we first discuss the diversity among Web information sources by investigating the user interfaces of some sources. Then we discuss the architecture of our meta-search engine prototype and its components.

3.1 The Diversity of Sources

At the beginning of this section, six concrete user interface examples are displayed to demonstrate the great diversity among heterogeneous information sources. Fig.1 displays the user interface of the NCSTRL[8] search engine. Fig.2 displays the user interface of the ACM-DL[9]. Although these two search engines are designed for searching computer science papers, we can see many differences between them. ACM-DL provides a more complicated user interface than NCSTRL, so users can input more specific queries by using ACM-DL. In the second CGI form of NCSTRL search engine, each input-box belongs to one specific bibliographic field (i.e., Author, Title, Abstract). While in ACM-DL, there are five check-boxes (each stands for a field, i.e. Title, Full-Text, Abstract, Reviews, and Index Terms.) and users can select one or some of them to limit the scope of the input terms in the input-box. These kinds of differences make query translation from a meta-search engine to a target source difficult.

---

[6] http://www.tu-darmstadt.de/iuk/global-info/sfm-7/

[7] http://www.allonesearch.com

[8] Networked Computer Science Technical Reference Library http://www.ncstrl.org

[9] ACM Digital Library http://www.acm.org/dl/newsearch.html

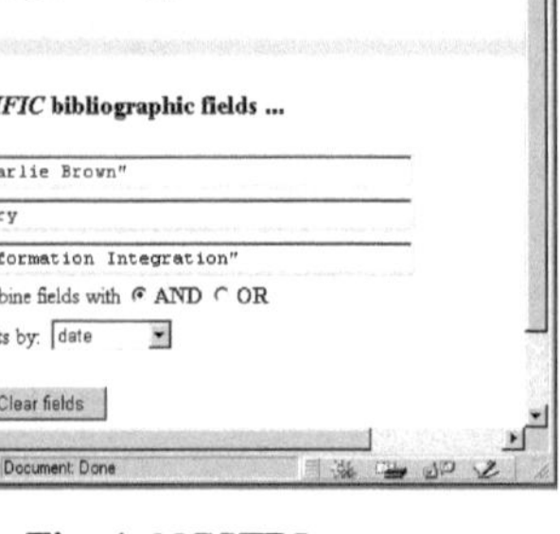

Fig. 1. NCSTRL

Fig. 2. ACM-digital library

Figures 3 and 4 show the user interfaces of two Internet sources for finding vehicles. Their interfaces are also quite different. Fig. 3 displays a rich-function query interface in which users can set many specific parameters for searching. Fig.4 displays three forms (for browsing, fast searching and advanced searching, respectively). Compared with figures 1 and 2, figures 3 and 4 are more difficult to integrate into a uniform interface because there are more differences between them.

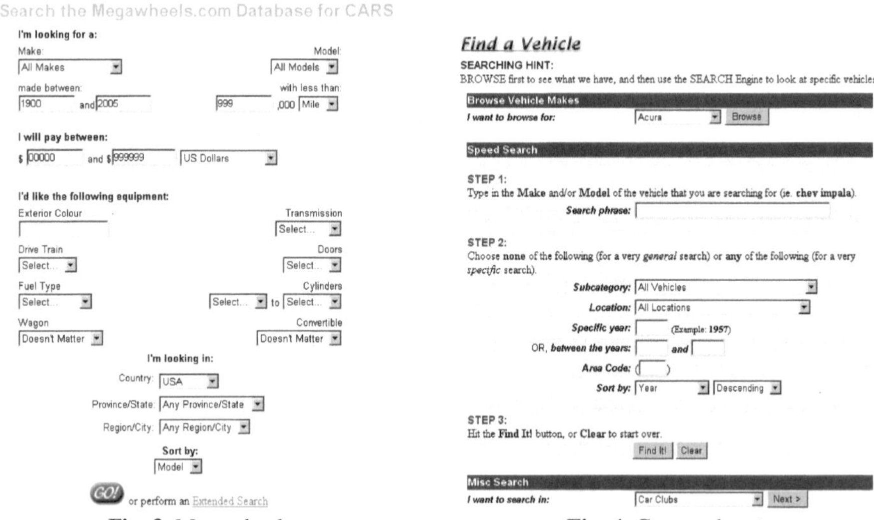

Fig. 3. Megawheels.com **Fig. 4.** Carsearch.net

Figures 5 and 6 display two sources for weather forecasting. These kinds of sources do not provide CGI-based query forms, meaning that users can only browse pages for information. However, these semi-structured pages can easily be queried with the help of wrappers. In section 3.2, we will discuss the wrappers. Most pages on the Web are semi-structured or non-structured, such as product information, personnel information, etc., so that meta-search engine must integrate not only search engines, but also these other kinds of sources.

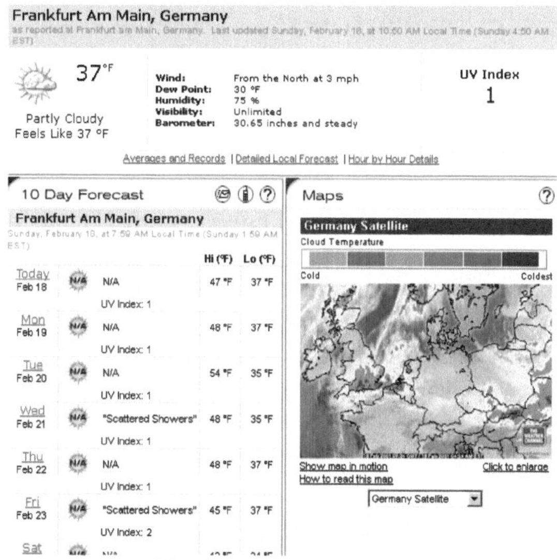

Fig. 5. www.weather.com

Frankfurt, Germany

Current Conditions	Forecast				
Updated 10:26 am local, 0926 GMT	SUN	MON	TUES	WED	THURS
mostly cloudy	p/cloudy	cloudy	snow	p/cloudy	snow
Temp: 37 F, 3 C Rel. Humidity: 86% Wind: W at 14 mph (23 kph) Sunrise: 07:12 am Sunset: 05:21 pm	HIGH 34 F 1 C LOW 21 F -6 C	HIGH 33 F 1 C LOW 24 F -4 C	HIGH 30 F -1 C LOW 22 F -6 C	HIGH 33 F 1 C LOW 32 F 0 C	HIGH 42 F 6 C LOW 30 F -1 C

Fig. 6. http://www.cnn.com/WEATHER/

In addition, the diversity of heterogeneous information sources also exists in other aspects:

1. Formats: For example, different search engines return their results using different date formats. Some systems use "September 8, 2000", others use "09/08/00", "08/09/00", "Sep. 2000", or "08092000", etc. Some sources use standard names and some use abbreviations (e.g. "kilometer", "km").
2. Naming: Different systems use different names for synonyms or homonyms. For examples, some systems use "all fields" to denote this field modifier, some use "anywhere". In the Dublin Core metadata set, there is only one

element for authors: "Creator". While in USMARC, there are two elements for authors: "Corporate author" or "Individual author".

3. Scaling: Different information retrieval systems have different ranking methods. For example, ACM-DL assigns the value 11 to an entry. While the Cora search engine assigns 0.9156 to another entry. How can you compare the relevance of these two entries?
4. Capability: Different retrieval models and query languages. Some sources support Boolean-based queries, some support vector-space-based queries, some support natural language queries. Some sources automatically drop stop-words (e.g. and, with, etc.). Some sources support fuzzy expansion, stemming, right-/left- truncation, or wildcards, and so on.
5. Interface designing: Some sources provide static HTML form user interface, some provide dynamical HTML form user interface, some provide HTML form user interface with JavaScript, and some provide java applet user interface. Some sources provide customizing services for users to personalize their user profiles. Some sources can return all results for a user query, while some sources demand users to visit their web sites more than one time to get complete results. Some sources can let users refine their queries after results come.

From Figures 1-6, we know that there are many discrepancies among the user interfaces of heterogeneous sources and it makes integration difficult. However, all the controls available in user interfaces can be divided by function into three groups:

(1) **Classification Selection Controls**, a classification selection control is a component on the user interface to a search engine, by selecting one or more items of which, users can limit their information needs to certain domains, subjects, categories, etc. For example, in Fig. 2, there is a choice control for users to limit their searches to a certain publication (which proceeding or journal) or all publications. In Fig.3, there are some classification selection controls, such as "Maker", "Model", "Country", "Transmission", etc.;
(2) **Result Display Controls**, A result display control can be used by users to control the formats, sizes or sorting methods of the query results. For example, in Fig. 1, there is a result sorting control by selecting which the retrieved results can be sorted by "author's name", "date" or "relevance". In Fig. 4, the results can be sorted by "Date of entry", "Location", "Make", "Mileage", "Model", "Price", or "Year". Some sources provide "Results grouping size" controls;
(3) **Query Input Controls**, All terms, term modifiers and logical operators of a search engine constitute a query input controls group, through which users can express their information needs (queries). A term is the content keyed into an input box on the user interface. A term modifier is used to limit the scope, the quality or the form of a term (e.g. <Title>, <Full-Text>, <Keywords>, <Abstract>, <Author>, <Exactly Like>, <Multiple Words>, <Using Stem Expansion>, etc.). A logical operator is used to logically combine two terms to perform a search, the results of which are then evaluated for relevance. For example, a logical operator can be <AND>, <OR> or <NOT>.

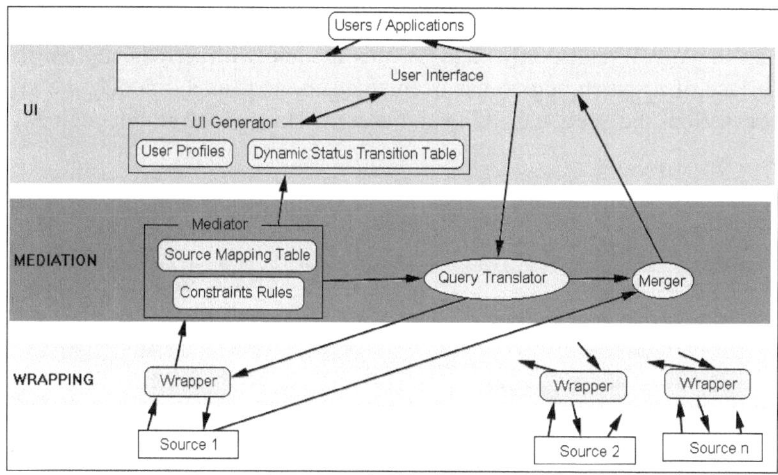

Fig. 7. Architecture of an adaptive meta-search engine

3.2 The Architecture of an Adaptive Meta-Search Engine and Its Components

Fig. 7 displays the architecture of our adaptive meta-search engine. It consists of three layers. The first one is the wrapping layer. Each wrapper describes the characteristics (such as input, output, domain, average response time, etc.) of a source and is responsible for the communication between the meta-search engine and this source. The second one is the mediation layer, which acts as an agent between users and the wrappers. The third one is the UI (user interface) layer that dynamically constructs the query form for users to input information needs. In the following, we will discuss all components in this architecture.

3.2.1 Wrappers – Mediator

The "Mediator [8] - Wrapper" architecture has been used by many information integration systems. The mediator manages the meta-data information on all wrappers and provides users with integrated access to multiple heterogeneous data sources, while each wrapper represents access to a specific data source. Users formulate queries in line with the mediator's global view, that is the combined schemas of all sources. Mediators deliver user queries to some relevant wrappers. Each selected wrapper translates user queries into source specific queries, accesses the data source, and translates the results of the data source into the format that can be understood by the mediator. The mediator then merges all results and displays them to users.

Due to the heterogeneity of Internet information sources, different kinds of wrappers should be employed for different kinds of sources, such as web search engines (e.g. Altavista), web databases (e.g. Lexis Nexis), online repositories (e.g. digital libraries), other meta-search engines (e.g. Ask Jeeves), semi-structured web documents (e.g. product lists), non-structured documents (e.g. Deja UseNet, e-mail installations), and so on. Each wrapper records the features of an integrated source. Because the information on the WWW constantly changes, this module will periodically check if the user interface of a search engine has been changed, and timely modify the information that describes the query capability and user interface of the search engine.

IBM

Image Specs Review Audio Video Coupon Compare

Description	Part#	Price	Usually Ships	Order
Thinkpad A21P 850MHz 128MB 32G 850MHz Pentium III Processor, 128MB RAM, 32GB Hard Drive, 15" TFT Display, 24X CD ROM, Windows 98	MPN31000	$3,579.95	Same Day	Add to Basket
Thinkpad A21P 850MHz 128MB 32G 850MHz Pentium III Processor, 128MB RAM, 32GB Hard Drive, 15" TFT Display, 8X DVD ROM, Windows 98	MPN31002	$3,679.95	Same Day	Add to Basket
Thinkpad T21 850MHz 128MB 32G 850MHz Pentium III Processor, 128MB RAM, 32GB Hard Drive, 14.1" TFT Display, 8X DVD ROM, Windows 98	MPN31004	$3,779.95	Same Day	Add to Basket

TOSHIBA

Image Specs Review Audio Video Coupon Compare

Description	Part#	Price	Usually Ships	Order
Satellite Pro 4600 850MHz 850MHz Pentium III, 128MB RAM, 20GB Hard Drive, 15.0" TFT Display, CD-RW, 10/100 LAN, Windows 2000	MPN61007	$2,989.95	CALL	Add to Basket
Tecra 8100 850MHz 256MB 20GB 850MHz Pentium III, 256MB RAM, 20GB Hard Drive, 14.1" Active Matrix Display, 6X DVD ROM, Windows 2000	MPN61004	$3,699.95	Same Day	Add to Basket

Fig. 8. An example Web page that can be queried by wrappers

In Fig.8, a screenshot of Web page for sailing laptop computers is displayed. From this page, we can extract the information of producers (IBM, TOSHIBA, HP, COMPAQ, etc.), descriptions (CPU, Memory and Storage, Display and Graphics, Multimedia, Communications, etc.), Part#, Prices, etc; and then use wrappers to record such information. The meta-search engine can use an SQL-like query language (e.g. "SELECT * FROM 'http://www.companyURL.com/...' where producer = 'IBM' and price < '$2000' ...") to express the information needs of users and extract the relevant information from the pages.

3.2.2 Constraints Rules - Source Mapping Table

Due to the various conflicts existing among heterogeneous sources, a meta-search engine must coordinate these conflicts for the purpose of both constructing a harmonious user interface, and making a more accurate query translation from the meta-search engine to the sources. Constraints rules are employed to record the conflict information between the controls of a source or between different sources. For example, a term belonging to the 'Date' field cannot be modified by the 'Sound like' qualifier, while a term belonging to the 'Author' field can be modified by this qualifier but not by 'Before' or 'After'. When the meta-search engine dynamically constructs the user interface, these constraints rules will be considered in order to eliminate various kinds of conflicts and to let users input their queries more easily and accurately. Fig. 9 displays three screen shots of the meta-search engine using constraints rules to construct its user interface. In Fig. 9(a), the 'author' field can only be modified by 'Exactly like', 'Sound like' and 'Spelled like' qualifiers. While in Fig. 9(b), the 'date' field can only be modified by 'Before' and 'After' qualifiers and the search terms are two choice controls (one for month and another for year) instead of an input-box.

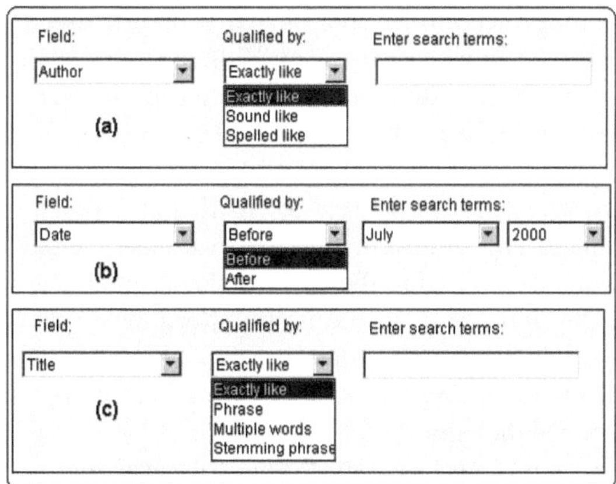

Fig. 9. Three examples of applying constraints rules to UI construction

A source mapping table is used to record the mapping situations between the items of a meta-search engine's choice controls (e.g. category selection controls) and the integrated search engines. For example, if users choose the item <Zoological Journal of the Linnean Society> from a 'Journal' choice control in the meta-search engine, then a car-oriented source will retrieve nothing. When a user has finished the query construction and then submits it, the meta-search engine's mediator will judge which search engines should be used to answer the user's query. If the number of relevant search engines is large, the running priority order will also be decided.

3.2.3 User Profiles – Dynamic Status Transition Table – UI Generator

Because the users of a meta-search engine come from all kinds of application areas, it is favorable for users to be able to personalize their user interfaces. For example, some people have interests only in the field of computer science. In this case, the meta-search engine has to provide users with functionality to customize the query interface, such as selecting relevant search engines and relevant category items of some general-purpose search engines. Users can also set up other parameters like sorting criteria, grouping size, quality of results (layout, file format, field selection, etc). When a user customizes the interface to a certain extent, the resulting interface will be limited to some specific search engines. User profiles are employed to record the configuration that users set.

A dynamic status transition table is used to record the information on user manipulations and user interface status. Depending on the control constraint rules, when a user finishes an action of clicking an item in a control, the system will check the status of all controls. If one condition of a constraint rule can be satisfied, then the items in the right part of the rule are disabled or enabled. Why do we use such a "dynamic status transition table" to manage the information on user interface status and user manipulations? The reason is that various constraints among the controls of a search engine or several search engines need to be coordinated when the user interface of a meta-search engine is being dynamically generated. If there is no such a table, the dynamic interface may be inconsistent with some control constraint rules or even in disorder when the user interface has been changed a lot. This dynamic status transition table can also be used to help users move back to a former status.

Although a static user interface is easy to create and maintain, it lacks flexibility and the interactive nature of an information retrieval dialogue between users and the system. The functionality offered to the users is also limited. During the information retrieval dialogue, the expression of the users' information needs is a self-refining process. Therefore, it is necessary to design a flexible and progressive query interface that is capable of supporting the iterative and self-refining nature of an interactive information retrieval dialogue.

For a progressive query interface, the starting page usually consists of some common controls just like the "Least-Common-Denominator" interface supported by most search engines. During the user query process, the query pages change according to users' needs until the query is finished. Therefore, this kind of query interface has the advantages of both a simple and a sophisticated query interface. Sometimes users only need a simple interface to input keyword(s) without any extra controls. But sometimes users want to input complicated queries and they will complain about the lack of input controls.

The user interface of an adaptive meta-search engine changes in accordance with both the user manipulation and control constraint rules. When users gradually express their information needs by manipulating the controls in the user interface to a meta-search engine, especially some choice controls, the number of search engines that can satisfy the information needs of users may decrease (according to the source mapping table). Suppose that only some of the integrated search engines may be relevant, then the system need not consider the irrelevant controls and items that cannot be supported by these search engines when dynamically constructing the next-step query page. Synthesizing an integrated interface will coordinate the conflicts arising from heterogeneous sources with differing query syntax. There are many differences between the user interfaces and query models of search engines for different domains. For example, it is difficult for a meta-search engine to provide a uniform interface that can be efficiently used by users searching for information on movies, news and architectural engineering. Each time users execute a query, their information needs are on a certain domain or subject. In addition, search engines for similar domains have many similarities in their user interfaces, for example, figures 1 and 2 (they share fields such as 'Author', 'Title', 'Abstract', etc.), figures 3 and 4 (fields such as 'Makers', 'Models', 'Cylinders', etc.), figures 5 and 6 (fields such as 'Temperature',

'Relative humidity', 'Wind', etc.). Therefore, based on the adaptive query model discussed before, an information integration system can facilitate the expression of both the query capabilities of information sources, and the information needs of users. Such an adaptive information system will have higher flexibility and better scalability than traditional ones.

3.2.4 Query Translator – Merger

A user query will be translated into the specific form that a search engine supports. This query translation will consider not only the syntax of the query expression, but also the constraints of term modifiers, logical operators and the order restrictions. This problem needs to be considered more elaborately; otherwise, fully exploiting the functions of heterogeneous search engines will be impossible. Therefore, we should try to make use of the functions of heterogeneous search engines as much as possible. Only thus can we improve the processing speed and achieve more accurate and complete results. This is exactly what meta-search engines should do.

When translating the user query into the format supported by the target source. Sometimes, query subsuming and results post-processing are employed to compensate for the functional discrepancies between a meta-search engine and sources. When the system translates the original query Q^o into the target query Q^t, one of the following three cases will occur:

1. In this case, Q^o can be completely supported by Q^t.
2. Some term modifiers or logical operators in Q^o cannot be supported by Q^t, but after relaxing them (i.e. broadening the scope of the limitation and therefore enabling that more results may be retrieved), the newly-generated Q^o can be supported by Q^t. In this case, the system dispatches the relaxed query, and when the results come, the results are post-processed according to the previous relaxing information. For the relaxed field modifiers, term qualifiers and logical operators, the system uses some filters to record such information and later use them to refine the results in order to compensate for the relaxing of constraints.
3. In this case, Q^o cannot be supported by Q^t even after relaxing some modifiers or logical operators. The system will break Q^o into several sub-queries, then translate and dispatch each sub-query separately. We use some special filters to record such decomposition information. When the corresponding results come, these "special filters" are employed to compose the results.

The "Merger" module translates the results from a remote search engine into the uniform format and performs some post-processing: (1) sorting and grouping all results according to certain criteria; (2) revisiting search engines (some search engines need to be accessed more than one time to get complete information); (3) dynamically reorganizing the displayed results when the results come from some slow-responding search engines; (4) processing the results according to the previous query decomposition; and so on.

4 Experiments

In order to test this applicability of the architecture for meta-search engines, three experiments on different kinds of user interfaces have been carried out. The first experiment (EXPM1) adopted a "Least-Common-Denominator" user interface that contains only a simple input box without constraint controls. (See Fig. 10. There are no modifiers for each keyword, but users can use quotation marks "" to denote phrases.) The second experiment (EXPM2) employed a static HTML user interface (See Fig. 11) containing major controls that may conflict with each other. Almost all current information integration systems employ one of these two kinds of user interfaces. The user interface of the third experiment (EXPM3) was a progressive, dynamically generated Java Applet user interface, in which almost all conflicts are automatically resolved using our approach. (See Fig. 12, which shows four snapshots of user interfaces in which a query has been input progressively. Fig. 12(a) is the initial interface and in Fig. 12(d), the query is completely input.)

"Alon Levy " Metadata "Information Integration" XML query [Search]

Fig. 10. User interface of the first experiment (EXPM1)

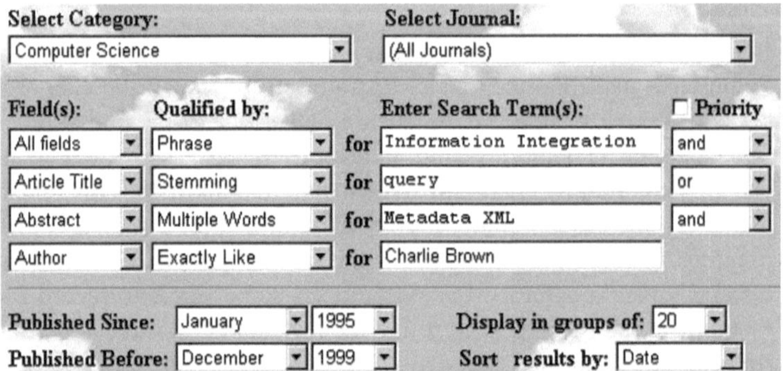

Fig. 11. User interface of the second experiment (EXPM2)

Now we briefly introduce the experimental environments: (1) Seven scientific publication oriented sources were chosen: ACM-DL, CORA, Elsevier, ERCIM, IDEAL, Kluwer, and NCSTRL. (2) Thirty papers were selected from the proceedings of the ACM Digital Libraries, SIGIR, SIGMOD, and VLDB annual conferences between 1995 and 1999. From these 30 papers we chose 45 keywords (including 20 phrases and 25 single words) and 5 authors' names. (3) Thirty-five queries were constructed from these 45 keywords and 5 authors' names according to the actual situation of selected papers. In these queries, most keywords were limited to certain fields. For example, ((**Author is "Charlie Brown"**) *AND* ((("**Information Integration" in All fields**) *AND* (**Title contains "query"**)) *OR* (("**Metadata", "XML"**) **in Abstract**))) in the Computer Science category of digital libraries, published during the period of 1995 to 1999, the results to be sorted by date. We judge that a hit is qualified if it is one of the 30 papers or relevant papers.

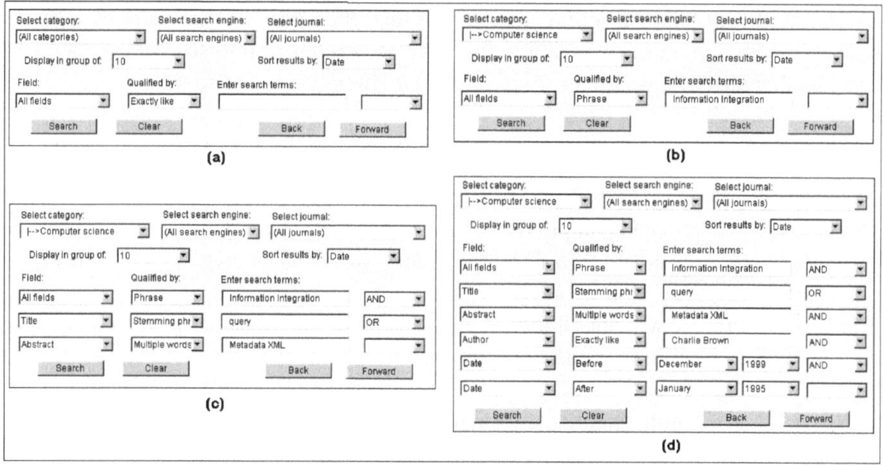

Fig. 12. User Interface of the third experiment (EXPM3)

Table 1. Experimental results

	Returned hits/query	Relevant hits /query	Precision (%)
EXPM1	130.5	1.5	1.1
EXPM2	5.9	1.3	22.0
EXPM3	1.5	1.3	86.7

Table 1 displays the results of the experimental results. The first experiment does not use any kind of constraint information, so it retrieves a lot of irrelevant information. The second experiment can use constraint controls, but it is not flexible enough for users to input queries that are closer to the formats understood by the sources. Because almost all conflicts between different sources or between the controls of the same source have been sufficiently coordinated, the third experiment achieves higher precision than the other two experiments. Table 2 compares the pros and cons of the three kinds of information integration systems with differing user interfaces.

5 Conclusions

Our experiments show that an information integration system with an adaptive, dynamically generated user interface, coordinating the constraints among the heterogeneous sources, will greatly improve the effectiveness of integrated information searching, and will utilize the query capabilities of sources as much as possible. Now, the adaptive meta-search engine architecture proposed in this paper has been applied to the information integration of scientific publications-oriented search engines. It can also be applied to other generic or specific domains of information integration, such as integrating all kinds of (especially specific-purpose) WWW search engines (or search tools) and online repositories with quite different user interfaces and query models. With the help of source wrapping tools, they can also be used to integrate queryable information sources delivering semi-structured or non-structured data, such as product catalogues, weather reports, software directories, and so on.

Table 2. Comparison of the three kinds of user interfaces for information integration

"Least-Common-Denominator" user interface (Fig. 10)	
Pros	1. It can be supported by all integrated sources; 2. It is simple for users to input information needs and for the system to map queries;
Cons	1. It will inevitably discard the rich functionality provided by specific information sources; 2. It is difficult for users to input complicated queries and retrieve more specific information.
Mixed user interface (Fig. 11)	
Pros	1. Users can express their information needs more accurately than in the "LCD" user interface;
Cons	1. It will increase the users' cognitive load and make the system hard to use for novice users; 2. The constraints between the user interfaces of heterogeneous sources may cause a user query to be inconsistent with a source and make the query mapping difficult; 3. Considering the instability of information sources on the Internet, to maintain it becomes difficult; 4. The static user interface lacks flexibility and makes the interaction between users and system difficult.
Adaptive, dynamically-generated user interface (Fig. 12)	
Pros	1. It has the advantages and avoids the disadvantages of both "LCD" and mixed user interfaces; 2. It will benefit the progressively self-refining construction of users' information needs; 3. Conflicts among heterogeneous sources can be coordinated efficiently; 4. User queries will match the queries supported by target sources as much as possible;
Cons	1. Its implementation is more difficult and time-consuming than the other two.

Such an adaptive meta-search engine can be used in the following spheres:

1. In enterprises, it can help people who are working on market research, decision support and competitive intelligence. By using it, enterprise analysts can simply formulate a single query in a uniform user interface to locate the information they need, rather than accessing several different internal and external sources separately. Therefore, they can easily monitor all marketing and commercial information concerning their businesses worldwide and can answer and respond very quickly to questions on specific subjects.
2. In organizations, researchers, librarians, and other information workers can profit from it.
3. It can serve individuals as a personal web agent.

Acknowledgements

Thanks to Barbara Lutes for discussion on this work.

References

1. Ambite J., Ashish N., Barish G., Knoblock C., Minton S., Modi P., Muslea I., Philpot A., Tejada S.: Ariadne: A system for constructing mediators for Internet sources. In: Proc. of the ACM SIGMOD. Seattle, WA, USA, June 1-4, 1998, pp. 561-563.
2. Adali S., Bufi C., Temtanapat Y.: Integrated Search Engine. Proceedings of the IEEE Knowledge and Data Engineering Exchange Workshop, KDEX97, Newport Beach, CA, USA. November 4, 1997, pp. 140-147.
3. Dreilinger D., Howe A.: Experiences with Selecting Search Engines Using Meta-Search. ACM Transactions on Information Systems. 15 (3), pp 195-222, July 1996.
4. Gravano L., Chang K., Garcia-Molina H., Paepcke A.: STARTS: Stanford Protocol Proposal for Internet Retrieval and Search. In Proceedings of the 1997 ACM SIGMOD, Tucson, AZ, USA. May 11-15, 1997, pp. 207-218.
5. Levy A., Rajaraman A., Ordille J.: Querying Heterogeneneous Information Sources Using Source Descriptions. Proceedings of the 22nd VLDB Conference Bombay, India, pp. 251-262, September 3-6, 1996.
6. Schmitt B., Schmidt A.: METALICA: An Enhanced Meta Search Engine for Literature Catalogs. In Proceedings of the 2nd Asian Digital Library Conference (ADL'99), Taipeh, Taiwan, November 8-9, 1999. pp 142-160.
7. Vassalos V., Papakonstantinou Y.: Describing and Using Query Capabilities of Heterogeneous Sources. In: Proc. 23rd VLDB Conf. Athens, Greece, Aug. 25-29, 1997, pp. 256-265.
8. Wiederhold G.: Mediators in the architecture of future information systems. IEEE Computer, 25(3), March 1992, pp. 38-49.
9. Yerneni R., Li C., Garcia-Molina H., Ullman J.. Computing Capabilities of Mediators. In: Proc. SIGMOD, Philadelphia, PA, USA, May 31-June 3, 1999, pp. 443-454.

A Semantic Approach to Integrating XML and Structured Data Sources

Peter McBrien[1] and Alexandra Poulovassilis[2]

[1] Department of Computing, Imperial College,
180 Queen's Gate, London SW7 2BZ
pjm@doc.ic.ac.uk

[2] Department of Computer Science, Birkbeck College, University of London,
Malet Street, London WC1E 7HX
ap@dcs.bbk.ac.uk

Abstract. XML is fast becoming the standard for information exchange on the WWW. As such, information expressed in XML will need to be integrated with existing information systems, which are mostly based on structured data models such as relational, object-oriented or object/relational data models. This paper shows how our previous framework for integrating heterogeneous structured data sources can also be used for integrating XML data sources with each other and/or with other structured data sources. Our framework allows constructs from multiple modelling languages to co-exist within the same intermediate schema, and allows automatic translation of data, queries and updates between semantically equivalent or overlapping heterogenous schemas.

1 Introduction

The presentation-oriented nature of HTML has been widely recognised as being an impediment to building efficient search engines and query languages for information available on the WWW. XML is a more effective means of describing the semantic content of WWW documents, with presentational information being specified using a separate language such as XSL. However, XML is still to some extent presentation-oriented, since the structuring of the data for a particular application is often made to suit the later presentation of that data. This is due to XML's hierarchical nature, with tags being nested inside each other, requiring document designers to make an *a priori* choice as to the ordering of the nesting. Whilst languages such as **DTD**s and **XML Schema** serve to structure XML documents, it is still the case that an essentially hierarchical model is being used.

For example, consider an application where a bank has records of customers, their accounts, and the bank site where the account is held. A site may have accounts belong to different customers and a customer may have accounts at several sites. Fig. 2(a) shows how one might list details of customers (called cust) in XML, detailing under each customer the account (called acc) and the site of the account. Alternatively, Fig. 2(b) shows how the same information

K.R. Dittrich, A. Geppert, M.C. Norrie (Eds.): CAiSE 2001, LNCS 2068, pp. 330–345, 2001.
© Springer-Verlag Berlin Heidelberg 2001

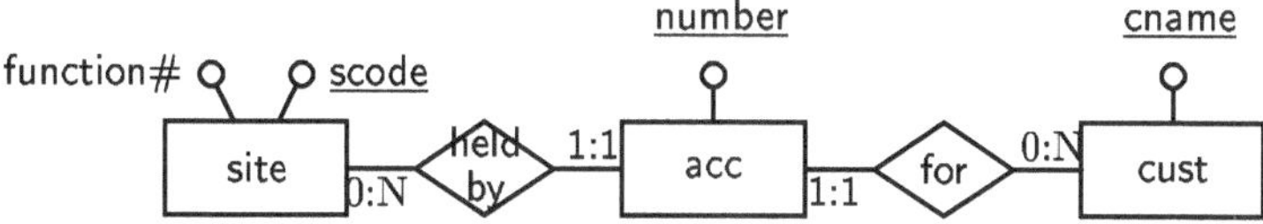

Fig. 1. ER schema of the Bank database

could be listed by site, with details of accounts, and the customer holding the account listed under each site.

The example appears to be that of a many-many relationship between customers and sites. Such choices of ordering as made in Fig. 2(a) and (b) do not arise in data models such as ER or UML, which are based on the classification of entities into types or classes, with relationships between them. For example, Fig. 1 shows an ER model for the same data as in Fig. 2(a) and (b). However, in XML the order of data can be significant and there may be semantic information embedded within XML documents which is assumed by applications but which is not deducible from the document itself. Looking at Fig. 2(a) for example, it might be the case that the first account listed for any customer is to be used for charging any banking costs to *e.g.* charges for customer Jones will be made to account 4411 and not to account 6676.

Apart from these variations in how the ER schema is navigated to produce a hierarchical data structure, there is also a choice as to whether to use XML elements or attributes. The examples given in Figures 2(a) and (b) use XML elements to represent data (as is usually the case in most of the literature) but for the 'leaf' nodes we could equally well use attributes. Fig. 2(c) takes this approach for the same data as shown in Fig. 2(a). XML (with DTDs) also supports a tuple-based representation of data as illustrated in Fig. 2(d) where duplication of data is avoided.

These variations in XML as to how the hierarchy is built, the possibility that ordering may or may not be significant, and the choice of using attributes or elements, has led us to investigate how a semantic data model can be used as the basis for integrating XML data sources with each other and/or with other structured data sources, rather than using XML itself. Note that this approach does not preclude the use of XML as the data transfer mechanism — it only precludes its use as the data modelling language.

The approach that we present in this paper extends our previous work on integrating structured data sources [16,12,11]. In this work, we have used as the common data model a low-level **hypergraph-based data model (HDM)**. We will see later in the paper that the separation in the HDM of data sets and constraints on data sets is useful for modelling XML data, since ordering of XML elements can be represented by extra node and edge information, leaving any other constraints on the data unchanged.

The remainder of the paper is structured as follows. In Section 2 we extend our previous work to show how XML can be represented in the HDM. This leads to the first contribution of the paper — providing a common underpinning for

```
⟨cust⟩
  ⟨cname⟩Jones⟨/cname⟩
  ⟨acc⟩
    ⟨number⟩4411⟨/number⟩
    ⟨site⟩
      ⟨scode⟩32⟨/scode⟩
    ⟨/site⟩
  ⟨/acc⟩
  ⟨acc⟩
    ⟨number⟩6976⟨/number⟩
    ⟨site⟩
      ⟨scode⟩56⟨/scode⟩
      ⟨function⟩Business⟨/function⟩
    ⟨/site⟩
  ⟨/acc⟩
⟨/cust⟩
⟨cust⟩
  ⟨cname⟩Frazer⟨/cname⟩
  ⟨acc⟩
    ⟨number⟩8331⟨/number⟩
    ⟨site⟩
      ⟨scode⟩32⟨/scode⟩
    ⟨/site⟩
  ⟨/acc⟩
⟨/cust⟩
```

(a) by customer, elements preferred

```
⟨site⟩
  ⟨scode⟩32⟨/scode⟩
  ⟨acc⟩
    ⟨number⟩4411⟨/number⟩
    ⟨cust⟩
      ⟨cname⟩Jones⟨/cname⟩
    ⟨/cust⟩
  ⟨/acc⟩
  ⟨acc⟩
    ⟨number⟩8331⟨/number⟩
    ⟨cust⟩
      ⟨cname⟩Frazer⟨/cname⟩
    ⟨/cust⟩
  ⟨/acc⟩
⟨/site⟩
⟨site⟩
  ⟨scode⟩56⟨/scode⟩
  ⟨function⟩Business⟨/function⟩
  ⟨acc⟩
    ⟨number⟩6976⟨/number⟩
    ⟨cust⟩
      ⟨cname⟩Jones⟨/cname⟩
    ⟨/cust⟩
  ⟨/acc⟩
⟨/site⟩
```

(b) by site, elements preferred

```
⟨cust cname="Jones"⟩
  ⟨acc number="4411"⟩
    ⟨site scode="32"/⟩
  ⟨/acc⟩
  ⟨acc number="6976"⟩
    ⟨site scode="56"
        function="Business"/⟩
  ⟨/acc⟩
⟨/cust⟩
⟨cust cname="Frazer"⟩
  ⟨acc number="8331"⟩
    ⟨site scode="32"/⟩
  ⟨/acc⟩
⟨/cust⟩
```

(c) by customer, attributes preferred

```
⟨cust cid="c1" cname="Jones"/⟩
⟨cust cid="c2" cname="Frazer"/⟩
⟨acc aid="a1" number="4411"
    cid="c1" sid="s1"/⟩
⟨acc aid="a2" number="6976"
    cid="c1" sid="s2"/⟩
⟨acc aid="a3" number="8331"
    cid="c2" sid="s1"/⟩
⟨site sid="s1" scode="32"/⟩
⟨site sid="s2" scode="56"
    function="Business"/⟩
```

(d) tuple-based

Fig. 2. Example XML data files for the Bank database

structured data models and XML, and hence the possibility of transforming between them and integrating them. In Section 3 we discuss how XML documents can be transformed into an ER representation, and from there into each other, thus providing a framework in which XML data sources can be integrated and queried in conjunction with each other and with other structured data sources. This leads to the second contribution of the paper — providing a method for transforming between ER and XML representations which allows complete control over whether the various elements of the XML representation have set or list-based semantics. In Section 4 we discuss related work. We give our concluding remarks in Section 5

2 Representing XML in the HDM

In [12] we showed how the HDM can represent a number of higher-level, structured modelling languages such as the ER, relational and UML data models. We also showed how it is possible to transform the constructs of one modelling language into those of another during the process of integrating multiple heterogeneous schemas into a single global schema By extending our work to specify how XML can be represented in the HDM, we are adding XML to the set of modelling languages whose schemas can be transformed into each other and integrated using our framework.

Structured data models typically have a set-based semantics *i.e.* there is no ordering on the extents of the types and relationships comprising the database schema, and no duplicate occurrences. XML's semi-structured nature and the fact that it is presentation-oriented means that lists need to be representable in the HDM, as opposed to just sets which were sufficient for our previous work on transforming and integrating structured data models. In particular, lists are needed because the order in which elements appear within an XML document may be significant to applications and this information should not be lost when transforming and integrating XML documents.

Thus we extend the notions of nodes and edges in HDM schemas (which respectively correspond to types and relationships in higher-level modelling languages) so that the extent of a node or edge may be either a set or a list. For reasons of space we refer the reader to our earlier work [16,12,11] for a full definition of the HDM. Here we give a simplified summary of it together with the extensions needed for representing XML:

A **schema** in the HDM is a triple (Nodes,Edges,Constraints). Nodes and edges are identified by their **schemes**, delimited by double chevrons $\langle\langle \ldots \rangle\rangle$. The scheme of a node n consists of just the node itself, $\langle\langle n \rangle\rangle$. The scheme of an edge labelled l between nodes $n_1, \ldots, n_m$ is $\langle\langle l, n_1, \ldots, n_m \rangle\rangle$. Edges can also link other edges, so more generally the scheme of an edge is $\langle\langle l, s_1, \ldots, s_m \rangle\rangle$ for some schemes $s_1, \ldots, s_m$. Two primitive transformations are available for adding a node or an edge to an HDM schema, S, to yield a new schema:

addNode(s, q, i, c)
addEdge(s, q, i, c)

Here, s is the scheme of the node or edge being added and q is a query on S which defines the extent of s in terms of the extents of the existing schema constructs (so adding s does not change the information content of S)[1]. i is one of set or list, indicating the collection type of the extent of s, and c is a boolean condition on instances of S which must hold for the transformation to be applicable for that particular instance. Optionally, list may take an argument which determines the ordering of instances of s.

Often the argument c will simply be true, indicating that the transformation applies for all instances of S, and in our previous work i has always been set. In [11] we allowed q to take the special value void, meaning that s can not be derived from the other constructs of S. This is needed when a transformation pathway is being set up between non-equivalent schemas *e.g.* between a component schema and a global schema. For convenience, we use addNode(s,q,i) as a shorthand for addNode(s,q,i,true), addNode(s,q) for addNode(s,q,set,true), and expandNode(s) for addNode(s,void,set,true). We use similar abbreviations for adding edges.

There are also two primitive transformations for deleting a node or an edge from an HDM schema S, delNode(s, q, i, c) and delEdge(s, q, i, c), where s is the scheme of the node or edge being deleted and q is a query which defines how the extent of s can be reconstructed from the extents of the remaining constructs (so deleting s does not change the information content of S), and i the collection type of s. c is again a boolean condition on instances of S which must hold for the transformation to be applicable for that particular instance. Similar shorthands as for the add transformations are used. There are similarly two primitive transformations for adding/deleting a constraint from an HDM schema, addConstraint($s, constraint$) and delConstraint($s, constraint$).

Supporting a list collection type does not actually require the HDM to be extended and the above primitive transformations on the HDM can be viewed as 'syntactic sugar' for the primitive transformations we used in previous work. In particular, lists can be supported by introducing a reserved node order, the extent of which is the set of natural numbers. For any scheme s whose extent needs to be viewed as a list, an extra unlabelled edge ⟨⟨_,s,order⟩⟩ is used whose instances assign an ordinality to each instance of s. The instances of ⟨⟨_,s,order⟩⟩ do not necessarily need to be numbered consecutively, and the ordering of instances of s can be relative to the ordering of instances other schemes.

2.1 Specifying XML in Terms of the HDM

Table 1 summarises how the methodology we described in [12] can be used to build an HDM representation of an XML document. In particular:

1. An XML element may exist by itself and is not dependent on the existence of any other information. Thus, each XML element e is what we term a

[1] We first developed our definitions of schema equivalence, schema subsumption and schema transformation in the context of an ER common data model [9,10] and then applied them to the more general setting of the HDM [16]. A comparison with other approaches to schema equivalence and schema transformation can be found in [10].

Table 1. Specifying XML constructs in the HDM

Higher Level Construct		Equivalent HDM Representation
Construct	**element (Elem)**	
Class	nodal, set	Node $\langle\langle \mathsf{xml}{:}e \rangle\rangle$
Scheme	$\langle\langle e \rangle\rangle$	
Construct	**attribute (Att)**	Node $\langle\langle \mathsf{xml}{:}e{:}a \rangle\rangle$
Class	nodal-linking, constraint, list	Edge $\langle\langle \_, \mathsf{xml}{:}e, \mathsf{xml}{:}e{:}a \rangle\rangle$ Links $\langle\langle \mathsf{xml}{:}e \rangle\rangle$
Scheme	$\langle\langle e, a \rangle\rangle$	Cons $\mathsf{makeCard}(\langle\langle \_, \mathsf{xml}{:}e, \mathsf{xml}{:}e{:}a \rangle\rangle, 0{:}1, 1{:}N)$
Construct	**nest-list (List)**	Edge $\langle\langle \_, \mathsf{xml}{:}e, \mathsf{xml}{:}e_s \rangle\rangle, \langle\langle \_, \langle\langle \_, \mathsf{xml}{:}e, \mathsf{xml}{:}e_s \rangle\rangle, \mathsf{order} \rangle\rangle$
Class	linking, constraint, list	Links $\langle\langle \mathsf{xml}{:}e \rangle\rangle, \langle\langle \mathsf{xml}{:}e_s \rangle\rangle$
Scheme	$\langle\langle e, e_s \rangle\rangle$	Cons $\mathsf{makeCard}(\langle\langle \_, \langle\langle \_, \mathsf{xml}{:}e, \mathsf{xml}{:}e_s \rangle\rangle, \mathsf{order} \rangle\rangle, 1{:}1, 0{:}N)$
Construct	**nest-set (Set)**	Edge $\langle\langle \_, \mathsf{xml}{:}e, \mathsf{xml}{:}e_s \rangle\rangle$
Class	linking, set	Links $\langle\langle \mathsf{xml}{:}e \rangle\rangle, \langle\langle \mathsf{xml}{:}e_s \rangle\rangle$
Scheme	$\langle\langle e, e_s \rangle\rangle$	

nodal construct [12] and is represented by a node $\langle\langle \mathsf{xml}{:}e \rangle\rangle$ in the HDM[2]. Each instance of e in an XML document corresponds to an instance of the HDM node $\langle\langle \mathsf{xml}{:}e \rangle\rangle$.

2. An XML attribute a of an XML element e may only exist in the context of e, and hence a is what we term a **nodal-linking** construct. It is represented by a node $\mathsf{xml}{:}e{:}a$ in the HDM, together with an associated unlabelled edge $\langle\langle \_,\mathsf{xml}{:}e,\mathsf{xml}{:}e{:}a \rangle\rangle$ connecting the HDM node representing the attribute a to the HDM node representing the element e. A constraint states that each instance of the attribute is related to at least one instance of the element (we use a shorthand notation for expressing cardinality constraints using the function makeCard).
3. XML allows any number of elements $e_1, \ldots, e_n$ to be nested within an element e. This nesting of $e_1, \ldots, e_n$ is represented by a set of edges $\langle\langle \_,\mathsf{xml}{:}e, \mathsf{xml}{:}e_1 \rangle\rangle, \ldots, \langle\langle \_,\mathsf{xml}{:}e,\mathsf{xml}{:}e_n \rangle\rangle$. Each such edge is an individual **linking** construct, with scheme $\langle\langle \_,\mathsf{xml}{:}e,\mathsf{xml}{:}e_s \rangle\rangle$.
   Each such edge may have list or set semantics. For list semantics, there is an extra edge between the edge $\langle\langle \_,\mathsf{xml}{:}e,\mathsf{xml}{:}e_s \rangle\rangle$ and the node order, and a cardinality constraint which states that each instance of $\langle\langle \_,\mathsf{xml}{:}e,\mathsf{xml}{:}e_s \rangle\rangle$ is related to precisely one instance of order.
4. XML allows plain text to be placed within a pair of element tags. If a DTD is present, then this text is denoted as PCDATA. We thus assume there is an HDM node called pcdata whose extent consists of plain text instances. An element e can then be associated with a fragment of plain text by means of an edge $\langle\langle \_,\mathsf{e},\mathsf{pcdata} \rangle\rangle$.

[2] Because it is possible to have present within the same HDM schema constructs from schemas expressed in different higher-level modelling notations, higher-level constructs are distinguished at the HDM level by adding a prefix to their name. This prefix is xml for XML constructs and er for ER constructs.

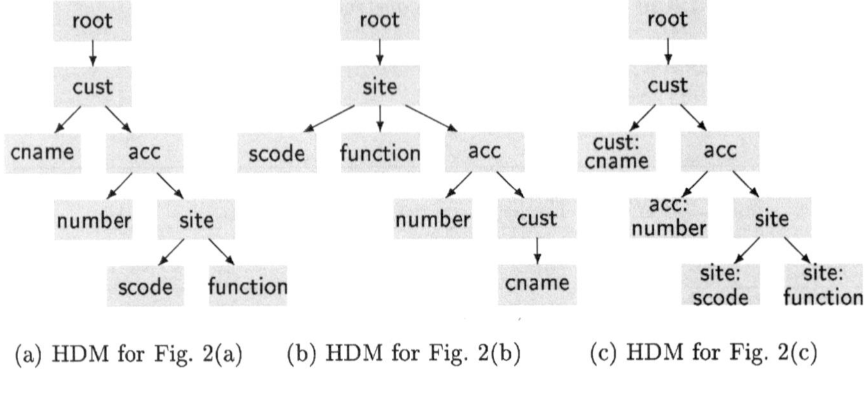

(a) HDM for Fig. 2(a) (b) HDM for Fig. 2(b) (c) HDM for Fig. 2(c)

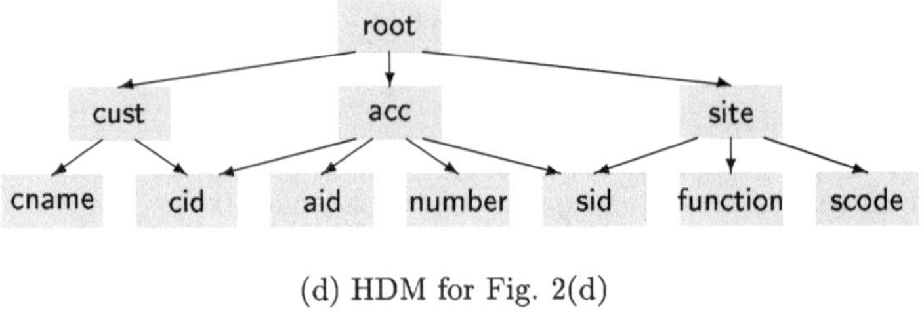

(d) HDM for Fig. 2(d)

Fig. 3. HDM schemas for the XML Documents (a)-(d)

To illustrate this representation of XML documents in the HDM, we show in Fig. 3 the HDM schemas corresponding to the XML documents of Fig. 2 (we have not shown the constraints and the links that each edge has to the order node). In the HDM schema (d), the common child node cid between site and cust, and aid between site and acc, arise if we assume the presence of a DTD in XML document (d) with ID attributes cid and aid on cust and acc, and corresponding IDREF attributes on site.

In general, elements within an XML document may be repeated, with identical attributes and nested elements within them. Hence elements and attributes are uniquely identified by their position within the document. In representing an XML document as an instance of an HDM schema, we thus assume that all nodes are assigned unique identifiers which are generated in some way from their position within the document. For example, the XML elements cust, cname and acc of Fig. 2(a) can be represented by the following instance of the schemes ⟨⟨root⟩⟩, ⟨⟨cust⟩⟩, ⟨⟨cname⟩⟩, ⟨⟨pcdata⟩⟩, and ⟨⟨acc⟩⟩ of the HDM schema in Fig. 3(a), where r1, c1, c2, cn1, cn2, a1, a2, a3 are unique identifiers:

⟨⟨root⟩⟩={r1}
⟨⟨cust⟩⟩={c1,c2}
⟨⟨cname⟩⟩={cn1, cn2}
⟨⟨pcdata⟩⟩={Jones, Frazer}
⟨⟨acc⟩⟩={a1,a2,a3}

```
⟨site⟩
  ⟨scode⟩32⟨/scode⟩
  ⟨acc⟩
    ⟨number⟩4411⟨/number⟩
    ⟨cust⟩
      ⟨cname⟩Jones⟨/cname⟩
    ⟨/cust⟩
  ⟨/acc⟩
⟨/site⟩
```

```
⟨site⟩
  ⟨scode⟩32⟨/scode⟩
  ⟨acc⟩
    ⟨number⟩8331⟨/number⟩
    ⟨cust⟩
      ⟨cname⟩Frazer⟨/cname⟩
    ⟨/cust⟩
  ⟨/acc⟩
⟨/site⟩
```

```
⟨site⟩
  ⟨scode⟩56⟨/scode⟩
  ⟨function⟩Business⟨/function⟩
  ⟨acc⟩
    ⟨number⟩6976⟨/number⟩
    ⟨cust⟩
      ⟨cname⟩Jones⟨/cname⟩
    ⟨/cust⟩
  ⟨/acc⟩
⟨/site⟩
```

Fig. 4. Transformed XML

The nesting relationships between the XML elements cust, cname and acc are represented by the schemes ⟪_,root,cust⟫, ⟪_,⟪_,root,cust⟫,order⟫, ⟪_,cust,cname⟫, ⟪_, ⟪_,cust, cname⟫,order⟫, ⟪_,cname,pcdata⟫, ⟪_, ⟪_,cname,pcdata⟫,order⟫, ⟪_, cust, acc⟫, and ⟪_, ⟪_, cust,acc⟫,order⟫. The instances of these schemes are shown below. Note that the one root r1 of the XML document contains two ordered customers c1 and c2, that c1 contains, in order, one cname cn1 and two accounts a1 and a2, and that c2 contains, in order, one cname cn2 and one account a3:

⟪_,root,cust⟫= {⟨r1,c1⟩, ⟨r1,c2⟩}
⟪_,⟪_,root,cust⟫,order⟫={⟨⟨r1,c1⟩,1⟩, ⟨⟨r1,c2⟩,2⟩}
⟪_,cust,cname⟫={⟨c1,cn1⟩, ⟨c2,cn2⟩}
⟪_,⟪_,cust,cname⟫,order⟫={⟨⟨c1,cn1⟩,1⟩, ⟨⟨c2,cn2⟩,1⟩}
⟪_,cname,pcdata⟫={⟨cn1,Jones⟩, ⟨cn2,Frazer⟩}
⟪_,⟪_,cname,pcdata⟫,order⟫={⟨⟨cn1,Jones⟩,1⟩, ⟨⟨cn2,Frazer⟩,1⟩}
⟪_,cust,acc⟫={⟨c1,a1⟩, ⟨c1,a2⟩, ⟨c2,a3⟩}
⟪_,⟪_,cust,acc⟫,order⟫={⟨⟨c1,a1⟩,2⟩, ⟨⟨c1,a2⟩,3⟩, ⟨⟨c2,a3⟩,2⟩}

An HDM representation which assumed set-based semantics for element containment would simply omit the schemes linking edges to ⟪order⟫.

2.2 Primitive Transformations on XML

In [12] we describe how, once the constructs of some higher-level modelling language have been defined in terms of the HDM constructs, this definition can be used to *automatically derive* the necessary set of primitive transformations on schemas expressed in that language. Thus, from the specification of XML given in Table 1 and described in the previous section, the primitive transformations shown in Table 2 can be automatically derived. The left-hand column of this table gives the names and arguments of the XML transformations and the right-hand column gives their implementation as sequences of the primitive transformations on the underlying HDM representation.

To illustrate the use of these primitive transformations on XML, we give below a sequence of primitive transformations that transform the document in

Table 2. Derived transformations on XML

Transformation on XML	Equivalent Transformation on HDM
$\text{renameElem}_{\text{xml}}(e,e')$	$\text{renameNode}(\langle\langle \text{xml}:e\rangle\rangle,\langle\langle \text{xml}:e'\rangle\rangle)$
$\text{addElem}_{\text{xml}}(e,q)$	$\text{addNode}(\langle\langle \text{xml}:e\rangle\rangle,q)$
$\text{delElem}_{\text{xml}}(e,q)$	$\text{delNode}(\langle\langle \text{xml}:e\rangle\rangle,q)$
$\text{renameAtt}_{\text{xml}}(a,a')$	$\text{renameNode}(\langle\langle \text{xml}:e:a\rangle\rangle,\langle\langle \text{xml}:e:a'\rangle\rangle)$
$\text{addAtt}_{\text{xml}}(e,a,q_{att},q_{assoc})$	$\text{addNode}(\langle\langle \text{xml}:e:a\rangle\rangle,q_{att})$; $\text{addEdge}(\langle\langle \_,\text{xml}:e,\text{xml}:e:a\rangle\rangle,q_{assoc})$; $\text{addConstraint}(x \in \langle\langle \text{xml}:e:a\rangle\rangle \rightarrow \langle \_, x\rangle \in \langle\langle \_, \text{xml}:e, \text{xml}:e:a\rangle\rangle)$
$\text{delAtt}_{\text{xml}}(e,a,q_{att},q_{assoc})$	$\text{delConstraint}(x \in \langle\langle \text{xml}:e:a\rangle\rangle \rightarrow \langle \_, x\rangle \in \langle\langle \_, \text{xml}:e, \text{xml}:e:a\rangle\rangle;)$ $\text{delEdge}(\langle\langle \_,\text{xml}:e,\text{xml}:e:a\rangle\rangle,q_{assoc})$; $\text{delNode}(\langle\langle \text{xml}:e:a\rangle\rangle,q_{att})$
$\text{addList}_{\text{xml}}(\langle\langle e, e_s\rangle\rangle,q,p)$	$\text{addEdge}(\langle\langle \_,\text{xml}:e,\text{xml}:e_s\rangle\rangle,q,\text{list}(p))$
$\text{delList}_{\text{xml}}(\langle\langle e, e_s\rangle\rangle,q,p)$	$\text{delEdge}(\langle\langle \_,\text{xml}:e,\text{xml}:e_s\rangle\rangle,q,\text{list}(p))$
$\text{addSet}_{\text{xml}}(\langle\langle e, e_s\rangle\rangle,q)$	$\text{addEdge}(\langle\langle \_,\text{xml}:e,\text{xml}:e_s\rangle\rangle,q)$
$\text{delSet}_{\text{xml}}(\langle\langle e, e_s\rangle\rangle,q)$	$\text{delEdge}(\langle\langle \_,\text{xml}:e,\text{xml}:e_s\rangle\rangle,q)$

Fig. 2(a) to that in Fig. 4. The HDM schema representation of the former is shown in Fig. 3(a) and of the latter in Fig. 3(b).

$$
\begin{array}{l}
\text{addList}_{\text{xml}}(\langle\langle \text{root,site}\rangle\rangle,\{\langle \text{r1}, x\rangle \mid \langle x\rangle \in \langle\langle \text{site}\rangle\rangle\},\ \text{after}(\langle\langle \text{root,cust}\rangle\rangle))\\
\text{addList}_{\text{xml}}(\langle\langle \text{site,acc}\rangle\rangle,\{\langle x, y\rangle \mid \langle y, x\rangle \in \langle\langle \text{acc,site}\rangle\rangle\},\ \text{after}(\langle\langle \text{site,function}\rangle\rangle))\\
\text{addList}_{\text{xml}}(\langle\langle \text{acc,cust}\rangle\rangle,\{\langle x, y\rangle \mid \langle y, x\rangle \in \langle\langle \text{cust,acc}\rangle\rangle\},\ \text{after}(\langle\langle \text{acc,number}\rangle\rangle))\\
\text{delList}_{\text{xml}}(\langle\langle \text{acc,site}\rangle\rangle,\{\langle x, y\rangle \mid \langle y, x\rangle \in \langle\langle \text{site, acc}\rangle\rangle\},\ \text{after}(\langle\langle \text{acc,cust}\rangle\rangle))\\
\text{delList}_{\text{xml}}(\langle\langle \text{cust,acc}\rangle\rangle,\{\langle x, y\rangle \mid \langle y, x\rangle \in \langle\langle \text{acc, cust}\rangle\rangle\},\ \text{after}(\langle\langle \text{cust,cname}\rangle\rangle))\\
\text{delList}_{\text{xml}}(\langle\langle \text{root,cust}\rangle\rangle,\{\langle \text{r1}, x\rangle \mid \langle x\rangle \in \langle\langle \text{cust}\rangle\rangle\},\ \text{before}(\langle\langle \text{root,site}\rangle\rangle))
\end{array}
$$

Notice that the above transformation consists of a 'growing phase' where new constructs are added to the XML model, followed by a 'shrinking phase' where the constructs now rendered redundant are removed. This is a general characteristic of schema transformations expressed within our framework.

The availability of a transformation pathway from one schema to another allows queries expressed on one schema to be automatically translated onto the other. For example, the following query on Fig. 3(a) finds the names of customers with accounts at site 32[3]:

$$
\begin{array}{l}
\{n \mid \langle r, c\rangle \in \langle\langle \text{root, cust}\rangle\rangle\ \wedge \langle c, n\rangle \in \langle\langle \text{cust, name}\rangle\rangle\ \wedge \langle c, a\rangle \in \langle\langle \text{cust, acc}\rangle\rangle\ \wedge\\
\quad \langle a, s\rangle \in \langle\langle \text{acc, site}\rangle\rangle\ \wedge \langle s, sc\rangle \in \langle\langle \text{site, scode}\rangle\rangle\ \wedge sc = 32\}
\end{array}
$$

By substituting deleted constructs appearing in this query by their restoring query *i.e.* by the 3rd argument of the $\text{delList}_{\text{xml}}()$ transformations above, it is possible to translate the query into the following equivalent query on Fig. 3(b) (see [11] for a general discussion of query/update/data translation in our framework):

[3] We do not consider here the issue of translating between different query languages, and assume a 'neutral' intermediate query notation, namely set comprehensions, into which queries submitted to local or global schemas can be translated as a first step.

$\{n \mid \langle r,c\rangle \in \{\langle \mathsf{r1},x\rangle \mid \langle x\rangle \in \langle\langle \mathsf{cust}\rangle\rangle\} \wedge \langle c,n\rangle \in \langle\langle \mathsf{cust}, \mathsf{name}\rangle\rangle \wedge$
$\quad \langle c,a\rangle \in \{\langle x,y\rangle \mid \langle y,x\rangle \in \langle\langle \mathsf{acc}, \mathsf{cust}\rangle\rangle\} \wedge$
$\quad \langle a,s\rangle \in \{\langle x,y\rangle \mid \langle y,x\rangle \in \langle\langle \mathsf{site}, \mathsf{acc}\rangle\rangle\} \wedge \langle s,sc\rangle \in \langle\langle \mathsf{site}, \mathsf{scode}\rangle\rangle \wedge sc = 32\}$

Any sequence of primitive transformations $t_1;\ldots;t_n$ is automatically reversible by the sequence $t_n^{-1};\ldots;t_1^{-1}$, where the inverse of an add transformation is a del transformation with the same arguments, and vice versa. Thus, the reverse transformation from Fig. 3(b) to Fig. 3(a) can be automatically derived to be:

$\mathsf{addList}_{\mathsf{xml}}(\langle\langle \mathsf{root,cust}\rangle\rangle,\{\langle \mathsf{r1},x\rangle \mid \langle x\rangle \in \langle\langle \mathsf{cust}\rangle\rangle\}, \mathsf{before}(\langle\langle \mathsf{root,site}\rangle\rangle))$
$\mathsf{addList}_{\mathsf{xml}}(\langle\langle \mathsf{cust,acc}\rangle\rangle,\{\langle x,y\rangle \mid \langle y,x\rangle \in \langle\langle \mathsf{acc,cust}\rangle\rangle\}, \mathsf{after}(\langle\langle \mathsf{cust,cname}\rangle\rangle))$
$\mathsf{addList}_{\mathsf{xml}}(\langle\langle \mathsf{acc,site}\rangle\rangle,\{\langle x,y\rangle \mid \langle y,x\rangle \in \langle\langle \mathsf{site,acc}\rangle\rangle\}, \mathsf{after}(\langle\langle \mathsf{acc,cust}\rangle\rangle))$
$\mathsf{delList}_{\mathsf{xml}}(\langle\langle \mathsf{acc,cust}\rangle\rangle,\{\langle x,y\rangle \mid \langle y,x\rangle \in \langle\langle \mathsf{cust,acc}\rangle\rangle\}, \mathsf{after}(\langle\langle \mathsf{acc,number}\rangle\rangle))$
$\mathsf{delList}_{\mathsf{xml}}(\langle\langle \mathsf{site,acc}\rangle\rangle,\{\langle x,y\rangle \mid \langle y,x\rangle \in \langle\langle \mathsf{acc,site}\rangle\rangle\}, \mathsf{after}(\langle\langle \mathsf{site,function}\rangle\rangle))$
$\mathsf{delList}_{\mathsf{xml}}(\langle\langle \mathsf{root,site}\rangle\rangle,\{\langle \mathsf{r1},x\rangle \mid \langle x\rangle \in \langle\langle \mathsf{site}\rangle\rangle\}, \mathsf{after}(\langle\langle \mathsf{root,cust}\rangle\rangle))$

This reverse transformation can now be used to translate queries on Fig. 3(b) to queries on Fig. 3(a).

We finally observe the similarity of Fig. 4 to the document in Fig. 2(b), which in fact has the same schema, Fig. 3(b). It is not possible to capture the non-duplication of site 32 in Fig. 2(b) using the above XML-to-XML transformation. It is however possible to transform Fig. 2(a) to Fig. 2(b) going via the ER model of Fig. 1, and we discuss how in Section 3 below.

3 Transforming between ER and XML

In [12] we showed how ER schemas can be represented in the HDM, and we refer the reader to that paper for details. Here we recall that ER entity classes are nodal constructs represented by HDM nodes, ER relationships are linking constructs represented by an HDM edge and a cardinality constraint, and ER attributes are nodal-linking constructs represented by a node, an edge linking this node to the node representing the attribute's entity class, and a cardinality constraint. The primitive transformations on ER schemas consist of the operations add, del, expand, contract and rename on the constructs Entity, Attribute or Relationship.

3.1 Automated Generation of a Canonical ER Schema from XML

Using the sets of primitive transformations on XML and ER schemas, an XML document or set of documents can be automatically transformed into a 'canonical' ER schema by applying the rules given below. For ease of reading, we have specified these rules at the level of the HDM, and the primitive transformations on the HDM, rather than at the higher level of the XML and ER constructs. As with the example in Section 2.2 above, the transformation consists of a 'growing phase' where new ER constructs are added to the schema, followed by a 'shrinking phase' where the XML constructs now rendered redundant are removed:

1. Each node representing an XML element (*i.e.* is named xml:e for some e and is not order or pcdata) generates an ER entity class of the same name e, with a key attribute also named e (this explicit key attribute is needed because there is no implicit notion of unique object identifiers in the ER model):

   addNode(⟨⟨er:e⟩⟩,⟨⟨xml:e⟩⟩)
   addNode(⟨⟨er:e:e⟩⟩,⟨⟨xml:e⟩⟩)
   addEdge(⟨⟨_,er:e,er:e:e⟩⟩, $\{\langle x,x\rangle \mid x \in$ ⟨⟨xml:e⟩⟩$\}$)
   addConstraint(makeCard(⟨⟨_,er:e,er:e:e⟩⟩,1:1,1:1))

   Note that for the purposes of this rule, the root of the document should also be considered as a node, so there will always be a root entity in the ER model. Each instance of the root entity will represent a distinct XML document.
2. Each node representing an XML attribute (*i.e.* is named xml:e:a for some e and a) generates an attribute of the ER entity class e[4]:

   addNode(⟨⟨er:e:a⟩⟩,⟨⟨xml:e:a⟩⟩)
   addEdge(⟨⟨_,er:e,er:e:a⟩⟩, ⟨⟨_,xml:e,xml:e:a⟩⟩)
   addConstraint(copyCard(⟨⟨_,xml:e,xml:e:a⟩⟩, ⟨⟨_,er:e,er:e:a⟩⟩))

3. Each node linked by an edge to ⟨⟨pcdata⟩⟩ has an attribute added called pcdata whose extent will consist of the instances of ⟨⟨pcdata⟩⟩ with which instances of the node are associated:

   addNode(⟨⟨er:e:pcdata⟩⟩, $\{x \mid \langle \_, x\rangle \in$ ⟨⟨_,xml:e,xml:e:pcdata⟩⟩$\}$)
   addEdge(⟨⟨_,er:e,er:e:pcdata⟩⟩, ⟨⟨_,xml:e,xml:e:pcdata⟩⟩)
   addConstraint(copyCard(⟨⟨_,xml:e,xml:e:pcdata⟩⟩, ⟨⟨_,er:e,er:e:pcdata⟩⟩))

4. Each nesting edge between two XML elements generates an edge between the two corresponding ER entity classes representing a relationship between them:

   addEdge(⟨⟨_,er:e,er:e_s⟩⟩, ⟨⟨_,xml:e,xml:e_s⟩⟩)

   For list-based nestings, each instance of the XML nesting edge will be linked to the order node. This information can be represented as an attribute order of the new ER relationship:

   addEdge(⟨⟨_,⟨⟨_,er:e,er:e_s⟩⟩,order⟩⟩, ⟨⟨_,⟨⟨_,xml:e,xml:e_s⟩⟩,order⟩⟩)

   Any constraint on the XML nesting edge should also be copied over onto the new ER relationship.
5. As a result of the above add transformations, all the XML constructs are rendered semantically redundant and can finally be progressively removed from the schema by applying a sequence of del transformations.

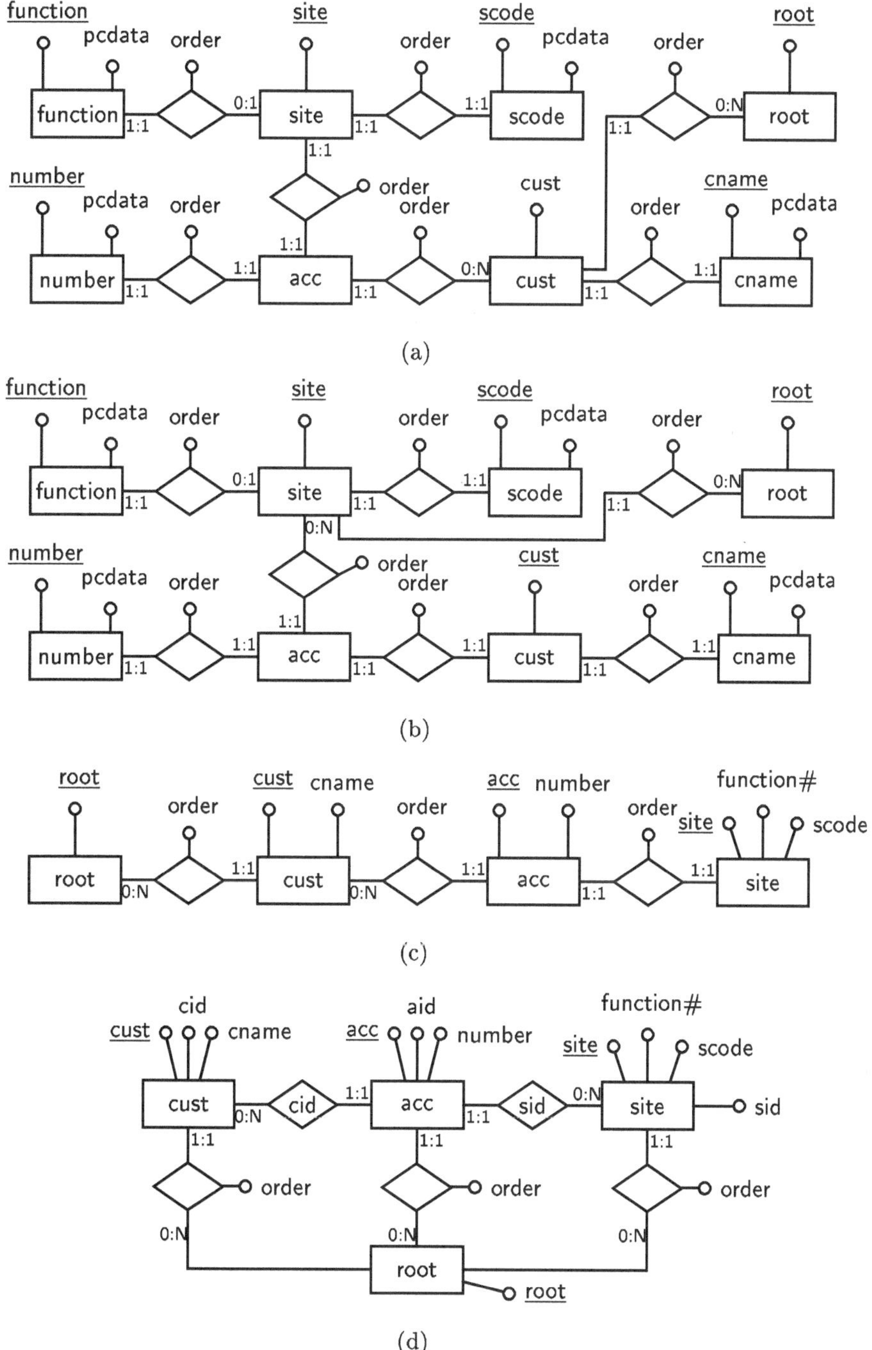

Fig. 5. Canonical ER representations of the XML documents (a)-(d)

The result of applying the above rules to the four HDM schemas representing the XML documents of Fig. 2 is shown in Fig. 5. We observe that each differs to a greater or lesser extent from the ER schema of Fig. 1 which we assumed was used to generate the original XML documents. We now study how to derive the original ER schema.

3.2 Manual Refinement of Canonical ER Schemas

Comparing Fig. 5(a) and (b) with Fig. 1, we notice the following characteristics of using XML to represent ER data, and hence possible refinements that may be made to the canonical ER schemas automatically generated by the method described in the previous subsection:

1. It is often the case that the ordering of elements in XML is not semantically significant, in which case the order attribute in the ER schema can be ignored. For example, if the order of site and number within acc is not significant in Fig. 2(a), then the order attribute in Fig. 5(a) can be removed from the relationship between acc and number, and from that between acc and site, as follows:

   $\mathsf{contractAttribute_{er}}(\langle\langle\langle\langle \_,\mathsf{acc},\mathsf{number}\rangle\rangle,\mathsf{order}\rangle\rangle)$
   $\mathsf{contractAttribute_{er}}(\langle\langle\langle\langle \_,\mathsf{acc},\mathsf{site}\rangle\rangle,\mathsf{order}\rangle\rangle)$

   In contrast if, as discussed in the introduction, the order of accounts within customers was significant, then the order attribute must remain on $\langle\langle \_,\mathsf{cust},\mathsf{acc}\rangle\rangle$.
2. Using XML elements to represent attributes in the original ER schema has resulted in entity classes being created in the new ER schema.
   For both Fig. 5(a) and (b) we see that function, number, cname and scode all have this property. The reason that this has occurred is because XML documents contain additional information regarding the position of constructs within the document, which is reflected by the order information being stored in the corresponding ER edges and the generation of a unique identifier for each XML element. If this ordering information need not be preserved in the ER schema (*i.e.* rule (1) above has been applied), these entity classes can be transformed into attributes, using a standard pattern of transformations which we give below for the case of the scode entity class:

   $\mathsf{addAttribute_{er}}(\langle\langle \mathsf{site},\mathsf{scode}\rangle\rangle, \{\langle x,z\rangle \mid \langle x,y\rangle \in \langle\langle \_,\mathsf{site},\mathsf{scode}\rangle\rangle \wedge$
   $\quad \langle y,z\rangle \in \langle\langle \mathsf{scode},\mathsf{pcdata}\rangle\rangle\}, \mathsf{copyCard}(\langle\langle \_,\mathsf{site},\mathsf{scode}\rangle\rangle,\langle\langle \mathsf{site},\mathsf{scode}\rangle\rangle))$
   $\mathsf{contractAttribute_{er}}(\langle\langle \mathsf{scode},\mathsf{pcdata}\rangle\rangle)$
   $\mathsf{contractAttribute_{er}}(\langle\langle \mathsf{scode},\mathsf{scode}\rangle\rangle)$
   $\mathsf{contractRelationship_{er}}(\langle\langle \_,\mathsf{site},\mathsf{scode}\rangle\rangle)$
   $\mathsf{contractEntity_{er}}(\langle\langle \mathsf{scode}\rangle\rangle)$

[4] Here, the function copyCard() translates the constraint on an XML edge to the same constraint on an ER edge.

The result of applying these transformations to function, number, cname and scode entities in Fig. 5(a) results in the ER schema shown in Fig. 5(c) *i.e.* the ER schema generated from the XML document of Fig. 2(c).

3. The order in which elements are nested within each other in an XML document effects the cardinality of the relationships between the corresponding ER entity classes.
   For example, in Fig. 5(a) sites are repeated for each customer, and thus each site entity is associated with only one acc (there will be two sites with scode 32 in our example, each with one account). This can be corrected by changing the key of site from the automatically generated, position-related, key to be scode instead. This involves renaming the site entity to site', creating a new site entity with the new scode key, copying the attributes and relationships across from site' to site, and finally removing site':

$\mathsf{renameEntity_{er}}(\langle\langle\mathsf{site}\rangle\rangle, \langle\langle\mathsf{site'}\rangle\rangle)$
$\mathsf{addEntity_{er}}(\langle\langle\mathsf{site}\rangle\rangle, \{x \mid \langle \_, x\rangle \in \langle\langle\mathsf{site'}, \mathsf{scode}\rangle\rangle\})$
$\mathsf{addAttribute_{er}}(\langle\langle\mathsf{site,scode}\rangle\rangle, \{\langle x, x\rangle \mid \langle x\rangle \in \langle\langle\mathsf{site'}, \mathsf{scode}\rangle\rangle\}, 1:1, 1:1)$
$\mathsf{addAttribute_{er}}(\langle\langle\mathsf{site,function}\rangle\rangle,$
$\quad \{\langle x, y\rangle \mid \langle z, x\rangle \in \langle\langle\mathsf{site'}, \mathsf{scode}\rangle\rangle \wedge \langle z, y\rangle \in \langle\langle\mathsf{site'}, \mathsf{function}\rangle\rangle\},$
$\quad \mathsf{copyCard}(\langle\langle\mathsf{site',function}\rangle\rangle, \langle\langle\mathsf{site,function}\rangle\rangle))$
$\mathsf{addRelationship_{er}}(\langle\langle\_,\mathsf{site,acc}\rangle\rangle,$
$\quad \{\langle x, y\rangle \mid \langle z, x\rangle \in \langle\langle\mathsf{site'}, \mathsf{scode}\rangle\rangle \wedge \langle z, \_, y\rangle \in \langle\langle\_, \mathsf{site'}, \mathsf{acc}\rangle\rangle\},$
$\quad \mathsf{copyCard}(\langle\langle\_,\mathsf{site',acc}\rangle\rangle, \langle\langle\_,\mathsf{site,acc}\rangle\rangle))$
$\mathsf{contractAttribute_{er}}(\langle\langle\mathsf{site',function}\rangle\rangle)$
$\mathsf{contractAttribute_{er}}(\langle\langle\mathsf{site',scode}\rangle\rangle)$
$\mathsf{contractAttribute_{er}}(\langle\langle\mathsf{site',site}\rangle\rangle)$
$\mathsf{contractRelationship_{er}}(\langle\langle\_,\mathsf{site',acc}\rangle\rangle)$
$\mathsf{contractEntity_{er}}(\langle\langle\mathsf{site'}\rangle\rangle)$

As already discussed, Fig. 5(c) is an intermediate stage of the transformations applied to Fig. 5(a) and only changing the key of entity classes from the automatically generated position-related key to another attribute remains to be done.

Finally, the 'flat' representation of information in the XML document of Fig. 2(d) results in the ER schema of Fig. 5(d). This is the closest to Fig. 1, with both the same attributes present and the same cardinality constraints. If they are not required for applications, we can simply drop the id nodes using $\mathsf{contractAttribute_{er}}(\langle\langle\mathsf{site,id}\rangle\rangle)$. We can also apply the key transformation rules to change the key from the position-related key to another attribute.

In summary, we have shown in this section how four XML documents d_1 to d_4 can be automatically transformed into ER schemas er_1 to er_4 which may then be manually transformed into a single ER schema er. To transform an XML document d_i to another XML document d_j the forward transformation $d_i \rightarrow er_i \rightarrow er$ can be applied, followed by the reverse transformation $er \rightarrow er_j \rightarrow d_j$.

4 Related Work

There has been much work on translation (a) between XML and structured data models, and (b) between different XML formats. Considering first (a), a common approach is to use some XML query language to build a new XML document as a view of another XML document. In common with other approaches to representing XML or semi-structured data, *e.g.* [15,2,1,3,17,5], we use a graph-based data model which supports unique identifiers for XML elements. One difference with our approach is how ordering is represented. Our use of the order node in the HDM graph allows list semantics to be preserved with the data if desired. Alternatively, the links to order can be ignored for a set semantics. Generally, the provenance of the HDM as a common data model for structured data gives it a rather different flavour to models with an XML or semi-structured data provenance. For example, (1) there are constraints in an HDM schema, and (2) nodal, linking and nodal-linking constructs are used to represent XML documents as opposed to just nodes and edges.

Considering the issue of translating between structured data models and XML, this has also received considerable attention [4,7,8,18,1,14,3]. In contrast to this previous work, what we have proposed here is a general method for translating between structured and XML representations of data, not tied to any specific structured data model. Specifying XML in the HDM has allowed the set of primitive transformations on XML to be automatically derived in terms of sequences of primitive transformations on the HDM. Both high and low-level transformations are automatically reversible because they require the specification of a constructing/restoring query for add/del transformations. These two-way transformation pathways between schemas can be used to auomatically translate data, queries and updates in either direction. Work in schema translation and matching [1,14,3] can be utilised to enhance our framework by automatically or semi-automatically deriving the constructing/restoring query in add/del transformations where possible.

5 Concluding Remarks

We have developed our previous work which supports the integration of structured data sources to also handle XML documents. We have demonstrated how XML documents can be automatically transformed into ER schemas, and have discussed how to further transform such schemas so that they are semantically closer to the original source database schema that the XML documents may have been generated from. This restructuring is based on the application of well-understood schema transformation rules.

The process of restructuring the ER schema would be a more complex task for data sources where the information has never been held in a structured form *i.e.* 'really' semi-structured data. In [19] techniques are presented for discovering structured associations from such data, and we are studying how our framework can be adapted to use such techniques.

In a longer version of this paper [13] we show how to represent in the HDM the additional information available in an XML DTD, if present. We are also studying how the approach proposed in this paper may be adapted to use XML Schema definitions [6] in place of DTDs, which will enable some of the restructuring rules for ER models to be automatically inferred.

References

1. S. Abiteboul, S. Cluet, and T. Milo. Correspondence and translation for heterogeneous data. In *Proceedings of ICDT'97*, 1997.
2. S. Abiteboul, *et al.* The Lorel query language for semistructured data. *Journal on Digital Libraries*, 1(1), 1997.
3. C. Beeri and T. Milo. Schemas for integration and translation of structured and semi-structured data. In *Proceedings of ICDT'99*, 1999.
4. V. Christophides, S. Cluet, and J. Siméon. On wrapping query languages and efficient XML integration. *SIGMOD RECORD*, 29(2):141–152, 2000.
5. A. Deutsch, M. Fernández, D. Florescu, A. Levy, and D. Suciu. A query language for XML. In *Proceedings of 8th International World Wide Web Conference*, 1999.
6. D.C. Fallside. XML schema part 0: Primer; W3C working draft. Technical report, W3C, April 2000.
7. M. Fernández, W-C. Tan, and D. Suciu. SilkRoute: Trading between relations and XML. In *Proceedings of 9th International World Wide Web Conference*, 2000.
8. D. Florescu and D. Kossmann. Storing and querying XML data using an RDBMS. *Bulletin of the Technical Committee on Data Engineering*, 22(3):27–34, September 1999.
9. P.J. McBrien and A. Poulovassilis. A formal framework for ER schema transformation. In *Proceedings of ER'97*, volume 1331 of *LNCS*, pages 408–421, 1997.
10. P.J. McBrien and A. Poulovassilis. A formalisation of semantic schema integration. *Information Systems*, 23(5):307–334, 1998.
11. P.J. McBrien and A. Poulovassilis. Automatic migration and wrapping of database applications — a schema transformation approach. In *Proceedings of ER99*, volume 1728 of *LNCS*, pages 96–113. Springer-Verlag, 1999.
12. P.J. McBrien and A. Poulovassilis. A uniform approach to inter-model transformations. In *Advanced Information Systems Engineering, 11th International Conference CAiSE'99*, volume 1626 of *LNCS*, pages 333–348. Springer-Verlag, 1999.
13. P.J. McBrien and A. Poulovassilis. A semantic approach to integrating XML and structured data sources. Technical report, Birkbeck College and Imperial College, November 2000.
14. T. Milo and S. Zohar. Using schema matching to simplify heterogeneous data translation. In *Proceedings of VLDB'98*, 1998.
15. Y. Papakonstantinou, H. Garcia-Molina, and J. Widom. Object exchange across heterogeneous information sources. In *Proceedings of ICDE'95*, 1995.
16. A. Poulovassilis and P.J. McBrien. A general formal framework for schema transformation. *Data and Knowledge Engineering*, 28(1):47–71, 1998.
17. R.Goldman, J. McHugh, and J. Widom. From semistructured data to XML: Migrating the Lore data model and query language. In *Proceedings of WebDB*, 1999.
18. J. Shanmugasundaram, K. Tufte, G. He, C. Zhang, D. DeWitt, and J. Naughton. Relational databases for querying XML documents: Limitations and objectives. In *Proceedings of the 25th VLDB Conference*, pages 302–314, 1999.
19. K. Wang and H. Liu. Discovering structural association of semistructured data. *IEEE Transactions on Knowledge and Data Engineering*, 12(3):353–371, 2000.

XML-Based Integration of GIS and Heterogeneous Tourism Information

Franz Pühretmair and Wolfram Wöß

Institute for Applied Knowledge Processing (FAW)
Johannes Kepler University Linz, Austria
{fpuehretmair, wwoess}@faw.uni-linz.ac.at

Abstract. With the tremendous growth of the Web, a broad spectrum of tourism information is already distributed over various Web sites. To fulfill the tourists request for an extensive data collection it is inevitable to make accumulated data from different sources accessible. In a first step towards a comprehensive integration of tourism data the official Austrian destination information and booking system TIScover is extended with a flexible data interchange adapter which allows interchange of structured data with other tourism information systems.
Beside the problem of distributed data sources tourists are also confronted with differences concerning information presentation on various Web sites. To cope with this problem our approach extensively uses maps for data presentation and federation of multiple structured and semi-structured tourism information sources on the Web. For this purpose touristic maps are generated dynamically including data resulting from database queries. This concept allows a clear and meaningful representation of up-to-date tourism information embedded in a geographical context.

Key words: geographic information systems (GIS), tourism information systems (TIS), information integration, electronic data interchange (EDI), meta data, extensible markup language (XML), scalable vector graphics (SVG).

1 Introduction

The aim of the official Austrian tourism information system TIScover [1], [2], [3] is twofold: First, tourists should be supplied with comprehensive, accurate and up-to-date tourism information on countries, regions, villages and all destination facilities they offer like hotels, museums or other places worth seeing. Second, it aims to attract tourists to buy certain tourism products either offline or even more important to allow tourists to buy them online.

The functionality provided by TIScover can roughly be categorized into three different components, the *public Internet* component, the *Extranet* and the *Intranet*. The *public Internet* component comprises that functionality of the system that is accessible to the public, the most important modules are Atlas and Booking. The *Extranet* provided by TIScover allows authorized tourism information providers, regardless whether it is a small guesthouse or a large local tourist office to update and extend their tourism information and products directly. Finally, the *Intranet*

K.R. Dittrich, A. Geppert, M.C. Norrie (Eds.): CAiSE 2001, LNCS 2068, pp. 346–358, 2001.
© Springer-Verlag Berlin Heidelberg 2001

component of TIScover which is accessible at the system provider's side only allows to configure the whole system in various ways. For example, it is possible to extend the geographical hierarchy, to specify expiration dates for reports and to define the default language for all system components.

Currently, TIScover manages a database of about two gigabyte of data, more than 500.000 Web pages (composed of more than one million files) covering among others 2.000 towns and villages and over 40.000 accommodations. The system has to handle up to 21 million page views as well as up to 90.000 information and booking requests per month.

Although, these figures illustrate that TIScover manages a fairly huge amount of tourism information, it is of course far from being complete. To be able to satisfy also requests for certain information which is not part of the TIScover database, but already available at other information systems, the IST project XML-KM (eXtensible Markup Language-based Mediator for Knowledge Extraction and Brokering) [4], [5], [6] was intended to provide a proper technical basis.

Resulting of many years of project experience in the field of tourism information systems, we identify the following requirements having great impact on the quality of a tourism information system focusing the accessibility to the users (tourists):

- integration of geographical data with tourism data and
- integration of distributed data sources.

It is obvious, that in the actual situation to search for and to locate interesting tourism information distributed over various Web sites takes a great effort, is time consuming and burdens tourists. The situation gets even worse since Web sites usually differ very much especially concerning information presentation and information access. For a satisfying result the user needs a fast and straightforward access to tourism information with a reliable and clear data representation. Typically, if tourists search for objects they are interested in, they often have to deal with large result sets characterized by complex tabular lists where it is difficult to obtain an objective view. In the case of integrating tourism data with a geographic information system (GIS), maps offer a unique possibility to combine geographical data with plain tourism data at the visualization level. Maps are a proper medium to support the human perception in a convenient way and use people's inherent cognitive abilities to identify spatial patterns and provide visual assistance concerning geographic objects and their locations.

Today most graphics on the Web are still in raster format. The new XML-based graphic specification language scalable vector graphics (SVG) [7] allows the presentation of dynamic and interactive vector graphics in the Web. SVG offers substantial advantages for tourism information systems because maps change from static raster graphics to interactive graphical representations allowing the presentation of the most extensive information possible thus satisfying the demands of the users.

To fulfill the tourists request for an extensive data collection it is inevitable to make data from distributed data sources accessible. In a first step towards a comprehensive integration of tourism data TIScover is extended with a flexible data interchange adapter which allows interchange of structured data with other tourism information systems. Information about data interchange formats and data structures is transformed into a meta data structure represented by XML (extensible markup language) [8], [9], [15] document type definitions (DTDs). This concept allows a flexible administration and maintenance of several data interfaces to other tourism information systems.

The contribution of the presented approach is twofold: First, to provide an extensive data collection to a tourist, distributed and heterogeneous tourism information is accumulated and integrated. Thus, flexible data interchange mechanisms are introduced. Second, tourism information is combined with geographical information to achieve efficient visualization of tourism data. This purpose is enabled by the new SVG format, which allows to generate interactive tourist maps dynamically.

The paper is organized as follows: Section 2 illustrates the new possibilities yielding from the integration of tourism data with geographic information systems. The basic concepts of XML and meta data based tourism information integration are presented in Section 3 including an overview about the objectives of the XML-KM project. Section 4 concludes the paper by discussing further improvements and pointing to future work.

2 XML-Based Integration of GIS and Tourism Data

Geographic information systems (GIS) is a rapidly expanding field enabling the development of applications that manage and use geographic information in combination with other media. Among others, this technology offers great opportunities to develop modern tourism applications using maps to present information to the user, thus exploiting the two dimensional capabilities of human vision and present the information in a compact and easy to read way [10].

GIS technology has broadened our view of a map. In contrast to paper-based maps, they are now dynamic representations of geographic data with the ability to visualize data according to the user's demands. The integrated GIS system generates maps based on data returned from databases. As a consequence changes of the GIS data and tourism data are automatically considered and valid after the generation of a new map.

Actual geographic information systems work with two fundamentally different types of geographic models - the raster model and the vector model. With the raster model objects are represented as a matrix of cells in continuous space. A point is one cell, a line is a continuous concatenation of cells and an area is represented as a field of contiguous cells. Today most of existing Web-based information systems use raster graphics for the representation of maps. In this case a satisfying user interaction (e.g., zooming) is not possible. In contrast to raster maps vector based graphics use x, y coordinates to define the position of graphical objects. This allows graphic transformations, like zooming, at the client without losses of quality or makes textual search mechanisms available. In addition a dynamic generation of maps is possible, which enables the presentation of up-to-date information to tourists.

Until now, there have been several initiatives to establish a vector standard [11], e.g., SVF (simple vector format) [12], which was the first attempt for vector representation in the World Wide Web, or Flash [13] which is the most common vector format, but it has never been admitted to be an official standard. Furthermore, Microsoft specified his own proprietary vector format VML (vector markup language) [14], which was an appropriate attempt but limited to Microsoft platforms.

VML was one of the basics to define the more generalized and advanced vector format SVG.

The SVG grammar comprises, generalizes and advances the preceding attempts to build an open standard recommended and developed from the World Wide Web Consortium [7] to describe two-dimensional vector graphics in XML. SVG enables the integration of three types of graphic objects: vector graphic shapes, text and images. Graphical objects can be grouped, styled, transformed and combined with other SVG objects. SVG offers the features of embedded interactivity (vector zoom, move operation, , etc.), animation, embedded fonts, XML, CSS and supports scripting languages with access to the DOM (document object model) to obtain full HTML (hypertext markup language) compatibility.

Even though SVG is an upcoming standard, there already exist several SVG tools like editors, viewers, converters and generators. But until now Web browsers do not have an embedded support for SVG, which is promised for the next generation of Web browser. In the meantime Adobe's SVG plug-in [16] allows to process SVG-based graphics within Web browsers.

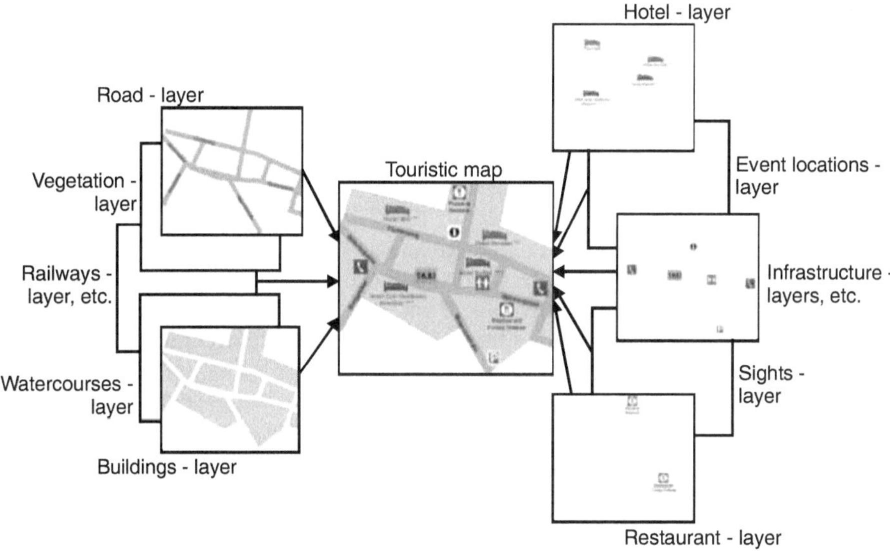

Fig. 1. Integration of themed layers.

The integration of geographical data, tourism information, textual data, images, and links are supported by SVG. Using vector graphics allows to interact, analyze, and to use screen-related functions, such as zooming and panning. It is easy to select features to be displayed, while ignoring unwanted data. Each kind of information within the map is located on a themed layer, which can be anything that contains similar features like roads, buildings, watercourses, hotels or sights. According to their needs users turn these layers on or off. To integrate GIS data and tourism data the traditional layer model (roads, buildings, vegetation, watercourses, railways, etc.) is enriched with additional tourism layers like hotels, restaurants, sights, event locations and further infrastructure layers (Figure 1).

SVG supports themed layers and the generation of user individual maps. In addition it is possible to integrate the following types of information and graphical symbols:

- object symbols representing the type of a touristic object,
- alphanumerical object descriptions (e.g., name and category of a hotel),
- colors, e.g., to visualize the availability of an hotel
- links to a homepage represented by a graphical object (e.g., a hotel's homepage).

In this approach tourism data stored in a tourism database still remains separated from GIS data stored in a GIS database. The XML-based integration of data takes place at the information visualization level. For the visualization of touristic objects it is necessary to add its geographical coordinates (latitude and longitude) which are then stored within the GIS database. The determination of the geographical coordinates of touristic objects is part of the administration expenses within the TIScover Extranet.

There are two different ways to determine object coordinates:

- *Manual positioning.* Authorized tourism information providers (e.g., hotel manager, tourist association) are responsible for the administration of geographical coordinates of those objects which are under their responsibility. The localization is done by marking the position of the object on the map. The corresponding geographic coordinates will be calculated and stored in the GIS database.
- *Automatic positioning.* Based on the postal address of touristic objects the geographic coordinates are calculated automatically [18].

Both approaches have advantages and disadvantages. The automatic approach will be faster, but it is less accurate and error prone because it is a prerequisite that the postal address of the touristic object is complete and that the address data within the GIS system matches to the data of the tourism system to enable accurate mapping. The manual approach is more precise because it is an WYSIWYG (what you see is what you get) approach since the administration as well as the map generation is based on the same map data. The problem of this alternative is a great administration effort.

After the geographic coordinates of touristic objects are determined, the objects can be visualized on touristic maps. Additionally, the geographic information system offers an advanced geographic search feature which allows to combine touristic search attributes like, object type, object name, category of hotels, etc. with geographic criterions like nearness, distance, location (city or province) or objects located inside a marked rectangular map region. Figure 2 describes the workflow of the data integration process.

The user starts a request, which is transmitted to the integrated GIS system. The GIS system submits the query to the spatial database to get the map data and to create a list of objects which are located in the queried area. Each object returned will be completed with tourism data like object name, URL of the object's homepage, category of the hotel, availability of hotel rooms, etc. Afterwards the transformation module converts the characteristics (layer extent, coordinate reference, shape, points, etc.) into SVG representations (polygon, line, path, etc.) and integrates the GIS data with the tourism data, identifies the layers, and defines the representation of the touristic objects (symbols, visualization of layers, object linking etc.). The data of the SVG map is provided in standard XML format and fulfills a DTD, which means that it is well-formed. Finally the touristic map is delivered to the client. On the client only queried layers are turned visible, but the users can turn layers on or off according to their needs.

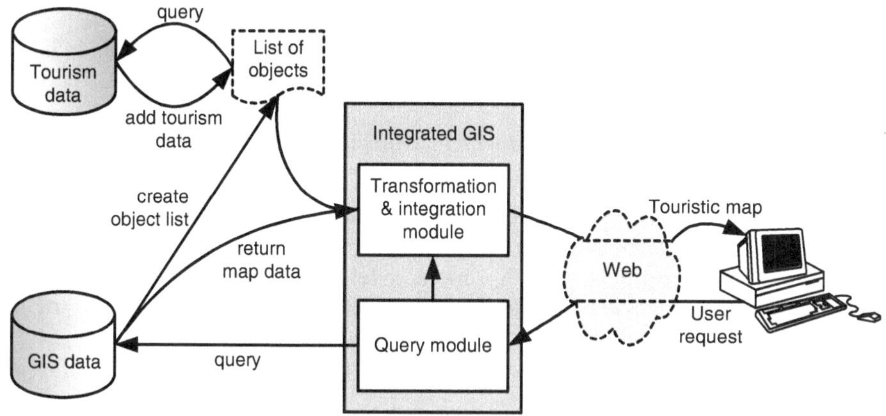

Fig. 2. GIS data integration workflow.

The results of the integration of GIS data and tourism data are touristic maps including the information where touristic objects are, how they can be reached, and which objects are located nearby. If the map representation of a touristic object is linked with the homepage of this object users acquire more detailed information about the object and / or in case of hotels features like online booking are provided.

A first prototype of the SVG based data integration and map representation is already implemented.

3. XML and Meta Data Based Information Integration

To fulfill the tourists request for an extensive data collection it is inevitable to make data from different sources accessible. In a first step towards a comprehensive integration of tourism data TIScover is extended with a flexible data interchange adapter which allows interchange of structured data with other tourism information systems (Figure 3).

Beside the main purpose of a global information integration this approach also focuses on data interchange (e.g., invoice data, order data, personal tourist data) between tourism information providers and other business partners without involving the TIScover system. For these data interchange requirements the existing functions of the TIScover Extranet are not sufficient or not adequate (Figure 4).

Especially the second data interchange scenario requires data interchange facilities which cope with the following demands:

- support of data interchange with a large number of business partners,
- high flexibility,

- lean administration and
- low costs.

In general, there are two contrary strategies to implement data interchange applications:

- Implementation of one data interface per EDI (electronic data interchange) communication partner. In the worst case n EDI partners require n different interfaces.
- Definition of a *common standard* which covers nearly all possible data interchange scenarios.

In the first case the implementation and maintenance of the whole set of interfaces is very time and cost intensive. In the second case a common standard results in a complex interface specification with a more or less large overhead. The question is: Why put up with the whole overhead if, for example, only a very simple data structure has to be transferred?

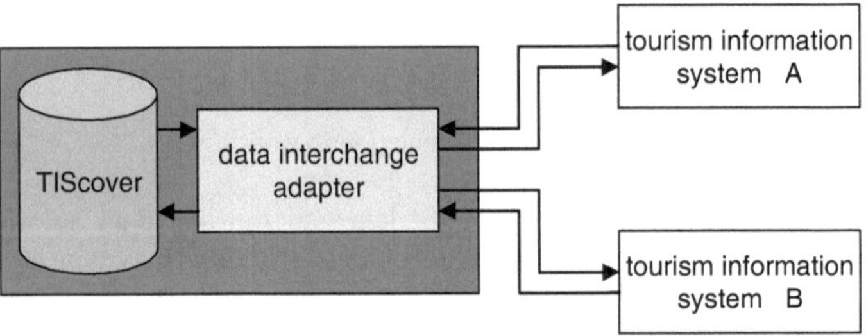

Fig. 3. Global information integration.

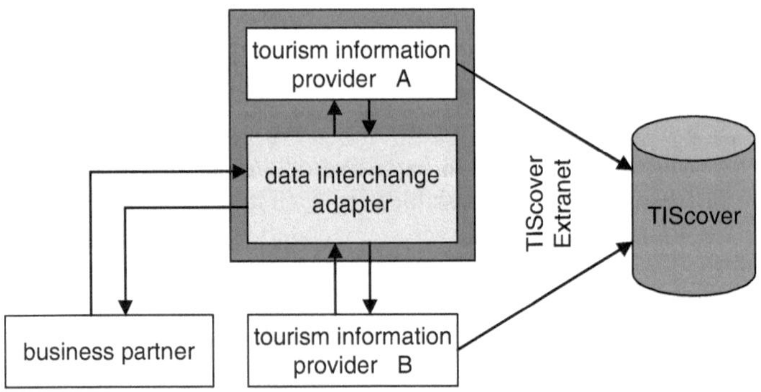

Fig. 4. Information interchange with other business partners.

The approach presented in this paper supports both strategies in order to combine the advantages of each of them. Knowledge about data structures and data formats is separated from the process of generation of destination files and electronic transmission. This separation achieves the purpose to make the data interchange application configurable. If any changes or extensions of the data interchange specification are necessary, it is sufficient to update the knowledge base. The implementation of the data interchange application remains unchanged. The required knowledge base is realized as a meta data structure.

3.1 Architecture of the Data Interchange Adapter

The general architecture of the data interchange adapter consists of three components (Figure 5):

- A *Knowledge base (meta data)* about data structures and data formats, represented by XML DTDs stored within a central database.
- The *mapping knowledge editor* is necessary for the administration of meta data information which is the representation of an individual data interface to an EDI communication partner.
- The *data interchange processor* in the first step receives input data which has to be transformed into the required destination format. In the next step the processor scans the meta data information corresponding to the required destination format. Based on this meta data the destination data file is generated dynamically. Data interchange is possible in both directions to and from a TIS application.

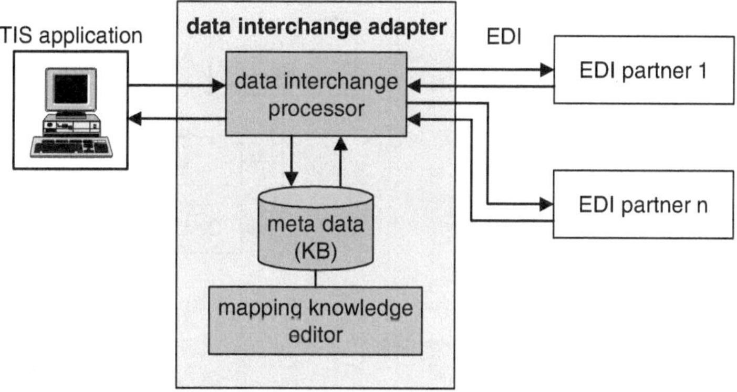

Fig. 5. General architecture of the data interchange adapter.

Beside several other key characteristics XML is a new and flexible concept for the specification of an EDI message. In contrast to HTML (hyperText markup language) which aims at the *presentation* of information XML focuses on *structuring* of information which is also the most important process of an EDI application. Hence, for example, it is possible to model existing EDIFACT message types [17] as XML messages.

Taking these considerations into account the latest implementation of the introduced data interchange adapter is based on XML DTDs (Figure 6). Within the XML based architecture source and destination files are in ASCII- (due to high compatibility) or XML-format. Meta data information for the specification of the transformation of a data source into its destination format is represented by XML DTDs.

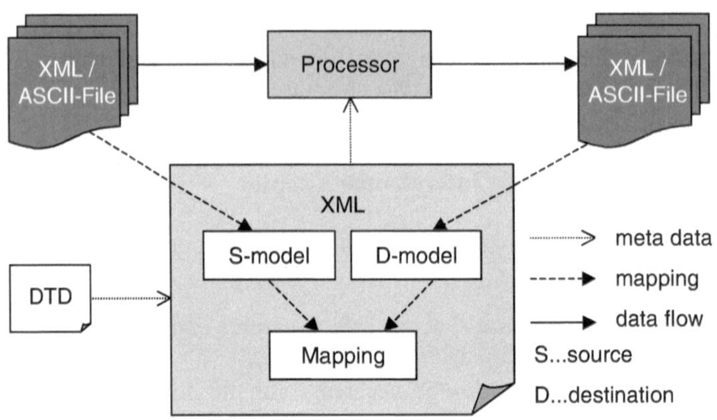

Fig. 6. XML based architecture.

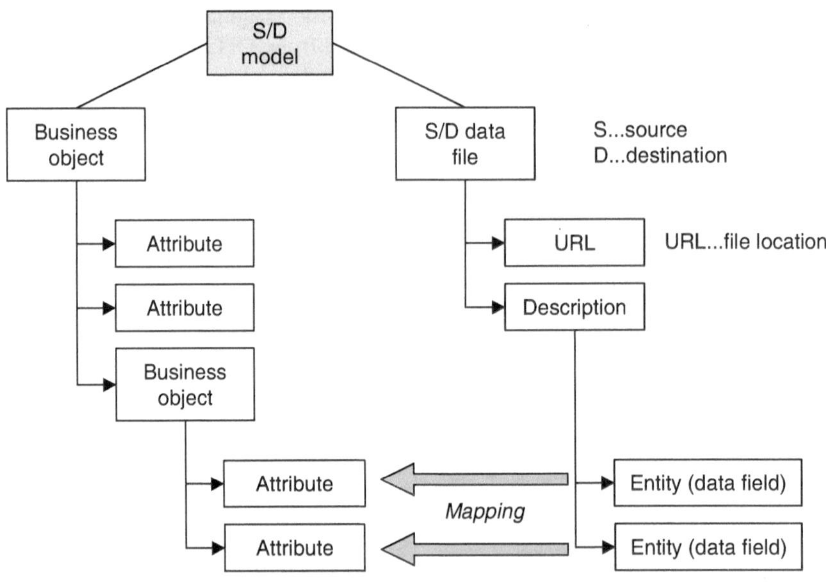

Fig. 7. Business object mapping.

A DTD also describes the mapping between the source model and the destination model each representing the data structures of the corresponding source/destination file. A model itself consists of business objects and references the corresponding source or destination file (Figure 7). Business objects are representations of the data structures within the source or the destination file. A business object is designed as a tree structure and therefore may itself consist of business objects and/or simple attributes (Figure 7). Each attribute of a source model knows its corresponding data field within the source file and each attribute of a destination model references the corresponding destination file, data field and data type.

The maintenance of the meta data information is done by using a mapping knowledge editor which allows to specify or update DTDs for source and destination models as well as the mapping between the represented data structures. Due to the central database system changes of the transformation specifications (meta data) are immediately valid for each interacting software system.

A further key aspect of the XML based architecture is the possibility to specify layout information of an XML document separated from the XML document itself. Hence, it is possible to generate different outputs of one XML document using different layout styles which are defined with XSL (XML stylesheet language) [8]. In this case it does not take great effort to provide different outputs of the same EDI process for different EDI partners.

3.2 The XML-KM Approach

The main advantages of XML based EDI are high flexibility and low implementation costs. In combination with XSL XML is important for structuring information *and* for its presentation.

A prototype of the XML and meta data based data interchange adapter is already finished. Towards a global and comprehensive integration of tourism information the introduced data interchange adapter is a first step, which makes flexible data interchange possible. Currently, XML is confined to syntactic exchange and thus can be regarded only as promising basis of semantically useful shared knowledge management [19], [20].

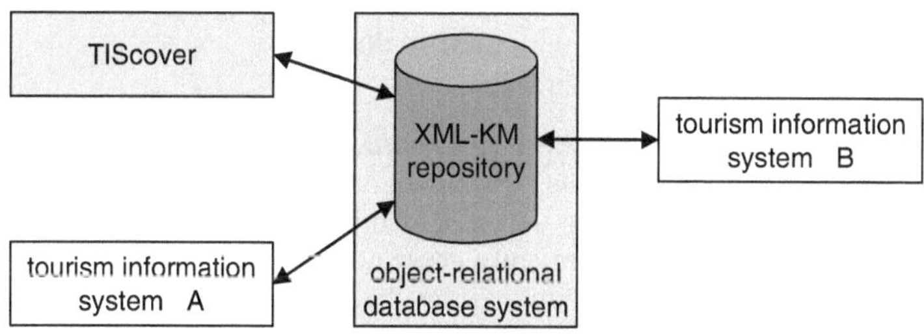

Fig. 8. Integration of tourism data in the XML-KM repository.

The main technical innovative points of the XML-KM project overcome some of these limitations. First, by adopting a semantic approach to the deployment of XML, combined with techniques to extract knowledge and to create conceptual views. Second, by developing concepts for the collection, fusion and dissemination of irregular, evolving information sources as opposed to the bilateral exchange. XML-KM extends XML DTDs with means to relate element-definitions and data types in ontologies using modeling means which will be harmonized with the efforts of the W3C working group on XML schemas [15]. This provides the basic framework to meaningfully collect information from diverse sources, integrate it into unified, domain-dependent schemas and to disseminate it according to the individual needs of diverse user groups or information systems.

Various data sources are integrated in the XML-KM repository (Figure 8), which provides facilities for querying integrated views of remote data and updating the repository. The XML-KM repository is built on top of an object-relational database system (Oracle 8).

The data to be integrated in the XML-KM repository can be structured or unstructured and can be managed by use of different technologies (databases, HTML pages, XML pages and text files).

Beside integration of tourism data in the XML-KM repository and integration of geographical data with tourism data the touristic part of the XML-KM project also focuses

- location based services in WAP (wireless application protocol) and
- dissemination of user obtained tourism newsletters on predefined schedules.

4 Conclusions

Due to the tremendous growth of the Web to search for and to locate interesting information distributed over various Web sites takes a great effort and is time consuming. This is especially true for tourism information systems. To fulfill the tourists request for an extensive data collection it is inevitable to make accumulated data from different sources accessible.

As discussed in Section 1 the following requirements have great impact on the quality of a tourism information system focusing the accessibility to the users:

- integration of geographical data with tourism data and
- integration of distributed data sources.

Actual tourism information systems are still characterized by a lack of integrated GIS data. The approach described in this paper integrates tourism and GIS data by using SVG the XML-based standard for vector and mixed raster-vector graphics for the generation of touristic maps. Dynamically generated vector maps based on SVG offer a wide spectrum of powerful functionalities like high performance zooming and printing as well as a flexible and open representation of interactive touristic maps in combination with the possibility to integrate other XML-related technologies. This kind of touristic maps gives tourist information providers the opportunity to present their tourism information to potential tourists in a clear, fast and powerful way. The flexibility of the presented concept and its implied technologies significantly improve the integration of tourism and GIS data to build dynamic interactive touristic maps.

Beside the integration of GIS data federation of tourism information sources is a primary goal. Thus, flexible and powerful data interchange mechanisms are required. TIScover is extended with a data interchange adapter which allows interchange of structured data with other tourism information systems. In contrast to many existing systems the presented approach separates knowledge about data structures and data formats from the process of generation of destination files and electronic transmission. This knowledge is transformed into a meta data structure represented by XML DTDs. The main advantage of this concept is that if changes of the data interchange specification to other tourism information systems are necessary, it is sufficient to update the corresponding meta data information within the XML DTDs. The implementation of the data interchange adapter remains unchanged.

Further work will be done primarily in the course of the IST project XML-KM. The goal of XML-KM which is strongly based on XML is to improve the data integration process in order to be able to collect and disseminate *knowledge* instead of just data. Through a rule-based XML-wrapper, information from corporate databases, HTML pages and office applications will be collected in the XML-KM repository. Through XML-based query tools, users will be able to receive personalized information in an appropriate format on various devices including computers, mobile phones based on WAP services and faxes.

References

1. Pröll, B., Retschitzegger, W., Wagner, R., Ebner, A.: Beyond Traditional Tourism Information Systems – TIScover. Journal of Information Technology in Tourism (ITT), Vol. 1, Inaugural Volume, Cognizant Corp., USA (1998)
2. Pröll, B., Retschitzegger, W., Wagner, R: TIScover – A Tourism Information System Based on Extranet and Intranet Technology. Proc. of the 4th Americas Conference on Information Systems (AIS 1998), Baltimore, Maryland (1998)
3. TIS Innsbruck, FAW Hagenberg: Homepage of TIScover, http://www.tiscover.com (2000)
4. Ebner, A., Haller, M., Plankensteiner, K., Pröll, B., Pühretmair, F., Starzacher, P., Tjoa, A. M.: TIS Business Workflow Specification (D5-1.0), XML-KM (IST 12030) (June 2000)
5. Bezares, J-L., Gardarin, G., Huck, G., Laude, H., Munoz, J-M., Pühretmair, F.: General Architecture Specification (D11-1.4), XML-KM (IST 12030) (April 2000)
6. The European Comission: Homepage of the IST-1999-12030 project: XML-based Mediator for Knowledge Extraction and Brokering, http://www.cordis.lu/ist/projects/99-12030.htm (2000)
7. W3C - The World Wide Web Consortium: Scalable Vector Graphics (SVG) 1.0 Specification, Candidate Recommendation, http://www.w3c.org/TR/SVG/ (2000)
8. Behme, H., Mintert, S.: XML in der Praxis. Professionelles Web-Publishing mit der Extensiblen Markup Language. Addison-Wesley Publishing, BRD (1998)
9. Bradley, N.: The XML companion. Addison-Wesley Publishing, Great Britain (1998)
10. Christodoulakis, S., Anastasiadis, M., Margazas, T., Moumoutzis, N., Kontogiannis, P., Terezakis, G., Tsinaraki, C.: A Modular Approach to Support GIS functionality in Tourism Applications, Proc. of the Int. Conf. on Information and Communication Technologies in Tourism (ENTER'98), pp. 63-72, Springer Verlag, Istanbul (1998)
11. Neumann, A., Winter, A.: Vector-based Web Cartography: Enabler SVG, Carto.net - Cartographers on the net, http://www.carto.net/papers/svg/index_e.html (2000)
12. Softsource: SVF (Simple Vector Format), http://www.softsource.com/svf/ (1996)
13. Macromedia: Macromedia Flash - Create animated vector-based Web sites, http://www.macromedia.com/software/flash/ (2000)
14. W3C - The World Wide Web Consortium: Vector Markup Language (VML), http://www.w3.org/TR/NOTE-VML.html (1998)
15. The World Wide Web Consortium (W3C). http://www.w3.org/XML/ (2000)
16. Adobe Systems Incorporated: Adobe SVG Viewer 2.0 (beta), http://www.adobe.com/svg/viewer/install/main.html (2000)
17. Schmoll, Thomas: Handelsverkehr elektronisch, weltweit: Nachrichtenaustausch mit EDI/EDIFACT. Markt & Technik Verlag, München (1994)
18. O'Neill, W., Harper, E.: Linear Location Translation within GIS. Proceedings of the ESRI User Conference (1997)

19. Huck, G., Fankhauser, P., Aberer, K., Neuhold, E.: Jedi: Extracting and Synthesizing Information from the Web. Proceedings of the COOPIS 1998 Conference, New York, IEEE Computer Society Press (1998)
20. Petrou, C., Hadjiefthymiades, S., Martakos, D.: An XML-based, 3-tier Scheme for Integrating Heterogeneous Information Sources to the WWW. Proceedings of the 10th International Workshop on Database and Expert Systems Applications (DEXA '99), IEEE Computer Society (1999)

Objects Control for Software Configuration Management

Jacky Estublier
LSR, Grenoble University
220 Rue de la Chimie, BP 53
38041 Grenoble 9 France
Jacky.Estublier@imag.fr

Abstract. A major requirement in Software Engineering is to reduce the time to market. This requirement along with a demand for product sophistication and better quality has led to larger teams which in turn dramatically increases the pressure for more concurrent work in a distributed context.

This paper, based on our experience in Software Configuration Management for large software systems, shows why object management in such a context requires specific facilities for the consistent management of objects in multiple copies, different locations and formats, accessed and changed simultaneously by many engineers.

We present the solutions we have developed with our partner Dassault Systèmes, for the definition and enforcement of consistent concurrent engineering work, including a number of measures showing that scalability and efficiency are really tough issues.

We argue that the scalability and efficiency constraints found in SCM can only be met by a new architecture of SCM systems and by the development of a middleware layer that should be common to all SCM tools, and also usable by other applications sharing the same concerns.

Keywords: Software configuration management, Version control, concurrent engineering, distribution, architecture.

1 Introduction

The strongest driving force in industry is the reduction of the time to market. In software engineering, the life cycle (the delay between two releases of a software product), has tended to decrease from 18 months to 4 months. Simultaneously, the size and complexity of the developed product substantially increase, as well as the required quality for these products.

Coping with these conflicting characteristics requires an increase in the number of persons involved in the teams, and an increase in the degree of concurrency. Today, one of the major CAE issues is concurrent engineering control. As will be shown, concurrency engineering control requires different characteristics from the underlying infrastructure, not available today.

K.R. Dittrich, A. Geppert, M.C. Norrie (Eds.): CAiSE 2001, LNCS 2068, pp. 359-373, 2001.
© Springer-Verlag Berlin Heidelberg 2001

The work presented here has been carried out in the area of software engineering, and more specifically in Software Configuration Management (SCM), but we believe that most, if not all, of our findings apply to all of CAE.

We will exemplify both the problem and the solution we propose with the experience we gained in addressing configuration management issues with our industrial partner Dassault Systèmes (DS). DS is the world wide leader in CAD/CAM/PDMII with its main software product CATIA. The fact that CATIA has over 4 millions lines and that it is developed simultaneously by 800 engineers with a 4 month life cycle, makes DS one of the world's major software producers. The numbers provided in this paper are measurements made on the real DS software.

Chapter 3 presents the issues, along with some illustrating measures, and shows that current technology does not satisfy our requirements. Chapter 4 presents our layered architecture; chapters 5, 6, 7 present each layer: Basic workspace manager, Synchronization manager and Concurrent Engineering manager

2 Object Management in Concurrent Engineering

The experience shows that we are a long way from the time when an engineer "owned" all the required objects, for the whole duration of an activity. Instead, at DS, each software engineer has direct access to all the needed objects; but due to the high degree of concurrency, each object is used by 50 to 100 different engineers simultaneously. Of course, changed objects must be kept private for the work duration. If N engineers change the same file concurrently, that file will have N+1 different values (also called cooperative versions). At DS, N = 3 in average[1] ; values for N greater than 6 are not uncommon, with maximum around 30.

A high degree of concurrency requires two or more persons to access AND change the same piece of information in a concurrent way. The database community addressed and solved that issue long ago, using the transaction concept. Unfortunately, the typical database transaction lasts a fraction of a second and involves few objects, while in design, the time span of a change is the task that leads to that change i.e. from a few hours to a few weeks or more; and the scope of a change can be huge (thousands of objects). As a consequence, the database serializability approach (changes are performed in sequence) cannot be used here. Concurrent engineering requires engineers to work and change different copies of the objects of interest.

As engineering is predominantly human-driven, the intellectual atom of change is the task (or activity), i.e. a significant body of changes/creations leading to a new "consistent" state of the product. In the meantime, the state of the object is "inconsistent", and the engineer requires his/ her job to be isolated from the other changes. Each engineer needs private copies of the objects. Conversely, the engineer knows that

---

[1] Averages are between 1 and 2 for kernel files, and around 3 for application modules

other changes are concurrently performed on the same objects, and asks for these changes to be incorporated when relevant.

In CAE, activities are performed using tools. These tools have been designed to work on "objects" (often files) in a predefined format and organization. Depending on the activity, the same object may need to be found in a different format and location.

Apparently files not to be changed can be shared, at least on a LAN; experience shows this is not realistic for three reasons:

- Efficiency. At DS, with 1500 machines, average sharing of 30 and high performance demand (compilations), only local copies of all files can provide enough efficiency.
- Name and directory. The same file can be located under different names and directories depending on the platform (NT or Unix) or product version (restructuring).
- Internal format. Some files need to be translated from a format to another depending on the tool and activity which use it.

These reasons means that each file, even when not changed, has to be physically copied to the machine and workspace where it is used. At DS, at any point in time, each atomic object (e.g. a source file) has between 10 and 100 physical copies; in each case with potential read and write privileges.

Altogether the DS software amounts close to one million files, each averaging 30 copies (thus about 30 millions files) not necessarily all identical. In addition complexity, reliability (source code cannot be lost) and efficiency are really critical issues. Object management in such a context raises many difficult issues.

Altogether, CAE requires objects to be:

- in multiples copies, locations and formats,
- resynchronised when found relevant, with
- high efficiency, reliability and availability.

Providing a solution to the first point can be simple; the challenge is to provide also ways to solve the last two points.

The issue for an SCM or a SE platform designer is to find a technology upon which the environment can be built. In practice we need database for storing the information and middleware to handle distribution.

3 Architecture

All SCMs to date are built following the same architecture: a central database contains all the objects and their versions, using the same schema (definition). All workspaces (a file tree in a file system) share the same database. The structure and content of a (complex) object in a WS is always the same (see Fig 1)

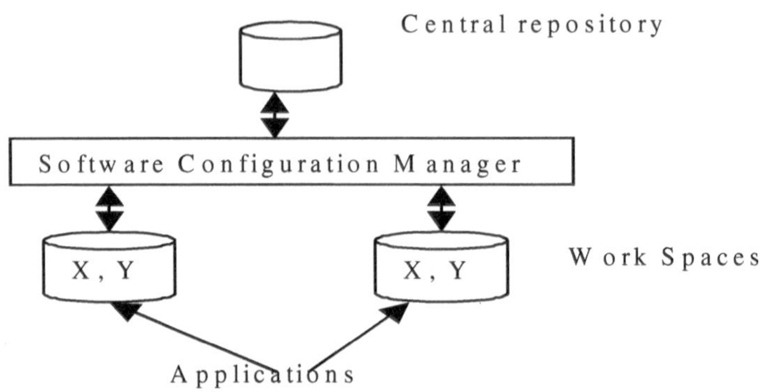

Fig. 1. Usual SCM Architecture

The first lesson learned in managing large software projects is that no central database can afford the needed efficiency and availability, even in the context of a LAN. For exemple, DS has an average of 2000 files per workspace. Synchronization operations involve on average a few hundred files i.e. a few MegaByte of information to transfer, merge, compress, format and so on. Consistency requires such heavy operations to be performed in transaction, but such transactions last several minutes during which objects are locked, and the server busy.

With an average of about 800 workspaces at any point in time, and such heavy transactions occurring on average 4 times a day per workspace (and most of them executed at about the same time i.e. on arriving at and leaving work), no DB can afford the required efficiency. Moreover, the measurements we took have shown that, even using a 100MB LAN, a transaction takes twice as long if the database and the WS are on different machines. Finally, availability requires that the whole company will not be stopped if a disk, a machine or the network crashes.

A major objective of this work was to find a truly distributed architecture, in which no repository is the main one, no one owns objects; each only contains object values. The satisfaction of the requirements also implies that we should be able to use different kind of repositories simultaneously (file systems and databases), with different schemas and on different kind of platforms (heterogeneity).

Availability, flexibility and efficiency all push us to consider as atomic (and often on the same machine): a WS, its manager and its associated repository. The architecture (Figure) we propose has as its basic element a workspace manager, organized in three layers. The first one is the basic Work Space manager (shaded). It deals with the problem of providing to each user (and his/her tools) the needed objects at the current location and in the right format. The second layer provides basic functions for transferring / synchronizing objects between any two workspaces. The third layer provides concepts for the consistency of concurrent activities, for structuring the workspaces, and for defining and enforcing cooperative work policies which satisfy some consistency requirements (see Fig 2). A fourth layer, not presented here, deals with general process support; see [1][5][6][19][20].

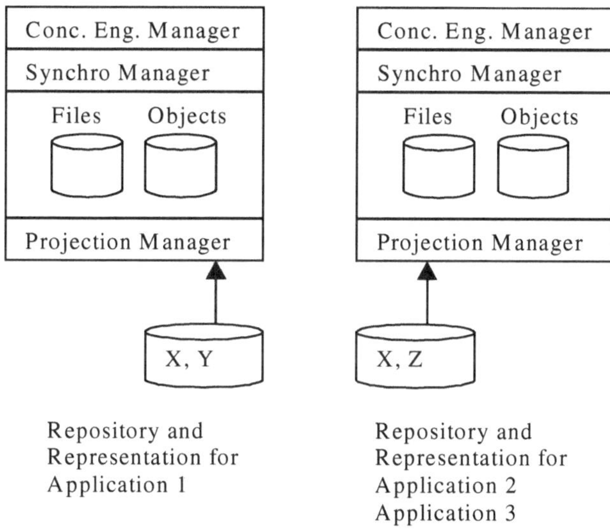

Fig. 2. Proposed Architecture

An object is an instance of a class in which the content and the behavior of each instance is defined. An object contains attributes. Attribute domains can be literals (i.e. strings, integers, file), or another object type; in the latter case we say the attribute is a composition attribute, and the object a complex or composed object.

This usual definition makes the identity: 1 object == 1 instance. This definition was extended by the concept of revision: an instance containing the snapshot of an object at a given point in time[2]. This is not sufficent; in our case, at each point in time, an object may have many copies (different instances) that may have different values simultaneously, and even different definitions. We need to refine these definitions.

A Workspace (WS) is a sub-set of a public repository[3] in which (part of) the objects of interest are stored.

A working object value is an object instance found in a WS. A object may simultaneously have different working values; one in each WS which contains it; but a WS can have one working value for a given object at the most.

---

[2] The concept of revision is more complex; many aspects are not discussed here. Some attributes in a revision are immutable (like files), others are common to a set of revision, others are mutable; still other attributes trigger the creation of a new revision on the attempt to change them and so on. A few more details are presented later

[3] A repository is a system which can store persistent information. It can be a file system or any kind of DB. It is public if access is performed by tools and users using the native repository functions without any need to be wrapped. It does not prohibit the control of these accesses, nor prevent undesired access, as long as this is done in a transparent way.

We call representation the format of an object instance at a given location. An object may have many different representations. A revision is a snapshot of a working value, in the WS manager specific representation, or a copy of another revision.

Revisions represent snapshots of an object's value both at a different point in time and in different places (workspaces).

An abstract object, or simply an object, is the sum of all its instances i.e. all its revisions plus its working values.

As defined in [10], three levels of versioning apply to each object: Historical versioning which is represented by a sequence of revisions in a WS (often called a branch). Cooperative versioning represented by the different branches of an object (one by WS) and logical versioning, represented by variants which are different objects sharing some logical or historical properties.

This paper discusses the different issues raised by the management of cooperatives versions of an object.

4 Basic WS Manager

The fundamental goal of the basic WS manager is to provide the objects in the representation required by the tools working in this WS, irrespective of the other possible representations, and let applications change the object in that specific format, (apparently) irrespectively of the changes made by other applications on other working values of that object.

A WS manager manages two areas. The first one is the public WS i.e. a part of a public repository containing the working values; the other one is the private WS area containing revisions. The private WS area is intended to contain the information required for the management of the public WS.

A WS type is defined by:

- Object type definitions, for objects allowed to be contained in this WS.
- Repository type (Unix FS, NTFS, Oracle, etc).
- Projection functions. They define the relationship between the working value and revisions of an object

For each type of object, the following projection functions must be provided:

- Projection: A function defining the mapping from a revision to a working value. In general it is a partial function; some attributes may not be mapped, others can be added. Example: attribute souceCode of object foo of type c, is mapped to file foo.c; attribute owner is mapped as the owner of the file, attribute protection is added.
- Reverse projection: A function defining the mapping from a working value to a revision. If the public repository data model is weaker than the private

one, this function requires conventions or additional information. Example: File foo.c is attribute sourceCode of object foo. But how are we to know that foo is a component of object X?.

- Change mapper: A function defining the mapping from a working value change to a revision change. If the public repository data model is weaker than the private one, this function, in theory, cannot exist. This function usually requires conventions and heuristics or changes to be performed through a specific interface. Example: If file foo.c is renamed or moved into another directory, what does that mean for object foo?

This contrasts with all approaches we know, including SCM systems, where the representation is unique. Existing SCM systems support a single WS type where these functions are predefined. Either there is no object model at all, thus the private model is the public repository model, or conventions are simple enough to compute the functions trivially. For example name identity (object foo of type c maps to file foo.c); a single composition relationship mapped into the relationship between directory and file (e.g. if foo.c is under directory X then object foo is a component of object X). The direct consequence is that (1) the object models are poor, and (2) WS type is unique[11].

In our system, WS types are formally defined, and projection functions are part of the type definition of each object type. For example, WS types proposed by SCM vendors can be defined easily which allows any product managed under "any" current SCM system to be integrated. It also provides for linking the work done on a product under an SCM system with the work done on the same product, under another representation, by another SCM system. This way we aim to address the virtual enterprise problem.

This approach is in complete opposition with current work in SCM (as well as our own previous work [9][11]) on at least the following:

- There is no longer any central repository. Experience has shown that no centralized approach can scale to very large projects (like the Dassault Systèmes one).
- There is no generic WS manager, because there is no generic projection function, and because efficiency requires the WS manager to be tailored to benefit from current company conventions.
- Basic WS managers are indeed low-cost simple SCM systems. They can be compatible with current habits and simple revision control tools like RCS or CVS.

Basic WS managers can be simple, but unlike RCS or CVS, our solution is scalable toward high functionality levels, huge software products and very large distributed teams. This is the topic of the next chapters.

5 Synchronization Manager

The synchronization manager defines how an object instance can be transferred and "synchronized" between two workspaces. Concurrent changes imply there is a way to reconcile different values of an attribute. We call that function the merge function. If we denote A1,... Ai the different values of attribute A, and Ai = Ci(A0) the value of A after change Ci is performed on A0, then a merge function M for A is such that M(Ai, Aj) = Ci(Cj(A0)) = Cj(Ci(A0)) = M(Aj, Ai).

This means the result of the merge is the same as if changes Ci and Cj were performed in sequence on A0, irrespective of the order. If an exact merge function existed for each attribute, concurrent engineering would always lead to consistent results! Unfortunately, for a given attribute, such a merge function either (1) exists, (2) is an approximation, or (3) does not exist at all. In our example, the components attribute has an exact merge, sourceFile merge is an approximate function, owner has no merge[15][16].

This approach contrasts with current work in SCM, at least with respect to the following aspects:

- Each WS manager ignores the formats and models used by the other WS managers; SCM system heterogeneity is possible.
- Only the relevant information is transferred, which is critical in distributed work (at DS, some complex objects total several Gigabytes!).
- We provide object merge instead of only file merge. Our customers are unanimous in saying that object merge is a major enhancement (the composition relationship merge is an exact merge).

6 Concurrent Engineering Manager

So far, nothing prevents us from to performing changes on any objects in each WS. Thus at a given point in time, each attribute of each object may have a different value in each WS in which it is present (at DS, each attribute would have 30 different values on average)!

Concurrent engineering control means ensuring that work performed collectively is "consistent". Unfortunately, for concurrent access to information, there is a single real consistency criteria: serializability (ACID transactions as found in databases) but is this case, concurrent changes are prohibited.

To which extent concurrent changes to an object are a relevant issue? A few measurements taken at DS answer the question. At DS, the average number of different values for a given file under work is around 3, with a maximum around 30.

Merges occur frequently, depending on the policy (see below) and kind of software. The average for an application file is 2 merges a year, for a kernel file 0.4 a year. These numbers may seem low, but averages are meaningless because most files are not changed or merged at all. Conversely, files under work are subject to many changes and merges. We have records of more than 200 merges a year for the same file, which means about 1 merge per work day. Globally, about one thousand merges occur each day at DS. Concurrent change control is a real critical issue.

These values apply only to files whereas, in this work and at DS, we deal with objects (files are atomic attributes in our object model). Our experience shows that concurrent changes to the same attribute (like file or composition) as well as changes to different attributes of the same object (like responsible, state, name, namefile, protection etc.) are very common. Merging must address both cases. For example, restructuring, renaming and changing files are common, independent activities. Raising the granularity from file to object makes appear new kinds of concurrent changes, which may produce new kinds of merges (typically composition changes). It is our claim that object concurrent change control subsumes traditional file control and provides homogeneous and elegant solutions to many difficulties which currently hamper concurrent software engineering.

6.1 The Group Approach

We call a group, a set of Work Spaces the goal of which is to make an object evolve in a "consistent" way. Each group contains a WS playing the role of reference repository for the group called the integration WS.

For consistency to be enforced in a group, the integration WS must behave as if ACID transactions are applied to it, with each WS playing the role of a local transaction cache.

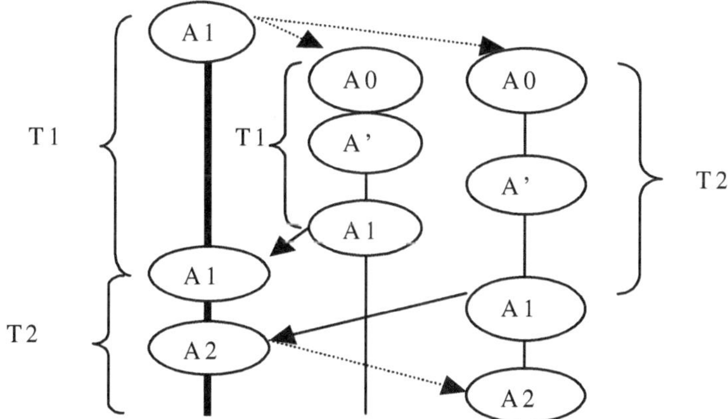

Fig. 3. Consistent Concurrent Changes

In Fig 3, both transactions T1 and T2 started with a copy of I/A0. Transaction T1 executed on I/A0 yields value W1/A1, while transaction T2 also performed on I/A0 yields value W2/A1. Changes T1 and T2 are consistent only if it is possible to compute the value I/A2 as the result of applying T2 changes to I/A1 instead of I/A0.

If this is possible, the group G = (I, W1, W2) behaves as if all changes (T1, T2...) are applied in sequence on I in any order; while they are really performed concurrently on the different WSs of that group.

Let us call Wi/O and Wi/O.A the working values of object O and attribute A of O in WS Wi. We have defined the following atomic functions:

Propose. P(Wi/O) informs that the current value for O in Wi is available to become the current value of the group.

Integrate. I(Wi/O) performs I/O = M (I/O, Wi/O).

Synchronize. S(O), when performed by Wi does Wi/O = M (Wi/O, I/O).

Reserve and **free**. R(O.A) and f(O.A). R(O.A) sets a lock on O.A for the whole group and f(O.A) releases that lock. Together they define a critical section for attribute O.A. At any point in time, a single WS of the group can be in the critical section for attribute A.

Free (O), releases all pending locks on attributes of O in that group.

6.2 Basic Policies in a Group

A basic concurrent engineering policy means defining and enforcing

- who has the right to perform a change on what (not anyone on anything at any time),
- what is to be merged and when,
- who has to merge.

Experience shows that merge control (including merge prohibition) is a central point when the merge function is missing or approximate. More specifically, file merge is an approximation which requires, potentially, manual work (for conflicts) and validatation if the result is consistent.

This becomes even more critical as soon as, in a group, the integration task is performed by a dedicated person. That person, the integrator, does not necessarily know the changes performed and would not be able to solve merge problems, nor it is his/her duty to fix the bugs that can result from the merge. At DS currently, each group has a integrator, and 15 integrators are working full time.

We claim that basic concurrent engineering policies can be defined as the valid sequences of operations performed on objects and attributes in a group. A definition of a few of the most useful policies follows.

Notations. In the following, let us denote by letter P, I, S, R and F commands Propose, Integrate, Synchronize Reserve and Free. C(O.A) denotes one or more change of attribute A; A,B the fact action B does not necessarily follow immediately A, (A1,A2 ... R) the repetition of the sequence in parenthesis if the R operation fails.

Finaly, letters in Roman caracters are executed by users in a WS, while in italic, they are executed by the integrator (i.e. from the integration WS).

Exclusion. O.A: (SR),C,PIF.

This sequence states the policy to be applied to attribute O.A in a group of WSs. It means that before changing A (C), a WS must first synchronize (S) then reserve (R) A. If the reserve fails, the couple SR must be repeated at a later time. After the changes, the WS executes proposes (P), integrates (I) and frees (F) on the object.

This policy implements an ACID transaction[4] by a two phase locking protocol. S copies the latest official value of the object to the local cache (WS), R sets a lock on the attribute before changing it and thus Integrate simply replaces in the reference repository (Integration WS) the old value by the new one (no merge), and F releases all locks. In practice any attribute for which no merge function exists should be managed in this way.

Delayed reserve: O.A: C,(R)SPIF.

This sequence states that changes on attribute A can be done before reserving it. As a consequence the synchronize operation (S) may require a merge (because meanwhile, a change on A may have been integrated). But since S is performed in the critical section (i.e. between R and F), operation I (integrate) will never require a merge (because since the reserve, only that WS can integrate a change on A). That policy has the property to never produce any merge in the integration WS. This is a consequence of any sequence in which S is before I in a critical section.

Delayed reserve and integration. O.A: C,RSP,I,F.

This sequence is similar to Delayed reserve, but I and F are performed by the integrator, at integrator convenience. Defered Integration deferred means the integrator is free to select which proposed change to integrate, when and in which order. Since the Free operation can also be delayed, it means the integrator has the opportunity, for example, to run test suites before validating the changes (F). With respect to the previous policy, flexibility is provided to integrators, not to developers.

Note that no merges are ever needed in the Integration WS. In other words, multiple concurrent changes are allowed but their integration is sequentialized and merges are performed in the WS which performed the changes, never in the integration WS. Integrators select the right change to integrate, and users can only synchronize on validated changes (F). This is a flexible but still pretty well controled evolution strategy.

Remember that a policy is defined on an attribute basis. For example, at DS, the management of source code objects in development groups is to use *Exclusion* for the filename attribute, *Delayed reservation and integration* for sourceCode attribute and *default* (no control) for the components attribute.

This approach contrasts with all other SCM cooperation strategies in the following ways:

- A policy can be defined independently for each attribute (default means no

[4] In our implementation, a WS is a (complex) object. Activities performed in a WS can be considered to be a transaction if the S,P,I operations involve the object WS instead of objects it contains. The rollback operation requires creating a revision (deep revision because of the transitive components relationship) of the WS before starting a transaction. This is why, in DS, some WSs have more than 2000 revisions

control). For instance, in a DS development group, it is possible, for the same object, to have three WS changing the file (sourceCode) another one changing that file name (fileName) and a third one changing which complex object it pertains to (components). Subsequent synchronization will consistently merge all these changes.

- A policy is declared on a group type basis i.e. all WSs in a group share the same policies; but each group can have different policies over the same objects. Typically, the development groups have more relaxed policies than the release group.
- A policy is formally defined (as the valid sequence of commands), and some "classic" policies have been studied so that their properties have been proved and used to optimize the implementation.

The integration WS plays the role of the coordinator for the whole group, it knows the policies and thus enforces and optimizes the controls, based on the properties of each policy.

6.3 Company Concurrent Work Policies

The integration WS behaves as the representative of the whole group, and can thus be a component of another higher level group (potentially of another type). In this way, the whole company can be organized in a hierarchy of WSs. At DS, three types of WS are defined, the depth of the WS tree ranges from 2 to 6, the number of concurrent WSs amounts to several hundred at any point in time. Traditionally, the WSs close to the root are the most stable (release WSs), while the leaf WSs are the most evolutive (Development WSs). Of course different policies apply.

It can be proven that the properties identified inside a single group hold for any sub-hierarchy with the single condition that reserving an attribute in a WS is considered as reserving that attribute in the integration WS of that group. It makes the reserve command transitive between groups which include the reserve command for that attribute in their policy.

The strict application of the previous strategy leads to a tree of groups, with data flow following the edges. Experience have shown this strategy supports the majority of the data flow, but is too limited. Data flow between siblings is needed, as well as data flow between almost any arbitrary group (as long as it is accepted by the process in place). Typical exemples are urgent fix procedures, where a fix "jumps" directly from a low level group to the top of hierarchy; or when a dependent group requires another sibling to make a bloking change.. In both cases the standard procedure would take days, which is sometimes not acceptable.

A major difficulty in SE is that almost any policy sometimes needs to be violated, for "good" reasons. This requires two features: first, the basic mechanisms must be able to perform un- standard operations, and second, these unusual procedures must be carefully defined and closely controlled; this is the goal of general process support, not presented in this paper.

We have demonstrated that synchronization between siblings is possible, still satisfying the coordination properties, provided some constraints, but at the cost of more expensive algorithms (many optimizations are no longer valid); demonstration is outside the scope of this paper.

Non sibling synchronization is also possible but breaks the coordination properties; for that reason, the underlying mechanisms keeps track of "abnormal" object synchronization (any synchronization that does not follow the tree); because it invalidates the assumptions used for optimization, it is required for performing subsequent merges (e.g. relationship merges require to store information), and is used for history, tracking and debugging.

These "simple" enhancements, as well as the facility allowing to "undo" complex operations like synchronize or integrate are responsible for significantly increasing the complexity of the whole system (remember that these commands can involve very complex objects, often the whole WS, which amounts to thousands of files).

7 Related Work

Object management, concurrent Engineering and distributions are hot topics. Nevertheless, it is a surprise to see that not much work have addressed seriously the issues of managing consistently copies of the "same" object having different values, different locations, different definitions and different format.

Databases have addressed the issue of consistency using the transaction concept, and different definitions (schema) with the multi-database and multi-view approaches. Some work has defined sub-databases, which look similar to SCM work spaces, [2][3]. But still there is a single (logical) version and a single consistency approach: serializability.

Middleware deal with object and distribution, but often objects do not move (CORBA 1 & 2) and in any case objects have a single (logical) value at any point in time. Service on top of Corba propose different coordination (flexible transaction features) and life cycle strategies; nevertheless there is not multiple version management and synchronization facility (yet).

The Web protocols, like HTTP also address some issues, but with different goals : a copy is sent with the assumption it will not be modified. WebDAV is proposing HTTP extensions for "distributed authoring" which includes attributes on objects, complex objects and locks [22][23] . Current proposed extensions to WebDAV [4] include versioning, workspace and configuration support . It is no surprise to note that these extensions are proposed by ClearCase, the leading SCM product [17][18]. Pushing this initiative further could lead to propositions similar to those contained in this paper[8].

Without surprise, SCM systems are those products coming closer to our reauirements. [6][14][13][11][21] but SCM systems, generally, manage files rather than objects, they know a single representation (the file) and evolution constraints are missing or unclearly stated[7][12].

8 Conclusion

Object management in SCM sets a number of very hard constraints. Our experience shows that the architecture followed by all vendors does not scale enough, does not provide sufficient efficiency, reliability and availability, and does not provide sufficiently high level services for concurrent engineering control. The dramatic increase in product size, concurrency factor and distribution will soon require a complete redesign of these systems.

We advocate for autonomous, heterogeneous and simple WS managers. They provide for the required availability, autonomy and efficiency. Heterogeneity affords for different format and projection conventions, and allows to reuse "standard" tools, preserving habits and investments. It also makes possible the incremental adoption of SCM strategies and policies, as well as scalability capabilities. All this is missing in today's SCM tools.

We claim that concurrent engineering requires the formalization of high level and flexible policies. We have proposed the group concept, and policy definition based on the valid sequence of 5 basic operations. This proposal contributes the idea that a policy is defined on a group basis, for each attribute individually, and may have properties that can be proved, and on which the system relies for performing substantial optimisations. Most policies of interest are available in standard (and recognized for optimisation) and any other can be user defined on the same basis.

Most of the complexity of the system comes from the fact that any policy may need to be violated. We believe it to be of critical importance (1) to define and control the processes where these violations are allowed to occur, and (2) propose a system where this is possible, even at the cost of performance degradation.

We think these propositions really contribute to build system, scalable both in term of amount of managed data, geographical location and level of services provided.

References

[1] ClearGuide: Product Overview". Technical report, Atria Software, Inc.

[2] A. Bjornersledl and C. Hullen. Version control in an Object-Oriented Architecture. In Won Kim and Frederick H. Lochowsky. editors. Objects-Oriented concepts, databases and application. Chapter 18, pages 451-485, Adisson-Wesley. 1990.

[3] E. Bratsberg. Unified Class Evolution by Object Oriented views. Proc of the 11th Conf on the relationship approach. LNCS N0645, Springer Verlag, Oct 1992.

[4] G. Clemm. Versioning Extensions to WebDav. Rational Software. May 1999. http://www.ietf.org/internet-drafts/draft-ietf-webdav-versioning-02.txt

[5] G. Clemm. The Odin System. SCM5, Seattle June 1995, pp241-263. Springer Verlag LNCS 1005.

[6] S. Dami, J. Estublier and M. Amiour. "APEL: a Graphical Yet Executable Formalism for Process Modeling". Automated Software Engineering journal, January 1998.

[7] S. Dart. "Concepts in Configuration Management Systems". Proc. of the 3rd. Intl. Workshop on Software Configuration Management. Trondheim, Norway, june, 1991.

[8] S. Dart. Content Change Management: Problems for Web Systems. In Proc SCM9, Toulouse, France, September 1999. pp1-17. Springer Verlag LNCS 1675.

[9] J. Estublier and R. Casallas. "The Adele Software Configuration Manager". Configuration Management, Edited by W. Tichy; J. Wiley and Sons. 1994. Trends in software.

[10] J. Estublier and R. Casallas. "Three Dimensional Versioning". In SCM-4 and SCM-5 Workshops. J. Estublier editor, September, 1995. Springer LNCS 1005, pp118-136.

[11] J. Estublier. "Workspace Management in Software Engineering Environments". in SCM-6 Workshop. Springer LNCS 1167. Berlin, Germany, March 1996.

[12] J. Estublier and S. Dami and M. Amiour. High Level Process Modeling for SCM Systems. SCM 7, LNCS 1235. pages 81--98, May, Boston, USA, 1997

[13] J. Estublier and R. Casallas. "Three Dimensional Versioning". In SCM-4 and SCM-5 Workshops. J. Estublier editor, September, 1995. Springer LNCS 1005.

[14] B. Gulla, E.A. Carlson, D. Yeh. Change-Oriented version description in EPOS. Software Engineering Journal, 6(6):378-386, Nov 1991.

[15] B. Gulla, J. Gorman. Experience with the use of a Configuration Language. In SCM-6 Workshop, Berlin, Germany, March, 1996. Springer Verlag LNCS1167, pp198-219.

[16] M. Hardwick, B.R. Dowine, M. Kutcher, D.L. Spooner, "Concurrent Engineering with Delta Files', IEEE Computer Graphics and Applications, January 1995, pp. 62-68.

[17] D. B. Leblang. and G.D. McLean. Configuration Management for large-scale software development efforts. In Proceedings of the workshop on Software Environments for Programming-in-the-Large. Pages 122-127. Harwichport, Massachussets, Jume 1985.

[18] D. B. Leblang. "The CM Challenge: Configuration Management that Works". Configuration Management, Edited by W. Tichy; J. Wiley and Sons. 1994. Trends in software.

[19] D.B. Leblang. Managing the Software Development Process with ClearGuide. SCM 7, LNCS 1235. pages=66, 80, May, Boston, USA, 1997

[20] J. Micallef and G. M. Clemm. "The Asgard System: Activity-Based Configuration Management". In SCM-6 Workshop, Berlin, Germany, March, 1996. Springer Verlag LNCS1167, pp175-187.

[21] Walter F. Tichy. Tools for software configuration management. In Proc. of the Int. Workshop on Software Version and Configuration Control, pp. 1-20, Grassau, January 1988

[22] Jim Whitehead. Goals for a Configuration Management Network protocol.In SCM9, LNCS 1675, pages 186-204, Toulouse September 1999.

[23] WebDav. HTTP extentions for distributed Authoring. RFC 2518. http://andrew2.andrew.cmu.edu/rfc/rfc2518.htm. February 1999.

Coordination Technologies for Managing Information System Evolution

Luís Filipe Andrade[1] and José Luiz Fiadeiro[2]

[1] ATX Software S.A.
Alameda António Sérgio 7 – 1 A,
2795 Linda-a-Velha, Portugal
landrade@oblog.pt

[2] LabMAC & Dept. of Informatics
Faculty of Sciences, University of Lisbon
Campo Grande, 1700 Lisboa, Portugal
jose@fiadeiro.org

Abstract. Information System Engineering has become under increasing pressure to come up with software solutions that endow systems with the agility that is required to evolve in a continually changing business and technological environment. In this paper, we suggest that Software Engineering has a contribution to make in terms of concepts and techniques that have been recently developed for Parallel Program Design and Software Architectures, which we have named *Coordination Technologies*. We show how such mechanisms can be encapsulated in a new modelling primitive – coordination contract – that can be used for extending Component Based Development approaches in order to manage such levels of change.

1 Introduction

More and more, companies live in a very volatile and turbulent environment in which both business rules and supporting technology change very quickly. In order to remain competitive, companies need their information systems to be easily adaptable to such changes, most of the time in a way that does not imply interruptions to critical services. Through the advent of the Internet and Wireless Applications, the New Economy is only fuelling this need even further: "... the ability to change is now more important than the ability to create e-commerce systems in the first place. Change becomes a first-class design goal and requires business and technology architecture whose components can be added, modified, replaced and reconfigured" [15].

Component-Based Development (CBD) [28] has been often proclaimed to be the ultimate approach for providing software solutions with the *agility* required to cope with such turbulence in business and technological environments. As put in [21], "Software developers have long held the belief that complex systems can be built from smaller components, bound together by software that creates the unique behav-

K.R. Dittrich, A. Geppert, M.C. Norrie (Eds.): CAiSE 2001, LNCS 2068, pp. 374–387, 2001.
© Springer-Verlag Berlin Heidelberg 2001

iour and forms of the system. Ideally, a new system can be built using mostly predefined parts, with only a small number of new components required... In a well designed system, the changes will be localised, and the changes can be made to the system with little or no effect on the remaining components".

However, it is also widely recognised that the promises brought in by CBD are not without several caveats. Our own experience in developing applications in one of the most volatile business areas – banking – indicates that *interactions* and *architectures* are at the core of the problems that still need to be addressed before CBD can deliver the goods. Many of the changes required on a system do not concern the computations performed by its components but the way they *interact*. Indeed, we often forget that the global behaviour of a system emerges both from the local behaviour of its components *and* the way they are interconnected. In dynamic business areas, the most frequent changes are likely to occur not at the level of the core entities of the domain (say, the notion of a bank account) but at the level of the business rules that determine how these core entities interact (say, how customers interact with their accounts). The old-fashioned technique of coding interactions in the components that implement the core services (say, reflecting policies on cash withdrawals directly on the withdrawal operation of an account) leads to systems that are very difficult to evolve because any change on the business rules triggers changes on the core components that it involves, their clients, possibly their client's clients, and so on.

Notice that object-oriented methods are very old-fashioned in this respect: interactions between objects are usually coded in the way messages are passed, features are called, and objects are composed, leading to intricate spaghetti-like structures that are difficult to understand, let alone change. Hence, new methodological concepts and supporting technology are needed that promote interactions to first-class entities, leading to systems that are "exoskeletal" in the sense that they exhibit their configuration structure explicitly [23]. Such an externalisation of the interconnections should allow for systems to be reconfigured, in a compositional, non-intrusive way, by acting directly on the entities that capture the interactions between components, leading to an evolution process that follows the *architecture* of the system [11,26].

Whereas interactions relate two or more components, the technology that is required for externalising them is also meaningful when applied to the evolution of single components. Indeed, another class of changes often required on a system concerns the adaptation of individual components [9], either existing ones (as in legacy systems) or newly acquired ones (as from third-party suppliers). In both cases, the implementation of the components is often not available and new ones cannot be brought in to replace them. A solution is to connect each component that needs to be evolved to an adapter such that the behaviour that emerges from the interaction between component and adapter fulfils the changes that were required initially. The advantage of making this interconnection explicit is that the adapter can be evolved independently of the component that it regulates, thus adding flexibility with respect to future changes.

In order to cope with the complexity of system evolution, we need more than mechanisms for managing change: we need to be able to determine *what* needs to be changed. Because change is most easily perceived and understood at the level of the application domain, the component model must be abstract enough to accommodate these changes without bringing in implementation details. This requires that the more traditional low-level notion of software component, and the design techniques used for its development (e.g., design patterns [17]), be "upgraded" to account for entities that are meaningful at the level of the application domain. This reflects the principle of "business centricity" upheld in [20] according to which software should be architected such that it reflects the way business organisations, policies and processes are defined by their own people. The ultimate goal of this approach is "100% maintenance" by which the authors mean "continuous quick evolution performed directly by the user rather than by the producer of the system".

This "isomorphism" between business component and software component architecture requires that changes perceived at the level of the business domain be mapped to the implementation levels in a compositional, non-intrusive way by keeping them local to the implementations of the higher-level parts that need to be changed. In summary, the architectural structure of the system – its gross decomposition in terms of high-level components and the interconnections that bind them together – should reflect the structure of the business domain, and mechanisms should be provided that enable evolution to be compositional over this structure.

In this paper, we assert that the entities that provide interconnections between components and the adapters of single components, all of which are called *coordination contracts* in the rest of the paper, should be seen as *coordination* mechanisms that one may *superpose*, in a non-intrusive way, on existing systems, supporting "plug-and-play" in system construction and evolution. The terms "coordination" and "superposition" in the previous sentence have a precise technical meaning and they constitute the kernel of the contribution that we think Software Engineering can make to enhancing the *agility* of information systems.

Through "coordination" we mean principles and techniques that have been recently developed for separating what in systems accounts for the computations that are performed and the interactions that are maintained [18]. Our belief is that CBD needs to be enhanced with primitives that provide for the explicit representation of the coordination mechanisms that regulate the way components behave and interact, and enable systems to evolve through the reconfiguration of the coordination mechanisms in place, without having to interfere with the way the computations performed by the individual components are programmed.

The term "superposition" refers to a mechanism, also called "superimposition" by some authors, that has been developed and applied in Parallel Program Design for extending systems while preserving their properties [7,10,16,22]. Our belief is that this basic mechanism is lacking in object-oriented development approaches and fills the gap that we have identified in OO development for supporting the externalisation of interactions and their evolution.

These are the two sides of the coin that we call *Coordination Technologies* [5] and with which we want to contribute to the debate that has been launched for identifying synergies between Software and Information System Engineering. In section 2, we provide further motivation on what these Coordination Technologies are and how they can contribute to increased levels of agility in the next generations of information systems. In section 3, we discuss an extension that we proposed in [4] for the UML in terms of a new semantic primitive that encapsulates Coordination Technology – coordination contracts or, simply, *contracts*. In section 4, we discuss the semantics and deployment of contracts. We conclude with an overview of what is available on contracts and what we are planning to make available in the near future.

2 Motivation

In order to motivate the principles and techniques that we wish to put forward through Coordination Technologies, consider the familiar world of bank accounts and clients who can make deposits, withdrawals, transfers, and so on. The notation that we will use in the examples is a shortened version of the Oblog language [www.oblog.com] that we have been developing for object-oriented modelling [3]. An example of a class specification is given below for bank accounts.

```
class Account
operations
          class
            Create(client:Customer,iAmount:Integer)
          object
            Deposit(amount:Integer);
            Withdrawal(amount:Integer);
            Balance() : Integer;
            Transfer(amount:Integer, target:Account);
body
          attributes
            number : Integer;
            balance : Integer := 0
          methods
            Deposit
              is set Balance := Balance+amount
            Withdrawal [enabling Balance≥amount]
              is set Balance := Balance-amount
            Transfer
              is { call target.Deposit(amount);
                   call self.Withdrawal(amount) }
end class
```

In Oblog, a class specification includes a section in which the interface operations are declared. We distinguish between class and object operations: the former are used for managing the population of the class as a whole and the latter apply to each specific instance. Each operation is declared with a list of input and output parameters and a specification of its required behaviour in terms of pre/post conditions (omitted in the example for simplicity). In the case of the bank account, the operations that were chosen are self-explanatory.

The body section of a class specification identifies the attributes that define the state of the instances as well as the implementations of the operations (called methods). Methods can be guarded with state-conditions like in Dijkstra's guarded commands. This is the case of `withdrawal` that is being restricted to occur in states in which the amount to be withdrawn can be covered by the balance.

This restriction on the availability of withdrawals is a good example of the limitations of object-oriented methods for supporting the flexibility that is required for managing change. Indeed, the enabling condition `Balance≥amount` derives more from the specification of a business requirement than an intrinsic constraint on the functionality of a basic business entity like `Account`. The circumstances under which a withdrawal can be accepted is likely to change as competition dictates banks to come up with new ways for customers to interact with their accounts. Different financial packages are concocted every now and then that provide different policies on withdrawals, making it impossible to predict how accounts will be accessed over their lifetime. Nevertheless, there are basic functional properties that should remain much more stable, reflecting the core "semantics" of the operation as a business transaction, like the fact that the amount is actually deducted from the balance.

One could argue that, through inheritance, this guard could be changed in order to model these different situations and future ones that may arise as a consequence of changes on business policies. However, there are two main problems with the use of inheritance for this purpose. Firstly, inheritance views objects as white boxes in the sense that adaptations like changes to guards are performed on the internal structure of the object. From the point of view of evolution, this is not desirable. This is because, on the one hand, changes on the internal structure are often difficult to locate and localise because they may trigger changes to the interface of the component as well and, thus, initiate a cascade of changes throughout the chain of clientship. On the other hand, inheritance may not be applicable when the component structure is simply not available, as in the case of legacy systems or third-party components.

Secondly, from the business point of view, the adaptations that make sense may be required on classes other than the ones in which the restrictions were implemented. In the example above, this is the case when it is the type of client, and not the type of account, that determines the nature of the guard that applies to withdrawals. The reason the restriction ended up coded as a guard on withdrawals results already from the bias that the OO mechanism of clientship introduces in the modelling of interactions. Typically, restrictions of this sort are coded up on the server-side, which is where the operation is usually implemented, even if they are meant to reflect access modes that result from different categories of clients.

Hence, it makes more sense for business requirements of this sort to be modelled explicitly outside the classes that model the basic business entities. Our proposal is that guards like the one discussed above should be modelled as *coordination contracts* that can be established between bank customers and accounts, but *outside* the specifications of these two classes as explicit representations of the business rules from which they derive. In fact, we provide mechanisms for such contracts to be *superposed* on existing implementations of clients and accounts, considered as black boxes, so that contracts can be added and deleted in a non-intrusive way *(plug and play)*, reflecting the evolution of the business domain.

3 Coordination Contracts

From a static point of view, a coordination contract defines an *association class* in the sense of the UML (i.e. an association that has all the attributes of a class). However, the way interaction is established between the partners is more powerful: it provides a *coordination role* that is closer to what is available for configurable distributed systems in general, namely through the use of architectural connectors [2]. A contract consists of a collection of role classes (the partners in the contract) and the prescription of the coordination effects (the glue in the terminology of software architectures) that will be superposed on the partners. In Oblog, contracts are defined as follows:

```
contract <name>
   partners <list-of-partners>
   invariant <the relation between the partners>
   constants
   attributes
   operations
   coordination <behaviour superposed by the contract>
   behaviour < local behaviour of the contract>
end contract
```

The instances of the partners that can actually become coordinated by instances of the contract are determined through a set of conditions specified as invariants. The typical case is for instances to be required to belong to some association between the partners.

The behaviour that is required to be superposed over that of the partners is identified under "coordination" as trigger/reaction clauses of the form

```
<name> : when <condition>
         do <set of actions>
         with <condition>
```

Each coordination clause has a name that can be used for managing the interference between all the clauses and the contract's own actions, e.g. in terms of precedence relations. This is similar to what happens in parallel program design languages like Interacting Processes [16]. The condition under "when" establishes the trigger of the clause. Typical triggers are the occurrence of actions or state changes in the partners. Under "do" we identify the reactions to be performed, usually in terms of actions of the partners and some of the contract's own actions, which constitute what we call the *synchronisation set* associated with the trigger. Finally, under "with", we can put further constraints on the actions involved in the synchronisation set, typically further preconditions. The intuitive semantics (to be further discussed in the following section) is that, through the "when" condition, the contract intercepts calls to the partners or detects events in the partners to which it has to react. It then checks the "with" conditions to determine whether the interaction can proceed and, if so, coordinates the execution of the synchronisation set. All this is done atomically, in a transactional mode, in the sense that either all the actions in the set are executed or none is.

An example can be given through the account packages already discussed. The traditional package, by which withdrawals require that the balance be greater than the amount being withdrawn, can be specified as follows, where by `Account` we mean the class specification discussed in the previous section but without the enabling restriction on withdrawals:

```
contract Traditional package
          partners x : Account; y : Customer;
          invariants ?owns(x,y)=TRUE;
          coordination
            tp:when y.calls(x.Withdrawal(z))
                do x.Withdrawal(z)
                with x.Balance() ≥ z;
end contract
```

Notice that, as specified by the invariant, this contract is based on an ownership association that must have been previously defined. This contract involves only one interaction. It relates calls placed by the customer for withdrawals with the actual withdrawal operation of the corresponding account. The customer is the trigger of the interaction: the interaction requires every call of the customer to synchronise with the withdrawal operation of the account but enables other withdrawals to occur outside the interactions, e.g. by other joint owners of the same account. The constraint imposed through the with-clause is the guard already discussed in the previous section, which is now externalised as part of the contract. Hence, it regulates only a specific class of interactions between customers and accounts: those that have subscribed to the particular contract *Traditional package*. In particular, each instance of the contract and, hence, the constraint, applies only to an identified pair of customer and account, meaning that other owners of the same account may subscribe to different contracts.

The notation involving the interaction in this example is somewhat redundant because the fact that the trigger is a call from the customer to an operation of the account immediately identifies the reaction to be performed. In situations like this, Oblog allows for abbreviated syntactical forms. However, in the paper, we will consistently present the full syntax to make explicit the various aspects involved in an interaction. In particular, the full syntax makes it explicit that the call put by the client is intercepted by the contract, and the reaction, which includes the call to the supplier, is coordinated by the contract. Again, we stress that the interactions established through contracts are atomic, i.e. the synchronisation set determined by each coordination entry of the contract is executed as a single transaction – either all the actions in the set are performed or none is. In particular, the client will not know what kind of coordination is being superposed. From the point of view of the client, it is the supplier that is being called.

As already explained, the purpose of contracts is to externalise interactions between objects, making them explicit in the conceptual models, thus reflecting the business rules that apply in the current state. Hence, contracts may change as the business rules change, making system evolution compositional with respect to the evolution of the application domain. For instance, new account packages may be introduced that relax the conditions under which accounts may be overdrawn:

```
contract VIP package
          partners x : Account; y : Customer;
          constants CONST_VIP_BALANCE: Integer;
          attributes Credit : Integer;
          invariants
            ?owns(x,y)=TRUE;
            x.AverageBalance() >= CONST_VIP_BALANCE;
          coordination
            vp:when y.calls(x.Withdrawal(z))
               do x.Withdrawal(z)
               with x.Balance() + Credit() ≥ z;
end contract
```

Notice that, on the one hand, we have strengthened the invariant of the contract, meaning that only a restricted subset of the population can become under the coordination of this new business rule, namely the customers that qualify as VIPs. On the other hand, the contract weakens the guard imposed on withdrawals, meaning that there are now more situations in which customers can withdraw money from their accounts: this is the added benefit of the more restricted class of customers.

In general, we allow for contracts to have features of their own. This is the case of the contract above for which an attribute and a constant were declared to model the credit facility. It is important to stress that such features (including any additional operations) are all private to the contract: they fulfil a local purpose in establishing the coordination that is required between the partners and are not available for interaction with other objects. Indeed, the contract does not define a public class and its instances are not considered as ordinary components of the system. This is one of the reasons why association classes, as available in the UML, are not expressive enough to model the coordination mechanisms of contracts. Although contracts allow for interactions to be made explicit in conceptual models, they should not be accessed in the same way as the classes that model the core business entities. Contracts do not provide services: they coordinate the services made available by the core entities. Contracts arc the subject of a different level of management of the business component system: that of the definition and evolution of its configuration.

Another shortcoming of association classes is that they do not enforce the synchronisation and atomicity requirements of the proposed coordination mechanisms. As an example, consider one of the latest products to appear in banking – "the flexible package". This is a mechanism via which automatic transfers are made between a checking account and a savings account of the same client: from savings to checking when the balance is not enough to satisfy a withdrawal, and from checking to savings when the balance goes above a certain threshold.

Like before, the application of traditional object-oriented techniques for adding this new feature to the system would probably raise a number of problems. The first one concerns the decision on where to place the code that is going to perform the transfers: the probable choice would be the checking account because that is where the balance is kept. Hence, the implementation of account would have to be changed. The "natural" solution would be to assign the code to a new association class between the two accounts but, again, current techniques for implementing association classes require the implementations of the participating classes to be changed because the associations are implemented via attributes.

Another problem is concerned with the handling of the synchronisation of the transfers. If the transfers are not coded in the methods of the accounts, there is no way in which the whole process can be dealt with atomically as a single transaction. Again, what is required is a mechanism via which we can superpose a layer of coordination that is separate from the computations that are performed locally in the objects and enforces the required interactions, including the synchronisation constraints. This is exactly the level of coordination that contracts provide.

```
contract Flexible package
          partners c, s : Account;
          attributes max, min : Integer;
          invariants c.owner=s.owner;
          coordination
            putfunds:
              when calls(c.Deposit(z))
              do { c.Deposit(z);
                  if c.Balance()+z > max then
                    c.Transfer(c.Balance()-max,s)};
            getfunds:
              when calls(c.Withdrawal(z))
              do { if c.Balance()-z < min then
                    s.Transfer(min-c.Balance(),c);
                  c.Withdrawal(z)};
end contract
```

Transfers between two accounts are superposed on the operations of the partners as part of the coordination activity. Notice that, in both cases, the actions that trigger the interactions are executed as part of the synchronisation sets that define the reactions. Obviously, the execution of these actions as part of the synchronisation sets does not trigger the contract again! It may, however, trigger other contracts that the components have subscribed. For instance, the savings account may be under the coordination of a VIP-contract as well, in which case the withdrawals that are triggered by a *transfer* coordinated by *getfunds* will be handled by *vp* as well.

4 Semantical Aspects

The intuitive semantics of contracts can be summarised as follows:

- Contracts are added to systems by identifying the instances of the partner classes to which they apply. These instances may belong to subclasses of the partners; for instance, in the case of the flexible package, both partners were identified as being of type *account* but, normally, they will be applied to two subclasses of account: a checking account and a savings account. The actual mechanism of identifying the instances that will instantiate the partners and superposing the contract is outside the scope of the paper. In Oblog, this can be achieved directly as in languages for reconfigurable distributed systems [24], or implicitly by declaring the conditions that define the set of those instances.

- Contracts are superposed on the partners taken as black-boxes: the partners in the contract are not even aware that they are being coordinated by a third party. In a client-supplier mode of interaction, instead of interacting with a mediator that then delegates execution on the supplier, the client calls directly the supplier; however, the contract "intercepts" the call and superposes whatever forms of behaviour are prescribed; this means that it is not possible to bypass the coordination being imposed through the contract because the calls are intercepted;
- The same transparency applies to all other clients of the same supplier: no changes are required on the other interactions that involve either partner in the contract. Hence, contracts may be added, modified or deleted without any need for the partners, or their clients, to be modified as a consequence;
- The interaction clauses in a contract identify points of *rendez-vous* in which actions of the partners and of the contract itself are synchronised; the resulting synchronisation set is guarded by the conjunction of the guards of the actions in the set and the condition indicated in the with-clause. The execution of the synchronisation set requires the execution of all the actions in the set.
- The implementation of the synchronisation sets that are determined by the contracts can either be obtained through the primitives available in the transaction language (e.g. calls to the actions participating in the synchronisation set can be made by the regulator), or a completely new implementation can be superposed by the contract body as an alternative to the implementations available in the partners. In the latter case, the properties of the interfaces have to be preserved, namely any pre/post conditions and invariants specified through contracts in the sense of Meyer [25]. This is a useful way of bypassing legacy code as pointed out in [9]. Its implementation requires, however, that the contract body (the regulator) be awarded some sort of priority over the roles at the execution level.
- The effect of superposing a contract is cumulative: more than one contract may be active at a given state of the system. Because the superposition of a contract consists, essentially, of synchronous interactions, the different active contracts will superpose their coordinating behaviour, achieving a cumulative effect in the sense that the synchronisation set to be executed is the union of all the synchronisation sets of the active contracts and its guard is given by the conjunction of all the guards and "with" clauses associated with the active contracts. For instance, in the example of the flexible package, the transfers can also be subject to contracts that regulate the discipline of withdrawals for the particular account and client.
- In the paper, we will not address the issue of managing which contracts are active at which states. This aspect is being addressed as part of the configuration language that is being developed for assisting in the process of controlling or programming the evolution of systems. This will include logical mechanisms for reasoning about possible interactions between contracts. Preliminary work in this direction is reported in [31].

For simplicity, the examples are based on binary relationships. However, contracts may involve more than two partners. In this case, the invariant and coordination clauses may refer to all partners. In fact, contracts may be seen to correspond to synchronisation agents, as presented in the Actors model [1], that coordinate the rules of engagement of various objects participating simultaneously in the same task.

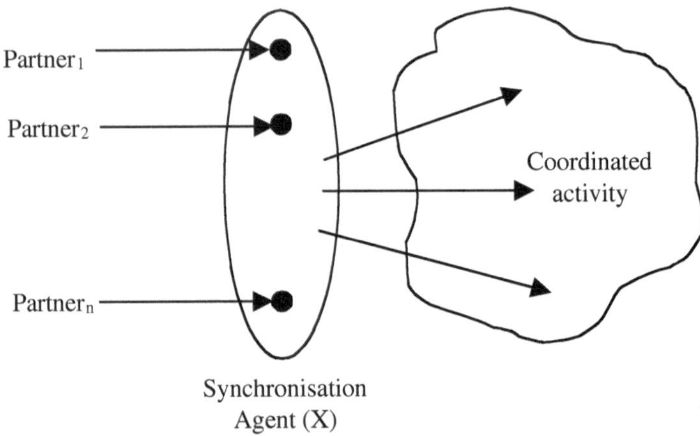

Notice that contracts in the sense of Meyer [25] fulfil a different, but complementary, role to the coordination contracts that we have described: their purpose is to support the development of object methods in the context of client-supplier relationships between objects. Therefore, they apply, essentially, to the "in-the-small" *construction* of systems rather than the "in-the-large" dimension, concerned with *architecture* and *evolution*, that is our target. In fact, in what concerns evolution, the use of Meyer's contracts in a non-architectural approach leads to some of the problems that we discussed in the introduction: by adopting clientship as the basic discipline for object interconnection, a bias is introduced in the way business rules get coded up, assigning to the supplier side the responsibility for accommodating changes that, from the point of view of the business rules, belong to the client. This was the case of the flexible withdrawals for VIP customers: by placing the contract on the supplier side (the account), the new rules are more easily modelled as specialisations of account whereas, in the application domain, they reflect specialisations of the client.

The intuitive semantics that we outlined above has been given both a mathematical and an implementational counterpart. The mathematical semantics draws on previous work on the categorical semantics of architectural and coordination principles. The presentation of this semantics is outside the scope of this paper. Please consult [12] for the formalisation of different kinds of superposition that lead to the identification of different kinds of contracts (regulators, monitors, etc); [13] for a formalisation of architectural connectors based on the previous formalisation of superposition, which includes the semantics of instantiation of the partners (roles); [14] for the coordination aspects as related to superposition and the application to software architectures; and [29,30,31] for the application to dynamic reconfiguration, including the definition of algebraic operations on architectural connectors that are directly applicable to contracts.

The fact that a mathematical semantics exists for justifying our principle of supporting evolution through the superposition of contracts does not mean that it can be put into practice directly over the technology that is available today. In [4,19] we have shown that, even if none of the standards for component-based software development that have emerged in the last few years (e.g. CORBA, EJB and COM) can provide a convenient and abstract way of supporting the proposed coordination principles as first-class mechanisms, an implementation can be given that is based on a design pattern that exploits some widely available properties of object-oriented programming languages such as polymorphism and subtyping. The relationship between this design pattern and the mathematical semantics is discussed in [6].

5 Concluding Remarks

In this paper, we have suggested that Information System Engineering can be significantly enhanced by adopting *Coordination Technologies* through which systems can be made more *agile* in responding to the turbulence that characterises current business and technological environments. More specifically, we described how a new semantic primitive – *coordination contracts* – can be added to current object-oriented approaches to fulfil some of the promises of Component Based Development in terms of supporting a new approach to business and technology architecture in which components can be dynamically added, modified, replaced and reconfigured [15,20,21,28].

Contracts draw from several mechanisms that have been available for sometime in Software Engineering:

- Contract-based development is based on the idea of separating *computation* from *coordination*, i.e. of making clear what components in a system provide the functionalities on which services are based, and what mechanisms are responsible for coordinating the activities of those components so that the desired behaviour emerges from the interactions that are established. This idea has been promoted by researchers in the area of Programming Languages who have coined the term "Coordination Languages and Models" [e.g. 18].
- The importance of having these coordination mechanisms available as first-class entities and as units of structure in system models and designs was inspired by the role played by connectors in software architectures [2,27]. This is why we proposed contracts as a semantic primitive enriching the notion of association class.
- The ability for objects to be treated as black-box components and, hence, for contracts to be dynamically added or removed from a system without having to recompile the partners, is achieved through the mechanism of superposition (or superimposition) developed in the area of Parallel Program Design [7,10,16,22].

An effective use of the proposed coordination mechanisms further requires language and tool support for the actual run-time superposition of contracts, either in reaction to customer demands or for the enforcement of business policies. Such policies have to be understood as "invariants" that apply to "evolution time" and not "computation time". That is, they constrain the way systems can be evolved in terms of their configuration. Naturally, they will also have an impact on the computations that are performed in the system because, as we have seen, the configuration determines the interactions from which the global computational behaviour of the system emerges. What is important to emphasise is that decisions on which contracts can be subscribed by which classes of partners and in which states of the system are an integral part of the modelling activity. This is the direction in which we are currently working.

References

1. G.Agha, ACTORS: A Model of Concurrent Computation in Distributed Systems, MIT Press 1986.
2. R.Allen and D.Garlan, "A Formal Basis for Architectural Connectors", ACM TOSEM, 6(3), 1997, 213-249.
3. L.F.Andrade, J.Gouveia, P.Xardone and J.Camara, "Architectural Concerns in Automating Code Generation", in Proc. TC2 First Working IFIP Conference on Software Architecture, P. Donohoe (ed), Kluwer Academic Publishers.
4. L.F.Andrade and J.L.Fiadeiro, "Interconnecting Objects via Contracts", in UML'99 – Beyond the Standard, R.France and B.Rumpe (eds), LNCS 1723, Springer Verlag 1999, 566-583.
5. L.F.Andrade and J.L.Fiadeiro, "Coordination: the evolutionary dimension", in Proc. TOOLS Europe 2001, Prentice-Hall, in print
6. L.F.Andrade, J.L.Fiadeiro, J.Gouveia, A.Lopes and M.Wermelinger, "Patterns for Coordination", in COORDINATION'00, G.Catalin-Roman and A.Porto (eds), LNCS 1906, Springer-Verlag 2000, 317-322
7. R.Back and R.Kurki-Suonio, "Distributed Cooperation with Action Systems", *ACM TOPLAS* 10(4), 1988, 513-554.
8. G.Booch, J.Rumbaugh and I.Jacobson, The Unified Modeling Language User Guide, Addison-Wesley 1998.
9. J.Bosch, "Superimposition: A Component Adaptation Technique", Information and Software Technology 1999.
10. K.Chandy and J.Misra, ParallelProgram Design - A Foundation, Addison-Wesley 1988.
11. H.Evans and P.Dickman, "Zones, Contracts and Absorbing Change: An Approach to Software Evolution", in Proc. OOPSLA'99, ACM Press 1999, 415-434.
12. J.L.Fiadeiro and T.Maibaum, "Categorical Semantics of Parallel Program Design", Science of Computer Programming 28, 1997, 111-138.
13. J.L.Fiadeiro and A.Lopes, "Semantics of Architectural Connectors", in TAPSOFT'97, LNCS 1214, Springer-Verlag 1997, 505-519.
14. J.L.Fiadeiro and A.Lopes, "Algebraic Semantics of Coordination, or what is in a signature?", in AMAST'98, A.Haeberer (ed), Springer-Verlag 1999.
15. P.Finger, "Componend-Based Frameworks for E-Commerce", Communications of the ACM 43(10), 2000, 61-66.
16. N.Francez and I.Forman, Interacting Processes, Addison-Wesley 1996.
17. E.Gamma, R.Helm, R.Johnson and J.Vlissides, *Design Patterns: Elements of Reusable Object Oriented Software*, Addison-Wesley 1995
18. D.Gelernter and N.Carriero, "Coordination Languages and their Significance", Communications ACM 35, 2, pp. 97-107, 1992.
19. J.Gouveia, G.Koutsoukos, L.Andrade and J.Fiadeiro, "Tool Support for Coordination Based Evolution", in Proc. TOOLS Europe 2001, Prentice-Hall, in print.
20. P.Herzum and O.Sims, *Business Component Factory*, Wiley 2000.

21. J.Hopkins, "Component Primer", Communications of the ACM 43(10), 2000, 27-30.
22. S.Katz, "A Superimposition Control Construct for Distributed Systems", ACM TOPLAS 15(2), 1993, 337-356.
23. J.Kramer, "Exoskeletal Software", in Proc. 16th ICSE, 1994, 366.
24. J.Magee and J.Kramer, "Dynamic Structure in Software Architectures", in 4th Symp. on Foundations of Software Engineering, ACM Press 1996, 3-14.
25. B.Meyer, "Applying Design by Contract", IEEE Computer, Oct. 1992, 40-51.
26. P.Oreizy, N.Medvidovic and R.Taylor, "Architecture-based Runtime Software Evolution", in Proc. ICSE'98, IEEE Computer Science Press 1998
27. D.Perry and A.Wolf, "Foundations for the Study of Software Architectures", ACM SIGSOFT Software Engineering Notes 17(4), 1992, 40-52.
28. C. Szyperski, Component Software: Beyond Object-Oriented Programming, Addison Wesley 1998.
29. M.Wermelinger and J.L.Fiadeiro, "Connectors for Mobile Programs", IEEE Transactions on Software Engineering 24(5), 1998, 331-341.
30. M.Wermelinger and J.L.Fiadeiro, "Towards an Algebra of Architectural Connectors: a Case Study on Synchronisation for Mobility", in Proc. 9th International Workshop on Software Specification and Design, IEEE Computer Society Press 1998, 135-142.
31. M.Wermelinger and J.L.Fiadeiro, "Algebraic Software Architecture Reconfiguration", in Software Engineering – ESEC/FSE'99, LNCS 1687, Springer-Verlag 1999, 393-409.

Using Metrics to Predict OO Information Systems Maintainability

Marcela Genero, José Olivas, Mario Piattini, and Francisco Romero

Department of Computer Science
University of Castilla-La Mancha
Ronda de Calatrava, 5
13071, Ciudad Real, Spain
{mgenero, jaolivas, mpiattin, fpromero}@inf-cr.uclm.es

Abstract. The quality of object oriented information systems (OOIS) depends greatly on the decisions taken at early phases of their development. As an early available artifact the quality of the class diagram is crucial to the success of system development. Class diagrams lay the foundation for all later design work. So, their quality heavily affects the product that will be ultimately implemented. Even though the appearance of the Unified Modeling Language (UML) as a standard of modelling OOIS has contributed greatly towards building quality OOIS, it is not enough. Early availability of metrics is a key factor in the successful management of OOIS development. The aim of this paper is to present a set of metrics for measuring the structural complexity of UML class diagrams and to use them for predicting their maintainability that will heavily be correlated with OOIS maintainability.

Keywords. object oriented information systems maintainability, object oriented metrics, class diagrams complexity, UML, fuzzy deformable prototypes, prediction models

1 Introduction

A widely accepted principle in software engineering is that the quality of a software product should be assured in the early phases of its life cycle. In a typical OOIS design at the early phases, a class diagram is first built. The class diagram is not merely the basis of modelling the persistent system data. In OO modelling, where data and process are closely linked, class diagrams provide the solid foundation for the design and implementation of OOIS.

As an early available, key analysis artifact the quality of the class diagram is crtucial to the success of system development. Generally, problems in the artifacts produced in the initial phases of system development propagate to the artifacts produced in later stages, where they are much more costly to identify and correct [2]. As a result, improving the quality of class diagrams, will therefore be a major step towards the quality improvement of the OOIS development. The appearance of UML [20], as standard OO modelling language, should contribute to this. Despite this, we have to be aware that a standard modelling language can only give us syntax and semantics to work with, but it cannot tell us whether a "good" model has been

K.R. Dittrich, A. Geppert, M.C. Norrie (Eds.): CAiSE 2001, LNCS 2068, pp. 388–401, 2001.
© Springer-Verlag Berlin Heidelberg 2001

produced. Naturally, even when language is mastered, there is no guarantee that the models produced will be good. Therefore, it is necessary to assess their quality.

The definition of the different characteristics that compose the concept of "quality" is not enough on its own in order to ensure quality in practice, as people will generally make different interpretations of the same concept. Software measurement plays an important role in this sense because metrics provide a valuable and objective insight into specific ways of enhancing each of the software quality characteristics. Measurement data can be gathered and analysed to assess current product quality, to predict future quality, and to drive quality improvement initiatives [27].

Quality is a multidimensional concept, composed of different characteristics such as functionality, reliability, usability, efficiency, maintainability and portability [15]. This paper focuses on UML class diagram maintainability, because maintainability has been and continues to be one of the pressing challenges facing any software development department. For our purpose we distinguish the following maintainability sub-characteristics:

- UNDERSTANDABILITY. The ease with which the class diagram can be understood.
- ANALYSABILITY. The capability of the class diagram to be diagnosed for deficiencies or to identify parts to be modified.
- MODIFIABILITY. The capability of the class diagram to enable a specified modification to be implemented.

But these maintainability sub-characteristics are external quality attributes that can only be measured late in the OOIS life cycle. Therefore it is necessary to find early indicators of such qualities based, for example, on the structural properties of class diagrams [4].

The availability of significant measures in the early phases of the software development life-cycle allows for better management of the later phases, and more effective quality assessment when quality can be more easily affected by corrective actions [3]. They allow IS designers:

1. a quantitative comparison of design alternatives, and therefore and objective selection among several class diagram alternatives with equivalent semantic content.
2. a prediction of external quality characteristics, like maintainability in the initial phases of the IS life cycle and a better resource allocation based on these predictions.

After performing a thorough review of several OO metric proposals [9],[18],[7],19], specially focusing in those that can be applied to class diagrams at a high level design stage we have proposed new ones [14] related to the structural complexity of class diagrams due to the usage of relationships (associations, dependencies, generalisations, aggregations). But proposing metrics it is not enough to assure that they really are fruitful in practice. Empirical validation is a crucial task for the success of software measurement [17],[13],[26],[1].

Our main motivation is to present metrics [14] for measuring UML class diagram structural complexity (internal quality attribute) and secondly demonstrate through experimentation that it can be used to predict UML class diagram maintainability (external quality attribute), which will strongly influence OOIS maintainability.

This paper is organised in the following way: In section 2 we will present a set of metrics for measuring UML class diagram structural complexity. In section 3 we describe a controlled experiment, carried out in order to build fuzzy deformable prototypes, using a new approach to Knowledge Discovery [21],[22], that characterise UML class diagram maintainability from the metric values. In section 4 we will use this prototypes to predict UML class diagram maintainability. Lastly, section 5 summarises the paper, draws our conclusions, and presents future trends in metrics for object modelling using UML.

2. A Proposal of Metrics for UML Class Diagrams

We only present here the metrics [14] that can be applied to the class diagram as a whole . They were called "Class Diagram-Scope metrics". Also we consider traditional ones like, the number of classes, the number of attributes, etc... We classify them in two categories: open-ended metrics, whose values are not bounded in an interval, and close-ended metrics whose values are bounded, in our case in the interval [0,1].

2.1 Open-Ended Metrics

- NUMBER OF CLASSES. (NC) is the total number of classes within a class diagram.
- NUMBER OF ATTRIBUTES. (NA) is the total number of attributes within a class diagram.
- NUMBER OF METHODS. (NM) is the total number of methods within a class diagram.
- NUMBER OF ASSOCIATIONS. (NAssoc) is defined as the total number of associations within a class diagram.
- NUMBER OF AGGREGATION. (NAgg) is defined as the total number of aggregation relationships within a class diagram (each whole-part pair in an aggregation relationship).
- NUMBER OF DEPENDENCIES. (NDep) is defined as the total number of dependencies relationship within a class diagram.
- NUMBER OF GENERALISATIONS. (NGen) is defined as the total number of generalisation relationships within a class diagram (each parent-child pair in a generalisation relationship).
- NUMBER OF GENERALISATIONS HIERARCHIES. (NGenH) is defined as the total number of generalisations hierarchies in a class diagram
- MAXIMUM DIT. The Maximum DIT in a class diagram is the maximum between the DIT value obtained for each class of the class diagram. The DIT value for a class within a generalisation hierarchy is the longest path from the class to the root of the hierarchy.

2.2 Close-Ended Metrics

- NUMBER OF ASSOCIATIONS VS. CLASSES. (NAssocVC) is defined as the ratio between the number of associations in a class diagram (NAssoc) divided by the total number of classes in the class diagram (NC).

- NUMBER OF DEPENDENCIES VS. CLASSES. (NDepVC) is defined as the ratio between the number of dependencies in a class diagram (NDep) divided by the total number of classes in the class diagram (NC).
- NUMBER OF AGGREGATIONS VS. CLASSES. (NAggVC) is defined as the ratio between the number of aggregations in a class diagram (NAgg) divided by the total number of classes in the class diagram (NC).
- NUMBER OF GENERALISATIONS VS. CLASSES. (NGenVC) is defined as the ratio between the number of generalisations in a class diagram (NGen) divided by the total number of classes in the class diagram (NC).

3. A Comprehensive Controlled Experiment to Build a Prediction Model for UML Class Diagram Maintainability

Taking into account some suggestions provided in [4],[5] about how to do empirical studies in software engineering, we carried out a controlled experiment with the goal of predicting UML class diagrams maintainability from metric values obtained at the early phases of OOIS life cycle.

3.1 Subjects

The experimental subjects used in this study were: 7 professors and 10 students enrolled in the final-year of Computer Science in the Department of Computer Science at the University of Castilla-La Mancha in Spain. All of the professors belong to the Software Engineering area and they have enough experience in the design and development of OO software. By the time the experiment was done all of the students had had two courses on Software Engineering, in which they learnt in depth how to build OO software using UML. Moreover, subjects were given an intensive training session before the experiment took place.

3.2 Experimental Materials and Tasks

The subjects were given twenty eight UML class diagrams of the same universe of discourse, related to Bank Information Systems. Each diagram has a test enclosed which includes the description of maintainability sub-characteristics, such as: understandability, analysability, modifiability. Each subject has to rate each sub-characteristic using a scale consisting of seven linguistic labels. For example for understandability we proposed the following linguistic labels:

Extremely difficult to understand	Very difficult to understand	A bit difficult to understand	Neither difficult nor easy to understand	Quite easy to understand	Very easy to understand	Extremely easy to understand

We allowed one week to do the experiment, i.e., each subject had carry out the test alone, and could use unlimited time to solve it.

After completion of the tasks subjects were asked to complete a debriefing questionnaire. This questionnaire included (i) personal details and experience, (ii) opinions on the influence of different components of UML Diagrams, such as: classes, attributes, associations, generalisations, etc... on their maintainability.

3.3 Experimental Design and Data Collection

The INDEPENDENT VARIABLES are those metrics proposed in sections 2.1 and 2.2.

The DEPENDENT VARIABLES are three of the maintainability sub-characteristics: understandability, analysability and modifiability measured according to subject's rating.

We decided to give our subjects as much time as they needed to finish the test they had to carry out. All tests were considered valid because all of the subjects have at least medium experience in building UML class diagrams and developing OOIS (this fact was corroborated analysing the responses of the debriefing questionnaire).

3.4 Construction of Fuzzy Deformable Prototypes to Characterise UML Class Diagram Maintainability

We have used an extension of the traditional Knowledge Discovery in Databases (KDD) [12]: the Fuzzy Prototypical Knowledge Discovery (FPKD) that consists of the search for fuzzy prototypes [28] that characterise the maintainability of an UML class diagram.

In the rest of this section we will explain each of the steps we have followed in the FPKD (see figure 1).

Selection of the Target Data. We have taken as a start set a relational database that contains 476 records (with 16 fields,13 represent metrics values, 3 represent maintainability sub-characteristics) obtained from the calculation of the metric values (for each class diagram) and the responses of the experiment given by the subjects.

Preprocessing. The Data-Cleaning was not necessary because we didn´t find any errors.

Transformation. This step was performed doing different tasks:

- SUMMARISING SUBJECT RESPONSES. We built a unique table with 28 records (one record for each class diagram) and 17 fields (13 metrics and 3 maintainability sub-characteristics). This table is shown in Appendix A). The metric values were calculated measuring each diagram, and the values for each maintainability sub-characteristics were obtained aggregating subjects´s rating using the mean of them.

– CLUSTERING BY REPERTORY GRIDS. In order to detect the relationships between the class diagrams, for obtaining those which are easy, medium or difficult to maintain (based on subject rates of each maintainability sub-characteristics), we have carried out a hierarchical clustering process by Repertory Grids. The set of elements is constituted by the 28 class diagrams, the constructions are the intervals of values of the subjects´rating. To accomplish an analysis of clusters on elements, we have built a proximity matrix that represents the different similarities of the elements, a matrix of 28 x 28 elements (the diagrams) that above the diagonal represents the distances between the different cycles. Converting these values to percentages, a new table is created and the application of Repertory Grids Analysis Algorithm returns a graphic as a final result (see figure 1).

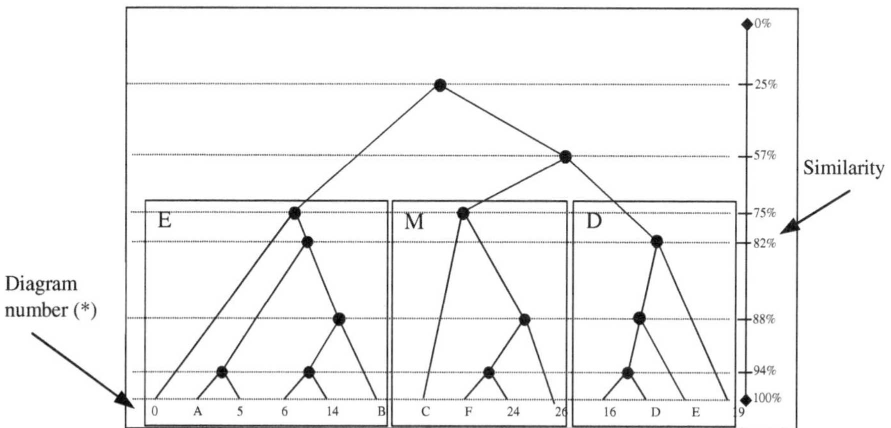

Fig. 1. Clustering results (E: Easy to maintain, M: Medium to maintain, D: Difficult to maintain)

(*) We have grouped some class diagrams assigning them one letter because they have 100% of similarity (see appendix A)

– DATA MINING. The selected algorithm for data mining process was summarise functions. Table 1 shows the parametric definition of the prototypes. These parameters will be modified taking into account the degree of affinity of a new class diagram with the prototypes. With the new modified prototype we will be able to predict the maintainability of a new class diagram.

– FORMAL REPRESENTATION OF CONCEPTUAL PROTOTYPES. The prototypes have been represented as fuzzy numbers, which are going to allow us to obtain a degree of membership in the concept. For the sake of simplicity in the model, they have been represented by triangular fuzzy numbers. Therefore, in order to construct the prototypes (triangular fuzzy numbers) we only need to know their centerpoints ("center of the prototype"), which are obtained by normalising and aggregating the metric values corresponding to the class diagrams of each of the prototypes (see figure 2).

Table 1. Prototypes "Easy, medium and difficult to manitain"

	Understandbility	Analisability	Modifiability
Difficult			
Average	6	6	6
Max.	6	6	7
Min.	6	5	6
Medium			
Average	5	5	5
Max.	5	6	5
Min.	4	4	4
Easy			
Average	2	2	3
Max.	3	3	3
Min.	2	2	2

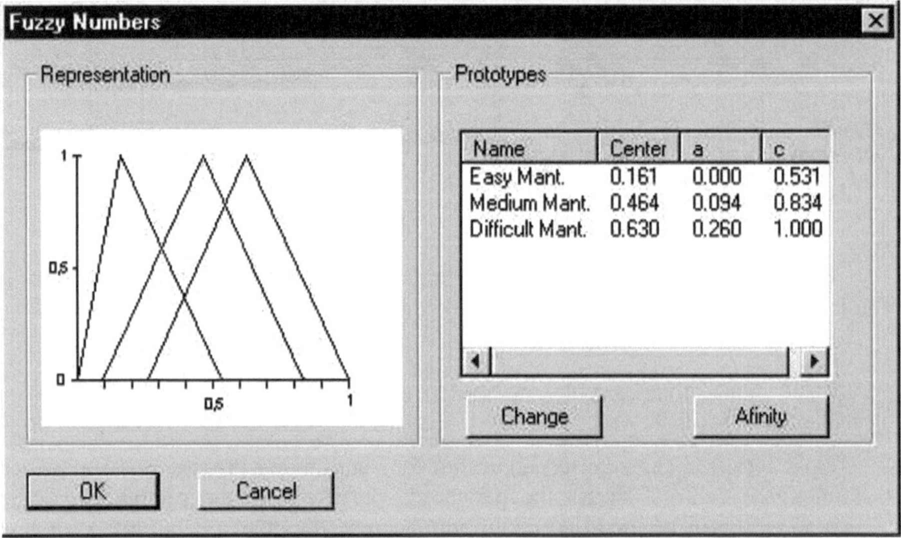

Fig. 2. Representation of the prototypes

– FORMAL REPRESENTATION OF CONCEPTUAL PROTOTYPES. The prototypes have been represented as fuzzy numbers, which are going to allow us to obtain a degree of membership in the concept. For the sake of simplicity in the model, they have been represented by triangular fuzzy numbers. Therefore, in order to construct the prototypes (triangular fuzzy numbers) we only need to know their centerpoints ("center of the prototype"), which are obtained by normalising and aggregating the metric values corresponding to the class diagrams of each of the prototypes (see figure 2).

3.5 Threats to Validity

Following several empirical studies [10],[5],[6] we will discuss the empirical study's various threats to validity and the way we attempted to alleviate them.

- CONSTRUCT VALIDITY. The degree to which the independent and the dependent variables accurately measure the concepts they purport to measure.
- INTERNAL VALIDITY. The degree to which conclusions can be drawn about the causal effect of independent variables on the dependent variables.
- EXTERNAL VALIDITY. The degree to which the results of the research can be generalised to the population under study and other research setting.

Threats to Construct Validity. The dependent variables we used are maintainability sub-characteristics: understandability, analysability and modifiability. We propose subjective metrics for them (using linguistic variables), based on the judgement of the subjects (see section 3.3). As the subjects involved in this experiment have medium experience in OOIS design and implementation we think their ratings could be considered significant. For construct validity of the independent variables, we have to address the question to which degree the metrics used in this study measure the concept they purport to measure. Our idea is to use metrics presented in section 2.1 and 2.2 to measure the structural complexity of an UML class diagram. From a system theory point of view, a system is called complex if it is composed of many (different types of elements), with many (different types of) (dynamically changing) relationships between them [25]. According to this, we think that the construct validity of our independent variables can thus be considered satisfactory. In spite of this, we consider that more experiments must be done, in order to draw a final conclusion to assure construct validity.

Threats to Internal Validity. The following issues have been dealt with:

- DIFFERENCES AMONG SUBJECTS. Using a within-subjects design, error variance due to differences among subjects is reduced. As Briand remarks in [5] in software engineering experiments when dealing with small samples, variations in participant skills are a major concern that is difficult to fully address by randomisation or blocking. In this experiment, professors and students had the same degree of experience in modelling with UML.
- KNOWLEDGE OF THE UNIVERSE OF DISCOURSE AMONG CLASS DIAGRAMS. Class diagrams were designed for the same universe of discourse, only varying the number of attributes, classes, associations, i.e. their constitutents parts. So that, the knowledge of the domain doesn't attempt to the internal validity.
- ACCURACY OF SUBJECT RESPONSES. Subjects assumed the responsibility for rating each maintainability sub-characteristics. As they have medium experience in OO software design and implementation, we think their responses could be considered valid. However we are aware that not all of them have exactly the

same degree of experience, and if the subjects have more experience minor inaccuracies could be introduced by subjects.

- LEARNING EFFECTS. All the tests in each experiment were put in a different order, to avoid learning effects. Subjects were required and controlled to answer in the order in which test appeared.
- FATIGUE EFFECTS. On average the experiment lasted for less than one hour, so fatigue was not very relevant. Also, the different order in the tests helped to avoid these effects.
- PERSISTENCE EFFECTS. In order to avoid persistence effects, the experiment was run with subjects who had never done a similar experiment.
- SUBJECT MOTIVATION. All the professors who were involved in this experiment have participated voluntarily, in order to help us in our research. We motivated students to participate in the experiment, explaining to them that similar tasks to the experimental ones could be done in exams or practice by students, so they wanted to take the most of the experiment.
- OTHER FACTORS. Plagiarism and influence between students really could not be controlled. Students were told that taking with each other was forbidden, but they did the experiment alone without any control, so we had to trust them as far as that was concerned.

Seeing the results of the experiment we can conclude that empirical evidence of the existing relationship between the independent and the dependent variables exists. But only by replicating controlled experiments, where the measures would be varied in a controlled manner and all the other factors would be kept constant, could really demonstrate causality.

Threats to External Validity. The greater the external validity, the more the results of an empirical study can be generalised to actual software engineering practice. Two threat of validity have been identified which limit the ability to apply any such generalisation:

- MATERIALS AND TASKS USED. In the experiment we tried to use class diagrams and tasks which can be representative of real cases, but more empirical studies taking "real cases" from software companies must be done.
- SUBJECTS. To solve the difficulty of obtaining professional subjects, we used professors and advanced students from software engineering courses. We are aware that more experiments with practitioners and professionals must be carried out in order to be able to generalise these results. However, in this case, the tasks to be performed do not require high levels of industrial experience, so, experiments with students could be appropriate [1].

In general in order to extract a final conclusion we need to replicate this experiment with a greater number of subjects, including practitioners. After doing replication we will have a cumulative body of knowledge; which will lead us to confirm if the presented metrics could really be used as early quality indicators, and could be used to predict UML class diagrams maintainability.

4 Prediction of UML Class Diagram Maintainability

Using Fuzzy Deformable Prototypes [21],[22], we can deform the most similar prototype to a new class diagram, and define the factors for a new situation, using a linear combination with the degrees of membership as coefficients. We will show an example of how to deform the fuzzy prototypes found in section 3.5. Given the normalised values corresponding to a new class diagram:

NC	NA	NM	NAssocR	NAssocVC	NAggR	NAggRVC	NAggH	NDepR	NDepVC	NGenR	NGH	MaxDIT
0.7	05	0.69	0.71	0.48	0.67	0.45	0.67	0.75	0.43	0.83	1	0.4

The final average is 0.64. The affinity with the prototypes is shown in figure 3.

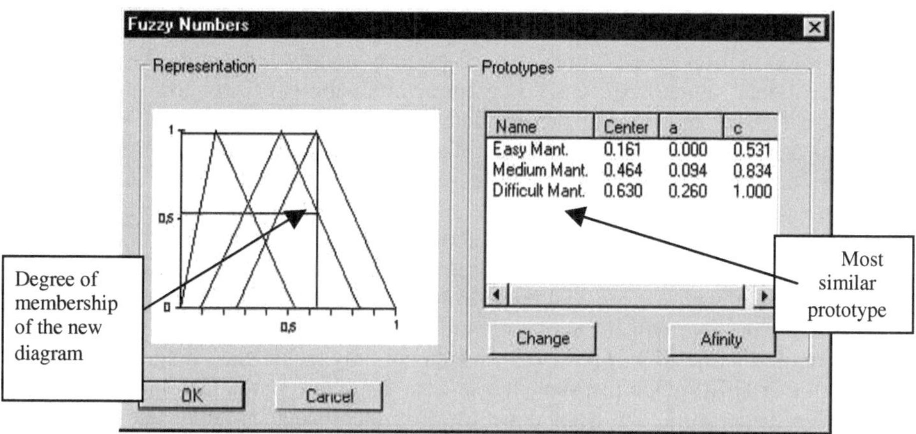

Fig. 3. Affinity of the real case with the prototypes

The most similar prototype for this new class diagram is "Difficult to maintain", with a degree of membership of 0.98. Then, the prediction is:

	Understandability	Analisability	Modifiability
Average	6	6	6
Maximum	6	6	7
Minimum	6	5	6

We want to highlight that this a first approach to predict UML class diagram maintainability, we need "real data" about UML class diagram maintainability efforts, like time spent in maintenance tasks in order to predict data that can be highly useful to software designers and developers.

5 Conclusions and Future Work

Due to the growing complexity of OOIS, continuous attention to and assessment of class diagrams is necessary to produce quality information systems. The fact that UML has emerged is a great step forward in object modelling. However this does not guarantee the quality of the models produced through the IS life cycle. Therefore, it is necessary to have metrics in order to evaluate their quality from the early phases in the OOIS development process.

In this paper we have presented a set of metrics for assessing the structural complexity of UML class diagrams, obtained at early phases of the OOIS life cycle.

We have also carried out a controlled experiment, with the objective of predicting UML class diagram maintainability based on the metrics values and the expert's rating of each of the maintainability sub-characteristics. The prediction model is an extension of the traditional KDD called FPKD and a novel technique which can be used for prediction based on Fuzzy Deformable Prototypes [21],[22]. This model have been used for different kinds of real problems, such as forest fire prediction, financial analysis or medical diagnosis, with very good results.

Nevertheless, despite of the encouraging obtained results we are aware that we need to do more metric validation, both empirical and theoretical in order to assess if the presented metrics could be really used as early quality indicators. Also could be useful "real data" about UML class diagram maintainability efforts, like time spent in maintenance tasks in order to predict data that can be highly fruitful to software designers and developers. But the scarce of such data continues to be a great problem we must tackle to validate metrics. In [8] suggested the necessity of a public repository of measurements experiences, which we think that could be a good step towards the success of all the work done about software measurement.

It will possible to that when more "real data" on systems developed using UML will be available, which is the challenge of most of the researchers in this area.

In future work, we will focus our research on measuring other quality factors like those proposed in the ISO 9126 (1999), which not only tackle class diagrams, but also evaluate other UML diagrams, such as use-case diagrams, state diagrams, etc. To our knowledge, little work has been done towards measuring dynamic and functional models [11],[23],[24]. As is quoted in [8] this is an area which lacks in depth investigation.

Acknowledgements

This research is part of the MANTICA project, partially supported by CICYT and the European Union (1FD97-0168). This research is part of the DOLMEN project supported by CICYT (TIC 2000-1673-C06-06) and the CIPRESES project supported by CICYT (TIC 2000-1362-C02-02).

Appendix A

The following table shows in each row the number of the class diagrams used in the experiment described in section 3, and in each column their metric values. Attached to some diagrams appear one letter. The diagrams which have the same letter mean that they have 100% of similarity.

	NC	NA	NM	NAssoc	NassocVC	NAgg	NAggVC	NAggH	NDep	NDepVC	NGen	NGH	MaxDIT	Understandability	Analisability	Modifiability
D0	2	4	8	1	0.5	0	0	0	0	0	0	0	0	1	1	1
D1 (A)	3	6	12	1	0.33	1	0.33	1	0	0	0	0	0	2	2	2
D2 (A)	4	9	15	1	0.25	2	0.5	1	0	0	0	0	0	2	2	2
D3 (A)	3	7	12	3	1	0	0	0	0	0	0	0	0	2	2	2
D4 (A)	5	14	21	1	0.2	3	0.6	2	0	0	0	0	0	2	2	2
D5	3	6	12	2	0.66	0	0	0	0	0	0	0	0	2	2	2
D6	4	8	12	3	0.75	0	0	0	1	0.25	0	0	0	2	3	3
D7 (B)	6	10	14	2	0.33	2	0.33	1	0	0	2	1	1	3	3	3
D8 (A)	3	9	12	1	0.33	0	0	0	1	0.33	0	0	0	2	2	2
D9 (B)	7	14	20	2	0.28	3	0.42	1	0	0	2	1	1	3	3	3
D10 (B)	9	18	26	2	0.22	3	0.33	1	0	0	4	2	1	3	3	3
D11 (B)	7	18	37	3	0.42	3	0.42	1	0	0	2	1	1	3	3	3
D12 (B)	8	22	35	3	0.37	2	0.25	1	1	0.12	2	1	1	3	3	3
D13 (A)	5	9	26	0	0	0	0	0	0	0	4	1	2	2	2	2
D14	8	12	30	0	0	0	0	0	0	0	10	1	3	2	3	3
D15 (C)	11	17	38	0	0	0	0	0	0	0	18	1	4	4	4	4
D16	20	42	76	10	0.5	6	0.3	2	2	0.1	10	3	2	6	6	6
D17 (D)	23	41	88	10	0.43	6	0.23	2	2	0.06	16	3	3	6	6	6
D18 (E)	21	45	94	6	0.28	6	0.28	2	1	0.04	20	2	4	6	5	6
D19	29	56	98	12	0.41	7	0.24	3	3	0.1	24	4	4	6	6	7
D20 (B)	9	28	47	1	0.11	5	0.55	2	0	0	2	1	1	3	3	3
D21 (F)	18	30	65	3	0.16	5	0.27	1	0	0	19	2	4	5	5	5
D22 (D)	26	44	79	11	0.42	6	0.23	2	0	0	21	5	3	6	6	6
D23 (F)	17	32	69	1	0.05	5	0.19	1	0	0	19	1	5	5	5	5
D24	23	50	73	9	0.4	7	0.3	3	2	0.08	11	4	1	5	6	5
D25 (E)	22	42	84	14	0.63	4	0.18	2	4	0.18	16	3	3	6	5	6
D26	14	34	77	4	0.28	9	0.64	2	0	0	7	2	4	4	5	5

References

1. Basili V., Shull F. and Lanubile F. Building knowledge through families of experiments. *IEEE Transactions on Software Engineering*, Vol. 25 No. 4, (1999) 435-437.

2. Boehm, B. *Software Engineering Economics*. Prentice-Hall (1981).
3. Briand L., Morasca S. and Basili V. Defining and Validating Measures for Object-Based high-level design. *IEEE Transactions on Software Engineering*. Vol. 25 No. 5, (1999) 722-743.
4. Briand L., Arisholm S., Counsell F., Houdek F., and Thévenod-Fosse P. Empirical Studies of Object-Oriented Artifacts, Methods, and Processes: State of the Art and Future Directions. *Technical Report IESE 037.99/E, Fraunhofer Institute for Experimental Software Engineering*, Kaiserslautern, Germany, (1999).
5. Briand L., Bunse C. and Daly J. A Controlled Experiment for evaluating Quality Guidelines on the Maintainability of Object-Oriented Designs. *Technical Report IESE 002.99/E, Fraunhofer Institute for Experimental Software Engineering*, Kaiserslautern, Germany, (1999).
6. Briand L., Wüst J., Daly J. and Porter D. Exploring the relationships between design measures and software quality in object-oriented systems. *The Journal of Systems and Software 51*, (2000) 245-273.
7. Brito e Abreu, F. and Carapaçua, R. (1994). Object-Oriented Software Engineering: Measuring and controllong the development process. *4th Int Conference on Software Quality*, Mc Lean, USA.
8. Brito e Abreu F., Zuse H., Sahraoui H. and Melo W. Quantitative Approaches in Object-Oriented Software Engineering. *Object-Oriented technology: ECOOP'99 Workshop Reader*, Lecture Notes in Computer Science 1743, Springer-Verlag, (1999) 326-337.
9. Chidamber S. and Kemerer C. A Metrics Suite for Object Oriented Design. *IEEE Transactions on Software Engineering*. Vol. 20 No. 6, (1994) 476-493.
10. Daly J., Brooks A., Miller J., Roper M. and Wood M. Evaluating Inheritance Depth on the Maintainability of Object-Oriented Software. *Empirical Software Engineering*, 1, Kluwer Academic Publishers, Boston, (1996) 109-132.
11. Derr K. *Applying OMT*. SIGS Books, New York. (1995).
12. Fayyad U., Piatetsky-Shapiro G. and Smyth P. The KDD Process for Extracting Useful Knowledge from Volumes of Data. *Communications of the ACM*, Vol. 39 No. 11, (1996) 27 – 34.
13. Fenton N. and Pfleeger S. *Software Metrics: A Rigorous Approach*. 2nd. edition. London, Chapman & Hall, (1997).
14. Genero, M., Piattini, M. and Calero, C. Early Measures For UML class diagrams. *L'Objet*. Hermes Science Publications, Vo.l 6 No. 4, (2000) 489-515.
15. ISO/IEC 9126-1.2. Information technology- Software product quality – Part 1: Quality model, (1999).
16. Henderson-Sellers B. *Object-Oriented Metrics - Measures of complexity*. Prentice-Hall, Upper Saddle River, New Jersey, (1996).
17. Kitchenham, B., Pflegger, S. and Fenton, N. Towards a Framework for Software Measurement Validation. *IEEE Transactions of Software Engineering*, Vol. 21 No. 12, (1995) 929-943.
18. Lorenz M. and Kidd J. *Object-Oriented Software Metrics: A Practical Guide*. Prentice Hall, Englewood Cliffs, New Jersey, (1994).
19. Marchesi M. OOA Metrics for the Unified Modeling Language. *Proceedings of the 2nd Euromicro Conference on Software Maintenance and Reengineering*, (1998) 67-73.
20. Object Management Group. *UML Revision Task Force. OMG Unified Modeling Language Specification, v. 1.3. document ad/99-06-08, (*1999).
21. Olivas J. A. and Romero F. P. FPKD. Fuzzy Prototypical Knowledge Discovery. Application to Forest Fire Prediction. *Proceedings of the SEKE'2000*, Knowledge Systems Institute, Chicago, Ill. USA, (2000) 47 – 54.

22. Olivas J. A. Contribution to the Experimental Study of the Prediction based on Fuzzy Deformable Categories, PhD Thesis, University of Castilla-La Mancha, Spain, (2000).
23. Poels G. On the use of a Segmentally Additive Proximity Structure to Measure Object Class Life Cycle Complexity. *Software Measurement : Current Trends in Research and Practice*, Deutscher Universitäts Verlag, (1999), 61-79,.
24. Poels G. On the Measurement of Event-Based Object-Oriented Conceptual Models. *4th International ECOOP Workshop on Quantitative Approaches in Object-Oriented Software Engineering*, June 13, Cannes, France, (2000).
25. Poels, G. and Dedene, G.. Measures for Assessing Dynamic Complexity Aspects of Object-Oriented Conceptual Schemes. *In: Proceedings of the 19th International Conference on Conceptual Modeling (ER 2000),* Salt Lake City, (2000), 499-512.
26. Schneidewind, N. Methodology For Validating Software Metrics. *IEEE Transactions of Software Engineering*, Vol. 18 No. 5, (1992) 410-422.
27. Tian J. Taxonomy and Selection of Quality Measurements and Models. *Proceedings of SEKE'99, The 11th International Conference on Software Engineering & Knowledge Engineering*, June 16-19, (1999) 71-75.
28. Zadeh, L. A.A note on prototype set theory and fuzzy sets. *Cognition* 12, (1982), 291–297.
29. H. Zuse. *A Framework of Software Measurement*. Berlin, Walter de Gruyter, (1998).

A Rigorous Metamodel for UML Static Conceptual Modelling of Information Systems

Régine Laleau[1] and Fiona Polack[2]

[1] CEDRIC-IIE(CNAM),
18 allée Jean Rostand, 91025 Evry, France
laleau@iie.cnam.fr

[2] Department of Computer Science
University of York, York, YO10 5DD, UK
fiona@cs.york.ac.uk

Abstract. Object-oriented specification and design approaches, such as the UML, are used in many sectors, including information systems development. One reason for the popularity of the UML is that it has notations for many types of system and all stages of development. However, this also makes it cumbersome and semantically imprecise. This paper looks at a UML for information systems specification. It both selects from and extends the UML1.3, defining the semantics of the IS UML in B-style invariants. The paper discusses the relationship of the work both to the current UML metamodels, and to proposals for extensions to the UML.

1 Introduction

The Unified Modeling Language (UML) is still evolving, but is widely used in commercial software engineering. The Object Management Group (OMG), which oversees the development of the UML has recognised the need to allow specialisation of the language for different types of system. The mechanisms of extension are not static or agreed, and Requests for Proposals (RFPs) have been issued, prior to evolution of a UML version 2.0.

The UML metamodels have varied over the years, but have all been general-purpose. The language contains concepts for abstract and concrete expression, and for many types of system (including, for example, real time and concurrent systems). One problem for users of UML is to determine the subset of notations which is relevant to their particular project, or to their particular level of abstraction.

This paper presents a variant of UML1.3 suitable for the specification of the static structure of an information system (IS). This derives from ongoing work on IS specification at CEDRIC-IIE, where an approach has been developed to derive formal specifications (in the B notation [1]) from object-oriented class and behavioural models [19,18,12][1].

[1] Metamodels for relevant behavioural notations have been defined, as well as the static model. Work on behavioural semantics is in progress.

K.R. Dittrich, A. Geppert, M.C. Norrie (Eds.): CAiSE 2001, LNCS 2068, pp. 402–416, 2001.
© Springer-Verlag Berlin Heidelberg 2001

1.1 Metamodel Presentation: The B Method

The IS UML semantics are presented as metamodels, which are compared to the OMG's UML 1.3 semantics (http://www.rational.com/uml). The presentation uses UML class diagrams, for the basic structure. The structure, detail and constraints are then expressed mathematically, as B invariant clauses.

B notations are used for a number of reasons:

- B underpins the IS specification work, on which this paper is based.
- B provides the needed power of expression. It includes standard operators on sets, functions and relations, and types such as sequences, allowing a precise, expression of invariants.
- A formal notation such as B supports reasoning, such as the verification of consistency of the metamodels. Meta-rules can also be defined to assist in the execution of proofs about the specification.
- A formal metamodel would assist development of advanced modelling tools. The formal meaning can be built in, such that diagrammatic models of a system are formally correct.

The B invariants use only notations in common usage in set theory and predicate logic. The intuitive meaning of each invariant is given in the text. This paper first considers the bases for modification of the UML. It then illustrates the metamodel for static structure, the rationale behind it and its relationship to UML1.3.

2 UML Specialisation and Extension

The UML1.3 variant illustrated here does not conform to the OMG's current UML extension mechanisms, because the UML mechanisms do not provide the relevant facilities [2,9,4].

The OMG structures its metamodel for UML1.3 using *packages*. However, an OMG package has limited power: it is not allowed to import another package in order to elaborate partially-defined features from that package [6, Sect. 2.1.2.8]. Since all the features of a concept must be stated in one package, it is not possible to separate out the UML specification features, or the components relevant to IS development using packages.

The OMG *profile* was proposed as a more flexible extension mechanism. The definition recommended for UML2.0 RFPs (see [9]) defines the profile as a subset of the UML metamodel, with additional well-formedness rules (in OMG's object constraint language, OCL). All profile concepts must be expressible in terms of existing UML metamodel concepts or stereotypes extending these concepts. An OMG profile cannot add structure (associations) or entirely-new concepts to the metamodel. Although the IS metamodel can express many of the required features in terms of existing UML1.3 concepts, it adds to and modifies the structure of the metamodel. It also modifies some semantics.

OMG also has a more robust extension approach, using the MetaObject Facility (MOF), but this is complicated by the fact that the UML metamodels are not MOF-compliant, and no rules are given for the syntactic or semantic interpretation of UML concepts under extension [9].

A number of alternative extension mechanisms have been proposed, motivated by a desire to shrink and focus the core UML [11,4,10]. These are more appropriate to this work. We intend to move towards an IS UML following the MOF-compliant pUML preface definition [6,9]. The pUML, like OMG, uses OCL to complement the diagrammatic metamodels[2]. However, OCL does not have a precise semantics; at this stage, we prefer to use a formal language with well-defined semantics for describing specialisation of and extensions to the UML notations.

3 Simplifying the UML for IS Specification

An IS manages large amounts of highly structured data.. A static model is essential, along with a large number of data constraints defining the integrity conditions of the IS. In a well-designed IS, operation preconditions maintain integrity by enforcing the data constraints. (This is explored formally in [14,15] and used in work on the formal derivation of databases [16]. Tool support is described in [17].)

This paper focuses on the semantics of class models for use in IS specification. Features of the general OMG metamodels which are rarely relevant to IS or to specification are omitted. These include attributes of metamodel classes concerning concurrency, real-time operation, visibility and scope. Relevant extracts of the OMG's UML1.3 metamodel are given in the Appendix to this paper.

Complex subclass hierarchies, used in the OMG metamodels to handle general system features, are collapsed and simplified. For example, the *Name* subclass in the *ModelElement* hierarchy of the UML1.3 fundamentals is demoted to an attribute of classes representing named constructs.

The simplification of the UML1.3 structures facilitates expression of some semantic features such as association qualifiers, and improves the readability of the metamodel diagrams. However, the collapsing of the OMG structures sacrifices some classification of semantic elements.

The IS metamodel (Fig. 1) brings together all relevant static UML features except n-ary associations and aggregation/composition. A n-ary association can always be replaced by binary associations and constraints. Aggregation associations imply constraints on the lifecycles of objects and the operations of the classes concerned. Work is in progress on this [24]. At this stage, OMG attributes recording aggregation/composition (and those relating to navigability) have been omitted from the metamodels.

Some structure is omitted for simplicity. For example, in the OMG metamodels, all generalisable features are subclasses of *GeneralizableElement* (Appendix,

[2] The formalisation of diagrammatic syntax and semantics dates from research tools for the formal review of diagrammatic specifications [25]. Initial UML work represented the whole UML metamodel in a formal notation, such as Z [8,5]. pUML now uses UML diagrams extended with OCL constraints [3,7].

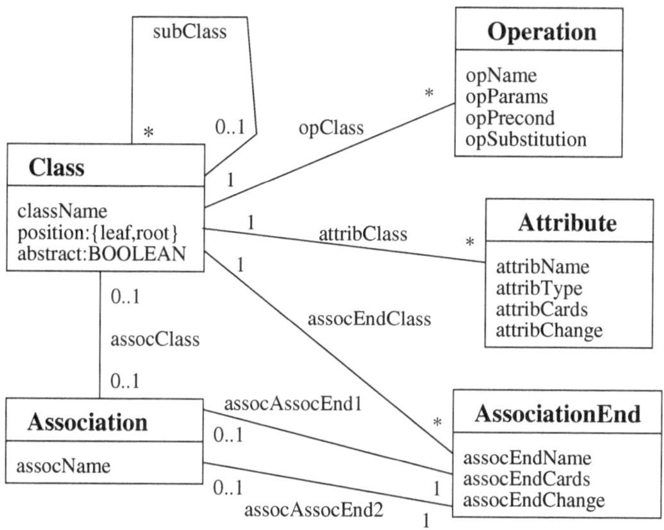

Fig. 1. Metamodel for the IS Class model

Fig. 5). For the IS class model, we restrict generalisation to classes, because the semantics of other generalisations (eg association) is not well defined.

Metamodel associations such as those between *Class* and *Attribute*, and between *Class* and *Operation*, are derived from the OMG metamodels by combining references to the various classes and then collapsing the hierarchies (Appendix, Fig. 5). The OMG's elaborate modelling of association class semantics, using multiple inheritance (Appendix, Fig. 6), is replaced by a simple association.

The OMG metamodel coverage of integrity constraints [21, p2-14] is not adequate for IS. This is addressed in Sect. 6, below.

4 Diagrammatic Metamodel for the IS Class Models

A class diagram provides the basic metamodel structure for the IS UML (Fig. 1). The discussion focuses on the meaning (the expressed semantics) and on the relationship between this metamodel and those of the OMG. The metamodel can be thought of as extending rather than replacing semantics of the OMG metamodels.

The attributes of the IS *Class* are implicitly inherited from the OMG's *GeneralizableElement* and *ModelElement*. Inheritance is expressed by the association *subClass*, a self-reference on *Class*. Any instance of *Class* is thus linked to its subclasses and superclass. For simplicity, multiple inheritance is not considered here.

The classes *Attribute* and *Operation* express the structure and behaviour of a class. A class may have many attributes and many operations, but each attribute or operation is directly linked to one class. Subclasses inherit the characteristics of their superclasses via the *subClass* association.

Operation has a name, a signature, a precondition and a substitution. It does not include attributes for concurrency, since this is conventionally handled in IS by a database management system.

The attributes of *Attribute* are from the OMG superclasses, omitting those relating to visibility and scope.

Association and *AssociationEnd* have name attributes (implicit inheritance from the OMG *ModelElement*). *AssociationClass* is modelled by the association *assocClass* between *Class* and *Association*. The semantics are elaborated below.

5 Formal Semantics: Extending the Diagrammatic Metamodels

The B invariants expressing the semantics of the IS UML class models can be typechecked (and, if required, used for further rigorous development, including refinement and proof).

In order to exploit the B tool support, the whole model, including the elements already expressed in diagrams, needs to be expressed in B. This section illustrates the precise expression of the semantics. The formal invariants representing the metamodel structure are derived using existing guidelines for representing object-oriented class diagrams in B [19,13]. The formal metamodel is extended with invariant clauses to clarify constraints among the components. The instantiation of classes to objects and associations to links is used to express multiplicity and other constraints.

This paper does not present the full definitions. The subsections illustrate typing information, the data structure, and finally the additional semantics required to express the full meta-language.

5.1 Type Structures for the Formal Semantics

Data types (written in upper case) can be equated to mathematical sets. The B Method incorporates integers (and integer-related proof rules), and has a variety of mechanisms for introducing new sets, at various levels of detail. In this presentation, most types are defined as given sets, about which most details remain unelaborated. A B given set may have enumerated values, or may simply be an (implicitly) non-empty, finite set [1, Sects. 2.2 and 4.17]. A given set called *TYPE* is defined here to represent UML- and user-defined data types.

The type of each class in the diagrammatic metamodel is a given set, as is that of the class *Object*, used to express multiplicity semantics. An invariant clause (Table 1) introduces sets of known instances of each class as a subset of the class type.

5.2 Structural Definitions Using B Relations and Functions

A binary relation is a set of ordered pairs, linking an element of the source set (the domain, dom) to an element of the target set (the range, ran) [1, p77-9].

Table 1. B invariants for sets of known instances of metamodel classes

class	$\subseteq$	*CLASS*
association	$\subseteq$	*ASSOCIATION*
attribute	$\subseteq$	*ATTRIBUTE*
operation	$\subseteq$	*OPERATION*
object	$\subseteq$	*OBJECT*
associationEnd	$\subseteq$	*ASSOCIATIONEND*

Table 2. Example mappings modelled by relation and functions

$\leftrightarrow$	*relation*	$*$ *to* $*$ *mapping*
$\rightarrow$	*function*	$*$ *to* 1 *mapping, all domain takes part*
$\rightarrowtail$	*injection*	1 *to* 1 *mapping, all domain takes part*
$\nrightarrow$	*partial function*	$*$ *to* 1 *mapping, not all domain takes part*
$\mapsto$	*partial injection*	1 *to* 1 *mapping, not all domain takes part*
$\twoheadrightarrow$	*partial surjection*	$*$ *to* 1 *mapping, not all domain, all range takes part*

The inverse of a relation ($relation^{-1}$) defines the reverse ordered pairs (ie. range first). The relation is a mathematical model of a many-to-many association.

Participation of domain and range elements can be constrained to mathematical functions [1, p86]. These can model any degree of association (Table 2).

Functions are used to model the associations in the metamodel (Table 3). Functions are also used to associate attributes to classes (Table 4). A total function ($\rightarrow$) from the set of known instances of the class to the type of an attribute models a mandatory, mono-valued attribute. Injections ($\rightarrowtail$) are used for mandatory attributes with a unique value. Partial functions ($\nrightarrow$) define optional, mono-valued attributes. There are no multi-valued attributes in the metamodel.

Most of the attribute types are OMG types [21, Sect. 2.7]. *NAME* is an implicit subset of the OMG's catch-all *STRING* type. The ranges of *attribCards* and *assocEndCards* are sets of natural numbers drawn from the powerset, $\mathbb{P}\,\mathbb{N}$. This allows any cardinality value to be expressed, with B's upper limit (*maxint*) equating to UML's *. The range of *opParams* is a sequence *seq*(*TPARAM*). This is an indexed structure (mathematically, it is an injection from natural numbers to the sequence type), and is used here to model an ordered list of parameters. The type *TPARAM* is a triple of a name, a basic UML type, and an indicator of whether the parameter is input (i) or output (o). (This simplifies the OMG representation, which additionally has *inout* and *return* parameters.)

$$TPARAM \in (NAME \times TYPE \times \{i, o\}) \; .$$

The range of *opPrecond* is the OMG type, *PREDICATE*. This type is not elaborated in UML1.3. However, the semantics could be defined more fully. For example, a predicate is only valid if it evaluates to a boolean (true or false). A semantically-relevant predicate should not always evaluate to the same value.

Table 3. B invariants representing associations in the IS metamodel

$subclass$	$\in$	$class \nrightarrow class$
$assocClass$	$\in$	$association \rightarrowtail\!\!\!\nrightarrow class$
$assocAssocEnd1$	$\in$	$association \rightarrowtail associationEnd$
$opClass$	$\in$	$operation \rightarrow class$
$assocAssocEnd2$	$\in$	$association \rightarrowtail associationEnd$
$attribClass$	$\in$	$attribute \rightarrow class$
$assocEndClass$	$\in$	$assocEnd \rightarrow class$

Table 4. B invariants associating attributes to IS metamodel classes

CLASS	*ATTRIBUTE INVARIANT*
Class	$className \in class \rightarrowtail NAME$
	$position \in class \rightarrow \{leaf, root\}$
	$abstract \in class \rightarrow BOOLEAN$
Attribute	$attribName \in attribute \rightarrow NAME$
	$attribType \in attribute \rightarrow TYPE$
	$attribCards \in attribute \rightarrow \mathbb{P}\ \mathbb{N}$
	$attribChange \in attribute \rightarrow BOOLEAN$
Operation	$opName \in operation \rightarrow NAME$
	$opPrecond \in operation \rightarrow PREDICATE$
	$opSubstitution \in operation \rightarrow STRING$
	$opParams \in operation \rightarrow seq(TPARAM)$
Association	$assocName \in association \nrightarrow NAME$
AssociationEnd	$assocEndName \in associationEnd \nrightarrow NAME$
	$assocEndCards \in associationEnd \rightarrow \mathbb{P}\ \mathbb{N}$
	$assocEndChange \in associationEnd \rightarrow BOOLEAN$

Two of the constraining invariants on the structure modelled in Fig. 1 express the fact that attributes (operations) of a class comprise those declared in that class and in all of its superclasses. For a given class, c:

$$allClassAttributes[\{c\}] = \bigcup\{attribClass^{-1}[\{c'\}] \mid c' \in subclass^*(c)\}$$
$$allClassOperations[\{c\}] = \bigcup\{opClass^{-1}[\{c'\}] \mid c' \in subclass^*(c)\}\ .$$

The notation subclass*, reflexive transitive closure, models the whole subclass hierarchy including the root class. A hierarchy without the root class is modelled by $subclass^+$, transitive closure. This is used in the invariant which prevents cycles of inheritance,

$$subclass^+ \cap id(class) = \varnothing\ .$$

5.3 Semantics of Class Models under Instantiation

The semantics of instantiated class models express, for instance, cardinalities and multiplicities. In the OMG metamodels, these are expressed diagrammatically [21, Fig. 2-16]; here they are modelled in B invariants.

The instantiation of a class is an object (Table 1), and each object is one instantiation of exactly one class:

$$objectClass : object \rightarrow class .$$

All the objects of a class, taking into account inheritance links, is defined as,

$$allClassObject = (subclass^*)^{-1} \mathbin{⨾} objectClass^{-1} .$$

The composition operator, ⨾, can be thought of as the navigation of the metamodel associations.

The definition of *object* is used to instantiate associations as sets of links among objects. There may be more than one association among the same objects:

$$linkAssociation : association \rightarrow (object \leftrightarrow object) .$$

The semantics of association ends requires that the ends of an association are different. This is stated as a partition of the set of association ends:

$$\begin{array}{l} \mathrm{ran}\ (assocAssocEnd1) \cup \mathrm{ran}\ (assocAssocEnd2) = assocationEnd \wedge \\ \mathrm{ran}\ (assocAssocEnd1) \cap \mathrm{ran}\ (assocAssocEnd2) = \varnothing . \end{array}$$

An additional invariant is required to constrain participation in the instantiation to objects of the classes implied by the association:

$$\begin{array}{l} linkAssociation \subseteq \\ ((assocAssocEnd1 \mathbin{⨾} assocEndClass \mathbin{⨾} allClassObject) \otimes \\ \ (assocAssocEnd2 \mathbin{⨾} assocEndClass \mathbin{⨾} allClassObject)) . \end{array}$$

This expression states that the links which instantiate an association are drawn from all pairs (constructed using the direct product, ⊗) of objects such that the first element comes from *assocAssocEnd1* and the second from *assocAssocEnd2*. The objects are those which instantiate the class (with all its superclasses) for the relevant association end.

An object which is the instantiation of an association class corresponds to one and only one link of the relevant association. The *card* operator returns the number of elements of a set, relation or function:

$$\begin{array}{l} \forall\, a \in \mathrm{dom}\ assocClass \bullet \\ card(linkAssociation(a)) = card(allClassObject(assocClass(a))) . \end{array}$$

This expression states that, for any association which has an association class, the number of instantiations of that association must be the same as the number of instantiations of the association class. This is more logical than the OMG semantics (using multiple inheritance): an association class is a simple class, which assists in the expression of any constraints, associations and inheritance structures in which it participates.

The multiplicity of association ends is expressed as a general constraint that the number of instantiated links must be a permitted cardinality of the association end:

$$\forall\, o \in object \bullet$$
$$card((linkAssociation(assocAssocEnd1(ae)) \cup (linkAssociation(assocAssocEnd2(ae))^{-1}))[\{o\}]) \in assocEndCards(ae)\ .$$

The expression refers to a specific association end, *ae*. It extracts and instantiates the association of which *ae* is the association end numbered 1. This gives a set of pairs of objects with those at the *ae* end first. The next part of the expression extracts and instantiates the other end of the association, giving a set of pairs of objects with those at the *ae* end second. To get the total number of objects taking part in the association at the *ae* end, the second pair is inverted (so that those at the *ae* end are first). The two sets of pairs are unified. The objects linked to each object of the class at the *ae* end are extracted (using a relational image of o, $[\{o\}]$) and counted using the *card* operator. The number of objects must be one of those defined in the set of numbers which is the value of the *assocEndCards* attribute of the association end, *ae*.

Instantiation is used to elaborate the semantics of attribute values and attribute cardinality. The instantiation of an attribute uses the given set, *VAL*, representing all possible values in the model. An object maps to many values, representing the instantiations of its attributes:

$$attribObjectValue : attribute \rightarrow (object \leftrightarrow VAL)\ .$$

When instantiated, an object can take only values for the attributes of its class (including superclasses):

$$\forall\, c \in class \bullet$$
$$\mathrm{dom}(\bigcup(attribObjectValue\ [allClassAttributes[\{c\}]])) \subseteq objectClass^{-1}[\{c\}]\ .$$

When any object is instantiated, the number of values assigned to each attribute a (including inherited attributes) must be a valid cardinality for that attribute:

$$\forall\, o \in object \bullet$$
$$\forall\, a \in (allClassAttributes(objectClass(o))) \bullet$$
$$card(attribObjectValues(a)[\{o\}]) \in attribCards(a)\ .$$

6 Static Integrity Constraints

In IS, complex relationships and restrictions on data are expressed as constraints. In general, these constraints are available to the developer when writing operations on the IS. In particular, cardinality, key and inter-association constraints can be used to automatically generate operation preconditions [14,15]. The OMG metamodels do not provide sufficient concepts to express the complexity and variety of IS constraints [21, p2-14]. The IS metamodel, of which Fig. 2 is an extract, is sufficient to express both general IS integrity concepts and data constraints specific to a particular IS. The structure specialises the OMG class

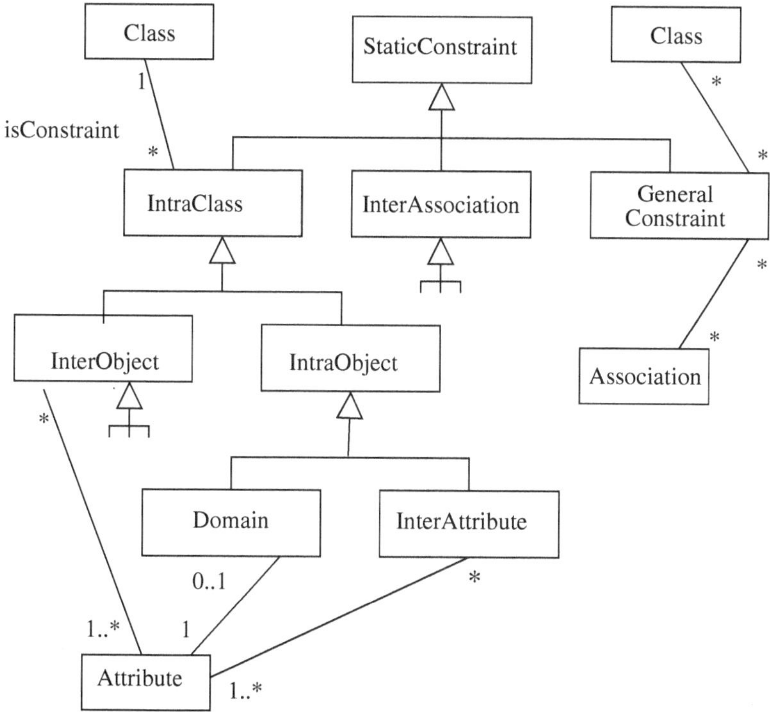

Fig. 2. Metamodel extract for the IS static constraints

StaticConstraint, and uses the IS metamodel classes *Class*, *Attribute* and *Association.*

Of the three top-level subclasses in Fig. 2, *GeneralConstraint* provides the structure for data constraints involving any combination of classes, attributes and associations. *InterAssociation* is considered below (Sect. 6.2). *IntraClass* defines constraints within one class. The subclass *IntraObject* can express constraints on domains and among attributes of one object. *InterObject* is considered below (Sect. 6.1).

6.1 *InterObject*: Keys and Dependency Constraints

It is common practice in IS specification to identify candidate keys, sets of attributes from a class which can uniquely identify any instance, and upon which other attributes are functionally dependent. To express these concepts, the IS metamodel in Fig. 1 is extended (Fig. 3) with *Key* and *FunctionalDependency*, two subclasses of *InterObject*. The association *isKey* is a specific link of the *isConstraint* association in Fig. 2

Invariants again capture and elaborate the semantics. The association *isKey* is a function from *key* to *class*, whilst *attribKey* is represented as a relation

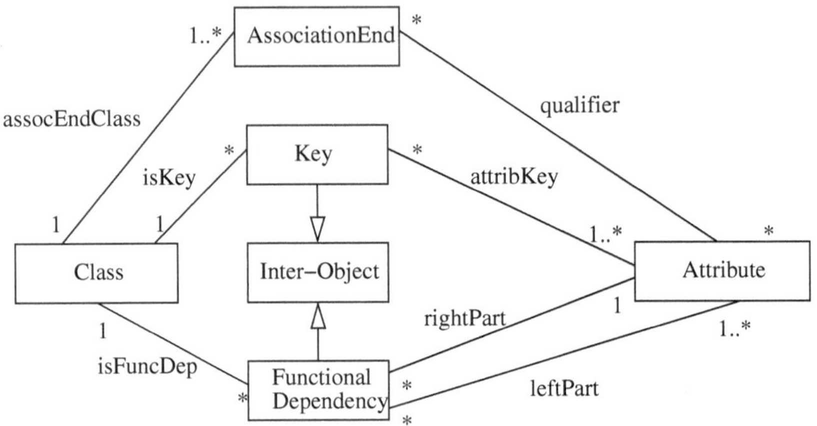

Fig. 3. Metamodel elaborating key and dependency constraints

between *key* and *attribute*. The additional constraint relates the attributes in the range of *attribKey* to the class in the range of *isKey* via the *key*.

$$\begin{array}{l} isKey \in key \rightarrow class \\ \wedge\ attribKey \in key \leftrightarrow attribute \\ \wedge\ \mathrm{dom}\ (attribKey) = key\ . \end{array}$$

The attributes of a key must be attributes of its class or superclasses:

$$\forall c \in \mathrm{ran}\ (isKey) \bullet attribKey[isKey^{-1}\ [\{c\}] \subseteq allClassAttributes[\{c\}]\ .$$

The uniqueness of keys requires that there is at most one object for each combination of values of the key of a class, and that all attributes in the key have values (the cardinality of the attribute cannot be 0). The constraints are defined on the set of attributes *attribKey[{K}]*, forming the key K of a class *cl*, and uses the fact that if the key value for two objects (*o1* and *o2*) are the same, then the two objects are in fact one object:

$$\begin{array}{l} \forall K \in key \bullet 0 \notin attribCards(attribKey[\{K\}]) \\ \quad \wedge\ isKey(K) = cl \Rightarrow (\forall o1,\ o2 \in allClassObject(cl) \bullet \\ \quad \forall a \in attribKey[\{K\}] \bullet \\ \quad (attribObjectValue(a)(o1) = attribObjectValue(a)(o2)) \\ \qquad \Rightarrow (o1 = o2))\ . \end{array}$$

The B invariants on qualifier and functional dependency are not illustrated here. The semantics of the qualifier are that an association can be qualified by a subset of attributes of the opposing class. Functional dependency semantics define that the attributes participating in the *leftPart* determine the value of those in the *rightPart*. These are disjoint sets of attributes.

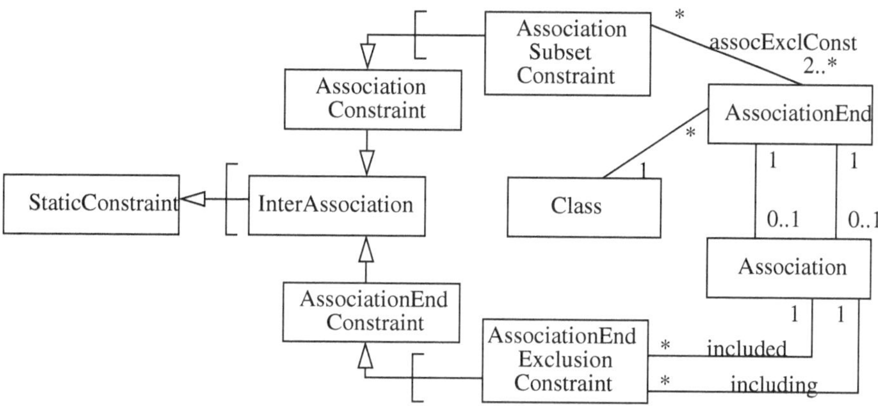

Fig. 4. Metamodel elaborating inter-association constraint semantics

6.2 *InterAssociation*: Participation Constraints

Inter-association constraints express constraints on participation (set exclusion, totality, subset) [20] among associations or association ends. (An example of such a constraint is that expressed above, stating that the ends of an association are different.)

Figure 4 extends Figs. 1 and 2, above. *InterAssociation* is specialised into *AssociationConstraint* and *AssociationEndConstraint*. One illustrative subclass is shown for each of these.

AssociationSubsetConstraint allows expression of the fact that links of one association may be a subset of links for another association. In metamodel terms, this is defined by the associations (*included* and *including*) between the constraint class and the *Association* class.

For association ends, the illustrated subclass is an exclusion constraint. This constraint is stated among two or more instances of *AssociationEnd*. Two invariants are shown.

If an exclusion constraint *ec* is defined between the association ends, the ends must be linked to the same class:

$$\forall\, ec \in AssociationEndExclusionConstraint \;\bullet\; card(assocEndClass \fatsemi assocExclConst[\{ec\}]) = 1\ .$$

If two association ends *ae1* and *ae2* participate in an exclusion constraint *ec*, then an object of the common class must belong to only one instance of the relevant associations:

$$\begin{aligned}
&\forall\, ec \in AssociationEndExclusionConstraint \;\bullet\\
&\forall\, ae1,\ ae2 \in assocExclConst[\{ec\}] \;\bullet\\
&\quad \mathrm{dom}\ (linkAssociation(assocAssocEnd1(ae1)^{-1}\\
&\quad \cup\ assocAssocEnd2(ae1)^{-1}))\\
&\quad \cap\ \mathrm{dom}\ (linkAssociation(assocAssocEnd1(ae2)^{-1}\\
&\quad \cup\ assocAssocEnd2(ae2)^{-1}\ )) = \varnothing\ .
\end{aligned}$$

7 Discussion

This paper presents metamodels, as class diagrams, for the main concepts required to model the static aspects of information systems in the UML. For reasons of space, it does not present all the invariants used to express the metamodels in B. Similarly, it does not present the specifications as a B machine. This would list, for example, all given sets in a *SETS* clause, and the conjoined invariants in an *INVARIANT* clause.

A B machine specification expresses both diagrammatic structure and the additional semantics of the metamodel. This is seen as a significant advantage over OMG's diagrams and OCL constraints. The OCL must be read in the context of the diagrams, and, although OCL support tools are emerging (see http://www.klasse.nl/ocl/index.htm), the mutual consistency of diagrams and OCL cannot be checked.

The diagrammatic and OCL semantics cannot define the creation, modification and deletion of model elements, whereas the B machine could be extended with operations to create, modify and delete semantically-correct components of the IS class metamodel. Such a machine would add semantic rules such as the requirement that a class cannot exist in isolation (ie from a hierarchy or association; the initial creation would have to be of two associated classes). The operations could be demonstrated (by proof using one of the two B Method tools[3]) to respect the invariant. A semantically-correct IS UML tool could then be formally generated from the B specification.

The paper only considers the semantics of UML class models for IS. The authors have also explored the behavioural model semantics, focusing on a subset of UML notations for state diagrams and collaboration diagrams.

Appendix

In UML1.3, the semantic models separate structure and behaviour (Fig. 5, drawing together semantics from UML1.3 Figs. 2.5 and 2.8 [21]). For example, the subclasses of *ModelElement* include *Features* and *GeneralizableElement*. Classes are a second-level specialisation of *GeneralizableElement*, whilst attributes and operations are second-level specialisations of *Features*. The UML1.3 metamodels have only indirect associations between *attribute* and *association* (Fig. 6, based on UML1.3 Figs. 2.6 and 2.8 [21]). This makes it difficult to express the semantics of structural features such as association qualifiers, which must be a set of attributes from the *Class* of the relevant *AssociationEnd*. Note that in UML1.1 [22], and in the UML Reference Manual [23, p161], association qualifiers are represented as an attribute of *AssociationEnd*. This does not permit expression of the semantics of qualification, via sets of attributes.

[3] Atelier-B, see http://www.atelierb.societe.com/ and The B Tool, see B-Core(UK) Ltd, http://www.b-core.com/

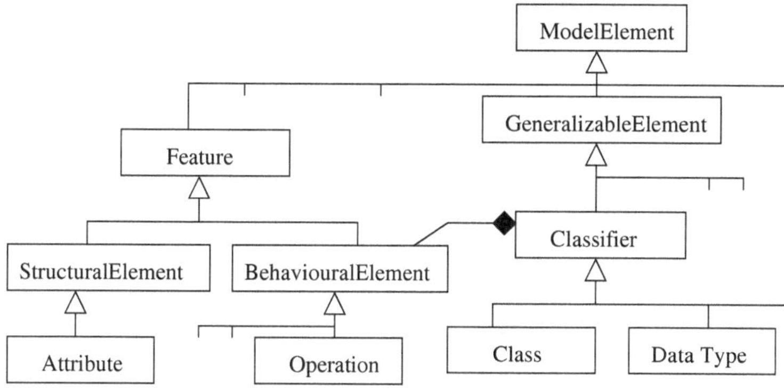

Fig. 5. Extracts of the OMG metamodels showing the use of subclasses to distinguish semantic element groups

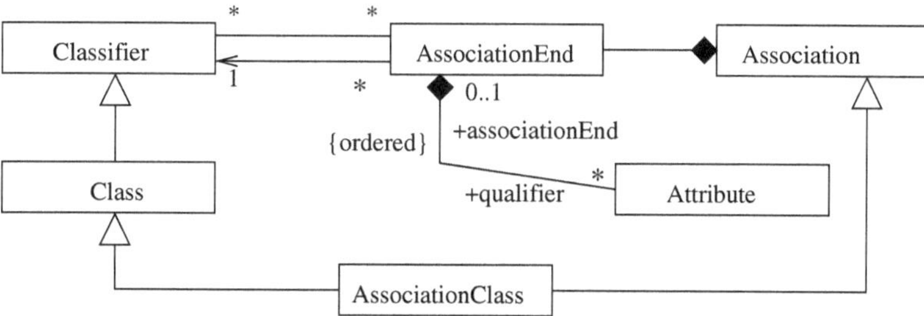

Fig. 6. Extracts of the OMG metamodels showing the semantics of associations and qualifiers

References

1. J-R. Abrial. *The B-Book*. CUP, 1996.
2. S. S. Alhir. Extending the Unified Modeling Language (UML). Technical report, January 1999. Available on line, http://home.earthlink.net/~ salhir/index.html.
3. J. Alvarez, A. Evans, and P. Sammut. MML and the metamodel architecture. In *Workshop on Transformations in UML, Fourth European Joint Conference on Theory and Practice of Software (ETAPS'01), Genova, Italy*, April 2001.
4. C. Atkinson and T Kühne. Strict profiles: Why and how. In *UML 2000 - The Unified Modeling Language. Advancing the Standard*, volume 1939 of *LNCS*, pages 309–322. Springer Verlag, 2000.
5. J-M. Bruel and R.B.France. Transforming UML models to formal specifications. In *UML'98*, volume 1618 of *LNCS*. Springer Verlag, 1998.
6. A. Clark, A. Evans, S. Kent, S. Brodsky, and S. Cook. A feasibility study in rearchitecting UML as a family of languages using a precise OO meta-modeling approach. Technical report, pUML Group and IBM, September 2000. Available on line, http://www.cs.york.ac.uk/puml/papers/.
7. A. Clark, A. Evans, S. Kent, and P. Sammut. The MMF approach to engineering object-oriented design languages. In *Workshop on Language Descriptions, Tools and Applications, LTDA2001*, April 2001.

8. T. Clark and A. Evans. Foundations of the Unified Modeling Language. In *Second Northern Formal Methods Workshop*, volume 1241 of *LNCS*. Springer Verlag, 1997.
9. S. Cook. The UML family: Profiles, prefaces and packages. In *UML 2000 - The Unified Modeling Language. Advancing the Standard*, volume 1939 of *LNCS*, pages 255–264. Springer Verlag, 2000.
10. S. Cook, A. Kleppe, R. Mitchell, B. Rumpe, J. Warmer, and A. Wills. Defining UML family members using prefaces. In C. Mingins and B. Meyer, editors, *Technology of Object-Oriented Languages and Systems, TOOLS'99 Pacific*. IEEE Computer Society, 1999.
11. D. D'Souza, A. Sane, and A. Birchenough. First-class extensibility for UML - packaging of profiles, stereotypes and patterns. In *UML 1999 - The Unified Modeling Language. Beyond the Standard*, volume 1723 of *LNCS*. Springer Verlag, 1999.
12. P. Facon, R. Laleau, A. Mammar, and F. Polack. Formal specification of the UML metamodel for building rigorous CAiSE tools. Technical report, CEDRIC Laboratory, CNAM, September 1999.
13. P. Facon, R. Laleau, and H. P. Nguyen. From OMT diagrams to B specifications. In M. Frappier and H. Habrias, editors, *Software Specification Methods. An Overview Using a Case Study*, pages 57–77. Springer, 2000.
14. P. Facon, R. Laleau, H. P. Nguyen, and A. Mammar. Combining UML with the B formal method for the specification of database applications. Technical report, CEDRIC Laboratory, CNAM, September 1999.
15. R. Laleau and N. Hadj-Rabia. Génération automatique de spécifications formelles VDM à partir d'un schéma conceptuel de données. In *INFORSIDS'95, Grenoble, France*, June 1995.
16. R. Laleau and A. Mammar. A generic process to refine a B specification into a relational database implementation. In *Proceedings, ZB2000: Formal Specification and Development in Z and B, York, August-September 2000*, volume 1878 of *LNCS*, pages 22–41. Springer Verlag, 2000.
17. R. Laleau and A. Mammar. An overview of a method and its support tool for generating B specifications from UML notations. In *Proceedings 15th IEEE Int. Conf. on Automated Software Engineering (ASE2000), Grenoble, France, September, 2000.*, 2000.
18. F. Monge. Formalisation du méta modèle des méthodes graphiques d'analyse et conception orientées objet. Master's thesis, IIE, CEDRIC Laboratory, CNAM, September 1997.
19. H. P. Nguyen. *Dérivation de spécifications formelles B à partir de spécifcations semi-formelles*. PhD thesis, CEDRIC Laboratory, CNAM, December 1998.
20. S. H. Nienhuys-Cheng. Classification and syntax of constraints in binary semantical networks. *Information Systems*, 15(5), 1990.
21. Object Management Group. *OMG Unified Modeling Language Specification(draft)*, April 1999. BetaR1 Release, available on line, www.rational.com/uml.
22. Rational Software Corporation. *Unified Modeling Language version 1.1*, July 1997. Available on line, www.rational.com/uml.
23. J. Rumbaugh, I. Jacobson, and G. Booch. *The Unified Modeling Language Reference Guide*. Addison-Wesley, 1998.
24. M. Sahbani. Formalisation du relations de composition et d'agrégation des modèles sémantiques. Master's thesis, IIE, CEDRIC Laboratoty, CNAM, September 1999.
25. L. T. Semmens, R. B. France, and T. W. G. Docker. Integrated structured analysis and formal specification techniques. *The Computer Journal*, 35(6):600–610, 1992.

Taxonomies and Derivation Rules in Conceptual Modeling

Antoni Olivé

Dept. Llenguatges i Sistemes Informàtics
Universitat Politècnica de Catalunya
Jordi Girona 1-3, C5-D207
08034 Barcelona (Catalonia)
olive@lsi.upc.es

Abstract. This paper analyzes the relationships between taxonomic constraints and derivation rules. The objectives are to see which taxonomic constraints are entailed by derivation rules and to analyze how taxonomic constraints can be satisfied in presence of derived types. We classify derived entity types into several classes. The classification reveals the taxonomic constraints entailed in each case. These constraints must be base constraints (defined in the taxonomy) or be derivable from them. We show how the base taxonomic constraints can be satisfied, either by the derivation rules (or the whole schema), or by enforcement. Our results are general and could be incorporated into many conceptual modeling environments and tools. The expected benefits are an improvement in the verification of the consistency between taxonomic constraints and derivation rules, and a guide for the determination of the taxonomic constraints that must be enforced in the final system.

1 Introduction

Taxonomies are fundamental structures used in many areas of information systems engineering and other fields [8]. In its most basic form, a taxonomy consists of a set of (entity or relationship) types and a set of *IsA* relations among them [11]. Extensionally, an *IsA* relation is a constraint between the populations of two types [3]. Usually, taxonomies include also other constraints, mainly disjointness and covering, which are needed to adequately represent domain knowledge [24].

Derived types and derivation rules have a long tradition in conceptual modeling, starting at least in the early eighties [22]. SDM [20] is recognized as one of the first languages that emphasized the need and support of derived types. Derivation rules were also included in the ISO framework [23]. CIAM was a methodology strongly based on derived types, with a temporal perspective [17]. Many other later conceptual languages include specific constructs for the definition of derivation rules. Among them, we mention the family of languages descendants of KL-ONE [11] called Description (or terminological) Logics [9]. The recent industry standard UML language [31] also allows derived types but, unfortunately, they are restricted to attributes and associations.

K.R. Dittrich, A. Geppert, M.C. Norrie (Eds.): CAiSE 2001, LNCS 2068, pp. 417–432, 2001.
© Springer-Verlag Berlin Heidelberg 2001

There are some relationships between constraints defined in a taxonomy, that we call taxonomic constraints, and derived types. As a very simple example, assume a taxonomy with entity types *Clerk*, *Engineer*, *Employee* and *Person*, and the constraint *Clerk IsA Employee*. Once defined a taxonomy, in general we have several options concerning derivability of its entity types, and the corresponding derivation rules. For example, we can make *Employee* base and *Clerk* derived. In this case, the derivation rule of *Clerk* must entail the constraint, with a rule like "A clerk is an employee such that its category is 'c'". Another option is to make *Employee* derived and *Clerk* base. In this case, the derivation rule of *Employee* may or may not entail the constraint. If the rule is like "An employee is a person that works in some company", then it does not entail the constraint. If the rule is "An employee is a person who is a clerk or an engineer" then it entails the constraint. However, in this case the rule entails also *Engineer IsA Employee*, and this constraint must be included in the taxonomy. Thus, we see that there are relationships from taxonomic constraints to derivation rules, and the other way around. The objective of this paper is to analyze such relationships at the conceptual level.

Knowledge of the relationships between taxonomic constraints and derivation rules is important for at least two purposes: (1) During conceptual modeling verification, to ensure that derivation rules entail some constraints defined in the taxonomy, and that the taxonomy includes all constraints entailed by derivation rules; and (2) During information systems design, to determine which taxonomic constraints need to be explicitly enforced (by database checks, assertions or triggers, or transaction pre/postconditions) and which are entailed by derivation rules, and need not to be enforced.

The main contributions of this paper are: (1) a classification of derived entity types; (2) an analysis of the constraints entailed by derivation rules and that must be defined in a taxonomy; and (3) an analysis of the possible ways of satisfaction of taxonomic constraints in presence of derived types. We assume a temporal conceptual schema and information base, and allow multiple specialization and classification.

We expect our contribution to be useful for the two purposes mentioned above. In particular, we want to stress the applicability to the enforcement of constraints. Current methods and techniques for the enforcement of taxonomic constraints do not take into account derived types [4, 27]. An illustrative example is [21], which details a wide range of techniques to express any kind of taxonomic constraints into standard DBMS constructs, but it is assumed that all entity types in the taxonomy are base. These methods could be extended easily to schemas that include derived types.

To the best of our knowledge, the relationships between taxonomic constraints and derivation rules have been studied, in conceptual modeling, only in the context of Description Logics (DL) [7, 9, 12] or similar [14]. This paper differs from most of that work in several aspects: (1) we deal with taxonomies of entity types in temporal conceptual schemas; (2) we assume a clearer separation between taxonomy (with taxonomic constraints) and derivability (with derivation rules); (3) we deal with derivation rules written in the FOL language, instead of a particular DL language [10]; (4) even if we use the FOL language, here we focus more on defining the relationships than on determining automatically them; we hope this will help in adapting and using the results reported here in several conceptual modeling languages; and (5) we give a special treatment to the determination of the taxonomic constraints that must be enforced by the designer.

The structure of the paper is as follows. Section 2 reviews taxonomic constraints and introduces the notation we will use. Section 3 develops a classification of derived entity types, based mainly on an analysis of their derivation rules. That analysis also allows us to determine some taxonomic constraints entailed by the rules. We discuss how these constraints must be reflected in the taxonomy. Section 4 analyzes how the constraints defined in a taxonomy can be satisfied, in presence of derived entity types. Section 5 summarizes the results and points out future work.

2 Taxonomies

A taxonomy consists of a set of concepts and their specialization (subsumption) relationships [11]. In a conceptual schema, we have a taxonomy of entity types and one or more of relationship types. In this paper we deal only with the former. Specializations have an intensional and an extensional aspect. Here, we focus on the extensional aspect, where a specialization is a constraint between the population of two entity types. Other constraints related to taxonomies are disjointness and covering. We call taxonomic constraints the set of specialization, disjointness and covering constraints defined in a schema. We introduce below the terminology and notation that will be used throughout the paper.

We adopt a logical and temporal view of the information base, and assume that entities and relationships are instances of their types at particular time points. In the information base, we represent by $E(e,t)$ the fact that entity e is instance of entity type E at time t. Similarly, we represent by $R(e_1,...,e_n,t)$ the fact that entities $e_1,...,e_n$ participate in a relationship instance of relationship type R at time t [17, 6, 30]. We say that $E(e,t)$ and $R(e_1,...,e_n,t)$ are entity and relationship facts, respectively.

2.1 Taxonomic Constraints

We denote by $S = E'$ *IsA* E a *specialization* constraint between entity types E' and E. For example, *Man IsA Person*. E' is said to be a subtype of E, and E a supertype of E'. An entity type may be supertype of several entity types, and it may be a subtype of several entity types at the same time. As usual, we require that an entity type may not be a direct or indirect subtype of itself. Informally, a specialization means that if an entity e is instance of E' at time t, then it must also be instance of E at t. In the language of first order logic, the meaning of a specialization $S = E'$ *IsA* E is given by the formula[1]:

$$E'(e,t) \rightarrow E(e,t)$$

We denote by $D = E_1$ *disjoint* E_2 a *disjointness* constraint (sometimes called *IsNotA*) between entity types E_1 and E_2. For example, *Man disjoint Woman*. The informal meaning is that the populations of E_1 and E_2 at any time are disjoint. Naturally, E_1 *disjoint* E_2 is equivalent to E_2 *disjoint* E_1. The formal meaning is given by the formula:

$$E_1(e,t) \rightarrow \neg E_2(e,t)$$

---

[1] The free variables are assumed to be universally quantified in the front of the formula.

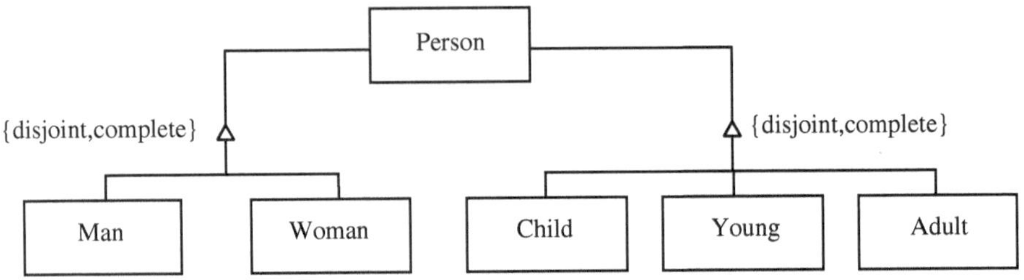

Fig. 1. Graphical representation in UML of *Person Gens Man, Woman*, and *Person Gens Child, Young, Adult.*

Finally, a *covering* constraint between an entity type *E'* and a set of entity types $E_1,...,E_n$, will be denoted by *C = E' covered* $E_1,...,E_n$. For example, *Person covered Man, Woman*. The informal meaning is that if *e* is instance of *E'* at time *t*, it must be also instance of at least one E_i at *t*. Formally:

$$E'(e,t) \rightarrow E_1(e,t) \vee ... \vee E_n(e,t)$$

Notice that, by itself, *E' covered* $E_1,...,E_n$ does not imply E_i *IsA E'*.

A specialization constraint is a covering constraint with only one covering type: *E' IsA E* ≡ *E' covered E* [24]. However, we will treat them separately, because both are widely used in practice.

We call *base* the taxonomic constraints explicitly defined in a schema. From the base constraints other may be *derived* using a set of inference rules. We are interested in derived constraints because derivation rules may entail them and, in this case, it will be necessary to check that such constraints may be derived from the base ones. We deal with derived constraints in Subsection 3.3.

All modern conceptual models and languages allow defining taxonomies, usually structured in generalizations [22, 4, 6]. A generalization *G = E Gens* $E_1,...,E_n$ corresponds to a set of specialization constraints E_i *IsA E*, for $i = 1,..,n$, with a common supertype *E*. *Gens* is a shorthand for *Generalizes*. A generalization is *disjoint* if their subtypes are mutually disjoint; otherwise, it is *overlapping*. A generalization is *complete* if the supertype *E* is covered by the subtypes $E_1,...,E_n$; otherwise it is *incomplete*. Many conceptual languages provide a graphical representation of generalizations. Figure 1 shows an example of two generalizations of *Person*, in the UML language [31].

2.2 Partitions

A partition is a modeling construct that allows us to define in a succinct way a set of taxonomic constraints. A *partition* is a generalization that is both disjoint and complete. We denote by *P = E Partd* $E_1,...,E_n$ a partition of entity type *E* into entity types $E_1,...,E_n$. *Partd* is shorthand for *Partitioned*. For example, *Person Partd Man, Woman*. A partition *P = E Partd* $E_1,...,E_n$ corresponds to the constraints:

$S_i = E_i$ IsA E, for $i = 1,..,n$
$C = E$ covered $E_1,...,E_n$
$D_{i,j} = E_i$ disjoint E_j, for $i,j = 1,..,n,\ i \neq j$.

We do not require here to structure a taxonomy in partitions. However, we give a special treatment to them due to their importance in conceptual modeling [26, 29].

3 Derivation Rules

An information system may know the population of an entity type in three distinct ways, which are reviewed below. Then, we classify derived entity types according to the form of their derivation rule, and show the entailed constraints in each case.

3.1 Derivability of Entity Types

The *derivability* of an entity type is the way how the information system knows the population of that entity type at any instant. According to derivability, an entity type may be *base*, *derived* or *hybrid* [5, 26]. We give below a few comments on each of them:

- *Base*. An entity type *E* is base when the population of *E* at time *t* is given directly or indirectly by the users by means of insertion and deletion events. An insertion event of entity *e* in *E* at *t* means that *e* is instance of *E* from *t*. A deletion event of *e* in *E* at *t* means that *e* ceases to be instance of *E* at *t*. The information system knows the population of a base entity type *E* at *t* by using a general *persistence* (or frame) *axiom*: the population is the set of entities that have been inserted at time $t_1 \leq t$ and have not been deleted between t_1 and *t*.
- *Derived*. An entity type *E* is derived when the population of *E* at *t* is given by a formal derivation rule, which has the general form:

$$E(e,t) \leftrightarrow \varphi(e,t)$$

- *Hybrid*. An entity type E is hybrid when the population of *E* is given partially by the users (insertion and deletion events) and partially by a formal derivation rule, which has the general form:

$$E(e,t) \leftarrow \varphi(e,t)$$

- We call such derivation rules *partial*, because they define only part of the population of *E*. The information system may know the population of *E* at any time by using the persistence axiom and the above rule.

Any hybrid entity type can be transformed easily into an equivalent derived one. This allows us to simplify our analysis, since we only have to deal with base and derived types. The procedure for the transformation is described in [5]. Here we show it by means of an example. Assume that *Building* is hybrid, with the partial derivation rule:

$$\text{Building}(b,t) \leftarrow \text{House}(b,t)$$

This means that houses are buildings, but that there may be other buildings besides houses. We define a new base entity type $Building_{ext}$, such that its population at time *t* is the set of entities that users have defined explicitly as being buildings. Usually, we will require the constraint $Building_{ext}$ *disjoint House*, but this is not mandatory. Now, *Building* is derived, with the derivation rule:

$$\text{Building}(b,t) \leftrightarrow \text{House}(b,t) \vee \text{Building}_{ext}(b,t)$$

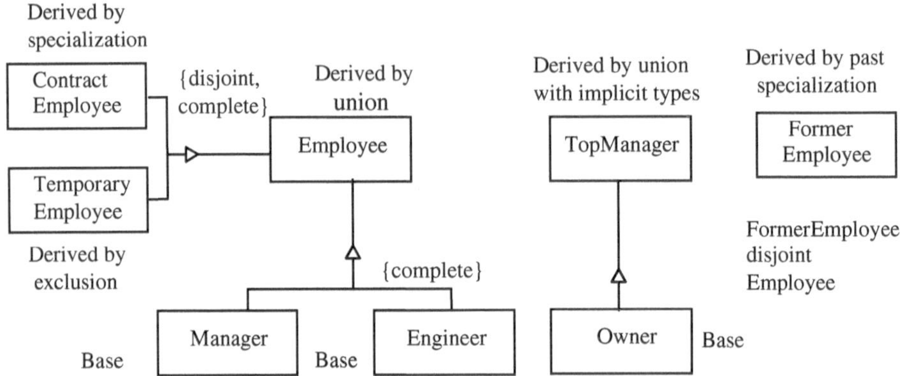

Fig. 2. Example of entity types with their derivability, and the taxonomic constraints entailed by their derivation rules.

3.2 Classification of Derived Entity Types

We are going to see that derived entity types can be classified into derived by: specialization, exclusion, past specialization, union and union with implicit types.

This classification is important to see the taxonomic constraints entailed by derivation rules. On the other hand, we think that the classification is useful for reasoning and implementation purposes (especially if entity types are materialized), in the same way as the classification of integrity constraints (key, referential, cardinalities, ...) eases reasoning about them and allows their efficient implementation.

For generality, we assume that derivation rules are written in the FOL language. However, the results shown below could be of interest to many conceptual modeling languages.

To illustrate the classification, we use as an example the following derived entity types and derivation rules:

DR1: Employee(e,t) $\leftrightarrow$ Manager(e,t) $\vee$ Engineer(e,t)
DR2: ContractEmployee(e,t) $\leftrightarrow$ HasContract(e,contract,t)
DR3: TemporaryEmployee(e,t) $\leftrightarrow$ Employee(e,t) $\wedge \neg$ ContractEmployee(e,t)
DR4: TopManager(m,t) $\leftrightarrow$ (Manager(m,t) $\wedge \neg\exists$super HasSubordinate(super,m,t))
$\vee$ Owner(m,t)
DR5: FormerEmployee(e,t) $\leftrightarrow$ Employee(e,t1) $\wedge$ Time(t) $\wedge$ t1 < t $\wedge \neg$ Employee(e,t)

In DR5, *Time(t)* is a special predicate whose facts are instants of the temporal domain. Figure 2 summarizes the classification of the above entity types and the taxonomic constraints (in UML) entailed by the derivation rules. Both results are obtained by the procedure described below.

The procedure is a simple four-step transformation of each derivation rule into an equivalent set of (logic programming) clauses [25]. The classification is based on the number of clauses thus obtained, and the structure of each of them, as we show in what follows.

Let $E(x,t) \leftrightarrow \phi(x,t)$ be a derivation rule. Assuming the semantics of the completion, the rule can be expressed by:

$$E(x,t) \leftarrow \phi(x,t)$$

In general, $\phi(x,t)$ can be any valid first order formula, which makes its analysis difficult. However, using the well-known (at least in the logic programming and deductive databases fields) procedure proposed in [25], we can transform it (first step) into an equivalent set of n clauses:

$$E(x,t) \leftarrow L_{i,1} \wedge \ldots \wedge L_{i,m} \quad \text{for } i = 1 \ldots n$$

where each of the $L_{i,j}$ in the body of the clause is a (positive or negative) literal. Other auxiliary clauses may be obtained as well, but they are irrelevant for our purposes. As usual, we require that the resulting clauses are safe and stratified [25, 13], thus guaranteeing that their computation is safe. We call *simple* a clause having exactly one literal of the form:

$$E(x,t) \leftarrow E_i(x,t)$$

In the example, transformation of DR1and DR4 yields the clauses (the transformation of the other rules is trivial):

DR1-1: Employee(e,t) ← Manager(e,t)
DR1-2: Employee(e,t) ← Engineer(e,t)
DR4-1: TopManager(m,t) ← Manager(m,t) ∧ ¬HasSuper(m,t)
DR4-2: TopManager(m,t) ← Owner(m,t)
Aux: HasSuper(m,t) ← HasSubordinate(super,m,t)

In the body of a clause $E(x,t) \leftarrow L_{i,1} \wedge \ldots \wedge L_{i,m}$, a literal $L_{i,j}$ may be:

- A positive entity fact $E_i(x,t_i)$ meaning the x is instance of E_i at time t_i.
- A positive relationship fact $R_i(x,y,t_i)$ including x as an argument. The position of x is irrelevant, and y is a vector of arguments. The meaning now is that x and y participate in a relationship of type R_i at time t_i.
- Other literals (positive facts not including x, negative, evaluable, *Time*). These are irrelevant for our current purposes. Note that in this paper we do not deal with aggregates.

There must be at least one positive $E_i(x,t_i)$ or $R_i(x,y,t_i)$ literal in the clause. This condition is implied by the safeness of clauses (recall that the head of the clauses is $E(x,t)$).

In conceptual modeling, participants p_i in relationships of type $R(p_1{:}E_1,\ldots,p_n{:}E_n)$ at time t must be instance of their corresponding entity type E_i at time t (referential integrity constraint). Therefore, assuming that E_i is the type corresponding to x, whenever literal $R_i(x,y,t_i)$ is true, $E_i(x,t_i)$ must be true also. This justifies our second step: for each positive relationship fact $R_i(x,y,t_i)$ appearing in the body of a clause, we add the corresponding literal $E_i(x,t_i)$. These literals are redundant with respect to $R_i(x,y,t_i)$, but they will allow us to find the entity types specialized by E.

In the example, this step can be applied to DR2. Assuming the relationship type *HasContract(Employee,Contract)*, we add *Employee(e,t)* to the body of DR2:

DR2': ContractEmployee(e,t) ← HasContract(e,contract,t) ∧ Employee(e,t)

In the third step we remove redundant positive entity fact literals. If we have a clause:

$$E(x,t) \leftarrow L_{i,1} \wedge ... \wedge E_i(x,t_k) \wedge \; ... \wedge \; E_j(x,t_k) \wedge \; ... \wedge \; L_{i,m}$$

and we have also a direct or indirect E_i *IsA* E_j in the schema, then we remove the redundant literal $E_j(x,t_k)$.

In the fourth step, we deal with the literals in the body of a clause which are positive entity facts including argument x and with a time argument t_i distinct from t, where t is the time argument of the head of the clause, $E(x,t)$. In general, such t_i may range in the interval $[1,...,t]$. The case $t_i > t$ is not permitted, because then $E(x,t)$ would not be computable at time t. We want to distinguish the case when $t_i = t$ and the case when $t_i < t$. For this purpose, given that $t_i \leq t \leftrightarrow t_i = t \vee t_i < t$, we replace each clause having a $E_i(x,t_i)$ literal:

$$E(x,t) \leftarrow L_{i,1} \wedge ... \wedge E_i(x,t_i) \wedge ... \wedge L_{i,m}$$

by the two equivalent clauses:

$$E(x,t) \leftarrow L_{i,1} \wedge ... \wedge E_i(x,t) \wedge ... \wedge L_{i,m} \qquad (t_i \text{ is replaced by } t)$$

$$E(x,t) \leftarrow L_{i,1} \wedge ... \wedge E_i(x,t_i) \wedge t_i < t \wedge ... \wedge L_{i,m}$$

If any of the clauses has a contradictory body, it is removed. After this step, all positive entity facts will have a time argument with t or with a t_i such that $t_i < t$. We call the former *current* facts, and the latter *past* facts.

The application of this step to DR5 gives:

DR5-1: FormerEmployee(e,t) $\leftrightarrow$ Employee(e,t) $\wedge$ t = t $\wedge$ Time(t) $\wedge$ t < t $\wedge$
$\neg$ Employee(e,t)
DR5-2: FormerEmployee(e,t) $\leftrightarrow$ Employee(e,t1) $\wedge$ Time(t) $\wedge$ t1 < t $\wedge$
$\neg$ Employee(e,t)

The first clause, DR5-1, is removed because of the contradiction $t = t$ and $t < t$. Such contradictions can be easily detected with the algorithms given in [18].

We now have sufficient machinery in place to classify derived entity types. We start by distinguishing two main cases: the derivation rule is transformed into a single clause (DR2',DR3,DR5-2) or into multiple clauses (DR1,DR4).

Single Clause. An entity type E is derived by *specialization* of $E_1,...,E_n$ if the above transformation of its derivation rule yields the single clause:

$$E(x,t) \leftarrow E_1(x,t) \wedge ... \wedge E_n(x,t) \wedge Res$$

with $n \geq 1$, and *Res* is a (possibly empty) residual set of literals, none of which is a positive entity fact $E_j(x,t)$. The above clause cannot be simple (that is, $n = 1$ and *Res* empty) because then E and E_1 would be redundant entity types.

If entity type E is derived by specialization of E_1, ..., E_n then the derivation rule of E entails the n specialization constraints: E *IsA* E_1, ..., E *IsA* E_n. If any of the literals in *Res* has the form $\neg E_p(x,t)$, then the derivation rule entails also the constraint: E *disjoint* E_p.

According to this definition, in our example *ContractEmployee* and *TemporaryEmployee* are derived by specialization of *Employee* (DR2' and DR3).

In conceptual modeling, many derived entity types are derived by specialization, and for some conceptual models, this is the only allowed type (for instance, NIAM [28] and Chimera [13])

A particularly interesting subcase of entity type derived by specialization is when the clause has exactly the form:

$$E(x,t) \leftarrow E'(x,t) \wedge \neg E_1(x,t) \wedge ... \wedge \neg E_m(x,t)$$

with $m \geq 1$. In this case, we say that E is derived by *exclusion*, because its instances are those of E' but excluding those of E_1 and ... and E_m. Now the derivation rule of E entails also:

- the disjointness constraints: *E disjoint* E_i, for $i = 1,..,m$.
- the covering constraint E' covered E, E_1, ..., E_m.

In the example, *TemporaryEmployee* is derived by exclusion (DR3). The set of constraints entailed by DR3 is:

TemporaryEmployee IsA Employee,
TemporaryEmployee disjoint ContractEmployee, and
Employee covered TemporaryEmployee, ContractEmployee.

Finally, we say that an entity type E is derived by *past specialization* of $E_1,...,E_n$ if the transformation of its derivation rule yields the clause:

$$E(x,t) \leftarrow E_1(x,t_1) \wedge ... \wedge E_n(x,t_n) \wedge Res$$

with $n \geq 1$, all $E_i(x,t_i)$ past, and where *Res* is a (possibly empty) residual set of literals, none of which is a positive entity fact $E_j(x,t)$. In this case the specializations *E IsA* E_i do not hold. If any of the literals in *Res* has the form $\neg E_p(x,t)$, then the derivation rule entails the constraint: *E disjoint* E_p.

The relationship between E_i and E that exists in this case is:

$$E(x,t) \rightarrow \exists t_k\, E_i(x,t_k) \wedge t_k < t$$

which can be seen as a past specialization. This specialization can be captured by the expression *E WasA* E_i, and could be considered a new kind of taxonomic constraint.

In the example, *FormerEmployee* is derived by past specialization of *Employee* (DR5-2). The constraint entailed by DR5-2 is *FormerEmployee disjoint Employee.*

Derivation by specialization or by past specialization are the two only possible cases when the original derivation rule is transformed into a single clause. The reason is that, as we have said before, after the first step the body of the clause must contain at least one $E_i(x,t_i)$ or $R_i(x,y,t_i)$ literal. If it is a $R_i(x,y,t_i)$ literal, then after the second step the body will include at least one $E_i(x,t_i)$ literal.

Multiple Clauses. Now we deal with the case when the original derivation rule is equivalent to a set of n clauses ($n > 1$):

$$E(x,t) \leftarrow L_{i,1} \wedge ... \wedge L_{i,m} \qquad \text{for } i = 1...n$$

An entity type E is derived by *union* of E_1, ..., E_n if the n clauses are simple; that is, they have the form:

$$E(x,t) \leftarrow E_1(x,t)$$
$$...$$
$$E(x,t) \leftarrow E_n(x,t)$$

If entity type E is derived by union of E_1, ..., E_n then the derivation rule of E entails the n specialization constraints: E_1 *IsA E*, ..., E_n *IsA E* and the covering constraint: *E covered* E_1, ..., E_n.

In the example, *Employee* is derived by union of *Manager* and *Engineer* (DR1-1 and DR1-2). The constraints entailed are: *Manager IsA Employee, Engineer IsA Employee* and *Employee covered Manager, Employee.*

Some conceptual models allow defining entity types derived by specialization and union. Among them, we mention PSM [19].

If any of the *n* clauses is not simple, we say that *E* is derived by *union with implicit types*. The non-simple clauses will be either a specialization or a past specialization, which can be seen as defining an implicit type. If some clause is simple, then the derivation rule entails a corresponding specialization constraint.

In the example, *TopManager* is derived by union of implicit types (DR4-1,DR4-2). The simple clause DR4-2 entails *Owner IsA TopManager*. Note that we can always transform a type derived by union with implicit types into one derived by union, with the introduction of new types. Thus, *TopManager* could be derived by union of *Owner* and *TopEmployeeManager*, defined as:

DR4-1': TopManager(m,t) ← TopEmployeeManager(m,t)
DR4-2': TopManager(m,t) ← Owner(m,t)
New: TopEmployeeManager(m,t) ← Manager(m,t) ∧ ¬HasSuper(m,t)

where *TopEmployeeManager* is the implicit type (derived by specialization).

In some special cases, the transformation yields *n* clauses with one or more common $E_i(x,t)$ literals. If this occurs, then *E* is classified as derived by specialization of E_i. For example, in the derivation rule:

PromotionCandidate(e,t) ↔ Engineer(e,t) ∧
(TemporaryEmployee(e,t) ∨ (HasSalary(e,sal,t) ∧ sal < Min))

where *PromotionCandidate* becomes derived by specialization of *Engineer*.

3.3 Taxonomy and Constraints Entailed by Derivation Rules

As we have just seen, a derivation rule may entail taxonomic constraints. Now, the question is: must these constraints appear in the taxonomy? Our answer is positive, because for verification, validation, implementation, and evolution purposes it is important that a taxonomy be as complete as possible. It is also important in order to ensure the consistency between the taxonomy and the derivation rules. In our example, this means to check that the taxonomic constraints depicted graphically in Figure 2 are included in the taxonomy.

A constraint entailed by a derivation rule will be very often a base one. In these cases, derivation rules and taxonomy match perfectly.

In other cases, however, a constraint entailed by a derivation rule is not a base one. In these cases, the constraint must be derivable from the base ones, using a set of inference rules. [3, 24] give the complete and sound set of inference rules for taxonomic constraints. If a constraint is not derivable using that set of inference rules, then either the taxonomy or the derivation rules must be modified.

4 Satisfaction of Taxonomic Constraints

In this Section, we analyze how the constraints defined in a taxonomy can be satisfied. We apply the results to the particular case of partitions.

4.1 Satisfaction of Base Constraints

An information base comprises all temporal base and derived facts that are true in the domain of the information system. An information base *IB* satisfies a constraint *IC* if *IC* is true in *IB*. There are several different meanings of constraint satisfiability in logic [15, 16], but the differences will not be important in this paper.

Satisfaction of integrity constraints can be ensured by the schema or by enforcement. We say that a constraint *IC* is satisfied by the *schema* (or intensionally) when the schema entails *IC*. That is, the derivation rules and the (other) constraints defined in the schema imply *IC* or, in other terms, *IC* is a logical consequence of the schema. In this case no particular action must be taken at runtime to ensure the satisfaction of *IC*. In a way, *IC* is redundant, but it may be important to keep it in the schema for verification, validation, implementation or evolution purposes. For example, if the schema includes:

S = LongPaper IsA Paper

DR: LongPaper(p,t) $\leftarrow$ Paper(p,t) $\wedge$ Length(p,pages,t) $\wedge$ pages > 50

then *DR* entails *S*, and therefore *S* is satisfied by the schema.

We say that a constraint *IC* is satisfied by *enforcement* (or extensionally) when it is not satisfied by the schema, but it is entailed by the information base. That is, *IC* is a condition true in the information base. In this case, the system has to enforce *IC* by means of checking and corrective actions, to be executed whenever the information base is updated. For example, if *Manager* and *Engineer* are base entity types and we have the constraint *D* = *Manager disjoint Engineer*, then *D* must be enforced. This means to check at runtime that no entity is classified in both types at the same time and, if it is, to reject the cause of the constraint violation or to perform some corrective action.

The enforcement of constraints is expensive, but it may be the only option available for some constraints. However, methods developed so far assume that all taxonomic constraints must be enforced [21]. We are going to show that, in some cases, taxonomic constraints may be satisfied by the schema and, thus, their enforcement is not necessary. We discuss each kind of taxonomic constraint in turn.

Let *S* = *E' IsA E* be a base specialization constraint. We distinguish four cases:

(1) if both *E'* and *E* are base, then *S* must be enforced.

(2) if *E'* is derived and *E* is base, then the derivation rule of *E'* must entail *S*. The rationale is that if the derivation rule of *E'*: $E'(e,t) \leftrightarrow \varphi(e,t)$ does not entail *S*, then we can transform it into the equivalent one $E'(e,t) \leftrightarrow (\varphi(e,t) \wedge E(e,t))$, which now entails *S*. Note that this transformation does not change the stratification of *E'*.

   In the particular case that *E'* is derived by specialization of *E*, the derivation rule of *E'* entails *S*.

(3) when *E'* is base and *E* is derived, *S* may be entailed by the schema; if it is not, then *S* must be enforced.

In the general case, it may be difficult to prove that *S* is entailed by the schema, see comment below. In many practical cases, however, it suffices to prove that *S* is entailed by the derivation rule of *E*. This happens, for instance, when *E* is derived by union of a set of types that includes *E'*.

(4) when both *E'* and *E* are derived, *S* may be entailed by the derivation rules of *E'* or *E*, or by the schema; if it is not, then *S* must be enforced.

As an example, consider the five specialization constraints of Figure 1. Assume that *Man* and *Woman* are base, and that the other entity types are derived. The derivation rules are:

DR1-1: Person(p,t) $\leftarrow$ Man(p,t)
DR1-2: Person(p,t) $\leftarrow$ Woman(p,t)
DR2: Child(p,t) $\leftarrow$ Person(p,t) $\wedge$ Age(p,a,t) $\wedge$ a $<$ 5
DR3: Young(p,t) $\leftarrow$ Person(p,t) $\wedge$ Age(p,a,t) $\wedge$ a $>$ 5 $\wedge$ a $<$ 18.
DR4: Adult(p,t) $\leftarrow$ Person(p,t) $\wedge$ Age(p,a,t) $\wedge$ a $\geq$ 18.

S_1 = *Man IsA Person* is entailed by DR1-1 (case 3). Similarly, S_2 = *Woman IsA Person* is entailed by DR1-2 (case 3). S_3 = *Child IsA Person*; S_4 = *Young IsA Person* and S_5 = *Adult IsA Person* are entailed by DR2, DR3 and DR4, respectively (case 4).

In cases (3) and (4) it may be necessary to prove that *S* is entailed by the schema. If it cannot be proved, then a safe approach is to enforce *S*. Automation of this proof requires the use of some reasoner [18, 15]. When derivation rules are written in a language based on Description Logics, such a reasoner is usually available [7, 9].

Let $D = E_1$ *disjoint* E_2 be a base disjointness constraint. We distinguish three cases:

(1) if both E_1 and E_2 are base, then *D* must be enforced.

(2) if E_1 is derived and E_2 is base, then the derivation rule of E_1 must entail *D*. The rationale is that if the derivation rule of E_1: $E_1(e,t) \leftrightarrow \varphi(e,t)$ does not entail *D*, then we can transform it into the equivalent one $E_1(e,t) \leftrightarrow (\varphi(e,t) \wedge \neg E_2(e,t))$, which now entails *D*. Note that this transformation does not change the stratification of E_1.

In the particular case that E_1 is derived by specialization of some type *E* and exclusion of some types including E_2, *D* is entailed by the derivation rule of E_1.

(3) when E_1 and E_2 are derived, *D* may be entailed by the schema; if it is not, then *D* must be enforced.

Continuing our previous example, consider now the disjointness constraint D_1 = *Man disjoint Woman*. Given that both entity types are base, D_1 must be enforced (case 1). In D_2 = *Young disjoint Child*, D_3 = *Young disjoint Adult* and D_4 = *Child disjoint Adult* both entity types are derived (case 3). It is easy to see that their derivation rules entail them. In this particular example, the algorithms described in [18] determine the entailment efficiently.

Let $C = E$ *covered* $E_1,\ldots,E_n$ be a base covering constraint. We distinguish only two cases:

(1) if all entity types are base, then *C* must be enforced.

(2) in any entity type is derived, *C* may be entailed by the schema; if it is not, then *C* must be enforced.

There are two particular subcases, frequently found in practice, for which C is entailed by a derivation rule. The first is when E is derived by union of $E_1,...,E_n$. The second is when some E_j is derived by specialization of E and exclusion of $\{E_1,...,E_n\}$ - $\{E_j\}$.

Continuing again our previous example, consider now the covering constraint C_1 = *Person covered Man, Woman*. Given that *Person* is derived by union of *Man* and *Woman*, C_1 is satisfied by the schema. The constraint C_2 = *Person covered Child, Young, Adult* is entailed by the derivation rules of *Child*, *Young* and *Adult*. Again, in this particular example, the algorithms described in [18] determine the entailment efficiently.

Note, on the other hand, that if the derivation rule of *Adult* were:

Adult(p,t) $\leftrightarrow$ Person(p,t) $\wedge \neg$ Child(p,t) $\wedge \neg$ Young(p,t)

then *Adult* would be derived by specialization of *Person* and exclusion of *Child* and *Young* and, therefore, C_2 would be entailed by this derivation rule.

4.2 Satisfaction of Partitions

Partitions can be classified according to the derivability of their supertype and that of their subtypes. All combinations are possible. The only exception, at least in conceptual modeling, is when all entity types are base, because in this case the supertype could be derived by union of its subtypes. The classification is useful in order to analyze how the partition constraints can be satisfied in each case. We discuss below the two most popular kinds of partitions.

Partition by Specialization. Let $P = E$ *Partd* $E_1,...,E_n$ be a partition. We say that P is a *partition by specialization* when all E_i, $i = 1,..,n$, are derived by specialization of E. We have an example in Figure 1: *Person Partd Child, Young, Adult*, with all subtypes derived by specialization of *Person*, as shown in the derivation rules DR2 - DR4 given above.

Let's see how the partition constraints may be satisfied in this kind of partition. First, the specialization constraints $S_i = E_i$ *IsA* E are satisfied by the schema. More specifically, they are entailed by the derivation rule of E_i.

Second, the disjointness constraints $D_{i,j} = E_i$ *disjoint* E_j should be satisfied by the derivation rules of E_i and E_j. In the general case, when both E_i and E_j are derived, the constraint *may* be entailed by the schema, but in this case it seems sensible to require it to be entailed and, in particular, to be entailed by the derivation rules. The reason is that it should not be difficult to write the derivation rules in a way that ensures disjointness.

In many cases, a partition by specialization has a subtype for each value of some single-valued attribute of E. In these cases, disjointness is naturally ensured. For example, the partition *Person Partd Single, Married, Divorced, Widower*, where *Person* has attribute *MaritalStatus* and the derivation rules are:

Single(p,t) $\leftrightarrow$ Person(p,t) $\wedge$ MaritalStatus(p,Single,t)

and similarly for the other subtypes. The difference in the argument *ms* of the literal *MaritalStatus(p,ms,t)* ensures disjointness.

Finally, the covering constraint $C = E$ *covered* $E_1,...,E_n$ should be satisfied by the derivation rules of $E_1,...,E_n$, for the same reason as above. In the example, this requires to check only that there is a subtype for each possible value of *MaritalStatus*.

When one of the subtypes, E_j, is derived by specialization of E and exclusion of $\{E_1,...,E_n\}$-$\{E_j\}$, then C is satisfied by the derivation rule of E_j.

In summary, partitions by specialization does not require enforcement of any partition constraint. They only require writing carefully the derivation rules, so that the disjointness and covering constraints are satisfied by the schema.

Partition by Union. Let $P = E$ *Partd* $E_1,...,E_n$ be a partition. We say that P is a *partition by union* when E is derived by union of $E_1,...,E_n$. We have also an example in Figure 1: *Person Partd Man, Woman*, with *Person* derived by union of *Man* and *Woman*, as shown in the derivation rule DR1 given above.

Let's see how the partition constraints may be satisfied in this kind of partition. First, the specialization constraints $S_i = E_i$ *IsA* E are satisfied by the schema. More specifically, they are entailed by the derivation rule of E.

Second, the disjointness constraints $D_{i,j} = E_i$ *disjoint* E_j can be satisfied, as indicated in subsection 4.1, either by the schema or by enforcement, depending on the derivability of E_i and E_j. In the example, *Man disjoint Woman* must be enforced.

Finally, the covering constraint $C = E$ *covered* $E_1,...,E_n$ is satisfied by the schema. More specifically, it is entailed by the derivation rule of E.

In summary, partitions by union may require the enforcement of only some disjointness constraints. All other partition constraints are enforced by the schema.

5 Conclusions

This paper has focused on the relationships between taxonomic constraints and derivation rules. The objectives were to see which taxonomic constraints are entailed by derivation rules, and must be included in a taxonomy, and to analyze how taxonomic constraints can be satisfied in presence of derived types.

Based on an analysis of their derivation rules, we have classified derived entity types into derived by specialization, by past specialization, by union or by union with implicit types. The analysis reveals the taxonomic constraints entailed in each case. These constraints must be base constraints (defined in the taxonomy) or be derivable from them. We have shown how the base taxonomic constraints can be satisfied, either by the derivation rules (or the whole schema), or by enforcement. Our analysis can be extended naturally to taxonomies of relationship types [1,2].

Our results are general and could be incorporated into many conceptual modeling environments and tools. The expected benefits are an improvement in the verification of the consistency between taxonomic constraints and derivation rules, and a guide (to the information system designer) for the determination of the taxonomic constraints that must be enforced in the final system.

We see our results as complementary to those of Description Logics [9, 12], mainly because we assume a different context: temporal conceptual schemas, taxonomies of entity and relationship types, and derivation rules written in the FOL language. On the other hand, we deal with the design problem of determining which constraints must be enforced.

The work reported here can be extended in several directions. The analysis of derivation rules given in Subsections 3.2 and 5.1 can be completed by taking into account aggregate literals. The classifications given in these subsections may be refined, and further entailed constraints may be defined. We have focused on a single derivation rule to see the constraints entailed by it, but it should be possible to consider also sets of such rules. The analysis of partitions given in Subsection 4.2 can be extended to other frequent kinds. Finally, the analysis can be extended to taxonomies of relationship types.

References

1. Analyti, A.; Constantopoulos, P.; Spyratos, N. "Specialization by restriction and Schema Derivations", Information Systems, 23(1), 1998, pp. 1 - 38.
2. Analyti, A.; Spyratos, N.; Constantopoulos, P. "Property Covering: A Powerful Construct for Schema Derivations", Proc. ER'97, LNCS 1331, Springer-Verlag, pp. 271-284.
3. Atzeni, P.; Parker, D.S. "Formal Properties of Net-based Knowledge Representation Systems", Procs. ICDE'86, Los Angeles, 1986, pp. 700-706.
4. Batini, C.; Ceri, S.; Navathe, S. Conceptual Database Design. An Entity-Relationship Approach. The Benjamin/Cummings Pub. Co., 1992, 470 p.
5. Bancilhon, F.; Ramakrishnan, R. "An Amateur's Introduction to Recursive Query Processing Strategies". Proc. ACM SIGMOD Int. Conf. on Management of Data, May 1986, pp. 16-52.
6. Boman, M.; Bubenko, J.A.; Johannesson, P.; Wangler, B. Conceptual Modelling. Prentice Hall, 1997, 269 p.
7. Bergamaschi, S.; Sartori, C. "On Taxonomic Reasoning in Conceptual Design", ACM TODS 17(3), 1992, pp. 385-422.
8. Borgida, A.; Mylopoulos, J.; Wong, H.K.T. "Generalization/Specialization as a Basis for Software Specification". In Brodie, M.L.; Mylopoulos, J.; Schmidt, J.W. (Eds.) "On Conceptual Modelling", Springer-Verlag, 1984, pp. 87-117.
9. Borgida, A. "Description Logics in Data Management", IEEE Trans. on Knowledge and Data Eng., 7(5), 1995, pp. 671-682.
10. Borgida, A. "On the Relative Expressiveness of Description Logics and Predicate Logics". Artificial Intelligence 82(1-2), 1996, pp. 353-367.
11. Brachman, R.J.; Schmolze, J.G. "An Overview of the KL-ONE Knowledge Representation System". Cognitive Science 9, 1985, pp. 171-216.
12. Calvanese, D.; Lenzerini, M.; Nardi, D. "Description Logics for Conceptual Data Modeling", In Chomicki, J.; Saake, G. (Eds.) Logics for Databases and Information Systems, Kluwer Academic Press, 1998, pp. 229-263.
13. Ceri, S.; Fraternali, P. Designing Database Applications with Objects and Rules. The IDEA Methodology. Addison-Wesley, 1997, 579 p.
14. Delcambre, L.M.L.; Davis, K.C. "Automatic Verification of Object-Oriented Database Structures". Procs. ICDE'89, Los Angeles, pp. 2-9.
15. Decker, H.; Teniente, E.; Urpí, T. "How to Tackle Schema Validation by View Updating". Proc. EDBT'96, Avignon, LNCS 1057, Springer, 1996, pp. 535-549.
16. Godfrey, P.; Grant, J.; Gryz, J.; Minker, J. "Integrity Constraints: Semantics and Applications", In Chomicki, J.; Saake, G. (Eds.) Logics for Databases and Information Systems, Kluwer Academic Press, 1998, pp. 265-306.

17. Gustaffsson, M.R.; Karlsson, T.; Bubenko, J.A. jr. "A Declarative Approach to Conceptual Information Modeling". In: Olle, T.W.; Sol, H.G.; Verrijn-Stuart, A.A. (eds.) Information systems design methodologies: A Comparative Review. North-Holland, 1982, pp. 93-142.
18. Guo, S.; Sun, W.; Weiss, M.A. "Solving Satisfiability and Implication Problems in Database Systems". ACM TODS 21(2), 1996, pp. 270-293.
19. Halpin, T.A.; Proper, H.A. "Subtyping and polymorphism in object-role modelling", Data and Knowledge Eng., 15, 1995, pp. 251-281.
20. Hammer, M.; McLeod, D. "Database Description with SDM: A Semantic Database Model", ACM TODS, 6(3), 1981, pp. 351-386.
21. Hainaut, J-L.; Hick, J-M.; Englebert, V.; Henrard,J.; Roland, D. "Understanding the Implementation of IS-A Relations". Proc. ER'96, Cottbus, LNCS 1157, Springer, pp. 42-57.
22. Hull, R.; King, R. "Semantic Database Modeling: Survey, Applications, and Research Issues", ACM Computing Surveys, 19(3), 1987, pp. 201-260.
23. ISO/TC97/SC5/WG3. "Concepts and Terminology for the Conceptual Schema and Information Base", J.J. van Griethuysen (ed.), March, 1982.
24. Lenzerini, M. "Covering and Disjointness Constraints in Type Networks", Proc. ICDE'87, Los Angeles, IEEE, 1987, pp. 386-393.
25. Lloyd, J.W. Foundations of Logic Programming. 2nd Edition. Springer-Verlag, 1987, 212 p.
26. Martin, J.; Odell, J.J. Object-Oriented Methods: A Foundation. Prentice Hall, 1995, 412 p.
27. Markowitz, V.M.; Shoshani, A. "Representing Extended Entity-Relationship Structures in Relational Databases: A Modular Approach", ACM TODS 17(3), 1992, pp. 423-464.
28. Nijssen, G.M.; Halpin, T.A. Conceptual Schema and Relational Database Design. Prentice Hall, 1989, 342 p.
29. Olivé, A.; Costal, D.; Sancho, M-R. "Entity Evolution in IsA Hierarchies", Proc. ER'99, LNCS 1728, Springer, 1999, pp. 62-80.
30. Olivé, A. "Relationship Reification: A Temporal View", Proc. CAiSE'99, LNCS 1626, Springer, 1999, pp. 396-410.
31. Rumbaugh, J.; Jacobson, I.; Booch, G. The Unified Modeling Language Reference Manual. Addison-Wesley, 1999, 550 p.

Using UML Action Semantics for Executable Modeling and Beyond

Gerson Sunyé, François Pennaneac'h, Wai-Ming Ho, Alain Le Guennec, and Jean-Marc Jézéquel

IRISA, Campus de Beaulieu, F-35042 Rennes Cedex, France
sunye, pennanea, waimingh, aleguenn, jezequel@irisa.fr

Abstract. The UML lacks precise and formal foundations for several constructs such as transition guards or method bodies, for which it resorts to semantic loopholes in the form of "uninterpreted" expressions. The Action Semantics proposal aims at filling this gap by providing both a metamodel integrated into the UML metamodel, and a model of execution for these statements. As a future OMG standard, the Action Semantics eases the move to tool interoperability, and allows for executable modeling and simulation. We explore in this paper a specificity of the Action Semantics: its applicability to the UML metamodel, itself a UML model. We show how this approach paves the way for powerful metaprogramming capabilities such as refactoring, aspect weaving, application of design patterns or round-trip engineering. Furthermore, the overhead for designers is minimal, as mappings from usual object-oriented languages to the Action Semantics will be standardized. We focus on an approach for expressing manipulations on UML models with the upcoming Action Semantics. We illustrate this approach by various examples of model transformations.

1 Introduction

The Unified Modeling Language provides modeling foundations for both the structural and behavioral parts of a system. The designer is offered nine views which are like projections of a whole multi-dimensional system onto separate planes, some of them orthogonal. This provides an interesting separation of concerns, covering four main dimensions of software modeling: functional (use cases diagrams express the requirements), static (class diagrams), dynamic (statecharts, sequence diagrams for the specification of behavioral aspects) and physical (implementation diagrams).

But the UML currently lacks some precise and formal foundation for several constructs such as transition guards or method bodies, for which it resorts to semantic loopholes in the form of "uninterpreted" expressions. The Action Semantics (AS) proposal [1], currently being standardized at the Object Management Group (OMG), aims at filling this gap by providing both a metamodel integrated into the UML metamodel, and a model of execution for these statements. Along the lines of what already exists in the IUT-T Specification and

K.R. Dittrich, A. Geppert, M.C. Norrie (Eds.): CAiSE 2001, LNCS 2068, pp. 433–447, 2001.
© Springer-Verlag Berlin Heidelberg 2001

Description Language (SDL) community [11], the integration of the Action Semantics into UML should ease the move to tool interoperability, and allow for executable modeling and simulation, as well as full code or test case generation.

The contribution of this paper is twofold. First, it gives an update on the forthcoming UML/AS standard, for which we contributed some of the precise semantic aspects. Second, relying on the fact that the UML metamodel is itself a UML model, we show how the Action Semantics can be used at the metamodel level to help the OO designer carry on activities such as behavior-preserving transformations [16] (see Sect. 3.1), design pattern application [5] (Sect. 3.2) and design aspects weaving [12] (Sect. 3.3). But before extending the description of the Action Semantics, let us dispel some possible misunderstanding here. Our intention is not to propose yet another approach for model transformation, pattern application or refactoring. What we claim here, is that the Action Semantics may (and should) go beyond the design level and be used as a metaprogramming language to help the implementation of existing approaches. Also, as we show in the following sections, it has strengths in both cases, as a design level and as a metamodel level programming language.

In all these design level activities, we can distinguish two steps: (1) the identification of the need to apply a given transformation on a UML model; and (2) the actual transformation of that model. Our intention is not to usurp the role of the designers in deciding what to do in step 1, but to provide them with tools to help automate the second step, which is usually very tedious and error prone. Further, when carried out in a ad hoc manner, it is very difficult to keep track of the *what, why* and *how* of the transformation, thus leading to traceability problems and a lack of reusability of the design micro-process. This can be seen in maintenance, when one has to propagate changes from the problem domain down to the detailed design by "re-playing" design decisions on the modified part of a model. Automation could also be very worthwhile in the context of product lines, when the same (or at least very similar) design decisions are to be applied on a family of analysis models (e.g. the addition of a persistence layer on many MIS applications).

Because Joe-the-OO-designer cannot be expected to write complex metaprograms from scratch, in order to develop his design, he must be provided with pre-canned transformations (triggered through a menu), as well as ways of customizing and combining existing transformations to build new ones. The main interest of using the UML/Action Semantics at the metamodel level for expressing these transformations is that we can use classical OO principles to structure them into *Reusable Transformation Components* (RTC): this is the open-closed principle [15] applied at the metamodel level. Furthermore, since the Action Semantics is fully integrated in the UML metamodel, it can be combined with rules written in the Object Constraint Language (OCL), in order to verify if a transformation (or a set of transformations) may be applied to a given context.

We can then foresee a new dimension for the distribution of work in development teams, with a few *metadesigners* being responsible for translating the mechanistic part of a company's design know-how into RTC, while most other designers concentrate on making intelligent design decisions and automatically applying the corresponding transformations.

The rest of the paper is structured as follows. Section 2 recalls some fundamental aspects of the UML and gives a quick update on the Action Semantics proposal currently under submission for standardization at the OMG. Section 3 shows the interest of using the Action Semantics at the metamodel level for specifying and programming model transformations in several contexts. Section 4 analyses the impact of this approach on OO development processes, and presents our experience with using it on several case studies. Section 5 describes related work.

2 Executable Modeling with UML

2.1 The Bare-Bone UML Is Incomplete and Imprecise

The UML is based on a four-layer architecture, each layer being an instance of its upper layer, the last one being an instance of itself. The first layer holds the living entities when the code generated from the model is executed, i.e. running objects, with their attribute values and links to other objects. The second layer is the modeling layer. It represents the model as the designer conceives it. This is the place where classes, associations, state machines etc... are defined (via the nine views, cited in the above introduction). The running objects are instances of the classes defined at this level. The third layer, the metamodel level, describes the UML syntax in a metamodeling language (which happens to be a subset of UML). This layer specifies what a syntactically correct model is. Finally the fourth layer is the meta-metamodel level, i.e. the definition of the metamodeling language syntax, thus the syntax of the subset of UML used as a metamodeling language. UML creators chose a four-layer architecture because it provides a basis for aligning the UML with other standards based on a similar infrastructure, such as the widely used Meta-Object Facility (MOF).

Although there is no strict one-to-one mapping between all the MOF meta-metamodel elements and the UML metamodel elements, the two models are interoperable: the UML core package metamodel and the MOF are structurally quite similar. This conception implies the UML metamodel (a set of class diagrams) is itself a UML model.

A UML model is said to be syntactically correct if the set of its views merge into a consistent instance of the UML metamodel. The consistency of this instance is ensured via the metamodel structure (i.e. multiplicities on association ends) and a set of well-formedness rules (expressed in OCL), which are logical constraints on the elements in a model. For instance, there should not be any inheritance cycle and a FinalState may not have any outgoing transitions.

But apart from those syntactic checks regarding the structure of models, UML users suffer from the lack of formal foundations for the important behavioral aspects, leading to some incompleteness and opening the door to inconsistencies in UML models. This is true, for instance, in state diagrams, where the specification of a guard on a transition is realized by a BooleanExpression, which is basically a string with no semantics. Thus, the interpretation is left to the modeling tool, breaking interoperability. But more annoying is the fact that models are not executable, because they are incompletely specified in the UML. This makes it impossible to verify and test early in the development process. Such activities are key to assuring software quality.

2.2 The Interest of an Action Semantics for UML

The Action Semantics proposal aims at providing modelers with a complete, software-independent specification for actions in their models. The goal is to make UML modeling executable modeling [7], i.e. to allow designers to test and verify early and to generate 100% of the code if desired. It builds on the foundations of existing industrial practices such as SDL, Kennedy Carter [13] or BridgePoint [17] action languages[1]. But contrary to its predecessors, it is intended that the Action Semantics become an OMG standard, and a common base for all the existing and to-come action languages (mappings from existing languages to Action Semantics are proposed).

Traditional modeling methods which do not have support for any action language have focused on separating analysis and design, i.e. the *what the system has to do* and the *how that will be achieved.* Whilst this separation clearly has some benefits, such as allowing the designer to focus on system requirements without spreading himself/herself too thinly with implementation details, or allowing the reuse of the same analysis model for different implementations, it also has numerous drawbacks. The main one is that this distinction is a difficult, not to say impossible, one to make in practice: the boundaries are vague; there are no criteria for deciding what is analysis, and what is not. Rejecting some aspects from analysis make it incomplete and imprecise; trying to complete it often obliges the introduction of some *how* issues for describing the most complex behavior.

As described above, the complete UML specification of a model relies on the use of uninterpreted entities, with no well-defined and accepted common formalism and semantics. This may be, for example, guards on transitions specified with the Object Constraint Language (OCL), actions in states specified in Java or C++.
In the worst case – that is for most of the modeling tools – these statements are simply inserted at the right place into the code skeleton. The semantics of execution is then given by the specification of the programming language. Unfortunately, this often implies an over-specification of the problem (for example, in Java, a sequential execution of the statements of a method is supposed), verification and testing are feasible only when the code is available, far too late in the development process.

Moreover, the designer must have some knowledge of the way the code is generated for the whole model in order to have a good understanding of the implications of her inserted code (for instance, if you are using C++ code generator

[1] all the major vendors providing an action language are in the list of submitters.

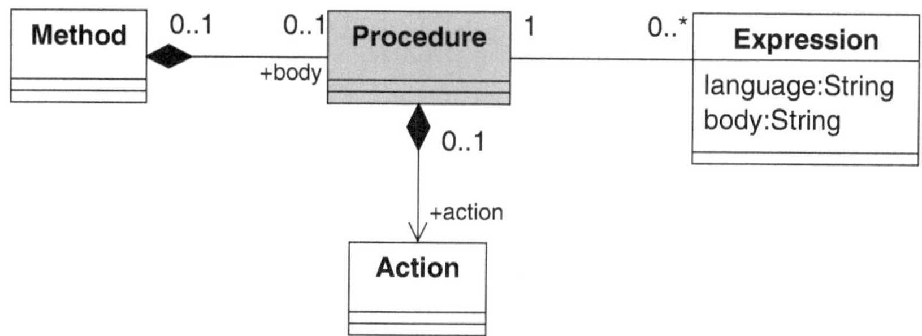

Fig. 1. Action Semantics immersion into the UML

of a UML tool, a knowledge of the way the tool generates code for associations is required in order to use such code in your own program).
At best, the modeling tool has its own action language and then the model may be executed and simulated in the early development phases, but with the drawbacks of no standardization, no interoperability, and two formalisms for the modeler to learn (the UML and the action language).

2.3 The Action Semantics Proposal Submitted to the OMG

The Action Semantics proposal is based upon three abstractions:

- A metamodel: it extends the current metamodel. The Action Semantics is integrated smoothly in the current metamodel and allows for a precise syntax of actions by replacing all the previously uninterpreted items. The uninterpreted items are viewed as a surface language for the abstract syntax tree of actions (see Fig. 1);
- An execution model: it is a UML model. It allows the changes to an entity over time to be modeled. Each change to any mutable entity yields a new snapshot, and the sequence of snapshots constitutes a history for the entity. This execution model is used to define the semantics of action execution;
- Semantics: the execution of an Action is precisely defined with a *life-cycle* which unambiguously states the effect of executing the action on the current instance of the execution model (i.e. it computes the next snapshot in history for the entity).

2.4 Surface Language

The Action Semantics proposal for the UML does not enforce any notation (i.e. surface language) for the specification of actions. This is intentional, as the goal of the Action Semantics is certainly not to define a new notation, or to force the use of a particular existing one. The Action Semantics was, however, conceived to allow an easy mapping of classical languages such as Java, C++ or SDL[2].

[2] Some of these mappings are illustrated in the AS specification document [4].

Thus, designers can keep their favorite language without having to learn a new one.

3 Metaprogramming with the Action Semantics

An interesting aspect of UML is that its syntax is represented (or metamodeled) by itself (as a class diagram, actually). Thus, when using a reflexive environment (where the UML syntax is effectively represented by a UML model), the Action Semantics can be used to manipulate UML elements, i.e. transform models.

In the following sections, we present three different uses for this approach: implementing refactorings, applying design patterns and weaving design aspects. Since the Action Semantics does not have an official surface language, we adopt an "OCL-based" version in our examples.

3.1 Refactorings

Refactorings [16] are behavior-preserving transformations that are used to restructure the source code of applications and thus make the code more readable and reusable. They help programmers to manipulate common language constructs, such as class, method and variable, of an existing application, without modifying the way it works. Since we believe that refactorings are an essential artifact for software development, we are interested in bringing them to the design level by means of a UML tool.

The implementation of refactorings in UML is an interesting task, since *design* refactorings – as opposed to *code* refactorings – should work with several design constructs shared among several views of a model. Furthermore, while code refactorings manipulate source code (text files), design refactorings should manipulate an abstract syntax tree.

This difference simplifies the implementation of some refactorings, for instance renaming elements, since an element is unique inside the abstract syntax tree. However, other refactorings (namely moving features) are more difficult to implement because we must take into account other constructs, such as OCL constraints and state charts.

The Action Semantics represents a real gain for refactoring implementation, not merely because it directly manipulates UML constructs, but also because of the possibility of combining it with OCL rules to write pre and post-conditions. More precisely, as refactorings must preserve the behavior of the modified application, they cannot be widely applied. Thus, every refactoring ought to verify a set of conditions before the transformation is carried out.

Below we present an example of a simple refactoring, the removal of a class from a package. This refactoring can only be applied if some conditions are verified: the class is not referenced by any other class and it has no subclasses (or no features). These conditions and the transformation itself are defined in the Action Semantics as follows:

```
Package::removeClass(class: Class)
pre:
    self . allClasses→ includes(class) and
    class . classReferences→isEmpty and
    ( class . subclasses→isEmpty or class.features→isEmpty)
actions:
    let aCollection := class . allSuperTypes
    class . allSubTypes→forAll(sub : Class |
        aCollection→forAll(sup : Class | sub.addSuperClass(sup)))
    class . delete
post:
    self . allClasses→ excludes(class) and
    class @pre.allSubTypes→forAll(each : Class |
        each.allSuperTypes.includesAll(class@pre.allSuperTypes)) and
    class @pre.features→forAll(feat : Feature | feat .owner = nil)
```

This refactoring calls two functions that are not present in the Action Semantics, *delete* and *addSuperClass()*, and should be defined elsewhere: the role of the former is to make the delete effective, by deleting the features, associations and generalization links owned by the class. The latter feature is described below. Its purpose is to create a generalization link between two classes:

```
Class :: addSuperClass(class: Class)
pre:
    self . allSuperTypes→excludes(class)
actions:
    let gen := Generalization.new
    gen.addLinkTo(parent, class)
    gen.addLinkTo(child, self)
post:
    self . allSuperTypes()→includes(class)
```

3.2 Design Patterns

Another interesting use for Action Semantics is the application of the solution proposed by a Design Pattern, i.e. the specification of the proposed terminology and structure of a pattern in a particular context (called instance or occurrence of a pattern). In other words, we envisage the application of a pattern as a sequence of transformation steps that are applied to an initial situation in order to reach a final situation, an explicit occurrence of a pattern.

This approach is not, and does not intend to be, universal since only a few patterns mention an existing situation to which the pattern could be applied (see [1] for further discussion on this topic). In fact, our intent is to provide designers with metaprogramming facilities, so they are able to define (and apply) their own variants of known patterns. The limits of this approach, such as pattern and trade-offs representation in UML, are discussed in [20].

As an example of design pattern application, we present below a transformation function that applies the Proxy pattern. The main goal of this pattern is to provide a placeholder for another object, called *Real Subject*, to control access to it. It is used, for instance, to defer the cost of creation of an expensive object until it is actually needed:

```
Class :: addProxy
pre:
   let classnames = self.package.allClasses→collect(each : Class | each.name) in
   (classnames→excludes(self.name+'Proxy') and
   classnames→excludes('Real'+self.name))
actions:
   let name := self.name
   let self.name := name.concat('Proxy')
   let super :=
      self.package.addClass(name,self.allSuperTypes(),{}→including(self))
   let real :=
      self.package.addClass('Real'.concat(name),{}→including(super),{})
   let ass := self.addAssociationTo('realSubject',real)
   self.operations→forAll(op : Operation | op.moveTo(real))
```

This function uses three others (actually, refactorings), that will not be precisely described here. They are however somewhat similar to the *removeClass()* function presented above. The first function, *addClass()*, adds a new class to a package, and inserts it between a set of super-classes and a set of subclasses. The second, *addAssociationTo()*, creates an association between two classes. The third, *moveTo()*, moves a method to another class and creates a "forwarder" method in the original class.

This transformation should be applied to a class that is to play the *role* of real subject[3]. Its application proceeds as follows:

1. Add the 'Proxy' suffix to the class name;
2. Insert a super-class between the class and its super-classes;
3. Create the *real subject* class;
4. Add an association between the *real subject* and the *proxy*
5. Move every method owned by the *proxy* class to the *real subject* and create a forwarder method to it (move methods).

As we have explained before, this is only one of the many implementation variants of the Proxy pattern. This implementation is not complete, since it does not create the *load()* method, which should create the *real subject* when it is requested. However, it can help designers to avoid some implementation burden, particularly when creating forwarder methods.

3.3 Aspect Weaving

Finally, we would like to show how Action Semantics can support the task of developing applications that contain multiple aspects. Aspects (or concerns) [14,21]

[3] Patterns are defined in terms of roles, which are played by one or more classes in its occurrences

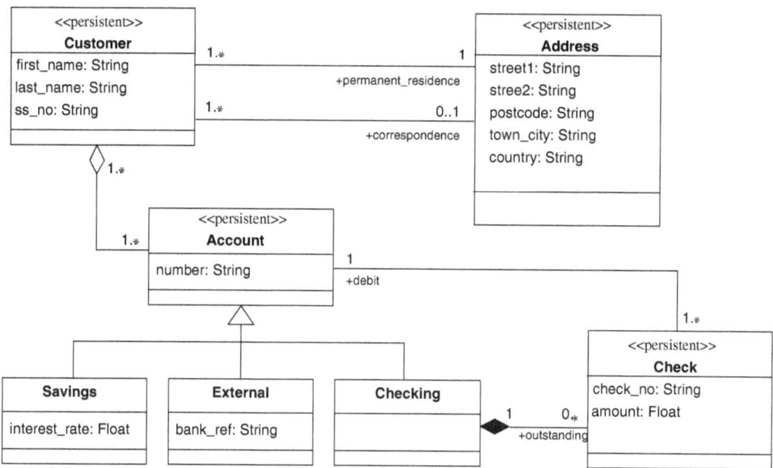

Fig. 2. Information management system for personal finance

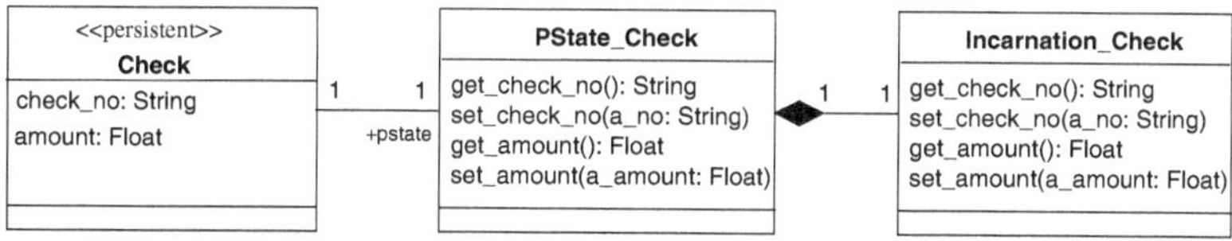

Fig. 3. Persistence proxies and access methods

refer to non-functional requirements that have a global impact on the implementation. The approach used in dealing with this is to separate these aspects from the conceptual design, and to introduce them into the system only during the final coding phase. In many cases, the merging of aspects is handled by an automated tool. In our example, we attempt to show how aspects can be weaved as early as the design level through model transformation [9], using the Action Semantics to write the transformation rules.

The class diagram in Fig. 2 illustrates a model of a bank's personal-finances information-management system. In the original system, the accounting information was stored in a relational database and each class marked with the "persistent" stereotype can be related to a given table in the database.
The aim of this re-engineering project is to develop a distributed object-oriented version of the user front-end to support new online access for its customers. One of the non-functional requirements is to map these "persistent" objects to the instance data stored in the relational database.
The task involves writing a set of proxy classes that hide the database dependency, as well as the database query commands. An example of the required transformation is illustrated by the model in Fig. 3. In this reference template, the instance variable access methods are generated automatically and database specific instructions are embedded to perform the necessary data access.

Since the re-engineering is carried out in an incremental manner, there is a problem with concurrent access to the database during write-back commits. The new application must cooperate with older software to ensure data coherence. A provisional solution is to implement a single-ended data coherence check on the new software. This uses a timestamp to test if data has been modified by other external programs. If data has been modified since the last access, all commit operations will be rolled back, thus preserving data coherence without having to modify old software not involved in this incremental rewrite. Fig. 4 shows the template transformation required. It adds a flag to cache the timestamp and access methods will be wrapped by timestamp-checking code.

The metaprogram needed to generate the proxy classes of figures 3 and 4 is composed of several operations. The first one is defined in the context of a Namespace (i.e. the container of UML modeling elements). It selects all classes that are stereotyped 'persistent' and sends them the *implementPersistent()* message:

```
Namespace:: implementPersistentClasses
actions:
  self . allClasses →select(each : Class| each.stereotype→notEmpty)→
    select(each : Class | each.stereotype→first .name = 'persistent')→
      forAll(each : Class | each.implementPersistent)
```

The *implementPersistent()* operation is defined in the context of a Class. This operation will first create two classes, *state* and *incarnation*, and then creates, in these classes, the access methods to its own stereotyped attributes. This operation is defined as follows:

```
Class :: implementPersistent
actions:
  let pstate :=
     self .package.addClass('PState_'.concat(pclass.name),{},{})
  pstate.addOperation('Load'); pstate.addOperation('Save')
  self .addAssociationTo(pstate, 1, 1)
  let incarnation :=
     self .package.addClass('Incarnation_'.concat(pclass.name),{},{})
  pstate.addCompositeAssociationTo(incarnation, 1, 1)
  let attrs := self . allAttributes →
     select(a : Attribute| a.stereotype→notEmpty)
  attrs→select(a : Attribute | a.stereotype→first .name = 'getset')→
     forAll(a : Attribute |
        pstate.createSetterTo(a); pstate.createGetterTo(a)
        incarnation.createSetterTo(a); incarnation.createGetterTo(a))
  attrs→select(a : Attribute | a.stereotype→first .name = 'get')→
     forAll(a : Attribute |
        incarnation.createGetterTo(a); pstate.createGetterTo(a))
  attrs→select(a : Attribute | a.stereotype→first .name = 'set')→
     forAll(a : Attribute |
        pstate.createSetterTo(a); incarnation.createSetterTo(a))
```

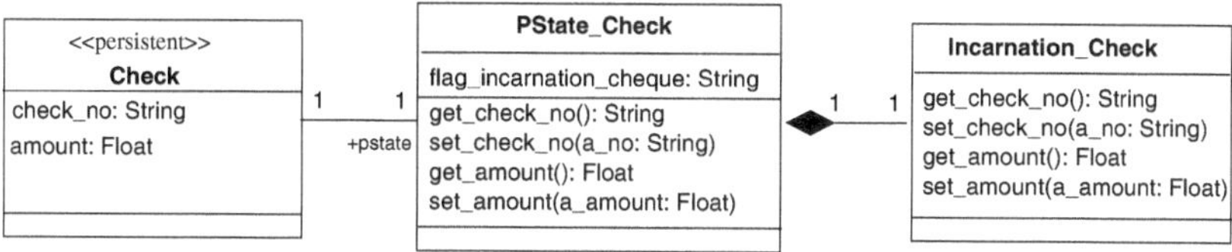

Fig. 4. Timestamp cache flag for concurrent data coherence

The creation of the access methods is implemented by the *createSetterTo()* and *createGetterTo()* operations. They are both defined in the Class context and implement a similar operation. They take an Attribute as parameter and create a Method for setting or getting its value. These operations use two other operations, *createMethod()* and *createParameter()*, which are explained above:

```
Class :: createSetterTo(att : Attribute)
actions:
   let newMethod := self.createMethod('set_'.concat(att.name))
   newMethod.createParameter('a_'.concat(attrib_name), att.type, 'in')
```

```
Class :: createGetterTo(att : Attribute)
actions:
   let newMethod := self.createMethod('get_'.concat(att.name))
   newMethod.createParameter('a_'.concat(attrib_name), att.type, 'out')
```

The *createMethod()* operation is also defined in the Class context. Its role is to create a new Method from a string and to add it to the Class:

```
Class :: createMethod(name : String)
actions:
   let newMethod := Method.new
   let newMethod.name := name
   self .addMethod(newMethod)
   let result := newMethod
```

Finally, the *createParameter()* operation creates a new parameter and adds it to a Method, which is the context of this operation:

```
Method::createParameter(name : String, type : Class, direction : String)
actions:
   let newParameter := Parameter.new
   let newParameter.name := name
   newParameter.setType(type)
   newParameter.setDirection(direction)
   self .addParameter(newParameter)
   let result := newParameter
```

The attractiveness of this approach is not immediately evident. Let us consider a different implementation for the persistent proxy of Fig. 3. In the case where there are composite persistent objects, it is possible to use a single per-

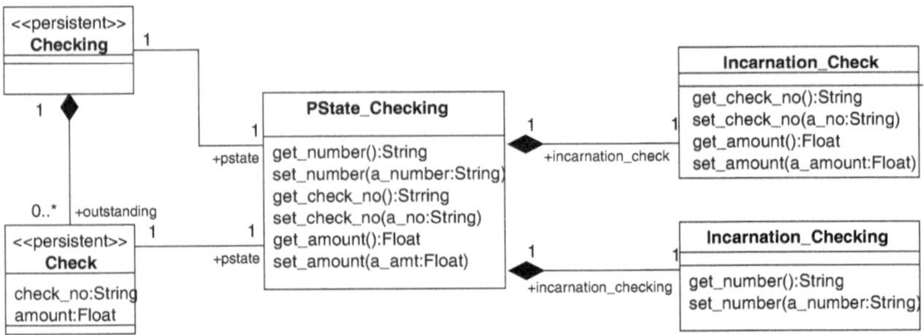

Fig. 5. Implementation template for shared proxy

sistent state proxy for a composite object and all its components (see Fig. 5). Through the use of metaprogramming, it is now possible to consider these different implementation aspects independently from the concurrency implementation. It enables the designer to conceptualize the modifications in a manageable manner. Making changes to a model by hand as a result of a change in an implementation decision is not a viable alternative as it is laborious and prone to error.

Therefore, it can be seen that metaprogramming using the Action Semantics can facilitate implementation changes at a higher abstraction level. It also leverages the execution machine for the Action Semantics by using it to perform the model transformation.

4 Impact in the Development Approach

The Action Semantics brings some possible changes to the traditional software development process. In other words, the Action Semantics is an important step towards the use of UML in an effective development environment, since it offers the possibility of animating early design models and evolving or refining them until their implementation. The development approach we propose here starts with an early design model, created by the designers from an analysis model. This model is completely independent of the implementation environment, it assumes an "Ideal World", where the processing power and the memory are infinite, there are no system crashes, no transmission errors, no database conflicts, etc. Since this model contains Action Semantics statements, it can be animated by the Action Semantics machine and validated. Once the validation is finished, the designers can add some environment-specific aspects to the design model (database access, distribution), apply design patterns and restructure the model using design refactorings.

A significant part of this information addition can be automated using the metaprogramming techniques we presented above (Sect. 3). More precisely, the designers are able to automatically apply a set of predefined transformations, defined by a special software developer, who we call the *metadesigner*. We see the metadesigners as experienced developers, who handle essential knowledge about the implementation environment. Their role is to use their knowledge to define and write model transformations and make them available to designers. The approach ends when the final design model is reached. After that, the implementation code can be generated.

4.1 Organizing Things

UML proposes an interesting extension mechanism called *Profiles*, allowing designers to expand the current metamodel. According to the UML 1.4 documentation (page 4-3), a UML Profile is:

(...) a predefined set of Stereotypes, TaggedValues, Constraints, and notation icons that collectively specialize and tailor the UML for a specific domain or process (e.g. Unified Process profile). A profile does not extend UML by adding any new basic concepts. Instead, it provides conventions for applying and specializing standard UML to a particular environment or domain.

Profiles can be used to organize Action Semantics metaprograms, whilst some other UML concepts, such as refinements and traces, can be used to manage the evolution of metaprograms (versioning). When the "refactoring" Profile is loaded into the metamodel, some operations are added to the UML concepts. For instance, *addClass()* is added to Package, *addSuperClass()* to Class, etc.

5 Related Work

Commercial UML tools often propose metaprogramming languages in order to manipulate models. This is the case, for instance, of Softeam's Objecteering, which uses J, a "Java-like" language [19]. This is also the case of Rational Rose and of Rhapsody from Ilogix, which use Visual Basic. The purpose of these languages is similar to the one we look for when using the Action Semantics as a metaprogramming language. However there are some notable differences:

To begin with, the Action Semantics only proposes an abstract syntax and no surface language. Thus, it is not a direct concurrent of J or Visual Basic, but it can provide a common underlying model for these languages.

Next, the navigation through the metamodel is done in a language-specific way, which does not reuse the standard OCL navigation facilities.

Finally, since there are discrepancies between the UML metamodel and the metamodels implemented by commercial tools, designers are not able to write truly universal UML transformations.

The lack of a surface language may be an annoying problem, since the choice of the surface language impacts on the practical use. While the choice of using a classic language (e.g. Java), thus possibly the same language at both the design and the meta-model level, is attractive, it may not be the most adequate. The adoption of a declarative language (e.g. OCL) may help bridge the gap between the specification of contracts for design transformations (specified with the OCL) and their implementation.

6 Conclusion

The Action Semantics provides designers with an expressive formalism for writing model transformations in an operational way, and a model for UML statement execution that fills some semantic gaps in the behavior aspects of a UML model. Moreover, since the UML metamodel itself is a UML model, the Action Semantics can be used as a powerful mechanism for metaprogramming. This particularity opens new perspectives for designers, compared to other model manipulations languages (e.g. J or VisualBasic), thanks to its perfect integration with the UML: all the features of the UML, such as constraints (pre or post-conditions, invariants), refinements or traces can be applied within the Action Semantics.

An implementation conforming to the current version of the Action Semantics specification is in development in UMLAUT[4], a freely available UML modeling tool. The complete integration of the Action Semantics and the UML in Umlaut provides an excellent research platform for the implementation of design patterns, refactorings and aspects.

Indeed, we are presently looking for more UML-specific refactorings with the purpose of extending our existent set, which is based on existing program restructuring transformations [2,3,6,10,16,18]. The use of the OCL and the Actions Semantics for defining refactorings opens two interesting perspectives. The first perspective concern the combinations of refactorings. As D. Roberts stated in his thesis [18], pre and post conditions can be used to verify if each single refactoring, within a set of combined refactorings, is (or will be) allowed to execute, before the execution of the whole set.

The second perspective concerns the property of behavior-preservation, that should be verified after the execution of a refactoring. In a similar way to other efforts in this area, we intend to propose a set of basic transformations and demonstrate how they ensure behavior-preservation. We will then be able to combine them in order to define higher-level transformations. We intend to define an approach based on the notion of *refactoring region*, which is a fraction of an instance of the UML metamodel (i.e. the underlying objects that represent a modeled system) that may potentially be modified by a transformation. With the help of bisimulation techniques we will be able to verify if the original and the transformed regions are equivalent. The notion of *snapshot history* - which is present in the Actions Semantics execution engine - provides excellent support for such a comparison.

We are also working on the representation of patterns trade-offs that could help the designer to define the variant she wants to apply. An introduction to this work can be found in [8]. The analysis of the Action Semantics syntax tree will help us to verify if a pattern was correctly implemented. More precisely, our tool will be able to verify if a given implementation of a pattern respects a set of predefined OCL constraints [20].

---

[4] http://www.irisa.fr/UMLAUT/

References

1. M. Cinnide and P. Nixon. A methodology for the automated introduction of design patterns. In *International Conference on Software Maintenance*, Oxford, 1999.
2. P. Bergstein. Maintainance of object-oriented systems during structural schema evolution. *TAPOS*, 3(3):185–212, 1997.
3. E. Casais. *Managing Evolutuin in Object Oriented Environments: An Algorithmic Approach.* Phd thesis, University of Geneva, 1991.
4. T. A. S. Consortium. Updated joint initial submission against the action semantics for uml rfp, 2000.
5. E. Gamma, R. Helm, R. Johnson, and J. Vlissides. *Design Patterns: Elements of Reusable Object-Oriented Software.* Professional Computing Series. Addison-Wesley, Reading, MA, 1995.
6. W. Griswold. Program restructuring as an aid to software maintenance, 1991.
7. O. M. Group. Action semantics for the uml rfp, ad/98-11-01, 1998.
8. A. L. Guennec, G. Sunyé, and J.-M. Jézéquel. Precise modeling of design patterns. In *Proceedings of UML 2000*, volume 1939 of *LNCS*, pages 482–496. Springer Verlag, 2000.
9. W. Ho, F. Pennaneac'h, and N. Plouzeau. Umlaut: A framework for weaving uml-based aspect-oriented designs. In *Technology of object-oriented languages and systems (TOOLS Europe)*, volume 33, pages 324–334. IEEE Computer Society, June 2000.
10. W. Hursch. *Maintaining Consistency and Behavior of Object-Oriented Systems during Evolution.* Phd thesis, Northeastern University, June 1995.
11. IUT-T. Recommendation z.109 (11/99) - SDL combined with UML, 1999.
12. R. Keller and R. Schauer. Design components: Towards software composition at the design level. In *Proceedings of the 20th International Conference on Software Engineering*, pages 302–311. IEEE Computer Society Press, Apr. 1998.
13. Kennedy-Carter. Executable UML (xuml), http://www.kc.com/html/xuml.html.
14. G. Kiczales, J. Lamping, A. Menhdhekar, C. Maeda, C. Lopes, J.-M. Loingtier, and J. Irwin. Aspect-oriented programming. In M. Akşit and S. Matsuoka, editors, *ECOOP '97 — Object-Oriented Programming 11th European Conference, Jyväskylä, Finland*, volume 1241 of *Lecture Notes in Computer Science*, pages 220–242. Springer-Verlag, New York, N.Y., June 1997.
15. B. Meyer. *Object-Oriented Software Construction.* Prentice-Hall, 1988.
16. W. F. Opdyke. *Refactoring Object-Oriented Frameworks.* PhD thesis, University of Illinois, 1992.
17. Projtech-Technology. Executable UML, http://www.projtech.com/pubs/xuml.html.
18. D. Roberts. *Practical Analysis for Refactoring.* PhD thesis, University of Illinois, 1999.
19. Softeam. UML Profiles and the J language: Totally control your application development using UML. In *http://www.softeam.fr/us/pdf/uml_profiles.pdf*, 1999.
20. G. Sunyé, A. Le Guennec, and J.-M. Jézéquel. Design pattern application in UML. In E. Bertino, editor, *ECOOP'2000 proceedings*, number 1850, pages 44–62. Lecture Notes in Computer Science, Springer Verlag, June 2000.
21. P. Tarr, H. Ossher, and W. Harrison. N degrees of separation: Multi-dimensional separation of concerns. In *ICSE'99 Los Angeles CA*, 1999.

Design and Implementation of a UML-Based Design Repository*

Rudolf K. Keller, Jean-François Bédard, and Guy Saint-Denis

Département d'informatique et de recherche opérationnelle
Université de Montréal
C.P. 6128, succursale Centre-ville
Montréal (Québec) H3C 3J7, Canada
{keller, bedardje, stdenisg}@iro.umontreal.ca
http://www.iro.umontreal.ca/~{keller, bedardje, stdenisg}

Abstract. The aim of this paper is to present the SPOOL design repository, which is the foundation of the SPOOL software engineering environment. The SPOOL design repository is a practical implementation of the UML metamodel, and is used to store detailed design-level information that is extracted from the source code of industrial systems. Its internal mechanisms and related tools provide functionalities for querying data and observing dependencies between the components of the studied systems, facilitating core tasks conducted in reverse engineering, system comprehension, system analysis, and reengineering. This paper discusses the architecture, the schema, the mechanisms, and the implementation details of the repository, and examines the choice of the UML metamodel. Experiences conducted with large-scale systems are also presented, along with related work and future avenues in design repository research.

Keywords. Design repository, Unified Modeling Language, data interchange, reverse engineering, system analysis, system comprehension, reengineering, system visualization.

1 Introduction

In information systems engineering and software engineering alike, repositories play an important role. Repositories at the design level, henceforth referred to as *design repositories*, are key to capture and manage data in domains as diverse as corporate memory management [10], knowledge engineering [15], and software development and maintenance. In this paper, we report on our experience in designing and

* This research was supported by the SPOOL project organized by CSER (Consortium for Software Engineering Research) which is funded by Bell Canada, NSERC (Natural Sciences and Engineering Research Council of Canada), and NRC (National Research Council Canada).

K.R. Dittrich, A. Geppert, M.C. Norrie (Eds.): CAiSE 2001, LNCS 2068, pp. 448–464, 2001.
© Springer-Verlag Berlin Heidelberg 2001

implementing a design repository in this latter domain, that is, in the realm of system comprehension, analysis, and evolution.

In order to understand, assess, and maintain software systems, it is essential to represent the analyzed systems at a high level of abstraction such as the analysis and design level. End user tools need access to this information, and thus, a design repository for storing the analyzed systems is required. Such a design repository should meet a number of requirements. First, it should be designed such that its schema will be resilient to change, adaptation, and extension, in order to address and accommodate easily new research projects. Second, it should preferably adopt a schema based on a standard metamodel, and offer extensibility mechanisms to cope with language-specific constructs. Third, in order to enable easy information interchange with other third-party tools, a flexible model interchange format should be used for importing and exporting purposes. Finally, the design of the repository should take into account scalability and performance considerations.

In the SPOOL project (*Spreading Desirable Properties into the Design of Object-Oriented, Large-Scale Software Systems*), a joint industry/university collaboration between the software quality assessment team of Bell Canada and the GELO group at Université de Montréal, we are investigating methods and tools for design composition [13] and for the comprehension and assessment of software design quality [14]. As part of the project, we have developed the SPOOL environment for reverse engineering, system comprehension, system analysis, and reengineering.

At the core of the SPOOL environment is the SPOOL design repository, which we designed with the four above-mentioned requirements in mind. The repository consists of the *repository schema* and the *physical data store*. The repository schema is an object-oriented class hierarchy that defines the structure and the behavior of the objects that are part of the reverse engineered *source code models*, the *abstract design components* that are to be identified from the source code, the *implemented design components*, and the *recovered and re-organized design models*. Moreover, the schema provides for more complex behavioral mechanisms that are applied throughout the schema classes, which includes uniform traversal of complex objects to retrieve contained objects, notification to the views on changes in the repository, and dependency accumulation to improve access performance to aggregated information. The schema of the design repository is based on an extended version of the *UML metamodel 1.1* [25]. We adopted the UML metamodel as it captures most of the schema requirements of the research activities of SPOOL. This extended UML metamodel (or SPOOL repository schema) is represented as a *Java 1.1* class hierarchy, in which the classes constitute the data of the *MVC-based* [3] SPOOL environment.

The object-oriented database of the SPOOL repository is implemented using *POET 6.0* [18]. It provides for data persistence, retrieval, consistency, and recovery. Using the precompiler of POET 6.0's *Java Tight Binding*, an object-oriented database representing the SPOOL repository is generated from the SPOOL schema. As POET 6.0 is *ODMG 3.0*-compliant [16], its substitution for another ODMG 3.0-compliant database management system would be accomplishable without major impact on the schema and the end user tools.

To deal with data interchange (importing source code into and exporting model information from the repository), *XMI* [17] technology is used. An import utility reads XMI-compliant files and maps the contained XML structures into objects of the repository's physical model. For exporting purposes, another utility traverses the objects of the repository and produces the corresponding XMI file. The adoption of standard technologies such as the UML and XMI enables easy information interchange between the SPOOL environment and other tools.

In the remainder of this paper, we first provide an overview of the SPOOL environment. Next, we describe the architecture of the SPOOL repository and detail its schema, discussing its top-level, core, relationship, behavior, and extension classes and relating it to the UML metamodel. Then, we describe two of the key mechanisms of the repository, that is, the traversal of complex objects and dependency management, and present one of the front-end user tools of the repository, the *SPOOL Design Browser*. Finally, we put our work into perspective, reporting on performance and interchange experiments, and discussing related and future work.

2 The SPOOL Environment

The SPOOL environment (Figure 1) uses a three-tier architecture to achieve a clear separation of concerns between the end user tools, the schema and the objects of the reverse engineered models, and the persistent datastore. The lowest tier consists of an *object-oriented database management system*, which provides the physical, persistent datastore for the reverse engineered source code models and the design information. The middle tier consists of the *repository schema*, which is an object-oriented schema of the reverse engineered models, comprising structure (classes, attributes, and relationships), behavior (access and manipulation functionality of objects), and mechanisms (higher-level functionality, such as complex object traversal, change notification, and dependency accumulation). We call these two lower tiers the SPOOL *design repository*. The upper tier consists of end user tools implementing domain-specific functionality based on the repository schema, i.e., *source code capturing*, and *visualization and analysis*.

In this section, we will describe the environment's techniques and tools for source code capturing and data interchange, as well as for visualization and analysis.

2.1 Source Code Capturing and Data Interchange

Source code capturing is the first step within the reverse engineering process. Its goal is to extract an initial model from the source code. At this time, SPOOL supports C++ and uses *Datrix* [6] to *parse* C++ source code files. With the deployment of the Datrix Java parser, SPOOL will soon be able to extend its support for reverse engineering Java source code. Datrix provides complete information on the source code in form of an ASCII-based representation, the *Datrix/TA intermediate format*. The purpose of this intermediate representation is to make the Datrix output independent of the

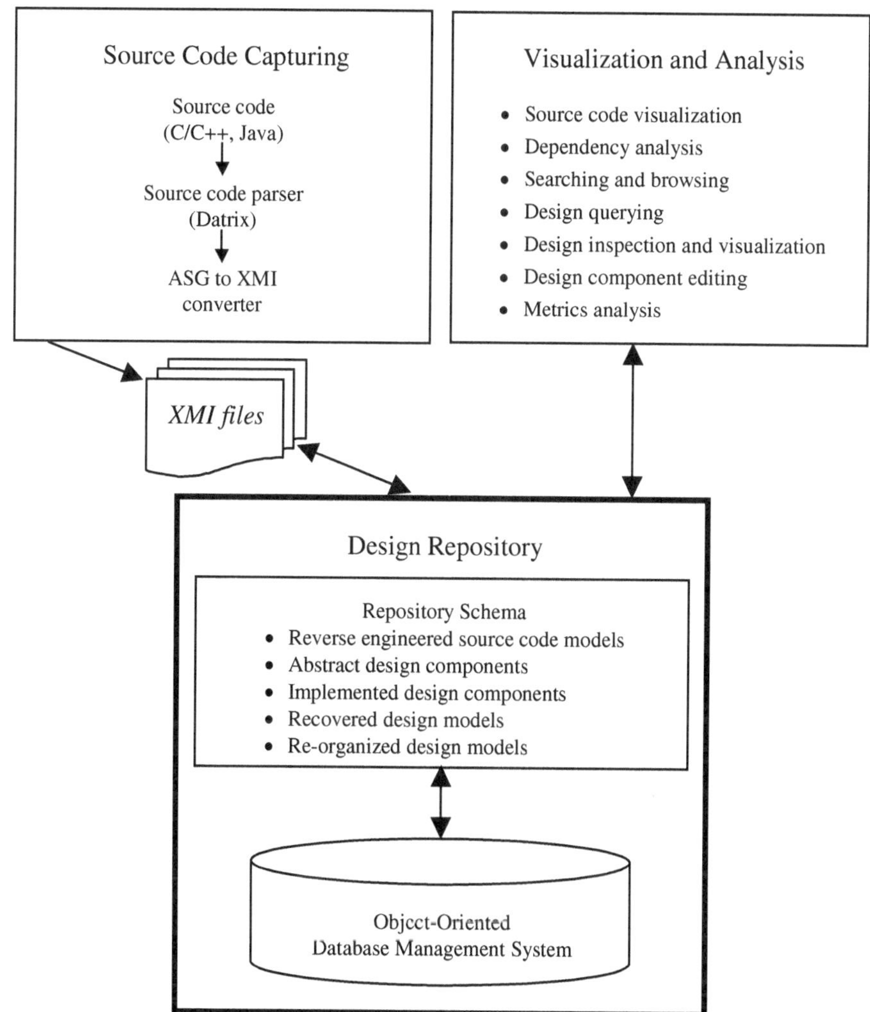

Fig. 1. Overview of the SPOOL environment

programming language being parsed. Moreover, it provides a data export mechanism to analysis and visualization tools, ranging from Bell Canada's suite of software comprehension tools to the SPOOL environment and to third-party source code comprehension tools. A *conversion* utility, built with *ANTLR* [1] and the *Datrix/TA* grammar description, assembles the nodes and arcs of the Datrix/TA source code representation files, applies some transformations to the resulting graphs (such as normalization of directory and file structures as well as addition of primitive data types) to map the Datrix/TA structures to those of the repository model, and generates XMI files. Thereafter, the resulting XMI files are read by an *import* utility, which leverages some components built for the *Argo* project [21] and which uses IBM's *xml4j* XML parser [12]. The importer constructs the objects of an initial physical model in the SPOOL repository. Another utility is used to export the content of the repository, translating the structures of the internal schema into a resulting XMI-compliant file. At the current state of development, we capture and manage in the repository the source code information as listed in Table 1.

Table 1. Source code information managed in the SPOOL repository

1.	*Files* (name, directory).
2.	*Classifier – classes, structures, unions, anonymous unions, primitive types* (char, int, float, etc.), *enumerations* [name, file, visibility]. Class declarations are resolved to point to their definitions.
3.	*Generalization* relationships [superclass, subclass, visibility].
4.	*Attributes* [name, type, owner, visibility]. Global and static variables are stored in utility classes (as suggested by the UML), one associated to each file. Variable declarations are resolved to point to their definitions.
5.	*Operations* and *methods* [name, visibility, polymorphic, kind]. Methods are the implementations of operations. Free functions and operators are stored in *utility* classes (as suggested by the UML), one associated to each file. *Kind* stands for *constructor, destructor, standard,* or *operator.*
5.1	*Parameters* [name, type]. The type is a *classifier.*
5.2	*Return types* [name, type]. The type is a *classifier.*
5.3	*Call actions* [operation, sender, receiver]. The receiver points to the class to which a request (operation) is sent. The sender is the classifier that owns the method of the call action.
5.4	*Create actions.* These represent object instantiations.
5.5	*Variable use* within a method. This set contains all member attributes, parameters, and local attributes used by the method.
6.	*Friendship relationships* between classes and operations.
7.	*Class and function template instantiations.* These are stored as normal *classes* and as *operations* and *methods*, respectively.

2.2 Visualization and Analysis

The purpose of design representation is to provide for the interactive visualization and analysis of source code models, abstract design components, and implemented components. It is our contention that only the interplay among human cognition, automatic information matching and filtering, visual representations, and flexible visual transformations can lead to the all-important why behind the key design decisions in large-scale software systems. To date, we have implemented and integrated tools (for details, see [14]) for

- *Source code visualization,*
- Interactive and incremental *dependency analysis* (see Section 4.2),
- Design investigation by *searching and browsing*, based on both structure and full-text retrieval, using the *SPOOL Design Browser* (see Section 4.3),
- *Design querying* to classes that collaborate to solve a given problem,

- *Design inspection and visualization* within the context of the reverse engineered source code models,
- *Design component editing*, allowing for the interactive description of design components, and
- *Metrics analysis* to conduct quantitative analyses on desirable and undesirable source and design properties.

3 Repository Architecture and Schema

The major architectural design goal for the SPOOL repository was to make the schema resilient to change, adaptation, and extension, in order to address and accommodate easily new research projects. To achieve a high degree of flexibility, we decided to shield the implementation of the design repository completely from the client code that implements the tools for analysis and visualization. The retrieval and manipulation of objects in the design repository is accomplished via a hierarchy of public Java interfaces, and instantiations and initializations are implemented via an *Abstract Factory* [9].

The schema of the SPOOL repository is an object-oriented class hierarchy whose core structure is adopted from the UML metamodel. Being a metamodel for software analysis and design, the UML provides a well-thought foundation for SPOOL as a design comprehension environment. However, SPOOL reverse engineering starts with source code from which design information should be derived. This necessitates extensions to the UML metamodel in order to cover the programming language level as far as it is relevant for design recovery and analysis. In this section, we present the structure of the extended UML metamodel that serves as the schema of the SPOOL repository. This includes the top-level classes, the core classes, the relationship classes, the behavior classes, and the extension classes.

3.1 Top-Level Classes

The top-level classes of the SPOOL environment prescribe a key architectural design decision, which is based on the *Model/View/Controller (MVC)* paradigm of software engineering [3, 9]. MVC suggests a separation of the classes that implement the end user tools (the views) from the classes that define the underlying data (the models). This allows for both views and models to be reused independently. Furthermore, MVC provides for a change notification mechanism based on the *Observer* design pattern [9]. The Observer pattern allows tools, be they interactive analysis or background data processing tools, to react spontaneously to the changes of objects that are shared among several tools. In SPOOL, the classes *Element*, *ModelElement*, and *ViewElement* implement the functionality that breaks the SPOOL environment apart into a class hierarchy for end user tools (subclasses of *ViewElement*) and a class hierarchy for the repository (subclasses of *ModelElement*). The root class *Element* prescribes the MVC based communication mechanism between ViewElements and ModelElements.

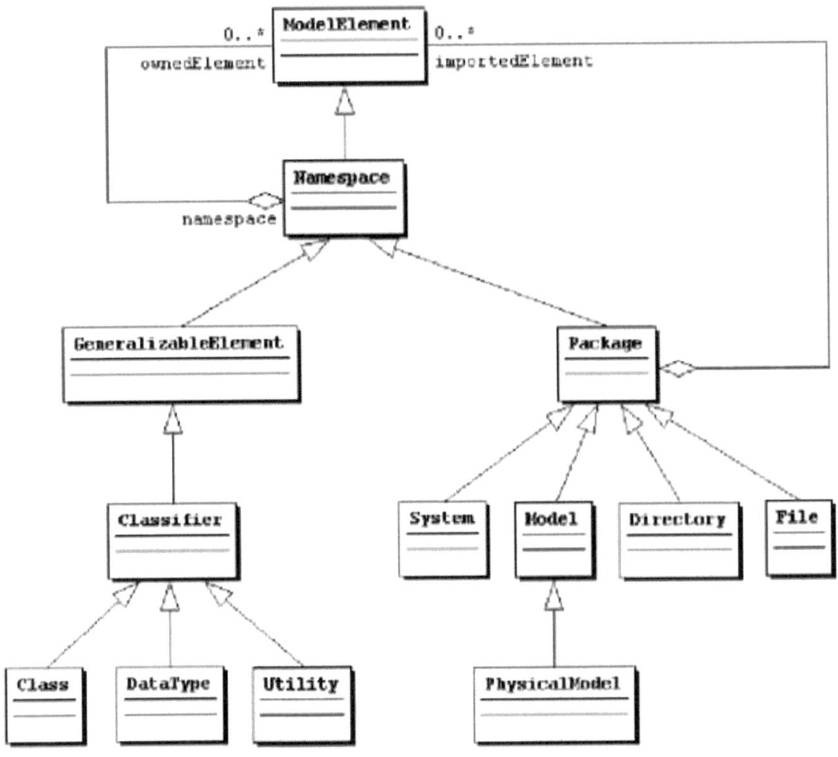

Fig. 2. SPOOL repository schema: Core classes

3.2 Core Classes

The core classes of the SPOOL repository schema adhere to a large extent to the classes defined in the core and model management packages of the UML metamodel. These classes define the basic structure and the containment hierarchy of the ModelElements managed in the repository (see Figure 2). At the center of the core classes is the *Namespace* class, which owns a collection of ModelElements. A *GeneralizableElement* defines the nodes involved in a generalization relationship, such as inheritance. A *Classifier* provides *Features*, which may be structural (*Attributes*) or behavioral (*Operations* and *Methods*) in nature. A *Package* is a means of clustering ModelElements.

3.3 Relationship Classes

"A relationship is a connection among model elements." [25] The UML introduces the notion of Relationship as a superclass of Generalization, Dependency, Flow, and Association for reasons of convenience, so that tools can refer to any connections among ModelElements based on the same supertype (for details, see [24]).

3.4 Behavior Classes

The behavior classes of the SPOOL repository implement the dynamics of the reverse engineered system. It is important to understand that the UML metamodel takes a forward engineering perspective and focuses on software analysis and design, rather than on the reverse engineering of source code. Therefore, the UML metamodel does not aim to encompass and unify programming language constructs.

The purpose of analysis and design is to specify what to do and how to do it, but it is the later stage of implementation in which the missing parts of a specification are filled to transform it into an executable system. However, the UML is comprehensive in that it provides a semantic foundation for the modeling of any specifics of a model. For example, the UML suggests State Machine diagrams (similar to Harel's Statechart formalism [11]) to specify the behavior of complex methods, operations, or classes. To cite another example, collaboration diagrams can be used to specify how different classes or certain parts of classes (that is, roles) have to interact with each other in order to solve a problem that transcends the boundaries of single classes.

In SPOOL, we look at a system from the opposite viewpoint, that is from the complete source code, and the goal is to derive these behavior specification models to get an improved understanding of the complex relationships among a system's constituents. For this purpose, we included in the SPOOL repository the key constructs of the behavior package of the UML metamodel, including the UML's Action and Collaboration classes. However, we modified certain parts to reduce space consumption and improve performance. For more information about the behavior classes of the SPOOL repository, see [24].

3.5 Extension Classes

The UML metamodel suggests two approaches to metamodel extension; one is based on the concept of TaggedValues and the other on the concept of Stereotypes. In SPOOL, we have only implemented the former approach since Stereotypes as defined in the UML metamodel would not scale to meet the performance requirements of the SPOOL repository.

4 Repository Mechanisms and Front-End

To be usable and reusable as the backend for a diverse set of interactive reverse engineering tools, the SPOOL repository implements a number of advanced mechanisms. The *traversal mechanism* defines how to retrieve objects of certain types from a complex object containment hierarchy. The *dependency mechanism* allows for compression of the vast amount of dependencies among ModelElements for fast retrieval and visualization. Finally, as a front-end to the repository and an interface to other SPOOL visualization tools, the *SPOOL Design Browser* is provided.

4.1 Traversal Mechanism

In SPOOL, the Namespace serves as a container for a group of ModelElements (Figure 2). Consequently, it defines methods that traverse complex object structures and retrieve ModelElements of a given type. For example, to identify all classes of a system, all files in all subdirectories of the system at hand must be checked for instances of the metatype Class. Unlike the objects in text-based repositories [8, 27], the objects in SPOOL's object-oriented database are typed and can be queried according to their types. SPOOL allows for the identification of the type of an object merely by using the Java *instanceof* operator or the reflective *isInstance* operation of the Java class *Class*. Hence, metaclass types can be provided as parameters to the retrieval methods of Namespace, which then recursively traverse the containment hierarchy of the namespace at hand and examine each ModelElement whether it is an instance of that type. If this is the case, the ModelElement is added to a return set, which is passed through the recursive traversal.

4.2 Accumulated Dependency Mechanism

An important requirement of the SPOOL repository is to provide information on dependencies between any pair of ModelElements within interactive response time. A straightforward approach to identify dependencies among ModelElements would be the traversal of the whole object structure at run-time. However, applied to reverse engineered software with directories that contain hundreds of files, this approach would require batch processing. A radically different approach would be to store each and every dependency among ModelElements as separate dependency objects, which would result in an unmanageable amount of dependency data. Hence, the solution that we adopted in SPOOL constitutes a trade-off between run-time efficiency and space consumption.

In SPOOL, we capture and accumulate dependencies at the level of Classifiers (for instance, classes, unions, or utilities). Accumulation refers to the fact that we store for each dependency its types together with the total number of primitive Connections on which each type is based. Given a pair of dependent Classifiers, we generate a so-called *AccumulatedDependency* object, which captures this information for the dependencies in the two directions. To be able to identify dependencies between higher-level namespaces, such as directories, files, or packages, without much lag time, we store the union of all AccumulatedDependencies of all contained Classifiers of a given Namespace redundantly with the Namespace. Hence, if we want to identify, for example, dependencies between the directories *Directory1* and *Directory2*, we only need to iterate over the set of AccumulatedDependencies of *Directory1* and look up for each element of the set whether the ModelElement at the other end of the element at hand (that is, the one which is not contained in the Namespace under consideration) has as one of its parent namespaces *Directory2*.

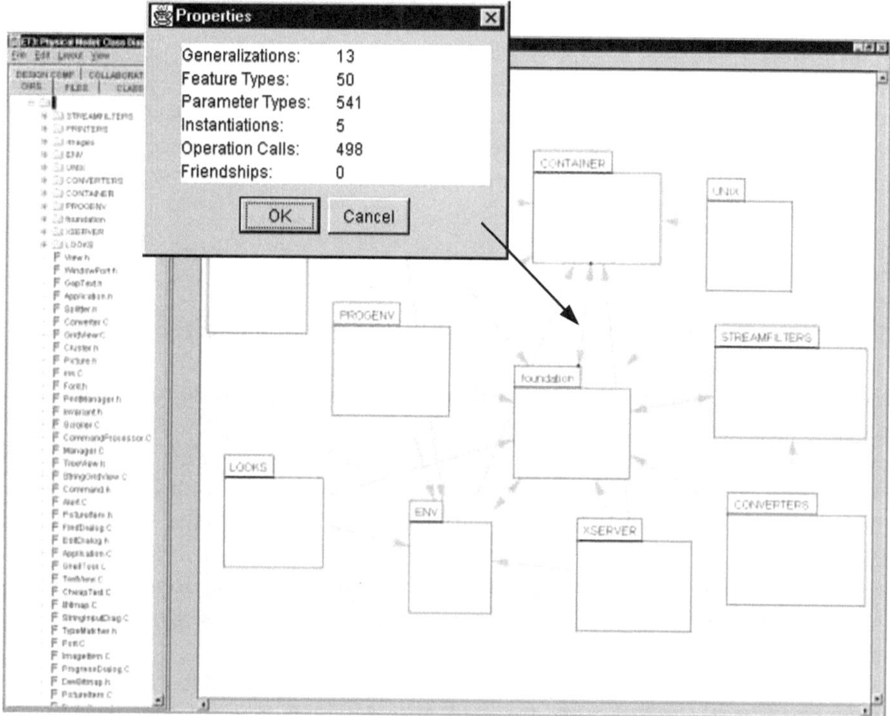

Fig. 3. SPOOL dependency diagram with dialog box for inspection of properties

Figure 3 shows a dependency diagram for the top-level directories of the system ET++ [26]. A property dialog box can be opened to inspect the nature of a specific dependency. In Figure 3, for instance, the dependency between the directories *CONTAINER* and *foundation* includes 13 generalization connections, 50 feature type connections (types of attributes and return types of operations and methods), 541 parameter type connections, 5 class instantiation connections (*CreateAction*), 498 operation call connections (*CallAction*), and 0 friendship connection. On demand, the dialog can also be invoked for each direction of a dependency. For more information about the accumulated dependency mechanism, see [24].

4.3 SPOOL Design Browser

The SPOOL environment provides a number of tools for design investigation and visualization. Among these is the *Design Browser*, which acts as a standard query engine to support design navigation. The SPOOL Design Browser offers predefined queries to the user, while making it possible to modify the set of predefined queries based on the traversal mechanism provided by the repository. It also manages the execution of queries, and displays the query results in a user-friendly way. For details on the Design Browser, refer to [22].

5 Experience and Perspectives

In this section, we first report on our experience with the SPOOL repository with respect to scalability and performance. Then, we discuss data interchange between the repository and external tools. Furthermore, we examine our choice of the UML metamodel and wrap up with a conclusion and a discussion of future work.

5.1 Scalability and Performance

Scalability and performance are critical for the success of source code investigation. Each step in the investigation process should be fast enough in order to avoid confusion and disorientation with the user, and the tools should be robust enough to accommodate industrial sized systems. In the following, we present two anecdotal experiments in which the performance of SPOOL queries was measured. The experiments were conducted on a 350MHz Pentium II machine with 256Mb of RAM running Windows NT 4.0. Two industrial C++ software systems were analyzed: *ET++ 3.0* [26], a well-known application framework, and *System A*, a large-scale system from the telecommunications domain provided by Bell Canada (for confidentiality reasons, we cannot disclose the real name of the system). The size metrics of these systems are shown in Table 2 (top section).

The first experiment consisted of measuring the times needed to execute a simple query which is predefined in the Design Browser but which is not directly supported by the repository schema, namely, the retrieval of all the *ModelElements* (directories, files, classes, C++ structures, C++ unions, C++ enumerations, operations, methods, and attributes) of a system. Table 2 (middle section) depicts the data for this query. The table shows that the first time the query is run, it takes longer (*Duration 1*) because, first, Poet needs to recreate the persistent objects that are stored on disk and, second, when loading a system, SPOOL caches some of the objects in internal hash tables. As soon as an element is "touched" by a query, it becomes available in memory, and the next time a query is accessing it, the execution is much faster (*Duration 2+*).

Table 2. Size metrics and durations of queries for two industrial systems

Size Metrics :	*ET++*	*System A*
Lines of code	70,796	472,824
Lines of pure comments	3,494	60,256
Blank lines	12,892	80,463
# of files (.C/.h)	485	1,900
# of classes (.C/.h)	722	3,103
# of generalizations	466	1,422
# of methods	6,255	17,634
# of attributes	4,460	1,928
Size of the system in the repository	19.3 MB	63.1 MB
# of *ModelElements*	20,868	47,834
Simple Query :	*ET++*	*System A*
Duration 1 (seconds)	22	47
Duration 2+ (seconds)	2	6
Template Method Query :	*ET++*	*System A*
# of occurrences found	371	364
Duration (seconds)	15	360

The above experiment shows that the retrieval of elements that are already referenced in the database is pretty fast. The execution of more complicated queries may take considerably longer. As a second experiment, we measured the time needed to retrieve all occurrences of the *Template Method* pattern [9] in the two systems. This query basically consists of the following five steps:

1. retrieve all classes in the system,
2. for each class, retrieve all methods,
3. for each method, retrieve all call actions,
4. for each call action, get the receivers,
5. for each receiver, look if the call action is defined in the same class and implemented in a subclass.

Table 2 (bottom section) shows the times needed to execute this query for the first time, assuming that all the *ModelElements* are already cached (a query that retrieves all *ModelElements* in the system was executed previously). These numbers are quite good considering that a considerable number of relations must be crossed in order to retrieve the desired information. The time needed to run a particular query may be higher, but these experiments suggest that only the complexities of the query and of the system are susceptible to increase execution time, whereas the access time to the *ModelElements* of the repository is relatively constant (mainly due to the use of hash tables).

5.2 Data Interchange

Before selecting XMI as the model interchange format for the SPOOL design repository, five interchange formats were considered and evaluated [23]. Advantages and disadvantages of the five formats were identified, where XMI came out as the strongest approach, mainly because it reuses existing solutions like the UML, XML, and MOF, has important industry support, and is generally complete. Other useful features that put XMI on top are partial or differential model exchanges, and general extension mechanisms. Working with XMI documents meant that we could benefit from readily available XML tools, components and expertise to develop our model importer, which was written in Java.

At the time of writing, we have completed our first experiments on exchanging XMI files with other software engineering tools such as *Rational Rose* [20]. We achieved good preliminary results, yet further experiments will be required, once more precise mappings between the supported programming languages and the respective repository schema constructs are available.

5.3 Discussion

The UML is hardly accepted in the reverse engineering community. Demeyer et al. have articulated some reasons for the "why not UML" [7]. We wholeheartedly agree that there is a lack of complete and precise mappings of programming languages to the UML. However, we consider this as a challenge for researchers, rather than a reason for abandoning the UML. With its Stereotype and TaggedValue extension mechanisms, the UML does provide constructs to capture the many details of source code written in different programming languages. The issue at hand is to define unambiguously how to map the various UML constructs to source code constructs and to provide tool support for the traceability in both directions. A second argument of Demeyer et al. against the UML is that it "does not include dependencies, such as invocations and accesses" [7]. This reflects a misconception in the software engineering community about the UML. All too often, the UML is looked at as a notation for structure diagrams only, and all other diagrams are rather neglected. Yet, the behavior package of the UML metamodel provides for a precise specification of the method internals. However, a critique against the UML may be that the behavioral package is too heavyweight to be directly applicable to software engineering. It is impossible to generate and store for each method a StateMachine object together with all its internal objects. In SPOOL, we implemented a shortcut solution for the representation of the bulk of the methods. We associated Actions directly to methods instead of generating StateMachines, which consist of Actions that are invoked by Messages. Refer to the UML for further details on the structure of StateMachines [2]. We do, however, allow StateMachines to be reverse engineered and stored for methods or classes of interest.

The UML has several advantages. First, the UML metamodel is well documented and based on well-established terminology. This is of great help to convey the semantics of the different modeling constructs to tool developers. Second, the metamodel is designed for the domain of software design and analysis, which is at the core of forward engineering and which constitutes the target of the reverse engineering process. The UML introduces constructs at a high level of granularity, enabling the compression of the overwhelming amount of information that makes source code difficult to understand. Third, the UML metamodel is object-oriented, meaning that the structure, the basic access and manipulation behavior, and complex mechanisms can be separated from end user tools and encapsulated in the repository schema. Fourth, the UML defines a notation for the metamodel constructs, thus providing reverse engineering tool builders guidelines for the visual representation of the model elements. Finally, since the UML has gained much popularity in industry and academia alike, tools and utilities supporting the UML and related formalisms such as XMI are becoming readily available. This proves highly beneficial in projects such as ours.

Other research efforts in repository technology include the design of the *Software Information Base (SIB)* and prototype implementation, as described by Constantopoulos and Dörr [4], and Constantopoulos et al. [5]. The SIB, as a repository system, is used to store descriptions of software artefacts and relations between them. Requirement, design, and implementation descriptions provide application, system, and implementation views. These descriptions are linked by relationship objects that express attribution, aggregation, classification, generalization, correspondence, etc. between two or more software components. Links may express semantic or structural relationships, grouping of software artefact descriptions into larger functional units, and even user-defined or informal relationships for hypertext navigation or annotations. The representation language used in the SIB is *Telos* [15], a conceptual modelling language in the family of entity-relationship models with features for increasing its expressiveness. Finally, the SIB comes with a set of visual tools for querying and browsing, which allow the user to search for software component descriptions that match specific criteria expressed as queries, or to navigate through the repository's content in an exploratory way within a given subset of the SIB.

Even if the Software Information Base and the SPOOL design repository share apparent similarities in their architecture and functionality, the SIB is mainly intended for the storage of user-written descriptions of software artefacts residing outside the system (links can be made from the descriptions to the physical components on external storage). The main goal of the SIB is to act as a large encyclopedia of software components, may they be requirements specifications, design descriptions, or class implementations in a specific programming language. These components are classified using a well-defined scheme, in a way that system developers may rapidly browse the contents of the SIB to find the building blocks they need for composing a new system with the help of registered parts. In contrast, the SPOOL repository stores information extracted from source code to help software engineers conduct metrics analysis and investigate the properties of object-oriented systems, such as the presence of design pattern instances and the existence of dependencies between classes, files, or directories. While a prototype of the SIB storing extracted facts from source code parsing has been developed, the retained facts remain of basic nature and serve as a start-up structure for the manual classification of classes, operations, and attributes. Furthermore, the SIB offers a mechanism to import artefact descriptions into its database; however, no export mechanism or data interchange facility is provided, assuming that information sharing with other tools is not seen as a primary objective. The SPOOL repository, in order to exchange information with other academic and commercial tools, comprises a stable and well-known internal datamodel and data interchange format. While both SPOOL and SIB were carefully designed with strict architectural and performance considerations, the main differences between them on the functionality side are best explained by their focus on reverse engineering and artefact reuse, respectively.

5.4 Conclusion and Future Work

In this paper, we presented the SPOOL design repository, the core part of the SPOOL environment. Based on the UML metamodel, its schema permits to store detailed information about the source code of systems, enabling users to conduct essential tasks of reverse engineering, system comprehension, system analysis, and reengineering. Its internal advanced mechanisms provide the core functionalities needed by the interactive visualization tools of the environment. XMI is used as model interchange format, easing information sharing between the SPOOL design repository and other software engineering environments. Our experience suggests that the choice of the UML and its metamodel was indeed a key factor in meeting the repository requirements stated at the outset of the project.

In our future work on the SPOOL design repository, we will aim to provide complete and precise mappings between the constructs of the UML-based SPOOL repository schema and the four programming languages C, C++, Java, and a proprietary language deployed by Bell Canada. We will also increase the information content of the SPOOL repository in respect to dynamic behavior. As discussed previously, a balance between space consumption and fast response time needs to be sought. One solution that we will investigate is parsing the source code of methods on the fly when querying, for example, control flow information. A third area of work will be to provide Web-based access to the repository, which will allow our project partners to remotely check in source code systems and immediately use SPOOL tools to query and visualize the repository content. Finally, we plan to investigate the UML-based repository approach beyond SPOOL in two other domains of our interest, that is, in corporate memory research [10], and in schema evolution [19].

Acknowledgments

We would like to thank the following organizations for providing us with licenses of their tools, thus assisting us in the development part of our research: *Bell Canada* for the source code parser *Datrix*, *Lucent Technologies* for their C++ source code analyzer *GEN++* and the layout generators *Dot* and *Neato*, and *TakeFive Software* for their software development environment *SNiFF+*.

References

1. ANTLR, ANother Tool for Language Recognition, 2000. <http://www.antlr.org>.
2. Booch, G., Jacobson, I., and Rumbaugh, J. *The Unified Modeling Language User Guide.* Addison-Wesley, 1999.
3. Buschmann, F., Meunier, R., Rohnert, H., Somerlad, P., and Stal, M. *Pattern-Oriented Software Architecture – A System of Patterns.* John Wiley and Sons, 1996.
4. Constantopoulos, P., and Dörr, M. "Component Classification in the Software Information Base", Object-Oriented Software Composition, Oscar Nierstrasz and Dennis Tsichritzis (Eds.), pp. 177-200. Prentice Hall, 1995.
5. Constantopoulos, P., Jarke, M., Mylopoulos, J., and Vassiliou, Y. "The Software Information Base: A Server for Reuse." VLDB Journal 4(1):1-43, 1995.
6. Datrix homepage, 2000, Bell Canada. <http://www.iro.umontreal.ca/labs/gelo/datrix/>.
7. Demeyer, S., Ducasse, S., and Tichelaar, S. "Why unified is not universal: UML shortcomings for coping with round-trip engineering." In *Bernhard Rumpe, editor, Proceedings UML'99 (The Second International Conference on the Unified Modeling Language).* Springer-Verlag, 1999. LNCS 1723.
8. Finnigan, P. J., Holt, R. C., Kalas, I., Kerr, S., Kontogiannis, K., Müller, H. A., Mylopoulos, J., Perelgut, S. G., Stanley, M., and Wong, K. "The software bookshelf." *IBM Systems Journal*, 36(4):564-593, 1997.
9. Gamma, E., Helm, R., Johnson, R., and Vlissides, J. *Design Patterns: Elements of Reusable Object-Oriented Software.* Addison-Wesley, 1995.
10. Gerbé, O., Keller, R. K., and Mineau, G. "Conceptual graphs for representing business processes in corporate memories." In *Proceedings of the Sixth International Conference on Conceptual Structures*, pages 401-415, Montpellier, France, August 1998.
11. Harel, D. "On visual formalisms." *Communications of the ACM*, 31(5):514-530, May 1988.
12. IBM-alphaWorks. XML Parser for Java, 2000. <http://www.alphaworks.ibm.com/tech/xml>.
13. Keller, R. K., and Schauer, R. "Design components: towards software composition at the design level." In *Proceedings of the 20th International Conference on Software Engineering*, Kyoto, Japan, pages 302-310, April 1998.
14. Keller, R. K., Schauer, R., Robitaille, S., and Pagé, P. "Pattern-based reverse engineering of design components." In *Proceedings of the Twenty-First International Conference on Software Engineering*, pages 226-235, Los Angeles, CA, May 1999. IEEE.
15. Mylopoulos, J., Borgida, A., Jarke, M., and Koubarakis, M. "Telos: Representing Knowledge About Information Systems." *ACM Transactions on Information Systems*, Vol. 8, No. 4, October 1990, pages 325-362.
16. Object Data Management Group (ODMG), 2000. On-line at <http://www.odmg.com>.
17. OMG. "XML Metadata Interchange (XMI)", Document ad/98-10-05, October 1998. On-line at <ftp://ftp.omg.org/pub/docs/ad/98-10-05.pdf>.
18. Poet Java ODMG binding, on-line documentation. Poet Software Corporation, San Mateo, CA, 2000. On-line at <http://www.poet.com>.

19. Pons, A., and Keller, R. K. "Schema evolution in object databases by catalogs." In *Proceedings of the International Database Engineering and Applications Symposium (IDEAS'97)*, pages 368-376, Montréal, Québec, Canada, August 1997. IEEE.
20. Rational Software Corporation, 2000. On-line at <http://www.rational.com>.
21. Robbins, J. E., and Redmiles, D. F. "Software architecture critics in the Argo design environment." *Knowledge-Based Systems*, (1):47-60, September 1998.
22. Robitaille, S., Schauer, R., and Keller, R. K. "Bridging program comprehension tools by design navigation." In *Proceedings of the International Conference on Software Maintenance (ICSM'2000)*, pages 22-32, San Jose, CA, October 2000. IEEE.
23. Saint-Denis, G., Schauer, R., and Keller, R. K. "Selecting a Model Interchange Format: The SPOOL Case Study." In *Proceedings of the Thirty-Third Annual Hawaii International Conference on System Sciences* (CD ROM, 10 pages). Maui, HI, January 2000.
24. Schauer, R., Keller, R. K., Laguë, B., Knapen, G., Robitaille, S., and Saint-Denis, G. "The SPOOL Design Repository: Architecture, Schema, and Mechanisms." In *Hakan Erdogmus and Oryal Tanir, editors, Advances in Software Engineering. Topics in Evolution, Comprehension, and Evaluation*. Springer-Verlag, 2001. To appear.
25. UML. Documentation set version 1.1, 2000. On-line at <http://www.rational.com>.
26. Weinand, A., Gamma, A., and Marty, R. "Design and implementation of ET++, a seamless object-oriented application framework." *Structured Programming*, 10(2):63-87, February 1989.
27. Wong, K., and Müller, H. (1998). Rigi user's manual, version 5.4.4. University of Victoria, Victoria, Canada. On-line at <ftp://ftp.rigi.csc.uvic.ca/pub>.

Why Enterprise Modelling? An Explorative Study into Current Practice

Anne Persson[1] and Janis Stirna[2]

[1] Department of Computer Science, University of Skövde, P.O. Box 408, SE-541 28 Skövde, Sweden
anne.persson@ida.his.se

[2] Department of Computer and Systems Sciences, Royal Institute of Technology and Stockholm University, Electrum 230, SE-16440, Kista, Sweden
js@dsv.su.se

Abstract. This paper presents an explorative study, which investigates the intentions behind current use of Enterprise Modelling (EM) in organisations. The intentions fall into two main categories: developing the business and ensuring the quality of the business. The results indicate that current methods are useful for these purposes.

1. Introduction and Background

Enterprise Modelling (EM) is an activity where an *integrated* and *negotiated* model describing different aspects of an enterprise is created. An Enterprise Model consists of a number of related "sub-models", each describing the enterprise from a particular perspective, e.g. processes, business rules, goals, actors and concepts/information/ data. Much research has been put into the development of Enterprise or Business Modelling methods, while the practice of *using* them has been more or less neglected by the research community.

EM method developers have suggested that their methods are applicable in a variety of contexts; e.g. BPR, strategic planning, enterprise integration and IS development. We investigate which intentions, *in fact*, are behind current use of EM in organisations. The paper is based on two qualitative studies using interviews, case studies, company observations, and literature studies. They were carried out in two separate research projects addressing EM *tool support* and *ways of working* (in particular the participative approach). The implications of the findings with regard to these two issues are further discussed by Persson and Stirna (2001).

2. Why Enterprise Modelling?

The goal hierarchy in Fig.1, resulting from analysing the interviews, shows the common objectives that organisations have for using EM. It contains two main branches. One deals with *developing the business*, e.g. developing business vision, strategies, redesigning business operations, developing the supporting information

K.R. Dittrich, A. Geppert, M.C. Norrie (Eds.): CAiSE 2001, LNCS 2068, pp. 465–468, 2001.
© Springer-Verlag Berlin Heidelberg 2001

systems, etc. The other deals with *ensuring the quality of the business,* primarily focusing on two issues: 1) sharing the knowledge about the business, its vision, the way it operates and 2) ensuring the acceptance of business decisions through committing the stakeholders to the decisions made. In the following two sections, the two branches will be discussed in more detail.

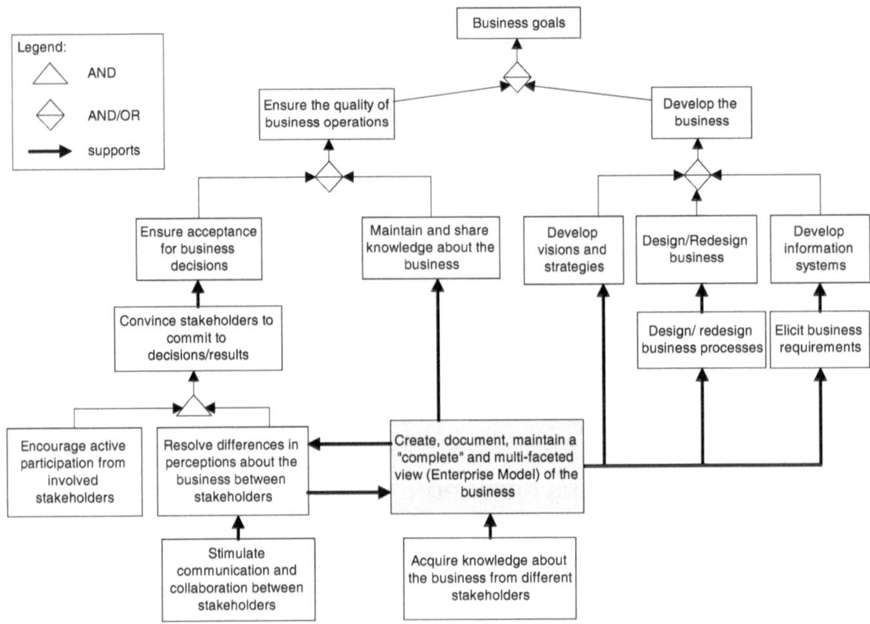

Fig. 1. Goal hierarchy of the most common intensions for using Enterprise Modelling

3. Business Development

Business development is one of the most common objectives for EM. It frequently involves change management – determining how to achieve visions and objectives from the current state in organisations. EM is used in this process with great success. Some specific issues addressed by EM are found in the following citation[1]:

"... questions like strategies, what type of market to participate in, how is the market structured, which are our clients, who are the other interested parties in the organisation, how should we structure our work sequencing, how do we structure our products comparing with the clients, do we sell everything to everyone. EM also aims to describe the reason for the organisation, the goals – to relate them to the strategies, to the business idea. EM continues all the way from the strategies through the processes, through the concepts – in order to arrive at a complete picture, or a picture that fits together."

Business process orientation is a specific case of business development – the organisation wants to restructure/redesign its business operations.

[1] Note that the quotations from this point onwards are excerpts from the interviews. Full transcripts of the interviews are available from the authors on request.

Also, EM is often used in the early stages of IS development. A common view is that EM can effectively be used for gathering business needs and high-level requirements. One experienced business consultant stated:

"In my experience, the most common modelling I have been doing, has been connected in some way to IT development. There has always been a superior decision of doing something in the IT sphere, which has led to the need to understand the business better and describe it much better, otherwise we can't build the right system. That is very often the situation. On the other hand I have not been very much involved in the rest of the IT development. I have just delivered the results – this is the business, this is how it's working, this is the information that needs to be handled. ... That's one situation. ...Another one is business process definition, where the idea as such has been to describe the business in terms of processes. Then other projects have sort of emerged. E.g. people see that some part of the business should be improved, or this part of the business is not supported by the IT at all."

4. Quality Assurance

Another common motivation to adopt EM is to ensure the quality of operations. Two important success factors for ensuring quality, mentioned by interviewees, were that stakeholders understand the business and they accept/are committed to business decisions. Recently, organisations have taken an increased interest in Knowledge Management, which is concerned with creating, maintaining and disseminating organisational knowledge between stakeholders. EM has a role to play here. Its aim is to create a multifaceted view of the business, which functions as a common platform for communicating between different stakeholders in the enterprise.

Sharing knowledge about the business becomes instrumental when two organisations merge or when different organisations collaborate in carrying out a business process. One part of the knowledge about a business is its terminology:

"I'm thinking about [organisation X and organisation Y] where they realised that they could use the same data. To be able to do that, they must use the same terms so that they could buy from and sell to each other ... and then it was quite clear that they needed modelling of their business concepts."

One of the Knowledge Management perspectives is keeping employees informed with regard to how the business is carried out. For example:

"...in those days ... when the company was expanding enormously, they increased by about 100% personnel each year, and it grew very rapidly over the globe. ... So how should we introduce [new people] to the [company E] world and teach [them] how to handle all the things in the [company E] community, etc. It's simply not possible, especially since we don't have good documentation of how we really operate, because everything went on so quickly, that [company E] had to change routines almost every year because of the expansion, etc. So their main motive actually for describing their processes was not to get a lot more efficient, because, maybe rightly, they thought that they were rather efficient, but as a tool to communicate to newly hired personnel, and to show people – this is how we think we are operating, do you have any ideas"

If Enterprise Models are to play a role in the maintenance and dissemination of business knowledge there is a need for supporting tools.

Most modern organisations subscribe to the view that the commitment of stakeholders to carry out business decisions is instrumental for achieving high quality business operations. To this effect, differences in opinions must be resolved which, in turn, requires that communication between stakeholders be stimulated. EM, particularly using a participative approach, is effective to obtain commitment from stakeholders.

"... if you want people actively involved and if you want people to go along with what is decided, then they have to be allowed to be involved from the beginning and not get decisions forced on them from management.."

"Active participation leads to commitment. So by achieving that, you make it impossible for people to escape commitment."

5. Conclusion and Future Outlook

We have discussed the role of EM in current practice and its perceived contributions to goals in an organisation. We show that EM in effect is used for a variety of purposes, such as creating visions and strategies, redesigning the business as well as for developing information systems. This seems to fit well with the intentions of EM method developers, which could indicate that available methods are in many ways useful. Apart from development purposes, EM seems also to play a role in ensuring the quality of the business. A recent challenge is the use of Enterprise Models in Knowledge Management. This challenge, among others, requires "mature" EM method and tool users.

In this paper we have focused on the true intentions behind the use of EM in organisations today. A further analysis of the implications of these findings on EM tool support and on ways of working has been carried out. This is discussed in detail in (Persson and Stirna, 2001). Some of the important conclusions from that analysis are:

- Participative EM is a strong way of truly committing stakeholders to business decisions, but should only be applied in consensus-oriented organisations.
- EM novices are poor judges of the applicability of participative EM. Nor can they assess which is the appropriate EM tool. Furthermore, they are not aware of their lack of knowledge in these respects, frequently causing EM projects to fail. These failures are often blamed on the methods and tools applied.
- Complex modelling tools tend to distract people from the issue at hand. In many cases, simple drawing tools can be just as effective if not more.
- EM activities require a modelling expert. Thus there is less need for method guidance facilities in tools. In fact, most modelling experts look for tools that provide as much freedom as possible.
- Tool vendors should play a more active role in getting to know what the practitioners need instead of overselling their products.

Current EM literature neglects the practical use of EM methods and related tools. It is important to bear in mind, however, that methods are only vehicles to take us somewhere. Modelling for the sake of modelling is not really useful. We have empirically found that method developers and researchers often forget this. The impression from our interviews is that practitioners feel the same way. More specifically, they feel that methods give very little guidance with regard to *how* and *why* methods and tools should be used in different situations.

References

Persson A., Stirna, J. (2001) , "An explorative study into the influence of business goals on the practical use of Enterprise Modelling methods and tools", Technical report no HS-IDA-TR-01-001, University of Skövde, Sweden.

A Glimpse into CM^3: Problem Management

Mira Kajko-Mattsson

Department of Computer and Systems Sciences, Royal Institute of Technology and Stockholm University, Electrum 230, SE-16440 Kista, Sweden
mira@dsv.su.se

Abstract. In this paper, we give a very short glimpse into a process model for software problem management. This model is the result of a long-term study of current generic process models and of industrial processes at ABB.

1. Introduction

CM^3: Problem Management is a process model for reporting and resolving software problems within corrective maintenance. It is the result of a long-term study of standard process models and of two industrial processes utilised at ABB Robotics and ABB Automation Products. Its primary role is to provide guidance to industrial organisations in the process of building and improving their problem management [4].

As depicted in Figure 1, *CM^3: Problem Management* has the following constituents: (1) *CM^3: Definition of Maintenance and Corrective Maintenance*, (2) *CM^3: Taxonomy of Activities* listing process activities [5], (3) *CM^3: Maturity Levels* indicating the degree of organisations' capability to manage problems and provide feedback to defect prevention and process improvement, (3) *CM^3: Conceptual Model* identifying problem management concepts [1], (4) *CM^3: Roles* designating roles of individuals involved in problem management [5], (6) *CM^3: Roadmaps* visualising the maintenance process [2-3], and, finally, (7) *CM^3: Maintenance Elements* providing explanations of and motivations for the above-mentioned constituents using the structure presented in Figure 2 [7].

Due to the restricted amount of space in this paper, we limit our presentation of *CM^3: Problem Management* to only the designation of *CM^3: Maturity Levels* (*Initial, Defined,* and *Optimal),* and some of their most important *CM^3: Maintenance Elements.* For more information about our process model, we recommend the interested readers to survey our other research contributions [1-7].

2. CM^3: Problem Management: Level 1 (Initial)

At the *Initial* level, the process is implicitly understood by maintenance engineers. Usually, this is due to the following reasons: (1) the process is not defined, (2) the process is defined, but not documented, (3) the process is defined and documented, but the documentation is either outdated or inconsistent, or (4) the process is defined and properly documented, but not consistently adhered to.

K.R. Dittrich, A. Geppert, M.C. Norrie (Eds.): CAiSE 2001, LNCS 2068, pp. 469–472, 2001.
© Springer-Verlag Berlin Heidelberg 2001

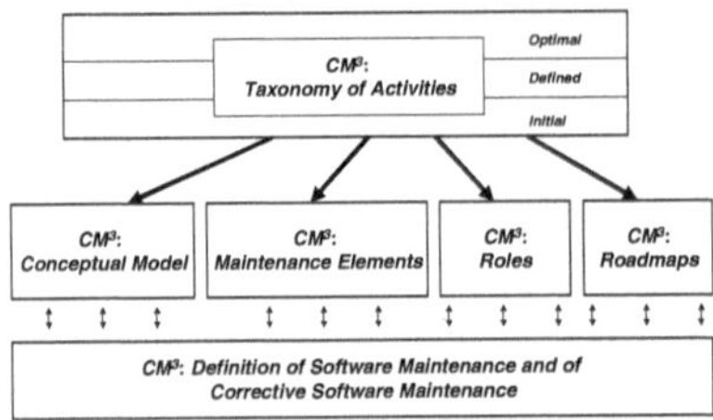

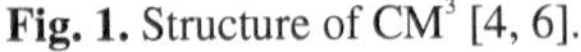

Fig. 1. Structure of CM^3 [4, 6].

Maintenance Element:	A statement of a maintenance element.
Goal:	The goal to achieve after having implemented the element.
Motivation:	Motivations for why the maintenance element should be implemented. Different problems and arguments are described, if any.
Measures:	A list of suggestions to be conducted in order to implement the maintenance element.

Fig. 2. CM^3: Maintenance elements [7].

At this level, the process offers little visibility, or gives a distorted view of its status. At the very most, the organisation has insight into the amount of software problems that have been reported and resolved. But, it cannot follow and follow up the problem resolution process, and make any kinds of reliable process analyses. The feedback provided by the process is not always meaningful. Success at this level may depend on the combination of the following factors: (a) competence of the maintenance engineers, (b) their dedicated overtime, (c) low staff turn-over rate, (d) the stability of the applications maintained, and (e) the repetitive nature of conducting similar tasks.

At this level, *CM^3: Problem Management* strongly recommends to consider the following maintenance process elements as a starting point on the organisations' voyage towards process excellence.

- *Software problems are communicated in writing*: to enable communication about problems in a formal, disciplined and tractable way, and to achieve control over all the maintenance requirements (reported problems in our case).
- *One problem is reported in one and only one problem report*: to facilitate communication on software problems, to enable follow and follow up of the problem resolution, and to enable problem validation and verification.
- *A problem submitter is identified*: to enable the delivery of the problem solution to the right customer.
- *Maintenance requirements are categorised*: to enable prioritisation of the maintenance requirements, to enable different kinds of process analyses, and to provide basis for assessing product quality.
- *Problems are classified as internal or external*: to be able to assess the quality of software products from the customer perspective, and to assess the effectiveness of quality assurance and quality control procedures.
- *Problems are classified as unique or duplicate*: to improve and measure productivity, and to provide correct statistics of the number of unique problems.

3. CM^3: Problem Management: Level 2 (Defined)

At the *Defined* level, a coarse-grained process model for managing software problems is defined and documented. The process covers the most rudimentary process phases and activities essential for managing software problems. These phases/activities offer visible milestones for following the progress and for making different kinds of intermediary decisions.

At this level, the process is consistently adhered to. Process compliance with its documentation is checked on a regular or event-driven basis. Although simple, the data and measurement provided by the process is meaningful, appropriately reflecting the process. Due to the coarse-grained process visibility, however, the assessment of effort and resources still corresponds to a rough estimation. Some of the maintenance elements applicable at this level are the following:

- *A template for how to structure a description of a software problem is institutionalised*: to develop a maximal support to problem submitters for describing their maintenance requirements (software problems). A proper description of a problem is the most important prerequisite for its effective resolution.
- *Correctness, consistency, and completeness of the problem report data is continuously checked and improved*: to maximise an objective understanding of a software problem, and to provide correct, consistent, and complete feedback for problem validation and verification, and for making different kinds of statistics and process analyses.
- *Sources (submitter and maintainer's) of problem description and problem report data are separated*: to enable efficient problem validation and verification, correct and reliable statistics, and to enable planning of maintenance work.
- *Problem management process, its phases, results and executing roles are documented:* to track the problem resolution process to their process phases, results, and the roles.
- *Process variances (allowed process deviations) are defined and institutionalised:* to enable process flexibility, and to provide feedback to process refinement and improvement.
- *The suggestion for a problem solution and the plan for its realisation is approved by a Change Control Board, before it gets implemented:* to assure that software is treated like a common organisational asset by commonly discussing and choosing the most optimal solution to the problem.

4. CM^3: Problem Management: Level 3 (Optimal)

At the *Optimal* level, the problem management process allows a fine-grained visibility into its status and progress. We have clear insight into every process step, and its results. The detailed process knowledge helps us make a thorough impact analysis during which the complexity and ramifications of a corrective change are recognised. This substantially helps assess the amount of effort and resources required for its resolution. In contrast to Level 2, the discrepancy between the planned and actual effort is strongly reduced due to the more detailed process feedback. The process does not only suffice for managing software problems, but also provides useful feedback to process improvement and defect prevention. The following maintenance elements apply at this level:

- *Causes of problem defects are defined and classified*: to be able to assess the product quality, and to enable a root cause analysis.
- *Traceability of change is ensured on the documentation/source code line level*: to correctly measure the modification size due to software problems, and to enable the tracking of all modifications to problem reports and vice versa.
- *Impact analysis is conducted*: to determine the complexity and ramifications (including ripple effect) of a software problem in order to correctly assess the amount of work, effort and resources required for its resolution.
- *A model for conducting a root cause analysis is defined and followed*: to identify the original sources of defects in order to provide feedback to process improvement and defect prevention.

5. Epilogue

In this paper, we have presented *CM³: Problem Management* process model. Our model does not only handle the resolution of software problems, but also provides a basis for quantitative feedback important for assessing product quality, crucial for continuous process analysis and improvement, and essential for defect prevention. Our aspirations are to provide a framework to the organisations building and improving their problem management processes, and to provide a pedagogical tool for universities and organisations in the process of educating and training their students and software engineers within the area of corrective maintenance.

References

[1] Kajko-Mattsson M. *Common Concept Apparatus within Corrective Software Maintenance*, In Proceedings, International Conference on Software Maintenance, IEEE Computer Society Press in Los Alamitos CA, 1999, pp. 287-294.

[2] Kajko-Mattsson, M., *Maintenance at ABB (I): Software Problem Administration Processes (The State of Practice),* Proceedings, International Conference on Software Maintenance, IEEE Computer Society Press: Los Alamitos, CA, Sep 1999.

[3] Kajko-Mattsson, M., *Maintenance at ABB (II): Change Execution Processes (The State of Practice),* In Proceedings, International Conference on Software Maintenance IEEE Computer Society Press: Los Alamitos, CA, Sep 1999.

[4] Kajko-Mattsson, M., *Corrective Maintenance Maturity Model (CM³),* Technical Report, No. 00-010, Department of Computer and Systems Sciences (DSV), Stockholm University and Royal Institute of Technology, December 2000.

[5] Kajko-Mattsson, Mira, *Taxonomy of Problem Management Activities*, In Proceedings 5th European Conference on Software Maintenance and Reengineering, IEEE Computer Society Press in Los Alamitos CA, 2001.

[6] Kajko-Mattsson, M., Forssander, S., Olsson, U., *Corrective Maintenance Maturity Model: Maintainer's Education and Training,* In Proceedings, International Conference on Software Engineering, , IEEE Computer Society Press in Los Alamitos CA, 2001.

[7] Kajko-Mattsson, M., *CM³: Problem Management: Maintenance Elements*, Technical Report, No. 2001-010, Department of Computer and Systems Sciences (DSV), Stockholm University and Royal Institute of Technology, December 2001.

A Method Engineering Language for the Description of Systems Development Methods (Extended Abstract)

Sjaak Brinkkemper[1], Motoshi Saeki[2], and Frank Harmsen[3]

[1] Baan R&D, P.O. Box 143, 3770 AC Barneveld, Netherlands
SBrinkkemper@Baan.nl

[2] Department of Computer Science, Tokyo Institute of Technology,
Ookayama 2-12-1, Meguro-Ku, Tokyo 152-8552, Japan
saeki@cs.titech.ac.jp

[3] Cap Gemini Ernst & Young Management Consultants
P.O. Box 3101, 3502 GC Utrecht, the Netherlands
nlharms4@nl.cgeyc.com

Abstract. We propose a Method Engineering Language, called MEL, as a formal representation language for the description of method fragments, i.e. the development processes, and the products and deliverables of a systems development method. The language allows representing the structures of method fragments, the applicable consistency rules, and a variety of method assembly operators, all of which the semantics are formally defined. The MEL language is illustrated by a simple example of a Sequence Diagram of UML.

1 Introduction

Systems development methods have always been described in an informal and ad-hoc manner. Interpretation and application of these methods is therefore subject to personal and circumstantial factors. In order to reason or manipulate with methods in an unbiased, neutral way a universal language for the representation of methods and tools is required and some languages have been developed and applied [1] [5]. In this paper we will introduce a dedicated language for the description and manipulation of methods. We have called this language simply *MEL: Method Engineering Language.*

In general, a specification language requires the expressive power to model the application domain in an effective manner, and should be practical to apply with respect to convenience, efficiency, and learnability [4]. For systems development methods in particular, the language should be able to support the representation and manipulation of 1) method fragments, i.e. method processes and method products, 2) development and project management aspects of methods, 3) conceptual definition and the supportive technical (i.e., tools) aspects of methods, and 4) constraints and rules concerning method fragments.

MEL (Method Engineering Language), which we have developed, is intended to strike a balance between formal meta-modelling languages and graphical techniques used for method modelling. The language is able to represent in an integrated manner

K.R. Dittrich, A. Geppert, M.C. Norrie (Eds.): CAiSE 2001, LNCS 2068, pp. 473–476, 2001.
© Springer-Verlag Berlin Heidelberg 2001

IS development methods both from the product perspective and the process perspective, as well as on various levels of decomposition and granularity. Moreover, MEL offers facilities to anchor method descriptions in an *ontology*, which is especially useful in the context of Situational Method Engineering. MEL provides operations to insert and remove fragments in and from the method base, to retrieve method fragments, and to assemble them into a situational method [2]. In the following sections the various constructs of MEL will be introduced.

2 Description Example in MEL

Methods and method fragments consist of product and process aspects. The product aspect specifies what products the developers should construct during a development process, while the process aspect navigates the developers to construct the products. The process descriptions suggest what activities and in what order are performed by the developers.

Let's consider an example of the simplified version of Sequence diagram of UML [6, p. 3-97]. Figure 1 depicts the descriptions of the product of Sequence diagram and the process for constructing a Sequence Diagram. Figure 1-(a) shows the former meta-modelled in a Class Diagram of UML, while the latter in Figure 1 (b) is modelled in a Flow Chart.

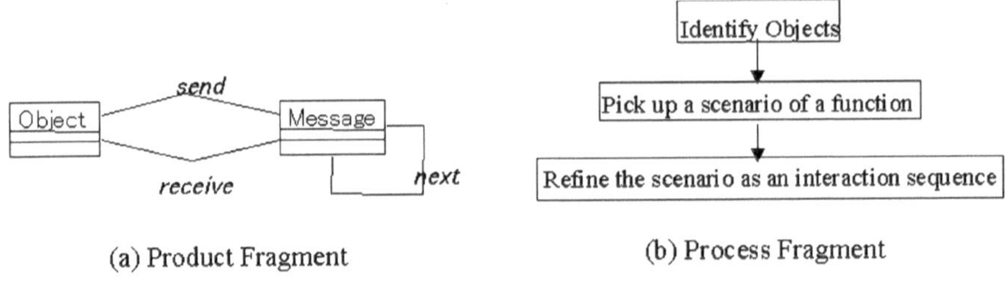

Fig. 1. Method fragment of sequence diagram

As shown in Figure 1 (a), the objects that are in the system or in the external environments send and/or receive messages and the diagram depicts the timing of message passing, i.e. a chain of messages based on timing order. The association *next* denotes the timing order of the messages. To construct a Sequence Diagram, we begin with identifying objects. The second step is to pick up a scenario for achieving a function of the system. The scenario is refined and structured in the third step.

The MEL description of the Sequence Diagram method fragment is given underneath. Descriptions that begin with the reserved word **PRODUCT** are product fragments. **LAYER** denotes the abstraction layer (granularity) of product fragments and, according to [4], there are four granularity levels defined: Concept, Diagram, Model, and Method. Association between concepts is enabled through **ASSOCIATION,** along with its cardinality and its roles. **PART OF** and **IS_A** specify aggregation and generalization relationships on method fragments respectively. We can express rules and constraints regarding method fragments with the **RULE** construct, which does not appear in the example, as well as with the **CARDINALITY** property type. Rules are first order logic formulas and only address the static aspects of product fragments.

As shown in the MEL specification, the process descriptions of method fragments are similar to those of conventional procedural programs. However, unlike to procedural programming languages, parallelism and non-determinism of development activities needs to be described. MEL has syntactic constructs to compose a complex process from activities, such as sequential execution, conditional branch, iteration, parallel execution and non-deterministic choice. In the example, the occurrences of hyphen (-) before activity names indicate sequential activities. Formally speaking, we have precedence relationship on activities specifying the execution order. Process fragments can be hierarchically decomposed into smaller ones in order to describe complex fragments in a comprehensive way. The construct **PART OF** relates a child process to a parent fragment in the hierarchical structure. The correlation of process and product fragments is established through the **REQUIRED** and **DELIVERABLES** section.

```
PRODUCT Sequence Diagram:
   ID Sequence Diagram;
IS_A Diagram;
LAYER Diagram;
SOURCE UML;
PART OF
   Analysis Model, Dynamic Model;
CREATED BY
   Construct a Sequence Diagram.

PRODUCT Sequence Diagram:
      LAYER Diagram;
      PART OF Use Case Model;
      NAME TEXT;

PRODUCT Object:
      LAYER Concept;
      PART OF Use Case Model;
      SYMBOL Rectangle;
      NAME TEXT;
      ASSOCIATED WITH
       {(send,), (receive,)}.

PRODUCT Message:
      LAYER Concept;
      PART OF Use Case Model;
      SYMBOL Arrow;
      NAME TEXT;
      ASSOCIATED WITH
    {(send,), (receive,), (next,)}.

   ASSOCIATION send:
      ASSOCIATES (Object, Message);
      CARDINALITY (1..1, 1..1).

   ASSOCIATION receive:
      ASSOCIATES (Object, Message);
      CARDINALITY (1..1,1..1).

   ASSOCIATION next:
      ASSOCIATES
       (Message, Message);
      CARDINALITY (0..1, 0..1).

PROCESS
Construct a Sequence Diagram:
      LAYER Diagram;
      TYPE Creation;
      PART OF
    Create an Analysis Model;
      REQUIRED {Interview results};
      REQUIRED OPTIONAL
    Current Information system;
      ( - Identify Objects ;
       - Pick up
           a Scenario of a Function ;
        - Refine the Scenario as
              an interaction sequences
      )
      DELIVERABLES
       {Sequence Diagram}.
```

3 Operational Constructs of MEL

The MEL operations for situational method engineering and method adaptation are the following:

1. Insertion and maintenance of fragments of methods and tools in the method base by a methods *administration* function, e.g. inserting and removing method fragments to and from the method base;
2. Selection of the method fragments for the project based on a characterisation of the project from a method base;
3. Adaptation of method fragments to suit the aims of the method engineer, e.g. the operations for including a method fragment into another one;
4. Assembly of the selected method fragments into a situational method, e.g. the operations for assembling two product fragments into a new product fragment by specifying the association or associations through which they should be connected, and the operations for creating a precedence relationship between two process fragments or between a process fragment and a set of process fragments. Since the method assembly process is considered as a procedure consisting of a sequence of MEL operations, it can be described as a process fragment.

A discussion and an extensive example of assembly operators can be found in [3].

4 Conclusion and Future Work

Departing from a set of requirements and purposes for method engineering languages, we have shown a part of MEL by using a simple example. It may be clear that the current textual format is not suitable as an end-user interface for a method engineer. Due to the complete formalisation of syntax and formal operational semantics of MEL [4], the language is unambiguous and suited to be serve as the underlying formalism for a Computer Aided Method Engineering (CAME) tool. This implementation has resulted in the MEL interpreter and editor contained in the *Decamerone* tool [3] environment for support of Situational Method Engineering.

References

1. Brinkkemper, S., and S.M.M. Joosten (Eds.), Method Engineering and Meta-Modelling. Special Issue. *Information and Software Technology*, vol. 38, nr. 2, pp. 259-305, 1996.
2. Brinkkemper, S., M. Saeki, and F. Harmsen, Meta-Modelling Based Assembly Techniques for Situational Method Engineering. *Information Systems*, vol. 24, No. 3, pp. 209-228, 1999.
3. Brinkkemper, S., M. Saeki, and F. Harmsen, *A Method Engineering Language for the Description of Systems Development Methods.* Baan R&D Report, November 2000.
4. Harmsen, F., *Situational Method Engineering*, Moret Ernst & Young, 1997.
5. Rolland, C., S. Nurcan, G. Grosz, A Decision Making Pattern for Guiding the Enterprise Knowledge Development Process. *Journal of Information and Software Technology*, Elsevier, 42(2000), p. 313-331.
6. The Unified Modeling Language Specification Ver 1.3, http://www.rational.com, 1999.

Flexible Support of Work Processes – Balancing the Support of Organisations and Workers

John Krogstie and Håvard D. Jørgensen

Norwegian University of Science and Technology and
SINTEF Telecom and Informatics, Norway
{jok,hdj}@informatics.sintef.no

1 Introduction and Motivation

Today's organisations are characterised by dynamic, only partially understood and error-prone environments. Thus systems must be developed, which support truly dynamic organisational processes, often involving co-operation across traditional organisational boundaries.

The limited success of workflow management systems (WMS) has partially been attributed to lack of *flexibility*. Consequently, flexible workflow is a hot research topic. Most work within this area looks at how conventional systems can be extended and enhanced in other words how static workflow systems can be made *adaptive*. Most of this work recognise that change is a way of life in most organisations, but a basic premise is still that work is repetitive and can be prescribed reasonably completely.

On the other hand, a lot of dynamic organisational activity is best described as emergent. In those activities the process model and structure is unclear at the start and emerges during enactment.

Traditional ERP systems tend to be quite inflexible, hardly adaptable at runtime, and primarily support the organisationally agreed processes. Existing workflow management systems have typically been focused on dealing with exceptions and have thus offered some support for adaptive processes. These types of systems, however, have typically overlooked emergent processes, which seem to encompass an increasing part of organised activity. An alternative approach is to support *active* models as a general technique for increasing the flexibility of computerised information systems for co-operative work support. What does it mean that a model is active? First of all, the representation must be *available* to the users of the information system at runtime. Second, the model must *influence the behaviour* of the computerised system. Third, the model must be *dynamic*, users must be supported in changing the model to fit their local reality, enabling tailoring of the system's behaviour.

On the other hand, only supporting the emergent work style of the individual knowledge worker is probably at times inefficient, because routine parts of the work can be prescribed and automated, and because sharing of explicit process models facilitates co-ordination, collaboration and communication between multiple parties. Thus there is a need for a balance between prescription and emergent representations.

K.R. Dittrich, A. Geppert, M.C. Norrie (Eds.): CAiSE 2001, LNCS 2068, pp. 477–481, 2001.
© Springer-Verlag Berlin Heidelberg 2001

The panel will describe and discuss different approaches to more or less flexibly support organisational, group and personal work processes, applying novel approaches including active models.

2 Position Statements of Panel Members

Stefan Jablonski (Stefan.Jablonski@informatik.uni-erlangen.de)
Universitaet Erlangen-Nuernberg

Flexible Support of Work Processes

We divide our contribution into three position statements. From our perspective they span a good amount of the issues related to the three topics

- flexibility
- work processes and
- workflow management.

The main theme throughout my position statements is that the need for work process support does not "automatically" connect to "workflow management". Therefore, we strongly advice to investigate first whether an application is suitable for workflow support at all. Not before the answer is "yes" it should be investigated how this support can be tailored to specific requirements of an application scenario; this investigations directs to the issue of flexibility for workflow management systems.

My three position statements are:

Uncritical application of workflow management

Our first position is that the workflow management techniques are applied uncritically. People do not sufficiently analyse application scenarios in order to find out whether it is really worth to be supported by workflow management. There is a huge range of different sorts of work processes. Some of them should never be supported by workflow management like emergencies in an hospital environment. This does not say that in such a case there might be something like an after-treatment that asynchronously maps the emergency to a workflow.

Wrong workflow models

Another problem we are facing very often is that workflow models are just wrong. They do not reflect the real application scenario but do impose processing constraints onto an application scenario that are not justified. Mostly these undesirable effects stem from the fact that workflow models are not expressive enough to allow adequate modeling.

Missing flexibility in workflow management approaches

Just this third position statement refers to the workflow management approaches. It is a fact that flexibility is not supported sufficiently by actual workflow approaches. We propose two mechanisms to introduce more flexibility to workflow management: flexibility by selection and flexibility by adaptation.

Paul Johannesson (pajo@dsv.su.se) Department of Computer and Systems Sciences, Stockholm University/KTH

Explicit Goal Representation as a Means for Flexible Process Modelling and Enactment

Current software systems for supporting process management, such as workflow management systems, are effective for predictable and repetitive processes. However, they are typically unable to adapt to a dynamic environment where unexpected situations have to be managed. It is possible to distinguish among four different types of exceptions that can occur during the execution of a process, [Eder95]: *basic failures*, which are failures of the supporting software system or its environment; *application failures*, which are failures of the applications that are invoked by the software system; *expected exceptions*, which are predictable deviations from the normal flow of the process; *unexpected exceptions*, which are mismatches between the actual execution of a process and its definition in the software system.

Unexpected exceptions are caused by process changes that could not have been anticipated at design time, e.g. a new government regulation. Thus, unexpected changes are the most difficult ones to handle in workflow systems. One approach for addressing the problem of managing unexpected exceptions is to explicitly include goals in process specifications and workflow systems – not only events and activities. Goals are typically more stable in a process than the events and activities that are carried out to achieve the goals. Therefore, including goals in process specifications will provide a stable frame around which activities can be ordered. A process specification will then consist of a collection of goals and for each goal a tentative and adaptable structure of activities to be carried out for achieving the goal. For this approach to work, it is required that the implemented workflow system makes goals visible to the user. Thus, goals are not only used in the early requirements specification phases – they should also become explicit in the application logic of a system as well as in its user interface. In this respect, the approach is similar to the philosophy behind Tropos, [Mylopoulos00], where intentional concepts are used in late software development phases.

[Eder95] Eder J, Liebhart W. The Workflow Activity Model WAMO. In: 3rd International Conference on Cooperative Information Systems. University of Toronto Press. 1995, pp. 87-98

[Mylopoulos00] Mylopoulos J, Castro J. Tropos: A Framework for Requirements-Driven Software Development. In: Information Systems Engineering, ed. Brinkkemper S. et. al., Springer 2000, pp. 261-274

John Krogstie (jok,@informatics.sintef.no) Norwegian University of Science and Technology and SINTEF Telecom and Informatics, Norway

From Supporting Organisations to Supporting Knowledge Worker in Organisations

Information and workflow systems are traditionally made to primarily support the goal of an organisation, and only secondarily the users of the systems. When the goals are well understood, this is a sensible approach, and production workflow and ERP systems are a good investment. On the other hand, a larger proportion of work is today done by what is termed symbolic analyst, whose work resources and work products are mainly symbol-structures. The main work pattern of these persons is knowledge intensive projects often in dynamically networked organisations. Symbolic analysts such as consultants, reporters, and researchers will typically have many tasks going on concurrently, and will be interested in a lot of different information there and then, much of which can not be anticipated fully beforehand. There might be a need for learning on the fly, but also for capturing interesting situations to feed back to the organisation supporting knowledge management and learning processes for a larger part of the organisation. Although many of their processes are emergent and creative, a large proportion of the time spend by these so-called knowledge workers, is on routine, administrative tasks. We also predict that many of these workers would like to be supported also in their more complex personal processes involving e.g. governmental agencies, banks, and insurance companies using the same system if possible, as the distinction of work and private life is blurred. This is similar to how many people nowadays use PDAs, where the calendar contains both private and professional tasks and appointments.

These developments have opened the need for emergent workflow systems where modelling is viewed as an integral part of the work, performed by the process participants, as they are the only actors who have sufficient knowledge of the process. The focus is on unique cases, especially knowledge intensive projects. Emergent models, as other active models, depend more on the individual, social, organisational and situational usage context than routine procedures modelled by software engineers or process designers. When process modelling and enactment happen in parallel, we cannot assume complete models. These challenges cannot be met by algorithms alone. Instead we must involve the users in resolving incompleteness and inconsistencies. In fact, it may not even be the case that emergent process models should be enacted by software, the interpretation of the models might be carried out by the users instead. Some challenges for emergent workflow systems for knowledge intensive project work are:

- Modelling must be done also by process participants, not just by external process experts. The end users must be motivated to externalise and share their knowledge about the process. We thus need to increase the users' benefits of keeping models up to date.
- Users don't want to start from scratch. In order to reuse modelling effort and learn from past experience, we need mechanisms that enable the harvesting of past models into reusable templates.
- How can we support co-ordination based on an emergent process model? We view enactment as "co-ordination by automated sequencing of tasks", and this should be combined with more flexible CSCW approaches.

Although the individual knowledge worker typically would prefer to be supported across his range of processes with the same tool, there are as indicated many differences between emergent workflow support and the traditional static and adaptive approaches. From this, it can be argued that combining conventional and emergent workflow into one system should not be a main objective. Though they share the core of active support based on process models, adaptive and emergent workflow is perhaps best kept in different systems and integrated through standards from the OMG, WfMC and IETF?

Matthias Weske (mathias.weske@acm.org) Eindhoven University of Technology

Flexible Workflow Management: Adapt and Bend, but Don't Break!

The dictionary tells that an artefact is flexible if it can "easily be changed to suit new conditions". This general definition of flexibility holds well in the workflow context, where artefacts correspond to workflow specifications and applications, and changes are the means to cope with new conditions of application processes, imposed by changes in their business or legal environments. Hence, it is not surprising that flexibility has been a major motivation for workflow management from its beginning: By explicitly modelling application processes consisting of activities, their causal and temporal ordering and the technical and organisational environment of their execution, the structure and embedding of a given application process can be changed conveniently. In particular, changes can be applied by modifying workflow specifications rather than by changing computer programs, which hard-coded application processes in the pre-workflow era. Hence, traditional workflow technology is well suited for modelling and controlling the execution of a wide range of application processes.

However, as turned out in recent years, traditional workflow technology does not suffice for advanced applications in today's dynamic and competitive environments in commerce, public administration, and in science and engineering. In these settings it is increasingly unlikely that an application process is modelled once to be executed repeatedly without any modifications. Hence, more advanced flexibility mechanisms to facilitate workflow modification at different instants with different scopes have to be developed. Recent work on flexible workflow management has generated a considerable body of literature, focusing on a variety of aspects. One of the hardest problems in this context is dynamic adaptation, i.e., the adaptation of running workflow instances to new workflow schemas. This functionality is required in settings where workflows are typically long running activities and where changes to the process environment occur frequently.

Besides conceptual research issues like correctness criteria for dynamic adaptations, flexibility requirements also raise a number of interesting workflow management systems design issues For example, the strict separation of a workflow's build time and its run time as found in many traditional workflow management system is not adequate for supporting dynamic adaptations effectively; interleavings of a workflow's build time and its run time are required and, thus, have to be facilitated by flexible workflow management systems.

Once the scientific results are consolidated and put into practice, the second interpretation of flexibility the dictionary tells can also be fulfilled: An artefact is flexible if it can "easily be bent without breaking", an interpretation which may be even more appropriate in the workflow context.

Author Index

Lecture Notes in Computer Science

For information about Vols. 1–1976
please contact your bookseller or Springer-Verlag

Vol. 1977: B. Roy, E. Okamoto (Eds.), Progress in Cryptology – INDOCRYPT 2000. Proceedings, 2000. X, 295 pages. 2000.

Vol. 1978: B. Schneier (Ed.), Fast Software Encryption. Proceedings, 2000. VIII, 315 pages. 2001.

Vol. 1979: S. Moss, P. Davidsson (Eds.), Multi-Agent-Based Simulation. Proceedings, 2000. VIII, 267 pages. 2001. (Subseries LNAI).

Vol. 1980: M. Agosti, F. Crestani, G. Pasi (Eds.), Lectures on Information Retrieval. Proceedings, 2000. XI, 311 pages. 2001.

Vol. 1981: J.M.L.M. Palma, J. Dongarra, V. Hernández (Eds.), Vector and Parallel Processing – VECPAR 2000. Proceedings, 2000. XVI, 580 pages. 2001.

Vol. 1983: K.S. Leung, L.-W. Chan, H. Meng (Eds.), Intelligent Data Engineering and Automated Learning – IDEAL 2000. Proceedings, 2000. XVI, 573 pages. 2000.

Vol. 1984: J. Marks (Ed.), Graph Drawing. Proceedings, 2001. XII, 419 pages. 2001.

Vol. 1985: J. Davidson, S.L. Min (Eds.), Languages, Compilers, and Tools for Embedded Systems. Proceedings, 2000. VIII, 221 pages. 2001.

Vol. 1987: K.-L. Tan, M.J. Franklin, J. C.-S. Lui (Eds.), Mobile Data Management. Proceedings, 2001. XIII, 289 pages. 2001.

Vol. 1988: L. Vulkov, J. Waśniewski, P. Yalamov (Eds.), Numerical Analysis and Its Applications. Proceedings, 2000. XIII, 782 pages. 2001.

Vol. 1989: M. Ajmone Marsan, A. Bianco (Eds.), Quality of Service in Multiservice IP Networks. Proceedings, 2001. XII, 440 pages. 2001.

Vol. 1990: I.V. Ramakrishnan (Ed.), Practical Aspects of Declarative Languages. Proceedings, 2001. VIII, 353 pages. 2001.

Vol. 1991: F. Dignum, C. Sierra (Eds.), Agent Mediated Electronic Commerce. VIII, 241 pages. 2001. (Subseries LNAI).

Vol. 1992: K. Kim (Ed.), Public Key Cryptography. Proceedings, 2001. XI, 423 pages. 2001.

Vol. 1993: E. Zitzler, K. Deb, L. Thiele, C.A.Coello Coello, D. Corne (Eds.), Evolutionary Multi-Criterion Optimization. Proceedings, 2001. XIII, 712 pages. 2001.

Vol. 1994: J. Lind, Iterative Software Engineering for Multiagent Systems. XVII, 286 pages. 2001. (Subseries LNAI).

Vol. 1995: M. Sloman, J. Lobo, E.C. Lupu (Eds.), Policies for Distributed Systems and Networks. Proceedings, 2001. X, 263 pages. 2001.

Vol. 1997: D. Suciu, G. Vossen (Eds.), The World Wide Web and Databases. Proceedings, 2000. XII, 275 pages. 2001.

Vol. 1998: R. Klette, S. Peleg, G. Sommer (Eds.), Robot Vision. Proceedings, 2001. IX, 285 pages. 2001.

Vol. 1999: W. Emmerich, S. Tai (Eds.), Engineering Distributed Objects. Proceedings, 2000. VIII, 271 pages. 2001.

Vol. 2000: R. Wilhelm (Ed.), Informatics: 10 Years Back, 10 Years Ahead. IX, 369 pages. 2001.

Vol. 2001: G.A. Agha, F. De Cindio, G. Rozenberg (Eds.), Concurrent Object-Oriented Programming and Petri Nets. VIII, 539 pages. 2001.

Vol. 2002: H. Comon, C. Marché, R. Treinen (Eds.), Constraints in Computational Logics. Proceedings, 1999. XII, 309 pages. 2001.

Vol. 2003: F. Dignum, U. Cortés (Eds.), Agent Mediated Electronic Commerce III. XII, 193 pages. 2001. (Subseries LNAI).

Vol. 2004: A. Gelbukh (Ed.), Computational Linguistics and Intelligent Text Processing. Proceedings, 2001. XII, 528 pages. 2001.

Vol. 2006: R. Dunke, A. Abran (Eds.), New Approaches in Software Measurement. Proceedings, 2000. VIII, 245 pages. 2001.

Vol. 2007: J.F. Roddick, K. Hornsby (Eds.), Temporal, Spatial, and Spatio-Temporal Data Mining. Proceedings, 2000. VII, 165 pages. 2001. (Subseries LNAI).

Vol. 2009: H. Federrath (Ed.), Designing Privacy Enhancing Technologies. Proceedings, 2000. X, 231 pages. 2001.

Vol. 2010: A. Ferreira, H. Reichel (Eds.), STACS 2001. Proceedings, 2001. XV, 576 pages. 2001.

Vol. 2011: M. Mohnen, P. Koopman (Eds.), Implementation of Functional Languages. Proceedings, 2000. VIII, 267 pages. 2001.

Vol. 2012: D.R. Stinson, S. Tavares (Eds.), Selected Areas in Cryptography. Proceedings, 2000. IX, 339 pages. 2001.

Vol. 2013: S. Singh, N. Murshed, W. Kropatsch (Eds.), Advances in Pattern Recognition – ICAPR 2001. Proceedings, 2001. XIV, 476 pages. 2001.

Vol. 2015: D. Won (Ed.), Information Security and Cryptology – ICISC 2000. Proceedings, 2000. X, 261 pages. 2001.

Vol. 2016: S. Murugesan, Y. Deshpande (Eds.), Web Engineering. IX, 357 pages. 2001.

Vol. 2018: M. Pollefeys, L. Van Gool, A. Zisserman, A. Fitzgibbon (Eds.), 3D Structure from Images – SMILE 2000. Proceedings, 2000. X, 243 pages. 2001.

Vol. 2019: P. Stone, T. Balch, G. Kraetzschmar (Eds.), RoboCup 2000: Robot Soccer World Cup IV. XVII, 658 pages. 2001. (Subseries LNAI).

Vol. 2020: D. Naccache (Ed.), Topics in Cryptology – CT-RSA 2001. Proceedings, 2001. XII, 473 pages. 2001

Vol. 2021: J. N. Oliveira, P. Zave (Eds.), FME 2001: Formal Methods for Increasing Software Productivity. Proceedings, 2001. XIII, 629 pages. 2001.

Vol. 2022: A. Romanovsky, C. Dony, J. Lindskov Knudsen, A. Tripathi (Eds.), Advances in Exception Handling Techniques. XII, 289 pages. 2001

Vol. 2024: H. Kuchen, K. Ueda (Eds.), Functional and Logic Programming. Proceedings, 2001. X, 391 pages. 2001.

Vol. 2025: M. Kaufmann, D. Wagner (Eds.), Drawing Graphs. XIV, 312 pages. 2001.

Vol. 2026: F. Müller (Ed.), High-Level Parallel Programming Models and Supportive Environments. Proceedings, 2001. IX, 137 pages. 2001.

Vol. 2027: R. Wilhelm (Ed.), Compiler Construction. Proceedings, 2001. XI, 371 pages. 2001.

Vol. 2028: D. Sands (Ed.), Programming Languages and Systems. Proceedings, 2001. XIII, 433 pages. 2001.

Vol. 2029: H. Hussmann (Ed.), Fundamental Approaches to Software Engineering. Proceedings, 2001. XIII, 349 pages. 2001.

Vol. 2030: F. Honsell, M. Miculan (Eds.), Foundations of Software Science and Computation Structures. Proceedings, 2001. XII, 413 pages. 2001.

Vol. 2031: T. Margaria, W. Yi (Eds.), Tools and Algorithms for the Construction and Analysis of Systems. Proceedings, 2001. XIV, 588 pages. 2001.

Vol. 2032: R. Klette, T. Huang, G. Gimel'farb (Eds.), Multi-Image Analysis. Proceedings, 2000. VIII, 289 pages. 2001.

Vol. 2033: J. Liu, Y. Ye (Eds.), E-Commerce Agents. VI, 347 pages. 2001. (Subseries LNAI).

Vol. 2034: M.D. Di Benedetto, A. Sangiovanni-Vincentelli (Eds.), Hybrid Systems: Computation and Control. Proceedings, 2001. XIV, 516 pages. 2001.

Vol. 2035: D. Cheung, G.J. Williams, Q. Li (Eds.), Advances in Knowledge Discovery and Data Mining – PAKDD 2001. Proceedings, 2001. XVIII, 596 pages. 2001. (Subseries LNAI).

Vol. 2037: E.J.W. Boers et al. (Eds.), Applications of Evolutionary Computing. Proceedings, 2001. XIII, 516 pages. 2001.

Vol. 2038: J. Miller, M. Tomassini, P.L. Lanzi, C. Ryan, A.G.B. Tettamanzi, W.B. Langdon (Eds.), Genetic Programming. Proceedings, 2001. XI, 384 pages. 2001.

Vol. 2039: M. Schumacher, Objective Coordination in Multi-Agent System Engineering. XIV, 149 pages. 2001. (Subseries LNAI).

Vol. 2040: W. Kou, Y. Yesha, C.J. Tan (Eds.), Electronic Commerce Technologies. Proceedings, 2001. X, 187 pages. 2001.

Vol. 2041: I. Attali, T. Jensen (Eds.), Java on Smart Cards: Programming and Security. Proceedings, 2000. X, 163 pages. 2001.

Vol. 2042: K.-K. Lau (Ed.), Logic Based Program Synthesis and Transformation. Proceedings, 2000. VIII, 183 pages. 2001.

Vol. 2043: D. Craeynest, A. Strohmeier (Eds.), Reliable Software Technologies – Ada-Europe 2001. Proceedings, 2001. XV, 405 pages. 2001.

Vol. 2044: S. Abramsky (Ed.), Typed Lambda Calculi and Applications. Proceedings, 2001. XI, 431 pages. 2001.

Vol. 2045: B. Pfitzmann (Ed.), Advances in Cryptology – EUROCRYPT 2001. Proceedings, 2001. XII, 545 pages. 2001.

Vol. 2047: R. Dumke, C. Rautenstrauch, A. Schmietendorf, A. Scholz (Eds.), Performance Engineering. XIV, 349 pages. 2001.

Vol. 2048: J. Pauli, Learning Based Robot Vision. IX, 288 pages. 2001.

Vol. 2051: A. Middeldorp (Ed.), Rewriting Techniques and Applications. Proceedings, 2001. XII, 363 pages. 2001.

Vol. 2052: V.I. Gorodetski, V.A. Skormin, L.J. Popyack (Eds.), Information Assurance in Computer Networks. Proceedings, 2001. XIII, 313 pages. 2001.

Vol. 2053: O. Danvy, A. Filinski (Eds.), Programs as Data Objects. Proceedings, 2001. VIII, 279 pages. 2001.

Vol. 2054: A. Condon, G. Rozenberg (Eds.), DNA Computing. Proceedings, 2000. X, 271 pages. 2001.

Vol. 2055: M. Margenstern, Y. Rogozhin (Eds.), Machines, Computations, and Universality. Proceedings, 2001. VIII, 321 pages. 2001.

Vol. 2056: E. Stroulia, S. Matwin (Eds.), Advances in Artificial Intelligence. Proceedings, 2001. XII, 366 pages. 2001. (Subseries LNAI).

Vol. 2057: M. Dwyer (Ed.), Model Checking Software. Proceedings, 2001. X, 313 pages. 2001.

Vol. 2059: C. Arcelli, L.P. Cordella, G. Sanniti di Baja (Eds.), Visual Form 2001. Proceedings, 2001. XIV, 799 pages. 2001.

Vol. 2064: J. Blanck, V. Brattka, P. Hertling (Eds.), Computability and Complexity in Analysis. Proceedings, 2000. VIII, 395 pages. 2001.

Vol. 2068: K.R. Dittrich, A. Geppert, M.C. Norrie (Eds.), Advanced Information Systems Engineering. Proceedings, 2001. XII, 484 pages. 2001.

Vol. 2072: J. Lindskov Knudsen (Ed.), ECOOP 2001 – Object-Oriented Programming. Proceedings, 2001. XIII, 429 pages. 2001.

Vol. 2073: V.N. Alexandrov, J.J. Dongarra, B.A. Juliano, R.S. Renner, C.J.K. Tan (Eds.), Computational Science – ICCS 2001. Part I. Proceedings, 2001. XXVIII, 1306 pages. 2001.

Vol. 2074: V.N. Alexandrov, J.J. Dongarra, B.A. Juliano, R.S. Renner, C.J.K. Tan (Eds.), Computational Science – ICCS 2001. Part II. Proceedings, 2001. XXVIII, 1076 pages. 2001.

Vol. 2091: J. Bigun, F. Smeraldi (Eds.), Audio- and Video-Based Biometric Person Authentication. Proceedings, 2001. XIII, 374 pages. 2001.

Vol. 2092: L. Wolf, D. Hutchison, R. Steinmetz (Eds.), Quality of Service – IWQoS 2001. Proceedings, 2001. XII, 435 pages. 2001.